New Approaches to Problems in Liquid State Theory

NATO Science Series

A Series presenting the results of activities sponsored by the NATO Science Committee. The Series is published by IOS Press and Kluwer Academic Publishers, in conjunction with the NATO Scientific Affairs Division.

General Sub-Series

A. Life Sciences	IOS Press
B. Physics	Kluwer Academic Publishers
C. Mathematical and Physical Sciences	Kluwer Academic Publishers
D. Behavioural and Social Sciences	Kluwer Academic Publishers
E. Applied Sciences	Kluwer Academic Publishers
F. Computer and Systems Sciences	IOS Press

Partnership Sub-Series

1. Disarmament Technologies	Kluwer Academic Publishers
2. Environmental Security	Kluwer Academic Publishers
3. High Technology	Kluwer Academic Publishers
4. Science and Technology Policy	IOS Press
5. Computer Networking	IOS Press

The Partnership Sub-Series incorporates activities undertaken in collaboration with NATO's Partners in the Euro-Atlantic Partnership Council – countries of the CIS and Central and Eastern Europe – in Priority Areas of concern to those countries.

NATO-PCO-DATA BASE

The NATO Science Series continues the series of books published formerly in the NATO ASI Series. An electronic index to the NATO ASI Series provides full bibliographical references (with keywords and/or abstracts) to more than 50000 contributions from international scientists published in all sections of the NATO ASI Series.
Access to the NATO-PCO-DATA BASE is possible via CD-ROM "NATO-PCO-DATA BASE" with user-friendly retrieval software in English, French and German (© WTV GmbH and DATAWARE Technologies Inc. 1989).

The CD-ROM of the NATO ASI Series can be ordered from: PCO, Overijse, Belgium.

Series C: Mathematical and Physical Sciences – Vol. 529

New Approaches to Problems in Liquid State Theory

Inhomogeneities and Phase Separation in Simple, Complex and Quantum Fluids

edited by

Carlo Caccamo
Dipartimento di Fisica,
Università di Messina, Italy

Jean-Pierre Hansen
Department of Chemistry,
University of Cambridge, U.K.

and

George Stell
Department of Chemistry,
State University of New York,
Stony Brook, NY, U.S.A.

Kluwer Academic Publishers

Dordrecht / Boston / London

Published in cooperation with NATO Scientific Affairs Division

Proceedings of the NATO Advanced Study Institute on
New Approaches to Old and New Problems in Liquid State Theory
Inhomogeneities and Phase Separation in Simple, Complex and Quantum Fluids
Patti Marina (Messina), Italy
7–17 July, 1998

A C.I.P. Catalogue record for this book is available from the Library of Congress

ISBN 0-7923-5670-5 (HB)

Published by Kluwer Academic Publishers,
P.O. Box 17, 3300 AA Dordrecht, The Netherlands.

Sold and distributed in North, Central and South America
by Kluwer Academic Publishers,
101 Philip Drive, Norwell, MA 02061, U.S.A.

In all other countries, sold and distributed
by Kluwer Academic Publishers,
P.O. Box 322, 3300 AH Dordrecht, The Netherlands.

Printed on acid-free paper

Printed in the Netherlands

TABLE OF CONTENTS

FOREWORD

The theory of simple and complex fluids has made considerable recent progress, due to the emergence of new concepts and theoretical tools, and also to the availability of a large body of new experimental data on increasingly complex systems, as well as far-reaching methodological developments in numerical simulations. This ASI aimed at providing a comprehensive overview of the most significant theoretical developments, supplemented by a few presentations of cutting-edge simulation and experimental work. The impact of the Institute in the overall landscape of Statistical Mechanics received an important recognition with its inclusion in the list of satellite events of STATPHYS20, the triennal international conference on Statistical Physics held in Paris in July 1998. These Proceedings contain the texts of the 13 Lecture Courses and 9 Invited Seminars delivered at Patti.

Two clear trends emerge from these Proceedings: first, the diversity of new and unexpected theoretical results relating to classic models of liquids, which have recently been subjected to fresh scrutiny; and secondly the parallel emergence of new concepts, models and methods, aimed at investigating complex fluids and phenomena, like the phase behaviour of fluids in pores, macromolecular assemblies, and the glass transition. Many of the new tools have their roots in traditional liquid state theory, and, in conjunction with fresh input from related fields, allow it wider applicability.

Modern studies that are able to assess the thermodynamic and structural properties of fluids obeying simple potential models over a wide range of thermodynamic states have revealed unexpected behaviour and theoretical challenges. For instance, the existence and location of both liquid-gas and ferroelectric phase transitions in dipolar fluids is exquisitely sensitive to molecular shape, details of dipole symmetry, and the presence or absence of dispersion forces.

Similarly the phase diagram of the simplest charged hard-sphere systems of anions and cations of equal size and charge number (the restricted primitive model) has emerged as a theoretical challenge. Although it is generally agreed that a liquid-gas critical point is to be expected in such

C. Caccamo et al. (eds.), New Approaches to Problems in Liquid State Theory, xiii–xvii.

a system, and a growing consensus that the criticality is Ising-like, its location is still subject to large uncertainties both in simulation studies and in analytic theories. Lectures and discussions clearly underlined the crucial importance of a proper account of the ion pairing process, already advocated by Bjerrum.

Although the structure of the hard-body fluid models is temperature independent, they too have recently yielded new challenges and surprises when studied over the whole range of size and shape parameters and densities accessible to them. In particular, the occurrence of fluid-fluid phase separation in binary hard-particle mixtures has proved to hinge sensitively upon the shapes of the particles, the disparity in size of the two species, and the degree of non-additivity of the distances of closest approach. For the case of binary hard-sphere mixtures, separation involving at least one ordered phase appears to always preempt demixing into two fluid phases except under metastable conditions. The behaviour of this model is an important key to understanding certain colloidal suspensions. The subtle dependence of phase separation involving liquid-crystal phases upon molecular shape for fluids of hard non-spherical particles is another example of a challenging entropic problem on which much progress has recently been made.

There has also been a breakthrough pertaining to the development of theories that are comprehensive enough to yield quantitatively satisfactory results for the thermodynamics and structure of a simple fluid both under typical liquid-state conditions and in the immediate vicinity of the liquid-gas critical point. The self-consistent Ornstein-Zernike approach (SCOZA) and the hierarchical reference theory (HRT) discussed at the ASI represent two approaches that are successful in this regard. Closely related is the significant progress recently achieved with traditional or improved integral equations to treat fluids of molecules with internal degrees of freedom, polymer solutions and melts, and the phase behaviour of model protein solutions.

Lattice-gas models also appear in these Proceedings, including several lattice models in interesting new roles that go considerably beyond standard applications. The tight-binding lattice-gas model of alkali fluids and the spin models used to describe glassy systems are two notable examples. The very recent work on supercooled liquids and glassy systems promises to bring to the phenomenological theory of such systems a set of quantitatively precise solvable models to serve as paradigms. The powerful replica method, borrowed from the theory of systems with quenched disorder (e.g. spin glasses) is now being successfully applied to the investigation of structural glasses, and a major example of recent progress is the precise formulation of an off-equilibrium fluctuation-dissipation theorem, which provides a powerful tool for the analysis of ageing dynamics.

A striking feature of much recent work reported in these Proceedings is the strong emphasis on liquids in the presence of solid surfaces, and, in particular, on liquids confined to micropores. One of the most remarkable phenomena with which contemporary theory is concerned is the wetting and drying of various kinds of surfaces. Our understanding of the molecular basis of the phenomenon of wetting has been steadily growing over the past decade, in part due to the increasingly sophisticated versions of density functional theory applied to the problem. But it is only in the past year or two that we have begun to understand the taxonomy of wetting properties in its full relation to the phase diagram of fluid behaviour in the bulk, the geometry of the solid substrate, and the range of the substrate-fluid forces. Closely related to this work are recent quantitative predictions for the structure and conformations of polymers and the effective interactions between charged colloidal particles near surfaces, or in confined geometries.

Until the 1990's, the statistical mechanical study of fluids adsorbed in micropores was almost entirely limited to pore geometry of well-characterized shape, especially slits, cylinders, and spheres. However, fundamental new results in the late 1980's and early 1990's dealing with mixtures of quenched and annealed (i.e. fluid) particles have led for the first time to a systematic means of dealing with fluids contained in a quenched porous matrix with a realistically disordered structure.

The expected volume available to the centre of a particle to be inserted into a fluid is one of a family of cavity functions, closely related to the probability distribution functions, that have played an important liquid-state role since Boltzmann first introduced them. Several speakers used these functions in different ways as the central objects of their geometrically-formulated theories, including the remarkably successful "fundamental measure" description of hard core fluids. Density functional theory is the generic tool in the field, and its foundations are carefully summarized in these Proceedings. It has a quantum mechanical version that has been developed to describe solutions to the Schroedinger equation, and the high level of sophistication recently attained in this direction in quantum mechanical time-dependent density functional studies is summarized in one of the lectures.

Two additional aspects, concerning quantum fluids, are presented in these Proceedings. A venerable problem on which a significant breakthrough was reported is the superfluidity of bulk and inhomogeneous ^{4}He, which is analyzed within the "shadow wave-function" approach. Also, the simulation of systems involving both classical and quantum degrees of freedom was convincingly illustrated by a presentation of the *ab initio* path integral technique, and its application to proton transfer in water.

Finally the power of classical Molecular Dynamics, combined with the latest methodological developments, to simulate increasingly complex fluids, is well illustrated by the Lectures devoted to surfactant self-assembly and membranes, paving the way to a quantitative molecular description of biological systems.

Though these contents have been masterly encompassed by the speakers in both lectures and invited seminars, the scientific scenario they outlined has been substantially enriched by contributed talks and short communications, and by the two poster sessions. The latter, in particular, have offered the youngest scientists a major opportunity to demonstrate their highly qualified involvement with the topics of the School, and to discuss the most debated issues with senior scientists.

The continuous and broad exchange of information between all the participants has probably been the most successful result of this ASI. It has certainly been favoured by the long afternoon break between the morning and the evening sessions, as well as by the two round tables and the panel discussion in which timely topics such as criticality in fluids, the most recent advances in computer simulation methodology, and hard sphere mixtures, have been hotly debated.

It is fair to say, however, that something else has contributed to creating the right atmosphere in which human contacts were favoured, namely: the secluded and calm site, merged in the enchanting landscape of the Gulf of Patti and of the Æolian Islands; the excursion to this volcanic archipelago has been, by unanimous consensus, one of the highlights of the meeting. Then, there was the splendid concert performed in the old church of historic Tyndaris, another highly appreciated social event of this ASI. Finally, the participants greatly enjoyed the comfortable hotel accommodation and facilities, and the Sicilian cuisine.

In concluding these introductory notes, we would like to express our gratitude, first of all, to the International Advisory Committee, in the persons of M. Baus, B.J. Berne, P.V. Giaquinta and H.N.W. Lekkerkerker, and to the Local Organizing Committee, in the persons of its Secretary, Prof. Maria C. Abramo, and of the other members: Prof. G. Pizzimenti; Drs. D. Costa and F. Saija, who skilfully managed all Web information and connections, and word processed the editorial material of the School and of these Proceedings; Mr. S. Rando, who handled all mailing lists.

We also warmly thank Dr. G. Pellicane, Mr. N. Crisarà for the technical assistance delivered in the conference room and for video recording all the talks, the hostesses of the ASI secretariat – Nunzi, Manuela and Sandra – and the Centro Elaborazione Statistica Informazione Scientifica Tecnologica (CESIST) of the University of Messina for technical support.

We would like to thankfully acknowledge: the substantial funding provided to the ASI by the NATO Scientific Affairs Division in Bruxelles; the patronage and sponsorship offered by the University of Messina; the financial support provided by the President of Regione Sicilia, by the President of the Sicilian Parliament, and by the following institutions: Gruppo Nazionale Struttura della Materia of CNR, Fondazione Bonino-Pulejo, Provincia Regionale di Messina, Azienda Autonoma Provincialc Incremento Turistico (AAPIT), Comune di Patti, Department of Physics of the University of Messina.

One of us (C.C.) wishes finally to express his gratitude to his wife, Maria, for her continuous encouragement since the idea of organizing this NATO-ASI was conceived.

Messina, November 1998

Carlo Caccamo, Jean-Pierre Hansen and George Stell

1. L.L. Lee
2. D. Di Caprio
3. F. Lado
4. N. B. Wilding
5. G. Stell
6. Y. Martinez-Raton
7. M. E. Fisher
8. L. Reatto
9. G. Pellicane
10. C. Caccamo
11. Sa. Dalgic
12. L.M. Saboungi
13. M.P. Tosi
14. J.O. Indekeu
15. G. Parisi
16. M. Dijkstra
17. S. Klapp
18. E. Bruno
19. N. Herrera-Pacheco
20. Y. Levin
21. J.P. Hansen
22. E. Kierlik
23. D. Levesque
24. M.M. Telo Da Gama
25. A. Robledo
26. V. Weiss
27. Se. Dalgic
28. K.P. Schneider
29. G. Pastore
30. J.S. Høye
31. R.J.F.L. De Carvalho
32. G.A. Martynov
33. M. Alvarez Fernandez
34. T. Coussaert
35. A. Malijevsky
36. A. Perera
37. Y. Rosenfeld
38. S. Jorge-Gonzáles
39. Z. Akdeniz
40. D. Costa
41. R. Evans
42. P. Jedlovszky
43. V. Paounov
44. M.P. Gaigeot
45. R. Roth
46. R.P. Sear
47. B.M. Mulder
48. S. Ramos
49. M.L. Klein
50. D. Goulding
51. M. Watzlawek
52. C. Bauer
53. B. Groh
54. A. Yethiraj
55. M. Björling
56. V. Balevicius
57. P. Gianinetti
58. K. Koga
59. N. Brilliantov
60. A. Reiner
61. J. Valleau
62. J.M. Romero Enrique
63. D. Pini
64. M.I. Rodriguez Ponce
65. L.R. Pratt
66. F. Saija
67. P. Tarazona
68. R. Van Roij
69. M. Aksimentiev
70. S. Grollau
71. J.M. Cordeiro
72. M. Benakli
73.
74. D. Price
75. M. Van Der Hoef
76. A. Giacometti
77. M. Tasinkevych
78. M. Reinaldo Falagán
79. T. Biben
80. V. Sokhan
81. V. Ballenegger
82. M.C. Abramo
83. J.F. Dufrêche
84. N.G. Almarza
85. H. Löwen
86. H. Xu
87. S. Dietrich
88. M.L. Rosinberg
89. G. Kahl

CHARGED AND POLAR LIQUIDS, PHASE TRANSITIONS AND METASTABILITY

UNDERSTANDING CRITICALITY:

Simple Fluids and Ionic Fluids

M. E. FISHER
Institute for Physical Science and Technology,
University of Maryland, College Park, Maryland 20742, U.S.A.

Critical behavior is thoroughly understood in Ising models *alias* lattice gases: But how far does that help our insight into the critical region of continuum models and real fluids? In particular, what causes some ionic solutions to exhibit van der Waals or "classical" critical exponents, while other, so-called "solvophopic" systems are of Ising-type?

Aspects of progress in answering these questions theoretically will be reviewed, highlighting the outstanding experimental, theoretical, and computational challenges. Theory and computation now show convincingly that continuum fluids with short-range or van der Waals, $1/r^6$, interaction potentials should exhibit Ising-type exponents ($\alpha \simeq 0.11, \beta \simeq 0.32_5, \gamma \simeq 1.24, \nu \simeq= 0.63$), etc., but with both analytic *and* singular corrections away from the critical point [1]: and experiments agree. But observations by Pitzer and others have shown that certain *ionic* solutions exhibit classical criticality ($\alpha = 0(discon.), \beta = \frac{1}{2}, \gamma = 1, \nu = \frac{1}{2}$) *or* "crossover" to Ising behavior only very close to criticality [2-6]. Why?

Various theoretical attempts have been made to understand these results [3,7,8] springing, ultimately, from the famous work of Debye and Huckel (DH) [9,10] in 1923 and the pioneering contribution by Bjerrum [11] who, in 1926, demonstrated the importance of *ion-pairing* at low temperatures which, in particular, proves crucial for the critical region. Subsequently, many attempts have been made using integral equations: but most of the standard approaches fail badly! The GMSA provides a passable account of the critical region but proves inferior [12] to recent extensions of the original DH self-consistent meanfield approach by Y. Levin, X-J. Li, B.P. Lee, and the author [13-19]. It must also be noted that the integral equations cast essentially no light on the issue of Ising-type *versus* classical critical behavior.

More concretely, in the context of the restricted primitive model (RPM), namely, equisized hard-spheres of diameter a carrying charges $\pm q$ in a uniform medium of dielectric constant D, the *full* Debye-Huckel (DH) ther-

C. Caccamo et al. (eds.), New Approaches to Problems in Liquid State Theory, 3–8.

modynamic theory (that recognizes ionic hard cores) has been extended by Levin and Fisher [13, 14] to include (+, -) ion pairing, the consequent *depletion* of the remaining "free" ions, and dipolar-pair/ionic-fluid solvation contributions. The results, while still predicting only classical criticality, are in quite satisfactory agreement with recent Monte Carlo simulations: see the discussion in [3(b)]. Furthermore, upper bounds on the free energy obtained by Gillan [20] enable one to prove [12], for both DH- and MSA-based [21-24] theories, that inclusion of ion pairing *and* of the associated free-ion depletion is *essential* at the relatively low temperatures where criticality occurs in the RPM, namely, $T_c^* \equiv k_B T_c D a/q^2 \lesssim 0.052$ (The so-called PMSA theories [23, 24] fail because they allow for pairing but *not* for depletion).

At higher temperatures, however, all the pairing theories fail to respect the Second Law, violating thermal convexity, owing to problems in implementing the "chemical picture" of ion pairing and in specifying appropriately the association constant, $K(T)$. Indeed, both Bjerrum's [11] and Ebeling's [21, 22] prescriptions prove unacceptable at intermediate and high temperatures, $T \gtrsim 2T_c$. To overcome these problems (and possibly improve the description near criticality) an "exact" thermodynamic formulation of chemical association has been developed [25]. This allows, in particular, the $T = \infty$, purely hard-core interaction limit to be correctly and explicitly achieved at all densities (and for *arbitrary* association constants!). Furthermore, by calculating the precise "re-summed" low-density expansions for the *extended primitive model* with *three* distinct hard-core-interaction diameters, a_{++}, a_{+-}, and a_{--} [26], new insights into (+,-) pairing have been obtained leading to better forms for $K(T)$. It should be stressed, however, that at low temperatures all physically sensible forms for $K(T)$ agree asymptotically [14].

One would like to broach the nature of ionic criticality in the RPM — Ising, classical, or something else? — or in its variants [3, 7, 15, 16] and extensions, by using a renormalization group approach. Within the usual ϵ-expansion, field theoretic formulations, however, that demands [3(a), 13] a fully adequate understanding at the mean-field level. Simply performing a Kac-Hubbard-Stratonovich transformation on the charge degrees of freedom leads to variants of a sine-Gordon model which then encounters serious difficulties: see, e.g., [27].

As an initial alternative, it is tempting [13] to employ a Ginzburg criterion which, ideally, should provide a semiquantitative measure of the domain of reliability of a mean-field or classical theory outside (or, perhaps, down to) the immediate critical vicinity. For that purpose, however, its is crucial to have a mean-field description of the *diverging* density fluctuations near criticality or, equivalently, of the *density-density* correlation function, $G_{NN}(\mathbf{r}; T, \rho)$, and, in particular, of the associated density correlation length(s) $\xi(T, \rho)$ ($\xi_{N,1}$, $\xi_{N,2}$, etc., [17-19, 26, 28]) where ρ is the overall ionic density. One would also like a parallel account of the associated (but *non*ordering) charge correlation function, $G_{ZZ}(\mathbf{r};T,\rho)$, near criticality. Until recently any such description has been lacking (except within a GMSA treatment [29] which has marked defects [3, 13, 17]).

However, by generalizing the DH charging approach, one can actually derive [17] a *free-energy functional* for slowly varying, locally neutral, ionic (and pair) densities, $\rho_\sigma(\mathbf{r})$, from which density-density correlations follow by functional differentiation. Not surprisingly, one finds that the inverse Debye Length, $\kappa_D(T,\rho)$, entering the generalized DH partial differential equation (or "linearized Poisson-Boltzmann equation") now becomes spatially dependent [30].

The calculated second-moment *density* correlation length, $\xi_{N,1}(T,\rho)$, is found to diverge like $1/(T\rho)^{1/4}$ when the overall ion density, ρ, approaches zero: furthermore, this predicted small-ρ behavior is *universal* (i. e. independent of the hard core diameter a) and proves to be *exact* ! (Recall that the DH result for the limiting *charge* correlation or "screening" length varies quite differently, namely, as $\xi_{Z,1} \approx 1/\kappa_D \sim (T/\rho)^{1/2}$). Indeed, by carefully carrying out the re-summed Mayer cluster diagram calculations for the correlation functions in the form developed by Meeron [31], one can obtain the exact density expansions for $\xi_{N,1}$, for the fourth-moment correlation length, $\xi_{N,2} \sim T^{1/8}/\rho^{3/8}$, and for the "true" or exponential decay length, $\xi_{N,\infty} \approx 1/2\kappa_D \sim (T/\rho)^{1/2}$. The exact magnitude of the divergence of $\xi_{N,1}$ when $\rho \to 0$, agrees precisely with the prediction of the generalized DH (GDH) theory [17, 18]. This is in contrast to an approximation developed subsequently by Schröer and Weiss [32] which does obtain the correct form for $\xi_{N,1}(T,\rho)$ when $\rho \to 0$ (unlike the GMSA or MSA [29]) but which predicts an amplitude about 5.7 greater than the exact (GDH) value! Note that the Ginzburg *critical crossover scale*, $t_G = |T_c - T_G|/T_c$, requires the magnitude of $\xi_{N,1}$ in the critical region raised to the *sixth* power [18, 29].

In the critical region, it is thus reasonable to use the GDH correlation length, $\xi_{N,1}$, to compute the crossover scale for departure from classical/mean-field behavior using the Ginzburg criterion [18]: On doing this one discovers that the restricted primitive model electrolyte should exhibit *Ising*-type criticality, contrary to some expectations or at least, hopes! [3, 18]. The reliability of this conclusion can be questioned theoretically although it agrees with inferences later drawn from simulations by Caillol, Levesque and Weis [33]. However, in other simulations by Valleau and Torrie [34] no indication of the corresponding Ising-type specific heat peak was seen above T_c. The experimental situation, even as regards the behavior of Pitzer's salt [5], has become less clear: at very least, an exceptional sensitivity to small variations in composition or to surface conditions, etc., has been uncovered [3, 4, 6]. Thus profound puzzles remain!

On the theoretical side, GDH theory can also be extended to the calculation of the *charge-charge* correlation function, $G_{ZZ}(\mathbf{r})$ [19, 28]. The analysis must now allow for the effects of a locally imposed *charge imbalance* [35]. The resulting expression satisfies both (i) the charge neutrality condition *and* (ii) the Stillinger-Lovett second-moment condition [36] for all T and ρ, and (iii) exhibits charge oscillations for densities above a "Kirkwood line" [29] in the (ρ,T) plane. This corrects the simple $e^{-\kappa r}/r$ form normally said to characterize DH charge correlations. Again, the low-density predictions can be checked diagrammatically [26] although now Meeron's analysis must be supplemented by the HNC re-summation theorem [37]. This analysis yields novel results for the fourth-moment charge correlation length, $\xi_{Z,2}(T,\rho)$, that go beyond the (correct) universal limiting predictions of GDH theory. At this point, then, GDH theory provides the only complete description of ionic correlation functions, as judged by (i)-(iii) together with (iv) exact behavior at low densities and (v) reasonable behavior near criticality.

In the applications of GDH theory to the critical region and to implement the Ginzburg criterion, ion-pairing and dipole-pair/ionic-fluid solvation were incorporated [18]. The same can be done when the charge correlation functions are calculated. In that case, however, one finds [19] that the Stillinger-Lovett relation is violated unless the "background dielectric constant", D is allowed a linear dependence on the pair density $\rho_2(T,\rho)$. For the RPM, however, that probably represents a defect of the pairing treatment which one might attempt to correct by allowing suitably for the effects of the dipoles on the free ions: the matter appears, however, to be more delicate than supposed in some treatment [38].

Finally, the role of the charge-density oscillations predicted by GDH [19] and earlier theories [39, 40] demands further study. The oscillations in $G_{ZZ}(\mathbf{r};T,\rho)$ are predicted to set in for $x \equiv \kappa_D a \gtrsim 1.18$; but when κ_D is evaluated using the (residual) free-ion density (i.e., allowing for depletion) one finds critical-point values of x_c close to unity [41]. Consequently, the "incipient charge ordering" which the oscillations represent, may well compete with the density fluctuations in the critical region (through, say, nonlinear terms in an appropriate, effective or LGW Hamiltonian describing criticality: see [3(a)] and [42]). Such behavior could, perhaps, invalidate the simple Ginzburg criterion estimates [18] and, possibly, account for the seemingly great sensitivity of some ionic critical points to small perturbations. At this point, however, these suggestions remain purely speculative.

Acknowledgments

The active collaboration of Yan Levin, Xiao-jun Li, Ben Lee, Daniel Zuckerman and Stefan Bekiranov and interactions with J.-P Hansen, C.W. Outhwaite, G. Stell, J.M.H. Levelt Sengers and W. Schröer have been much appreciated. Support from the National Science Foundation (through Grant Number CHE 96-14495) is gratefully acknowledged.

References

1. M.E. Fisher, Rev. Mod. Phys. **46**, 597 (1974) and references therein.
2. J.M.H. Levelt Sengers and J.A. Given, *Molec. Phys.* **80**, 989 (1993).
3. M.E. Fisher (a) *J. Stat. Phys.* **75**, 1 (1994); (b) *J. Phys. Cond Matt.* **8**, 1 (1996) and references therein.
4. W. Schröer *et al.*, *J. Phys. Cond. Matt.* **8**, 9321 (1996) and references therein.
5. K.S. Pitzer, *Accts. Chem. Res.* **23**, 333 (1990) and later work cited in [3, 4, 6].
6. But some puzzling recent observations on Pitzer's solution have, on the contrary, exhibited Ising-type behavior: S. Wiegand *et al.*, *J. Chem. Phys.* **106**, 2777 (1997); W. Schröer, J.M.H. Levelt Sengers and coworkers (private communication June 1997).
7. G. Stell (a) *J. Stat. Phys.* **78**, 197 (1994); (b) *J. Phys. Cond. Matt.* **8**, 9329 (1996).
8. J.S. Høye and G. Stell, *J. Molec. Liquids* [in press] (1997).
9. P.W. Debye and E. Huckel, *Phys. Z.* **24**, 185 (1923).
10. For a good modern account of Debye-Huckel theory, see e.g., D.A. McQuarrie, *Statistical Mechanics* (Harper Collins, New York, 1976) Chap 15.
11. N. Bjerrum, *Kgl. Dan. Vidensk. Selsk. Mat.-fys. Medd.* **7**, 1 (1926).
12. D.M. Zuckerman, M.E. Fisher, and B.P. Lee, *Phys. Rev. E* **56**, 6569 (1997).
13. M.E. Fisher and Y. Levin, *Phys. Rev. Lett.* **71**, 3826 (1993).
14. Y. Levin and M.E. Fisher, *Physica A* **225**, 164 (1996). See [12] for errata.
15. X.-J. Li, Y. Levin and M.E. Fisher, *Europhys. Lett.* **26**, 683 (1994).
16. Y. Levin, X.-J. Li and M.E. Fisher, *Phys. Rev. Lett.* **73**, 2716 (1994).
17. B.P. Lee and M.E. Fisher, *Phys. Rev. Lett.* **76**, 2906 (1996) and to be published.
18. M.E. Fisher and B.P. Lee, *Phys. Rev. Lett.*, **77**, 3561 (1996).
19. B.P. Lee and M.E. Fisher, *Europhys. Lett.* **39**(3) (1997) and to be published.
20. M.J. Gillan, *Molec. Phys.* **41**, 75 (1970).

21. W. Ebeling and M. Grigo, *Ann. Phys. (Leipzig)* **37**, 21 (1980).
22. M. Grigo and W. Ebeling, *J. Soln. Chem.* **13**, 321 (1984); but note numerical and other deficiences are present in this article and [21].
23. Y. Zhou, S. Yeh and G. Stell, *J. Chem. Phys.* **102**, 5785 (1995).
24. S. Yeh, Y. Zhou and G. Stell, *J. Phys. Chem.* **100**, 1415 (1996) and references therein.
25. M.E. Fisher and D.M. Zuckerman [to be published].
26. S. Bekiranov and M.E. Fisher, *Bull. Am. Phys. Soc.* **42** (1) 725 (1997) Q16 8; and to be published.
27. N.V. Brilliantov *et al.*, preprint posted as cond-mat/9706295, 30 June 1997.
28. M.E. Fisher, B.P. Lee and S. Bekiranov, Proc. Int. Conf. on "Strongly Coupled Coulomb Systems", Boston College, 3-10 August 1997, Ed. G.J. Kalman, K. Blagoev and J.M. Rommel (Plenum Publ. Corp, 1998) [in press].
29. R.J.F. Leote de Carvalho and R. Evans, *Molec. Phys.* **83**, 619 (1994); *J. Phys. Cond. Matt.* **7**, L575 (1995).
30. Somewhat analogous results appear in treatments of electrical double layers [see S.L. Carnie and G.M.Norrie *Adv. Chem. Phys.* **56**, 141 (1984) Sec. IV.B and references therein] via MPB theory as expounded by Outhwaite [40].
31. E. Meeron, *J. Chem. Phys.* **28**, 630 (1958).
32. V.C. Weiss and W. Schröer, *J. Chem. Phys.* **106**, 1930 (1997); W. Schröer and V.C. Weiss, *J. Chem. Phys.* **106**, 7548 (1997).
33. J.M. Caillol, D. Levesque, and J.J. Weis, *Phys. Rev. Lett.* **77**, 4039 (1996); *J. Chem. Phys.* **107**, 1565 (1997).
34. J. Valleau and G. Torrie, *J. Chem. Phys.* (1998) [in press].
35. One may compare with E. Trizac and J.-P. Hansen, *J. Phys. Cond. Matt.* **8**, 9191 (1996), Eq. (3.29).
36. F.H. Stillinger and R. Lovett, *J. Chem. Phys.* **48**, 3858 (1968).
37. J.M.J. van Leeuwen, J. Groeneveld, and J. DeBoer, *Physica* **25**, 792 (1959).
38. B. Guillot and Y. Guissani, *Molec. Phys.* **87**, 37 (1996).
39. J.G. Kirkwood, *J. Chem. Phys.* **2**, 351 (1934).
40. C.W. Outhwaite in *Statistical Mechanics: A Specialist Periodical Report*, Vol. **2**, ed. K. Singer (The Chemical Society, London, 1975).
41. It is also interesting that the simplest GDH result for $G_{ZZ}(\mathbf{r})$ predicts *undamped* charge oscillations, corresponding to "ionic crystallization", that set in for $x > x_X \simeq 6.652$ with an initial wavelength $\lambda_X \simeq 1.847a$ [19] although, clearly, the GDH theory cannot be trusted at the high values of ρ/T implied by this criterion.
42. V.M. Nabutovski *et al.*, *Phys. Lett.* **79A**, 98 (1980); *Molec. Phys.* **54**, 979 (1985).

EQUATION OF STATE AND CRITICAL BEHAVIOR OF SIMPLE FLUID MODELS

J. S. HØYE

Institutt for fysikk, NTNU; N-7034 Trondheim, Norway

Abstract.

The equation of state for fluids is considered starting with Van der Waals equation. Corrections from γ-ordering for long-range forces and its renormalization are dealt with and related to the mean spherical approximation (MSA) both for lattice gases and continuum fluids. The self-consistent Ornstein-Zernike approximation (SCOZA) is then considered and results so far for this approach are discussed. The results turn out to have good general accuracy. This also holds for the critical region which is discussed with respect to general properties, critical indices, and deviation from standard scaling.

1. Introduction

Equations of state (EOS) for fluids have been of interest for long. On one side this is a major problem in statistical mechanics. On the other side accurately predicting properties of real fluids is of great interest by applications. To obtain an EOS for fluids the famous Van der Waals equation was developed more than hundred years ago. This equation and modified versions of it are still used today by engineering applications. However, improving this equation has been a challenging problem in statistical mechanics. E.g. for low densities the virial expansion, which is useless for higher densities, can be used. Especially the critical region is difficult to treat more accurately. So theories usually have been restricted to various types of mean-field or Van der Waals like theories. Such theories are most inaccurate in the critical region, and they are in no way able to capture the singular behavior of thermodynamic quantities as the critical point is approached. This singu-

C. Caccamo et al. (eds.), New Approaches to Problems in Liquid State Theory, 9–29.

lar behavior is connected to density fluctuations whose correlation length grows towards infinity.

New insight and progress towards understanding the mechanism of critical behavior was obtained by Wilson and Kadanoff [1] by introduction of homogeniety from which the well-known scaling relations followed. Further on, renormalization group methods were introduced by Wilson after which Wilson and Fisher showed how critical exponents could be obtained as expansions in 4-d where d is dimension [2].

However, globally more accurate treatments of systems at thermal equlibrium was less developed. Some work was performed by the author in his thesis [3]. This work was based on the γ-ordering for long-range forces that was used by Hemmer and by Lebowitz, Stell, and Baer [4] to obtain corrections to Van der Waals equation. From this study it became clear that corrections to mean field theory beyond the high temperature contribution, were relatively small except in a region around the critical point. But in this region corrections to mean-field would diverge. To some extent this was rectified by a resummation or renormalization, and results that compared well with experimental results for Ar (argon) were obtained.

For the lattice case it became clear that the resummed γ-ordering was essentially the mean spherical approximation (MSA) [3]. So the critical properties with somewhat "irregular" isotherms were the same.

The MSA for fluids and modified versions of it were studied extensively by Chandler et al. [5], and accurate results were obtained outside the critical region. However, critical properties remained mean-field like. Also various other integral equation methods have been developed and refined to describe fluids [6].

Høye and Stell proposed to apply thermodynamic self-consistency in connection with the MSA for fluids [7]. For this approach, the self-consistent Ornstein-Zernike approximation (SCOZA), they developed equations for solving the self-consistency problem. Preliminary numerical results for supercritical temperatures suggested that SCOZA might produce good results of interest [8]. Recently this has been verified numerically in various cases, and high general accuracy is obtained also in the critical region [9, 10, 11, 12, 13, 14]. However, standard scaling is not obtained in this singular region, but a form of extended scaling that seems to represent features of real systems is found [10, 15]. This high accuracy can be surprising, but it seems to rely on the fact that the MSA form of the correlation function is close to its exact form. So the main defect of MSA itself is apparently its thermodynamic inconsistency which is remedied by SCOZA. Also one can note as argued by Høye and Stell [16], that in three dimensions the usual critical index η is close to its MSA value zero. Good results are also obtained when generalizing the SCOZA to a disordered system like the random field Ising model performing series expansions of the solution [17].

There is also another theory of global accuracy. This is the hierarchical reference theory (HRT) of Parola and Reatto [18]. This theory is based on renormalization group ideas that build in scaling while imposing a self-consistency condition, that although somewhat related, is different from the SCOZA onc. However, it results in a similar numerical problem of solving a non-linear partial differential equation of diffusion type.

In Sec. 2 we will start with Van der Waals or mean-field theory and focus upon some of its properties as an exact theory in the limit $\gamma \to 0$ where γ is the inverse range of interaction.

In Sec. 3 the first order γ-ordered contribution beyond mean field will be considered noting that it has only a perturbing effect also for realistic interactions of nearest neighbor type. But divergencies make it invalid in the critical region.

In Sec. 4 the γ-ordering of section 3 is resummed or renormalized such that the divergence follows the shift in the critical point location. By that thermodynamic self-consistency is improved greatly. And as found in Ref.[3] good results were obtained for Ar (argon) also in the critical region (disregarding "irregular" isotherms).

In Sec. 5 the renormalized γ-ordering is compared with the MSA. For lattice gases they are found to be eqvivalent while for continuum fluids the MSA is not properly "renormalized" and is thus different. This is a situation that is similar for other integral equation theories for fluids.

In Sec. 6 we impose thermodynamic self-consistency along with the MSA which then results in the SCOZA. As SCOZA imposes full thermodynamic self-consistency, the "irregularities" of the renormalized γ-ordering will be remedied and smoothed out.

In Sec. 7 numerical results obtained by application of SCOZA in various situations will be considered and discussed. So far only a limited number of situations have been considered in this respect. This limited number is partly connected to computational difficulties that had to be resolved in the subcritical region. However, the general picture is that SCOZA yields results of high accuracy quite generally for very different situations in one, two, and three dimensions. This we attribute to the analytic properties and general nature of the γ-ordering scheme which considers particles with a hard core plus an attractive tail as the interaction. Thus the method is not limited to a more specific interaction or dimension, and continuum as well as lattice gases or spin systems are included within the same framework.

In Sec. 8 focus will be put on the critical properties of SCOZA. Then both numerical and analytic results will be considered.

2. Van der Waals theory.

The famous Van der Waals EOS established more than one hundred years ago can be written

$$p = p_h - a\rho^2 \tag{1}$$

where p is pressure and ρ is number density. The physical interpretation is simple. The pressure p_h is due to excluded volume effects while the last term is due to the attractive interaction. Van der Waals used

$$\beta p_h = \rho f(x) \tag{2}$$

with

$$f(x) = 1/(1-x) \tag{3}$$

Here $x = \rho v$ where v represents the volume excluded by a particle, and $\beta = 1/k_B T$.

By use of statistical mechanics is was later shown that Van der Waals equation can be established as an exact limiting result for interactions that are infinitely weak and long-ranged [19]. Or more precisely, expression (3) is correct only for hard rods in one dimension. E.g. for hard spheres one can use the Percus-Yevick (PY) approximation

$$f(x) = \frac{1+x+x^2}{(1-x)^3}. \tag{4}$$

In this case $v = (\pi/6)d^3$ where d is hard sphere diameter. A more accurate expression is the Carnahan-Starling (CS) one where the term $-x^3$ is added to the numerator of (4). For lattice gases where the hard core is just the exclusion effect of one lattice site of unit volume, one has

$$\beta p_h = -\ln(1-\rho) \tag{5}$$

The quantity a in (1) is related to the attractive interaction $\psi(r)$ as

$$a = -(1/2)\tilde{\psi}(0) \tag{6}$$

where the tilde denotes Fourier transform. I.e. $\tilde{\psi}(0)$ is the integral over $\psi(r)$ with fourier variable $k = 0$.

3. γ-ordering

Due to the finite range of interaction there will be corrections to Van der Waals or mean field theory. E.g. for low density virial coefficients can be used, but for liquid densities they are useless. The work by Kac, Uhlenbeck, and Hemmer on an exactly solvable one-dimensional model initiated a new method [20]. They considered the Kac interaction $-\gamma \exp(-\gamma|x|)$. where the parameter γ is the inverse range of interaction. In the limit $\gamma \to 0$ Van der Waals EOS is obtained.

More generally the perturbing parameter γ was then used in connection with the Mayer graph expansion i statistical mechanics. It was found that the density expansion can be rearranged such that contributions will appear in increasing orders of γ [4]. By this rearranged graph expansion the repulsive hard (or soft) core interaction is regarded as a reference system whose properties are known. Furthermore the Mayer f-bond is separated into a hard core part and a part from the attractive interaction. The latter is further expanded into potential bonds

$$v(r) = -\beta\psi(r) \tag{7}$$

Formally one can write $\psi(r)$ in the form $\psi(r) = \gamma^d f(\gamma r)$ where d is dimensionality. The first order contribution to the free energy now comes from ring graphs with reference system hypervertices. The result is [3, 4]

$$I = -\frac{1}{2(2\pi)^d}\int [\ln(1-\tilde{\mu}(k)\tilde{v}(k)) + \tilde{\mu}(k)\tilde{v}(k)]\, d\mathbf{k} + I_{HT} \tag{8}$$

with

$$I_{HT} = \frac{1}{2}\int \rho^2 h_h(r) v(r)\, d\mathbf{r} = \frac{1}{2(2\pi)^d}\int (\tilde{\mu}(k) - \rho)\tilde{v}(k)\, d\mathbf{k}. \tag{9}$$

Here d is dimensionality, $\tilde{v}(k) = -\beta\tilde{\psi}(k)$, $h_h(r)$ is is the pair correlation function of the reference system, $\tilde{\mu}(k) = \rho + \rho^2\tilde{h}_h(k)$ represents the hypervertex that contains graphs from the reference system besides the usual ρ-vertex, and $I = -\beta\rho f$ where f is Helmholtz free energy per particle. The I_{HT}, which has been written as a separate term here, is the high temperature contribution. Contributions to pressure and chemical potential are obtained via

$$\beta p_1 = (1 - \rho\frac{\partial}{\partial\rho})I \quad , \quad \beta g_1 = -\frac{\partial I}{\partial\rho} \tag{10}$$

Expression (8) can be simplified noting that $\tilde{v}(k) = -\beta\tilde{f}(k/\gamma)$. Thus to leading order in γ one can put $\tilde{\mu}(k) = \tilde{\mu}(0) = \mu$. In this way I_{HT}, which represents a significant contribution (for realistic interactions), drops out. But I_{HT} is not crucial with respect to critical properties and will be incorporated later. With this (8) becomes (d=3)

$$I = -\frac{1}{2(2\pi)^3}\int [\ln(1-\mu\tilde{v}(k)) + \mu\tilde{v}(k)]\, d\mathbf{k}, \tag{11}$$

and further

$$\frac{\partial I}{\partial \rho} = \frac{1}{2}\mu' K \tag{12}$$

where $\mu' = \partial\mu/\partial\rho$ and

$$K = \frac{1}{(2\pi)^3}\int \frac{\mu(\tilde{v}(k))^2}{1-\mu\tilde{v}(k)}\,d\mathbf{k}. \tag{13}$$

Via the fluctuation theorem the μ can be directly related to the compressibility of the reference system

$$\frac{1}{\mu} = \frac{1}{\rho+\rho^2\tilde{h}_h(0)} = \frac{1}{\rho}\frac{\partial(\beta p_h)}{\partial\rho}. \tag{14}$$

So with (6) the lowest order result (1) is ($p = p_0$)

$$\frac{1}{\rho}\frac{\partial(\beta p_0)}{\partial\rho} = \frac{1}{\mu} - \tilde{v}(0). \tag{15}$$

One notes that (15) matches the correlation function $\tilde{h}(k)$ with its longe-range part to first order included. This contribution consisting of chain graphs with hypervertices is

$$\rho+\rho^2\tilde{h}(k) = \frac{\tilde{\mu}(k)}{1-\tilde{\mu}(k)\tilde{v}(k)}. \tag{16}$$

By including (11) in (15) an additional term will appear in the latter. With (10) and (12) this is

$$-\frac{\partial^2 I}{\partial\rho^2} = -\frac{1}{2}(\mu'' K + \mu' K') \tag{17}$$

and a mismatch will show up. In principle this will be corrected for by higher order terms. However, the more serious problem present originates in the divergence of the integrals for K and I. This takes place at the mean field critical point (and elsewhere along the spinodal curve) given by

$$1-\mu\tilde{v}(0) = 0. \tag{18}$$

Further (17) adds a positive term to (15) at the critical point since there $\mu' = 0$ and $\mu'' < 0$. This follows from the condition $\partial^2 p_0/\partial\rho^2 = 0$ which then also means that μ has its maximum at the critical density ρ_c. Thus this added term indicates a lowering of the critical temperature T_c, but the divergence make expressions invalid.

Now the divergence might be cured by including the I_{HT} piece in (8) such that T_c is moved away from the divergence. (The latter will though depend upon the choice of perturbing interaction inside the hard cores .) But this will only disguise the problems and singularities of the critical region that are intimately connected to long-range density correlations.

The numerical consequencies of including (11) were evaluated in [3] where it was applied to the Lennard Jones fluid with pair interaction

$$\varphi(r) = \frac{1}{r^{12}} - \frac{1}{r^6} \tag{19}$$

when using dimensionless units. The perturbing interaction was then chosen as $\psi(r) = \varphi(r)$ for $r > 1$ and $\psi(r) = 0$ for $r < 1$, and the repulsive part was replaced by an temperature dependent effective hard core that made its second virial coefficient exact. For the reference system hard spheres the Percus Yevick expression (3) was used. With this the mean-field theory (1) gives the critical parameters

$$T_c = 0.272 \qquad \rho_c = 0.268 \qquad z_c = \beta p/\rho = 0.360 \tag{20}$$

(with effective hard core volume $v = 0.481$ at T_c).

Due to the divergence the contribution (11) becomes useless in the critical region. However, one may expand the logaritm, and with a "suitable" number of terms (about 18) one can extrapolate to estimate T_c and ρ_c using the empirical rules of rectilinear diameter and power law for the curve of coexistence. This gave the estimates [3]

$$T_c = 0.251 \qquad \rho_c = 0.288 \qquad z_c = 0.293. \tag{21}$$

Here one notes the lowering of T_c, some shift in ρ_c, and the considerable lowering of the critical ratio z_c. These numbers can be related to experimental values for Ar which is represented well by a Lennard Jones fluid [21]. Scaled with interaction parameters used in Ref. [21] the latter were

$$T_c = 1.26/4 = 0.315 \qquad \rho_c = 0.318 \qquad z_c = 0.291. \tag{22}$$

Comparing (21) with this one notes that its z_c is essentially correct which contrasts the one of (20) and other mean-field results. But its T_c is clearly too low.

4. Critical renormalization.

From the results of the previous section it can be concluded that the first γ-ordered contribution has only a perturbing effect upon mean-field theory even for realistic interactions. Also this contribution is most dominant or important in the critical region as it decays away from it. A key problem in this connection is the attributed divergence. To try to partly remedy the divergence one can try to carry it along with the shift of the critical point. One then notes that at the (mean-field) critical point the last term of (17) vanishes as $\mu' = 0$ and K has maximum. Thus adding (17) to (15) one obtains a renormalized hypervertex μ_c such that

$$\frac{1}{\rho}\frac{\partial(\beta p)}{\partial \rho} \approx \frac{1}{\mu_c} - \tilde{v}(0) \tag{23}$$

where (when deleting the last term of (17)

$$\mu_c = \mu + \mu'' \mu \mu_c K/2 \tag{24}$$

The last expression can be identified with the short-range part of the correlation function integrated. I.e. its last term represents attaching a chain bond graph with two endpoints μ (and/or μ_c) to a hypervertex μ. Its second derivative μ'' then appears from the number of different ways to attach the chain bond to it [3, 4]. I.e. a graph with a number of n ρ-vertices yields the combinatorial factor $n(n-1)/2$. (Here the range of the vertex is again regarded as a δ-function. An exact treatment of it would involve four-body correlations.) Consistent with the renormalization of μ Eq. (13) is replaced by

$$\begin{aligned} K &= \frac{1}{(2\pi)^3}\int (\tilde{C}(k) - \tilde{v}(k))\, d\mathbf{k} \\ \tilde{C}(k) &= \frac{\tilde{v}(k)}{1 - \mu_c \tilde{v}(k)} \end{aligned} \tag{25}$$

The $\tilde{C}(k)$ is the Fourier transform of the chain bond. Note that the single v-bond is subtracted away in (25) as it will not contribute anyway when $v(0) = 0$ and is otherwise consistent with the neglect of I_{HT} in (8). With (24) or actually using [3]

$$\mu_c = \mu + \mu'' \mu_c^2 K/2 \tag{26}$$

the critical point can be evaluated directly from the divergence of the integrand in (25) (like (18) was used for mean field). Then the maximum value $\mu_c K_m$ of $\mu_c K$ is needed, and it was found to be

$$\mu_c K_m = 0.110. \tag{27}$$

Use of the PY expression (4) in (14) then gives

$$\mu = \rho(1-x)^4/(1+2x)^2 \tag{28}$$

from which μ'' is obtained. With the above inserted the maximum of μ_c can be obtained from Eq.(26). With $-\tilde{\psi}(0) = 8\pi/9$ the critical parameters became

$$T_c = 0.2545 \qquad \rho_c = 0.300 \qquad z_c = 0.286 \tag{29}$$

where the full EOS was needed to evaluate z_c. One sees that these values are similar to the extrapolated ones (21).

The major difference from Sec. 3 is that the renormalized γ-ordering yields results that are valid in the critical region. But near the critical point things still remain somewhat inconsistent, and isotherms become "irregular" such that for instance critical indices can not be defined. But nevertheless the singular behavior of the criticality of real fluids is clearly reflected. E.g. the coexistence curve for gas-liquid equilibrium is flattened on top due to the lowering of T_c found above. Also investigating the singular nature of the integral K in three dimensions one finds

$$\Delta(\mu_c\beta) = (const\,\Delta T + const\,(\Delta\rho)^2)^2 \tag{30}$$

for deviations from critical values. Inserted in (23), which will be the EOS via the fluctuation theorem, this gives for the critical isotherm and isochor

$$\Delta(\beta p) \sim (\Delta\rho)^\delta \quad \text{and} \quad \partial(\beta p)/\partial\rho = (\Delta T)^\gamma \tag{31}$$

respectively with critical indices $\delta = 5$ and $\gamma = 2$. These values are also the ones of the MSA to be dealt with in the next section.

The high temperature contribution I_{HT} in Eq.(8) can also be included. Formally it is of higher power in γ, but for realistic interactions it is of substantial magnitude at liquid densities due to the correlations of hard particles. To avoid direct interference with the critical renormalization performed above the I_{HT} can be regarded as part of the reference system, and its contribution can be included in p_h of Eq.(1) (via (10)). By that the previous quantity μ defined by (14) changes into a new quantity that we can call μ_v and which will replace μ in (14) and the equations that follow it. The main effect of this is a substantial shift in critical temperature, and it yields

$$T_c = 0.320 \qquad \rho_c = 0.330 \qquad z_c = 0.296 \tag{32}$$

that compares favorably with the experimental ones given by (22).

Concerning the renormalization it can be noted that replacing μ with μ_c in Eq.(11) will not be sufficient as the term $-\mu''(\mu_c K)^2/8$ should be added too. I.e. inserting expression (26) in (11) will count the appropriate second order contribution in γ twice. Thus this correction is needed to obtain proper results. (At critical point its value is about 0.002 as follows by use of (20), (26), (27), and (29). The maximum value of I in (11) is less than 1/4 of the value of $\mu_c K_m$, e.g. around 0.02. This shows the perturbative effect of higher order terms.) The remaining second order contribution that involves μ' is not included here, but $\mu' = 0$ at the mean-field critical point. The "irregular" isotherms mentioned above have their root in the last term of expression (17). Near the critical point (disregarding the shift in ρ_c) $\mu' \sim \Delta\rho$ and $K' \sim \Delta\rho$ (from Eqs.(25),(26), and (30)) such that $-\mu' K' \sim -(\Delta\rho)^2$, and for small $\Delta\rho$ this will dominate the variation in μ_c as follows from (30). When added to (23) this will make $\partial p/\partial\rho < 0$ on the critical isotherm, and this kind of inconsistency can not be removed in an easy way.

To conclude this section one notes that the renormalized γ-ordering yields an EOS that is near the exact one globally.

5. MSA

The mean spherical approximation (MSA) is the solution of the Ornstein-Zernike (OZ) equation with bondary condition [22]

$$\begin{aligned} h(r) &= \quad -1 \qquad && r < 1 \\ c(r) &= -\beta\varphi(r) && r > 1 \end{aligned} \tag{33}$$

for particles of unit diameter and pair interaction $\varphi(r)$. The hard core condition on $h(r)$ is exact while the condition on $c(r)$ is approximate. In the lattice case the conditions are similar, and in fact the MSA originates from the spherical model for spin systems [23]. Fourier transformed the OZ equation reads

$$\tilde{h}(k) = \tilde{c}(k) + \rho\tilde{c}(k)\tilde{h}(k) \tag{34}$$

Solved with respect to $\tilde{h}(k)$ this yields expression (16) if one identifies

$$\tilde{\mu}(k) = \rho/(1 - \rho\tilde{c}_0(k)) \qquad \tilde{v}(k) = \tilde{c}(k) - \tilde{c}_0(k) \tag{35}$$

such that $\tilde{\mu}(k)$ or $\tilde{c}_0(k)$ is the solution for PY hard spheres, and $\tilde{v}(k)$ defines the perturbing interaction. By this identification the MSA is nothing but the first order contribution in γ-ordering where the perturbing interaction inside hard cores is specified such that the exact core condition is fulfilled. This specification can be regarded as an optimization by which MSA plays a crucial role as the leading correction to mean-field theory. The MSA and modifications of it has been studied extensively, and generally it works well outside the critical region [5].

In the critical region properties of the MSA differ for lattice gases and continuum fluids. This can be understood from the renormalization described in the previous section. Consider then Eq.(26) for the renormalized hypervertex μ_c. For the lattice case Eqs.(5) and (14) yield $\mu = \rho(1-\rho)$, and thus $\mu'' = -2$ such that (26) can be written

$$\mu = \mu_c + \mu_c^2 K = \frac{1}{(2\pi)^3} \int \frac{\mu_c}{1 - \mu_c \tilde{v}(k)} \, d\mathbf{k}. \tag{36}$$

With $\mu = \rho(1-\rho) = (1-m^2)/4$ ($\rho = (1+m)/2$) where m is the magnetization for spin systems, this is nothing but the MSA core condition. Thus since they coincide the renormalized γ-ordering and MSA share critical properties as established in the previous section. Nevertheless, the MSA is somewhat inaccurate for the nearest neighbor Ising model on a simple cubic lattice say. This is related to the remaining thermodynamic inconsistency accompanying the relatively large shift in T_c as one then will find $\mu = 1.516..\mu_c$. I.e T_c is lowered by 34%. This contrasts somewhat the Lennard-Jones fluid where results (20) and (29) yield a lowering of only 6.5%.

Now consider MSA in the continuum case. The shift from $1/\mu$ to $1/\mu_c$ in Eq.(23) is the equivalent to adjusting the perturbing interaction inside the hard core. For small γ the MSA also gives a shift like (24), but with $-\mu''/2$ replaced with its low density value $8v = 4\pi d^3/3$ (d is hard core diameter) for all ρ. However, $-\mu''/2$ itself decreases rapidly with ρ (and even changes sign), and in accordance with the lowering of T_c one finds its value to be $0.16\,(8v)$ at ρ_c ($v = 0.48$). Thus the true shift in T_c will be much smaller than predicted by MSA compressibility route. So using the energy route thermodynamics will yield a mean-field type critical point away from the singularity. I.e. the two routes are clearly inconsistent, but the energy route will be the more accurate one. The reason is that the compressibility route requires one order higher in γ to obtain the same accuracy since it does not include the (longe-range) interaction when integrating.

From the above analysis it also follows that integral equations like HNC (Hypernetted chain) and PY usually do not improve upon MSA for fluids. They add contributions to next order in γ, but do not capture the crucial μ'' term discussed above. However, for ionic fluids the HNC is recognized to work well in the electrolyte regime. But this is connected to graph cancellations in symmetric electrolytes [24]. Both HNC and PY are unable to describe the critical region and a critical point can not be reached [25].

6. SCOZA

To improve the EOS beyond the renormalized γ-ordering is in no way obvious especially in the critical region. Høye and Stell proposed the self-consistent Ornstein-Zernike approximation (SCOZA) [7]. This method builds directly upon the MSA, and for continuum fluids they developed equations necessary for solving the self-consistency problem utilizing the explicit analytic solutions for the cases of one and two Yukawa terms in the interaction potential. The first ideas about a SCOZA approach were already made by Stell et al., but this was left unpublished due to difficulties to obtain numerical results [26]. Ideas along the same line were later dealt with by Høye, Lebowitz, and Stell in their work on polar and ionic fluids introducing the generalized MSA (GMSA). But thermodynamic self-consistency was less in focus there, although the notion SCOZA was introduced later [24, 27].

The idea with SCOZA is to impose thermodynamic self-consistency. I.e. based upon the pair correlation function one should have the same EOS both via the internal energy and the compressibility (fluctuation theorem). For the continuum case a third independent route can be imposed, namely the virial theorem [7].

The SCOZA is based upon the thermodynamic relation

$$\frac{\partial(\beta\chi^{-1})}{\partial\beta} = \frac{\partial^2(\rho u)}{\partial\rho^2} \tag{37}$$

where $\rho\chi^{-1}$ is the inverse compressibility $\partial p/\partial\rho$ and u is the internal energy per particle. (For spin systems one has the substitutions $\rho = (1+m)/2$ and $\chi^{-1} = \partial H/\partial m$ where H is magnetic field.) As both χ^{-1} and u can be expressed via the same pair correlation function $h(r)$ one obtains a partial differential equation the solution of which can be used to determine one parameter and thus enforce the same EOS via two independent routes. E.g. this parameter can be the use of an effective inverse temperature c_1 instead of β in the MSA problem (33). The advantage of this simple recipe is that the solution of the OZ equation will be as for the MSA problem itself. (But other more complex recipes will be of interest too.)

In Refs.[7] and [8] the quantity $S = -\beta\rho f$ where f is Helmholtz free energy per particle was the one to be solved for self-consistency. Then the SCOZA equation became $\partial S/\partial\beta = -\rho u$ with ρu eliminated in favour of $\beta\chi^{-1} = -\partial^2 S/\partial\rho^2$ as follows from the solution of the OZ equation. For supercritical temperatures this problem can be solved by standard methods, but serious difficulties were encountered below T_c. This is due to the bondary condition along the on beforehand unknown spinodal curve determined by $\rho\chi^{-1} = 0$, and where the "diffusion" constant of the linearized equation for deviations ΔS becomes infinite. Dickman and Stell then proposed to use Eq.(37), which is the second derivative of the one for S with respect to ρ [9]. In this way the bondary condition could be handled as then χ^{-1} or u can replace S as variable. But still the bondary condition has to be handled carefully to obtain accurate results.

In the lattice (or spin) system case the analytic structure from solution of the OZ equation is relatively simple. Then one has [9, 10, 11, 13]

$$\tilde{c}(\mathbf{k}) = c_0 + c_1\tilde{\psi}(\mathbf{k}) \tag{38}$$

where c_1 replaces β in (33) (and $\psi(r) = -\varphi(r), \psi(0) = 0$). With normalization $\tilde{\psi}(0) = 1$ the mean-field contribution to the internal energy becomes $\rho u_0 = -(1/2)\rho^2$. The core condition $h(0) = -1$ implies

$$\rho(1-\rho) = \frac{\rho}{1-\rho c_0} P(z) \tag{39}$$

where

$$P(z) = \frac{1}{(2\pi)^3} \int \frac{d\mathbf{k}}{1 - z\tilde{\psi}(\mathbf{k})} \qquad z = \frac{\rho c_1}{1-\rho c_0} \tag{40}$$

Here d is dimension, and the integration goes over one Brillouin zone. The part of internal energy due to correlations is then

$$\rho u_1 = -\frac{1}{2(2\pi)^d} \int (\rho + \rho^2 \tilde{h}(\mathbf{k}))\tilde{\psi}(\mathbf{k})\, d\mathbf{k} = -\rho(1-\rho)F(z) \tag{41}$$

where by use of Eqs.(39) and (40)

$$F(z) = \frac{P(z) - 1}{2zP(z)} \tag{42}$$

By use of the fluctuation theorem and OZ equation (34) we further have

$$\beta\chi^{-1} = \varepsilon^2/(\rho(1-\rho)) = (1 - \rho\tilde{c}(0))/\rho \tag{43}$$

where with Eqs.(38)-(40) the new quantity ε is

$$\varepsilon^2 = (1-z)P(z) \tag{44}$$

With the above inserted in (37) the SCOZA partial differential equation becomes

$$\frac{\partial \varepsilon^2}{\partial \beta} = -\rho(1-\rho)\left\{1 + \frac{\partial^2}{\partial \rho^2}[\rho(1-\rho)F(z)]\right\} \tag{45}$$

with bondary condition ε=1 for β=0, ρ=0, and ρ=1. For the continuum case one will end up with an equation of same type. But the relations for χ and u will be different [12, 14].

With the SCOZA approach one would expect improvement compared with MSA itself and renormalized γ-ordering, but the magnitude of it was not clear. For lattice gases these two latter approaches are the same as concluded in Sec. 5, but for continuum fluids the last one is more appropriate due to better thermodynamic consistency. This latter property is taken care of by SCOZA by its very definition. Thus it will incorporate the $-\mu''/2$ term discussed in Sec. 5, and which is crucial in Eq.(24) to determine correctly to leading order in γ the shifts in T_c and ρ_c. This can be easily understood from (45). On its right hand side one takes the second derivative of the internal energy with respect to density. This perturbs its left hand side where the compressibility sits and thus adjusts it to leading order in a way that Eq.(24) does. So in view of the γ-ordering analysis in Secs. 3 and 4 the SCOZA will give next order corrections to T_c and ρ. Compared with the relative magnitude of the first order correction reported in Sec. 5 one might expect something of order 1% for the Lennard Jones fluid, and of order 10% for the lattice gas. Along with such a shift the main effect of SCOZA will be to "smoothen" out the "irregularities" of isotherms as described in Sec. 4 to obtain well-defined results for critical properties and critical indices. So in view of this the SCOZA has a potential for promising results.

Finally it should be mentioned that SCOZA has been extended to D-dimensional spins and continuous spins [16, 28]. So far there are no numerical results published for these latter cases, and they will not be discussed further here, but explicit results for these latter cases will be of great interest. However, it should be mentioned that both in the spherical model limit ($D \to \infty$) and Gaussian model limit SCOZA yields exact results as it does in the $\gamma \to 0$ limit.

7. Numerical results of SCOZA

Initial numerical results of SCOZA were promising, and the first case considered was the lattice gas case that follows Eqs.(38)-(44). But to simplify the approximation [8]

$$F(z) = \gamma_r^3(1 - \varepsilon) \tag{46}$$

which has the proper singular behavior as $\varepsilon \to 0$ in three dimensions, was used. Here and below γ_r will denote inverse range of interaction to avoid confusion with the critical index γ for the inverse susceptibility along the critical isochore. With 20 gridpoints in the interval $0 < \rho < 0.5$ approaching T_c in diminishing steps (to avoid instabilities) the solution was obtained. Only the value $\gamma_r = 0.5$ was considered. A striking feature of the result obtained in the third and reported in the first work of Ref. [8] was the value $\gamma \approx 4/3$. On a $\varepsilon^{3/2}$ versus t plot a remarkably straight line was obtained with a minor bend at the $t \to 0$ end. Here $t = \beta_c - \beta$. However, with the $F(z)$ used, there were no other results with which to compare directly, so further investigations were not done. Later the more elaborate SCOZA for the one Yukawa fluid was done. Then values for γ varying from .98 to 1.14 for varying range of interaction was obtained. Whether these variations in γ were features of SCOZA or numerical uncertainties were not checked further, and the problem to go below T_c remained unsolved. So the value $\beta \approx 1/3$ of the critical index for the curve of coexistence reported in Ref.[8] was not the result of evaluation but followed from the scaling relation $\beta(\delta - 1) = \gamma$ with $\delta = 5$.

The above preliminary results indicated that effectively critical properties would depend upon the range of interaction, and one purpose of the work in Ref.[10] was to investigate this besides trying to solve the SCOZA problem below T_c. In the meantime Dickman and Stell were able to obtain numerical results below T_c [9]. The crucial step taken in this work was to use Eq.(37) as the basis for the SCOZA equation. Also, for the Ising model with nearest neighbor interaction on the three possible types of cubic lattices, they evaluated the correct $F(z)$ (Eq.(42)) that can be expressed in terms of elliptic integrals. A striking feature of their results was the high accuracy when compared with best estimates. E.g. their values for T_c were within 0.2% of such estimates, and the general behavior in the critical region gave effective critical exponents close to best estimates too.

The numerical procedure was improved by Pini who developed an unconditionally stable, accurate, and efficient one based upon previous experience on related work [18]. The investigations in Refs.[10] and [11] were then based upon use of programs developed by Pini. For simplicity approximation (46) for $F(z)$ was mostly used in Ref.[10]. As already tested out by Pini this had only a minor influence quantitativly. I.e. the use of $\gamma_r = 0.34$ gave almost the same results as the Ising model on a simple cubic lattice. This then also supported our expectation (from γ-ordering) that the precise form of the interaction besides its range has small influence upon the EOS.

By varying the γ_r in the range from 0 to 0.9 effective critical exponents that depended considerably upon this parameter, were found. The various exponents were defined as slopes of curves on logaritmic plots (e.g. $\log_{10} \chi^{-1}$ versus $\log_{10} t$ etc.), and mappings of these exponents for various γ_r were made. Typically one sees effective exponents that are relativly constant for a couple of decades as in the case studied in Ref.[9]. Then close to T_c a crossover to the universal limiting SCOZA values will take place. More SCOZA results, that further confirms the above picture, comparing with previous estimates for the three dimensional Ising model were obtained by Pini, Stell, and Dickman [13].

To further test the accuracy of SCOZA it was compared with exact results in one and two dimensions in Ref.[11]. For the Ising model with nearest neighbor interaction in one dimension the SCOZA is exact. So next nearest neighbor was added, and good results were again obtained as long as the second neighbor was small when antiferromagnetic. The latter also shows the limitation of SCOZA with $\tilde{c}(\mathbf{k})$ restricted to form (38). The situation with Kac interaction $-\gamma_r \exp(-\gamma_r |x|)$ on a lattice was also considered. In the limit $\gamma_r \to 0$ a result near the exact one was obtained for the exponential decay of the correlation function for large $|x|$ [29]. For the two dimensional

Ising model it is possible to compare with the exact result in zero field [30]. A special feature with this case is that SCOZA like MSA does not yield a sharp phase transition. In this respect this seemed like a rather bad case for SCOZA. However, numerical results came out differently, and e.g. the SCOZA internal energy is close to the exact one with small deviations (up to about2%) in the neighborhood of T_c of the exact solution. I.e. the phase transition is smoothed out somewhat, and maximum of the specific heat is almost precisely at the exact T_c. By extrapolation towards T_c the critical index γ is found to indicate something near its exact value 7/4.

To us the above results seem to strongly indicate that the MSA or OZ form of $c(r)$ used by SCOZA describes a pair correlation function near the exact one [13]. Thus thermodynamic self-consistency is crucial to improve the EOS by introducing an effective temperature in the MSA problem.

The continuum fluid with attractive interaction of Yukawa form was considered by Pini, Stell, and Høye [12]. Due to lack of previous accurate estimates the accuracy of its SCOZA approach remained somewhat unclear. But a clear feature was the flattening of the top of the curve of coexistence to make it similar in form to the one for real fluids. Further work in this respect was performed by Pini, Stell, and Wilding which extended the analysis to incorporate two Yukawa terms and comparing with Monte Carlo (MC) data [14]. The second Yukawa can be used for different purposes (including use of the virial theorem). One is to adjust the EOS at $\beta = 0$ to be the CS one instead of PY. (See Eq.(4) and below it.) Alternativly the CS EOS can be the reference and then use only one single Yukawa that is non-zero at $\beta = 0$. These two approaches made little difference. This we find similar to the lattice case; the precise form of $c(r)$ is not so crucial for accuracy of the EOS. However, the difference between the PY and CS EOS at $\beta = 0$ is quite significant when comparing with accurate MC data; the CS version is most accurate.

Another use of the two Yukawa version is to let both terms be represented in the interaction. E.g. in this way the Lennard-Jones fluid can be more accurately described. However, in view of γ-ordering the change of form of the interaction will only have a perturbing effect upon EOS. (An effective hard core may be introduced in this connection.) More generally it will be possible to use the EOS for more complex molecules as bondary condition for the SCOZA problem at $\beta = 0$.

8. Critical properties of SCOZA

In Ref.[10] the critical properties of SCOZA were analysed numerically. A typical feature was the presence of effective critical indices that varied with the range of interaction, but sufficiently close to the critical point common asymptotic values were obtained. In standard notation these were

$$\begin{array}{lll} \gamma = 2 & \delta = 5 & \alpha = 0 \\ \gamma' \approx 1.4 & \beta \approx 0.35 & \alpha' \approx -0.1. \end{array} \tag{47}$$

For continuum fluids these values will remain the same. As pointed out in Ref.[13] the last three estimates are to be replaced with equalities [15]. The indices above do not fulfill standard scaling relations that should be expected for exact results. So SCOZA fails somewhat within something like $|t| < 10^{-3}$ which is similar to the error in T_c for the Ising model [9]. However, outside this region results seem accurate and reliable. This indicates that standard scaling is not fulfilled by the exact result either outside this region. Thus in Ref.[10] isotherms (for the lattice gas case) using scaled variables were drawn, but due to deviations from scaling SCOZA gave two sets of closely spaced curves instead of just two curves. These were related to to experimental data on CO_2 (and other fluids) that were earlier plotted by Green *et al.* in terms of scaled variables [31]. Apparently the points fall along two curves with deviations regarded as experimental uncertainties. But by closer inspection it was possible to discern different CO_2 curves for different temperatures, and these follow the SCOZA curves reasonably well disregarding experimental uncertainties. Thus this feature indicates a deviation from standard scaling not noticed before.

Concerning the various indices the value $\delta = 5$ follows in an obvious way from the SCOZA Eq.(45) along with (46) ($\varepsilon \to 0$). In the first one of Ref.[8] a family of scaling solutions with $\gamma = 1$ was found, but numerical results told otherwise giving $\gamma = 2$ [9]. By renewed analysis this could be understood [32]. (Also from the numerical results in the third work of Ref.[8] this can be seen, but this was not paid attention to as it could just as well be numerical uncertainty.) Thus the scaling solution picked out by SCOZA is simply the β-independent one $\varepsilon \sim (\Delta\rho)^2$, so another β-dependent contribution must add to it. The value of α follows from $\alpha = 1 - \gamma/2$ as $\chi^{-1} \sim \varepsilon^2 \sim t^\gamma$ and $u \sim \varepsilon \sim t^{1-\alpha}$.

The asymptotic value for the index β is more spectacular as it is near best estimates and is clearly a new feature compared with the mean-field and spherical model value 1/2. The values for γ' and α' follow via the scaling relations $2\beta + \gamma' = \beta(\delta + 1) = 2 - \alpha'$. (The HRT of Ref.[18] produces nearly the same values for these indices.) The negative value of α' must be regarded as a defect of SCOZA as the exact one is expected to be non-negative. However, away from the critical point the effective γ' decreases, β increases slightly, and α' increases to obtain $\alpha' > 0$.

For the Ising model with nearest neighbor interaction the effective exponent $\gamma \approx 1.25$ was found in Ref.[9] in good agreement with best estimates. In Ref.[10] the same is obtained for $\gamma_r^3 = 0.34$, but other values of γ_r^3 yield different values of this exponent like $\gamma \approx 4/3$ for $\gamma_r^3 = 0.5$ as mentioned in Sec. 7, and $\gamma \to 1$ as $\gamma_r \to 0$. Thus we find reason to speculate whether the good agreement with $\gamma \approx 1.25$ is accidental, or would similar best estimates for more long-ranged interactions follow effective SCOZA values. To our knowledge this has not been investigated, although the universality hypothesis says that the true γ is fixed. As the SCOZA results give effective γ that varies with γ_r, one can ask which of them correspond to the true universal value of γ as $t \to 0$. Typically for $\gamma_r^3 > 0.1$ the crossover from effective indices starts around $t \approx 10^{-2}$ and asymptotic values are obtained only in the narrow region inside $t \approx 10^{-5}$. The situation is similar with the exponent δ.

With long-range forces there will be crossover to effective exponents that approach the mean-field ones. Roughly this happens when $\gamma_r^3 < 10^{-2}$ where the curves of each exponent as functions of $\log_{10} t$ or $\log_{10} \Delta\rho$ approach the same form except for a translation that depends upon γ_r. The region with asymptotic values of the exponents is inside the tiny region $t \approx c(\gamma_r^3)^2$ where c is a constant. This is in accordance with the Ginzburg criterion for such crossover [34]. Note that MSA has a similar crossover, but SCOZA gives a much more narrow region that is reduced about two decades compared to the MSA one [10].

Further accurate numerical evaluations by Høye, Pini, and Stell very close to the critical point strongly suggested the presence of of some scaling solution due to the typical form of curves obtained [15]. On basis of this numerical evidence the SCOZA equation was examined for the possibility to find an extended form of scaling, and results of this investigation were given and commented in Ref.[13]. Along with these results we here give a few additional details of this investigation planned to be reported in a separate paper [15]. Then consider the SCOZA equation (45) where one can use variables $t = \beta - \beta_c$ and $m = 1 - 2\rho$. Near the critical point t, m, and ε are small, so (46) can be used for $F(z)$. With $\varphi = (1 - m^2)\varepsilon$ and erazing quantities not of interest for the further analyses the SCOZA equation becomes

$$\frac{\partial(\phi^2)}{\partial t} = -1 + \frac{1}{2}\frac{\partial^2 \phi}{\partial m^2} \tag{48}$$

which has an obvious solution $\phi = \phi_0 = m^2$ that is the trivial member of the family of scaling solutions dealt with in Ref.[8]. But as the t-dependence has dropped out it can not fulfill the required bondary conditions, so there will be corrections to it, and one can write

$$\phi = \phi_0 + \psi. \tag{49}$$

Numerically a power law behavior $m = at^q$ (a is a constant) for the spinodal curve was demonstrated. So more generally one can put

$$\phi_0 = (m - at^q)^2 \tag{50}$$

to be used in Eq.(49) ($a \neq 0$ for $t > 0$). With this form the requirement $\psi \ll \phi_0$ will be fulfilled, and Eq.(48) can be linearized with respect to ψ with scaling variable $u/|t|^{1/4}$ provided $q = 3/4$. For $t < 0$ (i.e. with $a = 0$) the solution of interest is $\psi = -c(3t + m^4)$ (c is a constant). Here it can be mentioned that in Eq.(29) of Ref.[13] the m^4 term is missing in the quantity $\psi \propto t\psi_+ = t$ while the $-2at^q m$ term from ϕ_0 in (50) will be part of its quantity $t\psi_-$. The ϕ_0 determines subcritical properties, and Helmholtz free energy obtains the form

$$f \sim -t^{7/4}m + m^6, \tag{51}$$

when deleting constants not of interest here. With this one obtains the critical indices

$$\beta = 0.35 \qquad \gamma' = 1.4 \qquad \alpha' = -0.1 \tag{52}$$

as reported in Ref.[13].

It can be of interest to extend this analysis of SCOZA to interactions such that $c(r) \sim 1/r^{d+2-\eta}$ as $r \to \infty$ where η is the usual index for the decay of the correlation function such that $2 - \eta = d(\delta - 1)/(\delta + 1)$ (d is dimension). Using the known $\delta = 15$ for the two dimensional Ising model results into $\beta = (36/35)/8$, $\gamma = 9/5$, and $\alpha' = -2/35$ which are near the exact values 1/8, 7/4, and 0 respectively. Thus SCOZA aims for something near the exact solution although the related index γ is worse off as $\gamma = 7$.

To improve SCOZA and its critical properties further, one has the possibility to add a longe-range tail to it as indicated at the end of conclusion in Ref.[13]. In accordance with the scaling hypothesis one has near the critical point [33]

$$c(r) \to [h(r)]^\delta \underset{t \to 0}{\to} const\,/r^{d+2-\eta} \quad (r \to \infty) \tag{53}$$

This term which here is not a part of the interaction itself, can be interpreted as a contribution from graphs with a number of δ chain bonds (see Eq.(25)) in parallell. With this interpretation the critical index δ will be an odd number (as a leading graph with δ even would vanish at the critical point like the last term of (17) does). Although this is correct in two dimensions, best estimates in three, however, indicates a non-integer index δ. Anyway, it may be possible to argue that the coefficient A above can be related to derivatives (with respect to ρ) of a generalized hypervertex μ_c. The lowest remaining power δ in (53) at the critical point should determine critical properties as it will be the most dominating term and thus interfere most with the MSA solution and its inconsistencies in the critical region. However, solving SCOZA with (53) included in some way (plus possible lower powers of $h(r)$ that vanish at the critical point) will be a demanding task whose outcome one so far can only speculate about.

9. Conclusion

We have presented a systematic construction of the EOS from mean-field theory via γ-ordering and MSA to SCOZA. Numerical results obtained with SCOZA so far have been considered and discussed. The general feature is that SCOZA yields globally accurate EOS for fluids and lattice gases, and it describes the critical region well in many respects.

References

1. B. Widom, J. Chem. Phys. **43**, 3892 (1965); **43**, 3898 (1965); L. Kadanoff, Physics **2**, 263 (1966).
2. K. G. Wilson, Phys. Rev. B **4**, 3174(1971); K. G. Wilson and J. B. Kogut, Phys. Rev. C **12**, 2 (1974); K. G. Wilson and M. E. Fisher, Phys. Rev. Lett. **28**, 240 (1972).
3. J. S. Høye, thesis NTH, Trondheim (1973).
4. P. C. Hemmer, J. Math. Phys. **5**, 75 (1964); J. L. Lebowitz, G. Stell, and S. Baer, J. Math. Phys. **6**, 1282 (1965).
5. H. C. Andersen and D. Chandler, J. Chem. Phys. **57**, 1918 (1972); H. C. Andersen, D. Chandler, and J. D. Weeks, J. Chem. Phys. **57**, 2626 (1972); H. C. Andersen, D. Chandler, and J. D. Weeks, J. Chem. Phys. **56**, 3812 (1972); D. Chandler and J. D. Weeks, Phys. Rev. Lett. **25**, 149 (1970).
6. C. Caccamo, Physics Reports **274**, 1 (1996).
7. J. S. Høye and G. Stell, Mol. Phys. **52**, 1071 (1984). See also Sec. IV of J. S. Høye and G. Stell, J. Chem. Phys. **67**, 439 (1977).
8. J. S. Høye and G. Stell, Int. J. Thermophys. **6**, 561 (1985); C. B. Jensen, F. Rinnan, and J. S. Høye, unpublished results (1985); J. S. Høye, unpublished results (1977).
9. R. Dickman and G. Stell, Phys. Rev. Lett. **77**, 996 (1996).
10. A. Borge and J. S. Høye, J. Chem. Phys. **108**, 4516 (1997).
11. J. S. Høye and A. Borge, J. Chem. Phys., **108**, 8830 (1998).
12. D. Pini, G. Stell, J. S. Høye, Int. J. Thermophys., in press.
13. D. Pini, G. Stell, and R. Dickman, Phys. Rev. E, **57**, 2862 (1998).
14. D. Pini, G. Stell, and N. B. Wilding, Mol. Phys., in press.
15. J. S. Høye, D. Pini and G. Stell, unpublished.
16. J. S. Høye and G. Stell, Physica A **244**, 176 (1997).
17. E. Kierlik, M. Rosinberg, and J. Tarjus, J. Stat. Phys. **89**,215 (1997); E. Kierlik, M. Rosinberg, and J. Tarjus, preprint (1998).
18. A. Parola and L. Reatto, Phys. Rev. Lett. **53**, 2417 (1984); Phys. Rev. A **31**, 3309 (1985); A. Parola, A. Meroni, and L. Reatto, Phys. Rev. Lett. **62**, 2981 (1989); A. Parola and L. Reatto, Phys. Rev. A **44**, 6600 (1991); D. Pini, A. Parola, and L. Reatto, J. Stat. Phys. **72**, 1179 (1993).
19. J. Lebowitz and O. Penrose, J. Math. Phys. **7**, 98 (1966). J. S. Høye, Phys. Rev. B **9**, 2390 (1974).
20. M. Kac, G. E. Uhlenbeck, and P. C. Hemmer, J. Math. Phys. **4**, 216 (1963); G. E. Uhlenbeck, P. C. Hemmer, and M. Kac, J. Math. Phys. **4**, 229 (1963); P. C. Hemmer,

M. Kac, and G. E. Uhlenbeck, J. Math. Phys. **5**, 60 (1964).
21. M. Klein and M. S. Green, J. Chem. Phys. **39**, 1367 (1963).
22. J. L. Lebowitz and J. Percus, Phys. Rev. **144**, 251 (1966).
23. T. H. Berlin and M. Kac, Phys. Rev. **86**, 821 (1952). H. W. Lewis and G. H. Wannier, Phys. Rev. **88**, 682 (1952); **90**, 1131 (1953).
24. G. Stell, G. N. Patey, and J. S. Høye, Adv. Chem. Phys. **48**, 183 (1981).
25. R. J. Baxter, J. Chem. Phys. **49**, 2770 (1968); J. S. Høye, E. Lomba, and G. Stell, Mol. Phys. **75**, 1217 (1992); *ibid.* **79**, 532 (1993); L. Belloni, J. Chem. Phys. **98**, 8080 (1993).
26. G. Stell, J. L. Lebowitz., S. Baer, and W. Theumann, unpublished (1965). See comment in Ref. 4 of Ref. [16] above.
27. J. S. Høye, J. L. Lebowitz, and G. Stell, J. Chem. Phys. **61**, 3253 (1974); J. S. Høye and G. Stell, J. Chem. Phys. **67**,524 (1977).
28. J. S. Høye and G. Stell, Physica A **247**, 497 (1997).
29. M. Kac and E. Helfand, J. Math. Phys. **4**, 1078 (1963).
30. L. Onsager, Phys. Rev. **65**, 117 (1944).
31. M. S. Green, M. Vicentini-Missoni, and J. M. H. Levelt Sengers, Phys. Rev. Lett. **18**, 1113 (1967).
32. J. S. Høye, G. Stell, and G. Tarjus, unpublished notes.
33. G. Stell, Phys. Rev. Lett. **20**, 533 (1968); Phys. Rev. B **1**, 2265 (1970).
34. M. E. Fisher and Y. Levin, Phys. Rev. Lett. **71**, 3826 (1993); R. F. J. Leote de Carvalho and R. Evans, J. Phys. Condens. Matter **7**, L575 (1995); V. L. Ginzburg, Sov. Phys. Solid State **2**, 1824 (1960).

PHASE SEPARATION AND CRITICAL PHENOMENA IN SIMPLE FLUIDS AND IN BINARY MIXTURES

L. REATTO
INFM and Dipartimento di Fisica,
Università degli Studi di Milano,
Via Celoria 16 I-20133 Milano, Italy

1. Introduction

For simple fluids liquid state theories of the standard form [1] take into account in a satisfactory way the interparticle correlations at short range and the short wavelength density fluctuations. Under two conditions density fluctuations at long wavelength become relevant. One is in the neighbourhood of a critical point where critical fluctuations become dominant. The other is inside regions of coexisting phases in which large scale fluctuations lead to phase separation. With respect to this second aspect the convexity of the free energy, which is not guaranteed by integral equations or perturbative expansions, can be restored if the equations are supplemented by *ad hoc* procedures, such as the equal tangent construction. More serious is the problem with critical fluctuations. No standard liquid state theory has been produced which has the minimal requirements for an acceptable theory in the critical region: scaling with non-trivial critical exponents.

On the other side the essence of the critical behaviour of a one component fluid or of a mixture of neutral species is considered as solved by renormalization group (RG) techniques. The standard RG techniques [2] cannot be applied directly to realistic models of fluids because no lattice is present and the interatomic interaction is essentially singular at short distance. Operations like discretization of space or coarse graining, operations which are preliminary to the application of RG techniques, are such that several questions remain unanswered by RG, mostly regarding non-universal quantities like: the critical parameters as a function of the interatomic potential, the extension of the asyntotic behaviour with a description of the crossover to

C. Caccamo et al. (eds.), New Approaches to Problems in Liquid State Theory, 31–46.

some other behaviours or the precise nature of the order parameter at the transitions of a multicomponent fluid.

This led us to the search of a theory which is able to keep information on all length scales and which is able to recover the RG results in the appropriate regime maintaining at the same time an accurate description of the short range correlations. Notice that this is by no means limited to liquids but it is in this case that the necessity of such a theory was particularly acute. One such theory, and the only one as far as we know, is what has been called the hierarchical reference theory [3, 4] (HRT) of fluids. This will be discussed in this lecture as well as some of its applications to the liquid-vapor transition of a one component system and to the phase transitions in binary mixtures.

2. The hierarchical reference theory of fluids

We consider a one component fluid of classical particles interacting via a pairwise spherical symmetric potential $v(r)$. Realistic models for the interatomic potential contain a strongly repulsive part (v^R) at short distance which is responsible for excluded volume effects and a weaker part (w) at longer range which is important for the cohesive effect leading to the liquid-vapor transition. Therefore we split the potential into a sum of two terms

$$v(r) = v^R(r) + w(r). \tag{1}$$

Standard liquid state theories are very efficient in treating the excluded volume effects due to $v^R(r)$ and we assume the properties of the reference system, i.e. the system with just the interaction $v^R(r)$ at the same temperature T and average density ρ of the physical system, known as an input. Well known schemes to take into account the effects of $w(r)$ are the so called high temperature expansion, which to lowest order is just mean field, or the random-phase approximation (RPA). What is peculiar of HRT is the way in which the effects of $w(r)$ are turned on. We want to take into account how the density fluctuations of the reference system are modified as an effect of $w(r)$ but we want to be selective in wavevector k starting from large k and successively taking into account the effect of smaller k in order to build up the RG scheme. Infinite order resummation of the expansion of the thermodynamic potential in power series of $w(r)$ is needed in order to build the non-analyticities typical of the critical regime. To this end we found it necessary to set up the full high-temperature expansion of the Helmholtz free energy $A(\rho, T)$, thus generalizing to all orders the known first two terms. As a second step a partial infinite order resummation is perfomed in such a way that there is loop ordering in the perturbative expansion. The lowest order is simply the familiar RPA and the other terms give the correction to RPA arising from the non-linear coupling between density fluctuations.

The generating functional for the turning on of $w(r)$ selective in wavevector can be obtained in two equivalent ways [4]. One is based on the introduction in the formal perturbative series for the free energy and for the correlation functions of an infrared cut-off Q over all momentum integrations. This procedure is closely related to Wilson's momentum shell integration technique [5]. This generates a cut-off dependent quantity $\mathcal{A}^Q$ which has a direct physical meaning only in the limit $Q \to 0$ where $\mathcal{A}^0$ represents the Helmholtz free energy of the system. The other possibility is to introduce a cut-off dependent potential [3]

$$v^Q(r) = v^R(r) + w^Q(r) \tag{2}$$

where $w^Q(r)$ in Fourier space has the form

$$w^Q(k) = \begin{cases} w(k) & for \quad k > Q \\ 0 & \quad k < Q \end{cases} . \tag{3}$$

One can verify from the perturbative series that at all orders a cut-off on the potential as in (3) is the same as a cut-off on fluctuations with the exception of the trivial zero loop contribution (the RPA). This means that there is a simple relation between the free energy $A^Q$and the two-body direct correlation function $c_2^Q(k)$ of the partially coupled system (2) and the related quantities $\mathcal{A}^Q$ and $\mathcal{C}^Q(k)$ for the fully interacting system with a cut-off on fluctuations:

$$\begin{aligned} -\frac{\beta \mathcal{A}^Q}{V} = -\frac{\beta A^Q}{V} &+ \frac{1}{2}\rho^2\left[\Phi(k=0) - \Phi^Q(k=0)\right] \\ &- \frac{1}{2}\rho\left[\Phi(r=0) - \Phi^Q(r=0)\right], \end{aligned} \tag{4}$$

$$\mathcal{C}^Q(k) = c_2^Q(k) + \Phi(k) - \Phi^Q(k) \tag{5}$$

where V is the volume and Φ is defined by

$$\Phi = -\beta w \tag{6}$$

and similarly for Φ^Q. No zero loop contributions are present in the higher order direct correlation functions $c_n^Q\,(n > 2)$ so no distinction between the two definitions is present.

The previous equivalence is important because the system with the cut-off potential (2) corresponds to a well defined system, even if with a potential with a somewhat artificial interaction, for which all quantities have a well defined physical meaning. This will be important at the moment to devise approximations and in the derivation of the evolution equation. This evolution equation is obtained as follows. For an arbitrary value of Q one can take the system with interaction (2) as reference system and consider $\delta w^Q(k) = v^{Q-dQ}(k) - v^Q(k)$, where δQ is an infinitesimal shift of the cut-off, as the perturbation. The fact that this perturbation is infinitesimal allows us to obtain the *exact* evolution equation of the properties of the system as the cut-off is changed. It is more convenient to write the evolution equations for $\mathcal{A}^Q$ and $\mathcal{C}^Q$ because these are more well behaved than the corresponding A^Q and c_2^Q. These equations read

$$-\frac{d}{dQ}\left(\frac{-\beta\mathcal{A}^Q}{V}\right) = \frac{1}{2}\int_{k=Q}\frac{d\Omega_k}{(2\pi)^3}\ln\left[1+\Phi(k)/\mathcal{C}^Q(k)\right], \tag{7}$$

$$-\frac{d\mathcal{C}^Q(k)}{dQ} = \int_{p=Q}\frac{d\Omega_p}{(2\pi)^3}c_3^Q\left(\vec{k},\vec{p},-\vec{k}-\vec{p}\right)F_2^Q\left(\vec{k}+\vec{p}\right)c_3^Q\left(-\vec{k},-\vec{p},\vec{k}+\vec{p}\right)$$
$$\frac{\left[\mathcal{F}^Q(p)\right]^2\Phi(p)}{1+\mathcal{F}^Q(p)\Phi(p)} + \frac{1}{2}\int_{p=Q}\frac{d\Omega p}{(2\pi)^3}c_4^Q\left(\vec{k},-\vec{k},\vec{p},-\vec{p}\right)\frac{\left[\mathcal{F}^Q(p)\right]^2\Phi(p)}{1+\mathcal{F}^Q(p)\Phi(p)}, \tag{8}$$

where the integrals are on the shell $\left|\vec{k}\right| = Q$, $|\vec{p}| = Q$, and the Ornstein–Zernike relation takes the form $\mathcal{F}^Q(k) = -\left[\mathcal{C}^Q(k)\right]^{-1}$ and $F_2^Q(k) = -\left[c_2^Q(k)\right]^{-1}$. Similar evolution equations can be written in an explicit way for c_n^Q for arbitrary n. These together with (7) and (8) represent an exact hierarchy of equations to be integrated from $Q=\infty$ to $Q=0$. The initial conditions are

$$\frac{-\beta\mathcal{A}^{Q=\infty}}{V} = \frac{-\beta A^R}{V} + \frac{1}{2}\rho^2\Phi(k=0) - \frac{1}{2}\rho\Phi(r=0). \tag{9}$$

$$\mathcal{C}^{Q=\infty}(k) = c_2^R(k) + \Phi(k) \tag{10}$$

and $c_n^{Q=\infty} = c_n^R$ $(n>2)$. Looking at (9) and (10) it transpires which is the meaning of $\mathcal{A}^Q$ and $\mathcal{C}^Q$: these represent the free energy and the direct correlation function for the fully interacting system (1) when the effect of the Fourier components of w for $k<Q$ are taken into account at mean field and RPA level , respectively, whereas the components with $k>Q$ are fully accounted for. As we change Q some additional Fourier components are correctly taken into account at the expenses of the simplified treatment.

3. The critical region

Close to a critical point when we investigate the long-wavelength properties of the system some major simplifications occur in the exact hierarchy. $\mathcal{F}^Q(k)$ is, apart for a factor of density, the static structure factor so that the divergence of the isothermal compressibility at the critical point corresponds to a divergence of $\mathcal{F}^Q(k)$ as $k \approx Q \to 0$. This means that in (7) and (8) and in all the other equations of the hierarchy the unity can be neglected in comparison with $\mathcal{F}^Q(k)\,\Phi(k)$ as well as $\ln \Phi(k)$ can be neglected with respect of $\ln \mathcal{F}^Q(k)$. At this point the hierarchy acquires a *universal* form, i.e. it is independent on the form of $v(r)$, and the interaction only appears as a boundary condition where the asymptotic equations have to be matched with the complete ones. For example the first equation becomes

$$-\frac{d}{dQ}\left(\frac{-\beta A^Q}{V}\right) = \frac{\Omega_d}{2\,(2\pi)^d} Q^{d-1} \ln\left[\mathcal{F}^Q(Q)\right] \tag{11}$$

where Ω^d is the surface of the unit sphere in d dimension. These asymptotic HRT equations are equivalent to those for a scalar field theory developed by Nicol and co-workers [6] which represent a refinement of the RG differential approach of Wegner and Houghton [7] . The direct correlation functions at cut-off Q correspond to the running coupling constants at length scale Q^{-1} and the two-body function is related to the field theory propagator. Notice that in our equations also the odd terms are present because we do not have particle-hole symmetry. From this equivalence it is clear that the critical behaviour predicted by HRT can be analyed with tools developed in the framework of RG and that the universal properties will be the same. By introducing suitable rescaled quantities the usual RG analysis in terms of fixed points, perturbation expansion around the fixed points or dimensionality expansion can be performed. For instance, it is enough to analyse the first equation (11) to deduce the correct critical exponents to first order in $\varepsilon = 4 - d$ (e.g. $\eta = 0$ and $\gamma = 1 + \varepsilon/6$). By using the first two equations of the hiearchy one can obtain the critical exponents correctly to order ε^2 (e.g. $\eta = \varepsilon^2/54$). Here we are using the standard notation for the critical exponents (see M.E. Fisher's lectures).

More useful for liquids is a direct analysis of the system in three dimensions by introducing a non-perturbative closure. This is obtained by the assumption that $\mathcal{F}^Q(k)$ has an Ornstein-Zernike (OZ) form, i.e. it is an analytic function of k also at the critical point so that we can write at small k

$$\mathcal{C}^Q(k) = -\left[\mathcal{F}^Q(k)\right]^{-1} = \mathcal{C}^Q(k=0) - bk^2 = \frac{\partial^2\left(-\beta\mathcal{A}^Q/V\right)}{\partial\rho^2} - bk^2, \quad (12)$$

where in the third passage the compressibility sum rule for the partially coupled system has been taken into account . If, consistently with the OZ hypothesis, b is assumed to be a smooth function of the state variables, the closure (12) inserted in (11) produces a closed partial differential equation (PDE) in the ρ and Q variables. In the framework of field theory our OZ approximation is known as local potential approximation (LPA) because it corresponds to neglecting the momentum dependence of the renormalized coupling constants.

It is important to notice that if we expand the free energy in a Taylor series around a given density ρ_0, this PDE gives rise to an infinite set of ordinary differential equations for the coefficients of this expansion. One can verify that this full set of equations coincides with the zero momentum limit of the exact HRT hierarchy. This explicitly shows the non-perturbative nature of what can be considered as the simplest closure of HRT.

From the PDE (11,12) one can obtain the critical exponents which turn out to be

$$\gamma = 2\nu = 1.378..., \beta = 0.345..., \alpha = -0.07 \quad (13)$$

and, by construction, $\eta = 0$. These exponents are not very accurate but they represent truely non-classical exponents and γ and β differ from the correct values by less than 10%. In addition, scaling is satisfied by the free energy and other thermodynamic properties and by the correlation functions. It is interesting to point out the properties [8] of the HRT in the O-Z approximation below T_c. The initial condition at $Q = \infty$, when T is below the mean field critical temperature, has an unstable portion corresponding to a van der Waals loop. As the integration with respect to Q proceeds, the unstable portion is progressively reduced until it disappears at $Q = 0$. When $T > T_c$ the free energy is convex in T with finite curvature at all densities. When $T < T_c$ the free energy has the correct convexity but now there is a finite range of densities in which the free energy is rectilinear in density. This corresponds to a perfectly flat pressure isotherm, i.e. it signals the coexistence of phases. I stress that the flat isotherm is produced as a result of integration over density fluctuations of all length scales and no *ad hoc* procedure is used. It should be noticed however one limitation of HRT in the OZ approximation in the region of the first order phase transition: the spinodal line coincides with the phase boundary so that the isothermal compressibility is infinite at the locus of the first order phase transition. Some analysis shows that the OZ approximation is adequate in the region of coexisting phases in giving a finite isothermal compressibility at coexistence only when the dimensionality is 4 or above.

4. A partial differential equation for liquids

The next step is to embed the description outlined in the previous Section of the critical region in a proper theory of the fluid state able to deal with the universal as well as the non-universal properties of the system. If we limit ourself to the first equation of the HRT hierarchy, this requires the introduction of a closure for $\mathcal{C}^Q(k)$ which reduces to (12) at small k. In order to build up the closure we recall that the direct correlation function c_2^Q of the partially coupled system is related to $\mathcal{C}^Q(k)$ by (5) and that c_2^Q has a corresponding radial distribution function $g^Q(r)$. For simplicity we write down the equations only in the case the reference potential $v_R(r)$ is the hard sphere one of diameter d. Our ansatz [8] is the following:

$$\mathcal{C}^Q(r) = c_2^R(r) + \lambda^Q \Phi(r) \quad , \quad r > d \tag{14}$$

$$g^Q(r) = 0 \quad , \quad r < d. \tag{15}$$

Condition (15) is just the *exact* core condition for the Q-system. $c_2^R(r)$ is the direct correlation function of the reference system. The parameter λ^Q in (14) is determined by the condition that the compressibility sum rule $\left(\mathcal{C}^Q(k=0) = \partial^2\left(-\beta\mathcal{A}^Q/V\right)/\partial\rho^2\right)$ is satisfied. The HRT equation implied by the closure (14-15) is a non-linear partial differential equation in Q and ρ. When supplemented by suitable initial condition at $Q=\infty$ and by boundary conditions at $\rho=0$ and at large density it represents a closed equation whose critical behaviour should coincide with that discussed in the previous Section. It is easy to recognize the relationship between this approximate HRT and some known liquid state equations. If we take $\lambda = 1$ at $Q = 0$ Eqs(14-15) represent [10] just the optimized RPA (ORPA), an approximation very similar to the well known mean spherical approximation (MSA). If we take Eqs(14-15) at $Q = 0$ this is equivalent to the self-consistent OZ approximation [11] (SCOZA). Notice the difference between HRT in this OZ approximation and SCOZA. In this last approximation λ is determined via a condition of thermodynamic consistency when the full perturbation w is introduced. SCOZA gives a non trivial critical behaviour but unfortunately its asymptotic behaviour is not very good: there is an asymmetry between $T > T_c$ and $T < T_c$ and scaling is not obeyed. In HRT the compressibility sum rule is satisfied not only by the fully interacting system but also all along the process of the turning on process of the perturbation potential w. This is a much more stringent condition.

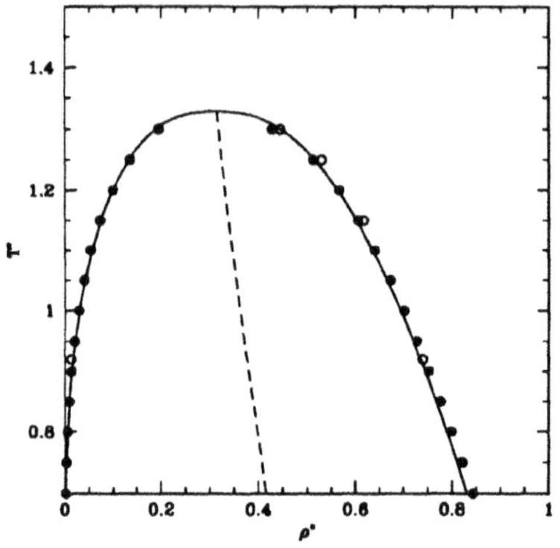

Figure 1. Coexistence curve and diameter for the LJ fluid in the $\rho^* - T^*$ plane. HRT (full curve), MHNC (o), and simulation (•) results.

The Eqs(14,15) have been generalized to the case of a continuous repulsive potential by a closure inspired by the soft MSA (SMSA) of liquid-state theory [12]. Some of the results [13] are shown in the Figures: in Fig. 1 we give the coexistence curve for the Lennard-Jones (LJ) potential, in Fig.2 the static structure factor $S(k)$ for three thermodynamic states is displayed and in Fig.3 we show the reduced isothermal compressibility $S(0) = \rho k_B T\, \mathcal{K}_T$ along the critical isocore. We notice that the HRT equation can be solved for temperature arbitrary close to T_c even if, given the discrete steps $\Delta\rho$ and ΔQ used in the integration, this limits the accuracy of the solution very close to the critical point. There is no intrinsic limit on how close one can be to the critical point and the limit derives from the computational effort that one is propared to perform. The numerical solution gives results which agree with the behaviour of the asymptotic analysis but this behaviour sets in quite close to T_c and ρ_c. For instance there is a rather wide region ($10^{-1} - 10^{-3}$ in reduced temperature $t = (T_c - T)/T_c$) in which $\mathcal{K}_T$ does not follow the asymptotic exponent ($\gamma = 1.378..$) of our approximation but it has a lower effective exponent which is closer to the correct behaviour ($\gamma = 1.24...$) and also the absolute values of $\mathcal{K}_T$ agree well with

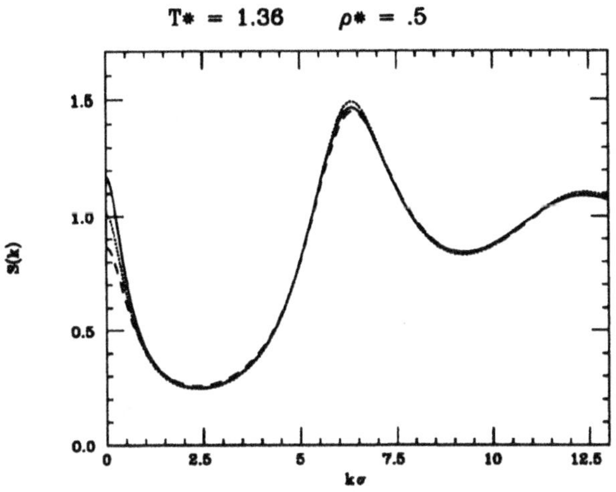

Figure 2. Structure factor for the state $T^* = 1.36, \rho^* = 0.5$. HRT (full curve), HRT as appropriate for a soft repulsive core (dot-dashed curve), MHNC (dashed curve).

the known values. Finally, the HRT $g(r)$ and $S(k)$ are rather accurate also far from the critical point and in the dense fluid, at a level similar to that of good liquid state theories. We conclude that HRT gives a viable global theory of fluids, from the critical region to the dense fluid regime.

5. Binary mixtures

The HRT hierarchy can be formally generalized [14] to mixtures, the only difference with respect to one-component fluids being the presence of species indices in the interactions and correlation functions. The two-body potential is written as a sum of a reference part and a remainder: $v_{ij}(r) = v_{ij}^R(r) + w_{ij}(r)$. After having introduced a cut-off Q in the Fourier components of all the tails w_{ij} as in Fig. (3), we can derive exact evolution equations for the free energy and the correlation functions which describe the inclusion of fluctuations in excess of the RPA ones on length scales up to Q^{-1}. For instance, the first equation of the hierarchy, the one for the modified free energy, reads

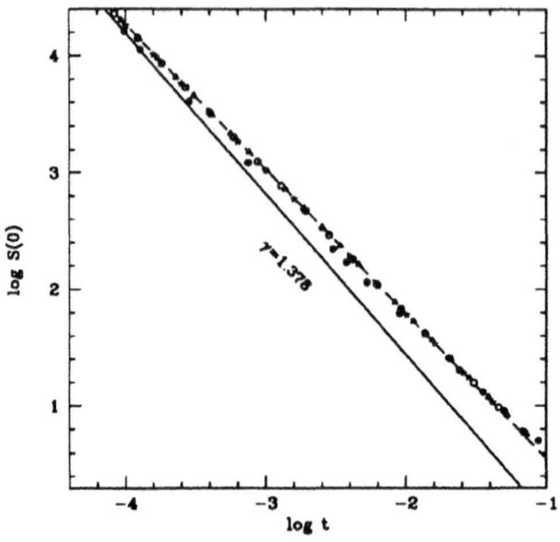

Figure 3. Dimensionless isothermal compressibility along the critical isotherm. Black dots: HRT results for a Lennard-Jones potential. Experimental results in Xe (open triangles), Kr (asteriks), Ar (open dots). Dotted line: power law with exponent $\gamma = 1.24$. Continuous line: $\gamma = 1.378$ (HRT result).

$$-\frac{d}{dQ}\left(\frac{-\beta A_Q}{V}\right) = \frac{1}{2}\int_{k=Q}\frac{d\Omega_k}{(2\pi)^3}\ln\det\left\{\mathbf{1} + \left[\mathcal{C}^Q(k)\right]^{-1}\cdot\mathbf{\Phi}(k)\right\}. \quad (16)$$

$\mathcal{C}$ and $\mathbf{\Phi} \equiv -\beta\, w$ are the matrix of the modified direct correlation functions and of the perturbative potentials. (16) holds for a mixture of an arbitrary number of species but our discussion focuses on the binary case, in which case we have 2x2 matrices. Notice that in order to derive this hierarchy we need not to specify the nature of the phase transitions that we want to describe: our equations are valid far from the critical region as well as in this region, independently on the character of the transition (liquid-vapor, demixing or of mixed character). The nature of the order parameter will arise from the analysis of Eq.(16). This is very different from the common approach which starts from the definition of the local order parameter and the introduction of an effective Hamiltonian which governs its fluctuations. A second observation is that $w_{ij}(r)$ need not to represent a purely attractive interaction but it can contain also some repulsive part as long as these have a well defined Fourier transform.

A key role in the analysis of criticality is played by the generalized compressibility sum rules

$$C_{ij}^{Q}(k=0) = \partial^2 \left(-\beta \mathcal{A}^Q/V\right)/\partial\rho_i\partial\rho_j \tag{17}$$

which are an extension of the one enbodied in (12). This condition together with an OZ assumption similar to (12) is enough for the study of the asymptotic critical behaviour. The non-universal behaviour and the properties of the mixture outside the critical region can be obtained by a closure similar to (14,15) for the one component system.

I review here the main results of this theory. At a critical point al least one of the second derivatives of the Helmholtz free energy has to vanish. Due to relations (17) this requires that det $\left[C^{Q=0}(k=0)\right]$ vanishes at the critical point so that at least one eigenvalue of the matrix (17) vanishes, The corresponding eigenvector defines a particular direction in the density - concentration plane specified by an angle φ. The linear combination

$$\Psi_1 = \delta\rho\cos\varphi + \delta c\sin\varphi \tag{18}$$

represents the combination of density and concentration which, via linear response theory, has the strongest fluctuations. Ψ_1 represents the natural choice as order parameter of the transition. We use standard notation for the total density $\rho = \rho_1 + \rho_2$ and concentration $c = \rho_2/\rho$, ρ_1 and ρ_2 being the number density of the two species, so that $\delta\rho = \delta\rho_1 + \delta\rho_2$ and $\delta c = (\rho_1\delta\rho_2 - \rho_2\delta\rho_1)/\rho$. Special cases of the order parameter correspond to $\varphi = 0$ (pure density fluctuation) and $\varphi = \pi/2$ (pure concentration fluctuation). In general, however, the order parameter turns out to be neither a pure density nor concentration but a mixture of the two. As an example we report in Fig.4 the critical lines for a neon-krypton mixture, modelled by LJ potentials, in the (ρ, c) plane and the arrows give a visual representation of the order parameter angle φ. For simplicity, in the figure the order parameter is computed at mean field level, i.e. when $Q = \infty$.

The asymptotic critical properties of a binary mixture have been studied [14] within HRT. This allowed a microscopic justification of the phenomenological scaling theory for mixtures [15]. In addition it gave evidence for interesting crossover effects that bear on some features of experiment. The HRT with the closure (17) gives rise to a partial differential equation in three variables: ρ, c and the cut-off parameter Q. In the critical region the HRT equation becomes identical to RG equation for an effective action involving two fluctuating quantities

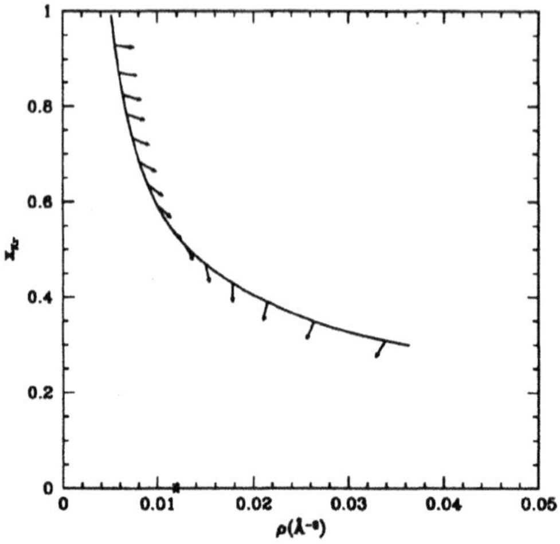

Figure 4. **Mean field critical lines and direction of the order parameter in the density-concentration plane for a Neon–Krypton mixture. The critical line originating from the critical point of pure Ne (marked in the figure) stops almost immediately upon addition of Kr.**

$$S[\Psi_1, \Psi_2] = \int d^3x \left[\frac{1}{2}|\nabla\Psi_1|^2 + r\Psi_1^2 + u\Psi_1^4 + g^2\Psi_2^2 + wg\Psi_1^2\Psi_2\right]. \qquad (19)$$

$\Psi_1(\vec{x})$ is the local order parameter (18) and Ψ_2 is a further combination $\Psi_2 = \delta\rho\,\cos\varphi - \delta c\,\sin\varphi$ of density and concentration fluctuation. Notice the presence in S of one odd term which couples the fluctuations of the order parameter Ψ_1 with those of Ψ_2 and the absence of the "kinetic" term for Ψ_2. The critical phenomena are governed by a fixed point which turns out to have a non-zero ($w^* \neq 0$) coupling between Ψ_1 and Ψ_2. Analysis of the RG flow equations around the fixed point shows that the critical exponents are related to those of the one-component system, i.e. those of the Ising universality class, via Fisher's renormalizion [15]. However, this asymptotic critical behaviour is usually masked by the presence of strong crossover effects due to the competition of another fixed point, which is unstanble but strongly attractive, with $w^* = 0$ so that in the crossover region there is a decoupling between the Ψ_1 and the Ψ_2 fluctuations. This implies for instance, that the isothermal compressibility $\mathcal{K}_{T,c}$ at fixed concentration asymptotically diverges weakly with temperature with an exponent -$\alpha/(1-\alpha)$ but in the crossover region $\mathcal{K}_{T,c}$ remains flat. In general, we find that the asymptotic behaviour sets in so close to the critical point that only the crossover region is experimentally accessible. This agrees with what has been experimentally observed [16]. If the coupling w becomes large, then the asymptotic region expands at the expenses of the crossover region. Favorable situations [17] for this to happen are mixtures near critical end points or with phase diagrams close to a change of topology.

The complete description [18],[19] of a mixture has been obtained on the basis of the closure

$$C_{ij}^{Q}(r) = c_{ij}^{R}(r) + \lambda_{ij} w_{ij}(r) \tag{20}$$

which shows the similarity with (14) for a one component fluid. c_{ij}^{R} are the direct correlation functions of the reference system. So far we have not implemented any core condition because this is expected to introduce only small changes in the phase diagram. In fact, it is our experience with the one component system that the core condition (15) is important when computing correlation functions but it has only a marginal role on thermodynamic properties. We have studied few mixtures modelled in terms of the LJ interatomic potential, like neon-krypton, neon-argon and argon-xenon. Also a symmetric LJ mixture with strength ratio $\varepsilon_{12}/\varepsilon_{11} = 0.75$ has been studied. These computations cannot be considered as exhaustive but they are preliminary in character. We wanted to show that HRT is a viable theory for mixtures and to put into evidence some of the main qualitative features.

Of some interest is the result for the argon-xenon mixture. The projecton of the critical line in the density-concentration plane is shown in Fig.5. Comparison with the mean field result shows that the non linear interaction between fluctuations in this case strongly modifies the phase diagram. For instance, the maximum value of the critical density is increased by fluctuations by almost a factor of two. The mixture appears to be of type I for which the two critical points of the pure components are joined by a line of critical points [20]. Not shown in the figure is a separate branch of critical points of mixing-demixing character which is present at densities above 20 nm^{-3}.From the direction of the arrows giving the order parameter, it is clear that the critical fluctuations have a strong component in concentration. In fact the interaction parameters of the system are such that it is close to a change to the different topology of mixtures of type II for which the critical points of the two pure substances do not fall on the same line of critical points but they are disjointed as in Fig.(4). As mentioned above this is a favorable situation for the detection of the asymptotic crit-

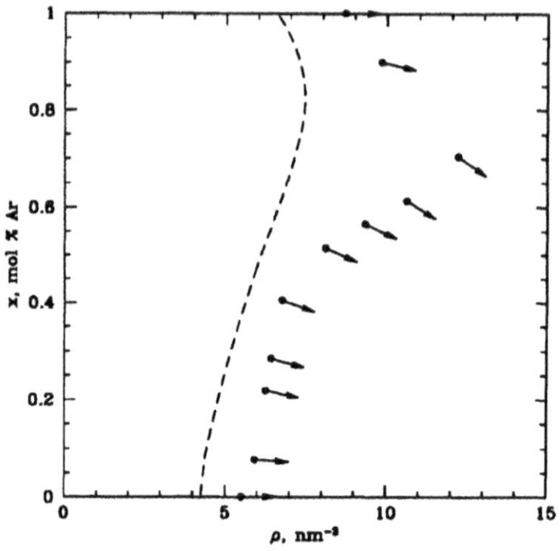

Figure 5. Density - concentration projection of the liquid-vapor critical line in an Ar-Xe mixture. Dotted line: mean-field theory. Circles: HRT results. The arrow represents the direction of the order parameter. The open circle marks the critical point of the isothermal section shown in Fig. 6.

ical behaviour due to an enhancement of the crossover temperature which is maximum near the hump of the critical line in the region of 20% xenon concentration. In order to see how HRT describes the region of first order phase transition, an isothermal section of the pressure-concentration plane is shown in Fig. 6a where states within the coexistence region are marked by squares. Thermodynamic stability requires that these points collapse on the same curve when plotted in the pressure-chemical potential plane. As shown in Fig. 6b the numerical solution satisfies this constraint.

6. Conclusion

HRT is a powerful scheme for dealing with equilibrium fluctuations over all length scales. HRT has the merits of liquid state theories in treating the short range local order. At longer range a proper critical behaviour is recovered if the system is close to a critical point. Truncation of the HRT hierarchy at the first equation already gives valuable results for one compo-

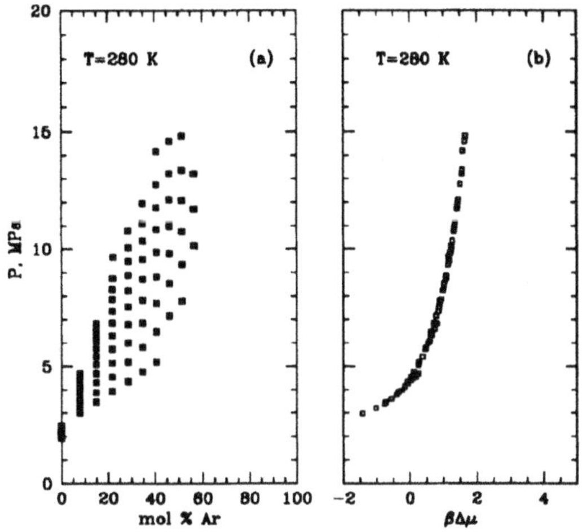

Figure 6. (a): Isothermal section of the concentration - pressure plane. Full symbols show the thermodynamic states inside the coexistence region. (b): Collapse of points of panel (a) when plotted in the chemical potential - pressure plane.

nent and two component fluids. One of the major merits of this scheme is that the region of first order phase transition is a direct output of the theory and no *ad hoc* construction built *a posteriori* from results for metastable and unstable states is used. This is a great advantage, particularly for mixtures.

It should be feasible to go beyond the first equation and this should improve the values of the critical exponents and the treatment of the first order phase transition. However, given the complexity of the higher order equations of HRT, it is rather unlikely that within the HRT scheme it will be possible to reach an accuracy in the values of the critical exponents comparable with that of the most sophisticated RG schemes. On the other hand the HRT can be used as a tool to generate, from the microscopic interactions, the effective coupling parameters which enter the phenomenological action to which field theoretical techniques are applied. Some preliminary study in this direction suggests that an accurate mapping of the microscopic interaction into an effective action requires a Ψ^6 model and that a Ψ^4 model is not enough.

Some of the interesting applications of HRT will require some modifications of the present approximations or even some extension of the equations. Suitable changes are needed in order to treat Coulombic mixtures and so far only the Kosterlitz-Thouless transition has been studied for the restricted primitive model for electrolytes at low densities in two dimensions. Some progress has been obtained in extending the basic HRT hierarchy in the case of a quantum fluid. In another direction it will be necessary to modify the present closures for the direct correlation functions in order to treat the effective interatomic potentials that typically one encounters in the approximate treatment of some soft matter system like protein solutions.

Acknowledgments

Let me thank in the first place Alberto Parola whose role has been fundamental in the invention and development of the HRT scheme. Then I should mention and thank Alberto Meroni, Davide Pini and Maria Tau who have greatly contributed to some of the developments presented in this lecture.

References

1. See for instance: J-P Hansen and I.R. McDonald , "*Theory of Simple Liquids*" (London Academic), (1986).
2. See for instance: J. Zinn-Justin, "*Quantum Field Theory and Critical Phenomena*" (Oxford: Claredon) (1983).
3. A. Parola and L. Reatto, *Phys. Rev. Lett.* **53**, 2417 (1984); Phys. Rev. A **31**, 3309 (1985).
4. An ample review of HRT is contained in: A. Parola and L. Reatto, *Adv. Phys.* **44**, 211 (1995).
5. K.G. Wilson and J.B. Kogut, *Phys. Rep.* C **12**, 75 (1974).
6. J.F. Nicoll, T.S. Chang and H.E. Stanley, *Phys. Rev. Lett.* **33**, 540 (1974).
7. F.J. Wegner and A. Houghton, *Phys. Rev.* **A8**, 401 (1973).
8. A. Parola, D. Pini and L. Reatto, *Phys. Rev.* , **E48**, 3321 (1993).
9. A. Meroni, A. Parola and L. Reatto, *Phys. Rev.* **A42** 6104 (1990).
10. H.C. Andersen and D. Chandler, *J. Chem. Phys.* **57**, 1918 (1972).
11. J.S. Hϕye and G. Stell, *J. Chem. Phys.* **67**, 439 (1977); J.S. Hϕye, in these Proceedings.
12. G. Zerah and J.P. Hansen, *J. Chem. Phys.* **84**, 2336 (1986).
13. M. Tau, A. Parola, D.Pini and L. Reatto, *Phys. Rev.* **E52**, 2644 (1995).
14. A. Parola and L. Reatto, *Phys. Rev.* **A44**, 6600 (1991).
15. M.E. Fisher, *Phys. Rev.* **176**, 257 (1968); R.B. Griffiths and J.C. Wheeler , *Phys. Rev.* **A2**, 1047 (1970).
16. J.V. Sengers and J.M.H. Levelt Sengers, *Annu. Rev. Phys. Chem.* **37**, 189 (1986).
17. D. Pini, A. Parola and L. Reatto, *J. Phys.* CM **9**, 1417 (1997).
18. L. Reatto and A. Parola, *J. Phys.* CM **8**, 9221 (1996).
19. D. Pini, A. Parola and L. Reatto, *Int. J. Theor. Phys.* (in press).
20. J.S. Rowlinson and F.L. Swinton, " *Liquids and Liquid Mixtures*" (Butterworths, London 1982).

CRITICAL BEHAVIOUR OF THE RESTRICTED PRIMITIVE MODEL (RPM)

D. LEVESQUE
Université Paris Sud
Laboratoire de Physique Théorique et Hautes Energies
Bâtiment 211, Université Paris-Sud
91405 Orsay Cedex, France

1. Introduction

The coulombic interactions have an important role in the physical characteristics of numerous fluids such as molten salts or ionic solutions. The last ones are mixtures of molecular solvents and ionized molecules which, in the simplest case, are ternary mixtures of a solvent and two species of ionized atoms, for example, water, Cl^- and Na^+. In these mixtures, phase transitions of demixing take place between phases of different ionic concentrations. They are often found in the mixtures of polar and ionic complex molecules where the intermolecular interactions can be divided into several contributions: short range repulsive interactions, van der Waals interactions, interactions between induced dipoles due to polarization effects, ... and coulombic interactions. All these interactions can contribute to the demixing process but some of these phase equilibria are considered to result mainly from the coulombic interactions between ionized molecules [1].

This hypothesis can make the simplest model of the electrolyte solutions, namely the restricted primitive model (RPM), relevant for the study of such a demixing. The RPM is a mixture of charged hard spheres of equal diameter, one half carrying charges $+q$, the other half charges $-q$ in a uniform dielectric continuum.

Theoretical approaches [2, 3] and numerical simulations [4, 5] have established that the RPM presents at low temperatures and densities a phase transition of gas-liquid type. This gas-liquid coexistence is generally believed similar to the phase equilibrium mentioned above in real electrolytes between phases of different ionic concentrations. This point of view is supported by the mapping, based on corresponding states arguments, between

C. Caccamo et al. (eds.), New Approaches to Problems in Liquid State Theory, 47–59.

the coexistence curve of the RPM and those of ionic molecular mixtures. Contrary to the case of simple liquids, this coexistence curve is characterized, contrary to the case of simple liquids, by a strong asymmetry, due to a different variation of the gas and liquid densities at equilibrium upon the temperature. The strong association of ions with opposite charges in pairs or clusters is also a characteristic of the gas phase near the coexistence line below the critical temperature [6]. These pairs and clusters are well defined due to the low values of the gas density at these thermodynamic states.

The long range of the coulombic interaction, the features of the gas-liquid RPM transition and several experimental data obtained on the demixing transition in ionic fluids, make possible that the critical behaviour of charged fluids, in spite of the screening effect, is different from that of Ising universality class and belongs to the "classical" or mean-field universality class. In the experiments, "classical" or Ising-like behaviours and crossover from mean-field to Ising behaviour seem having been observed [7, 8].Theoretical agurments in favour of the Ising criticality of the RPM [9] have been given and the question of the crossover between the two critical behaviours in ionic fluids has been investigated [10].

Recently, a finite size scaling analysis of simulation results, obtained in the vicinity of the critical point for the Lennard-Jones (LJ) system in 2 and 3 dimensions, has been realized [11, 12, 13]. This analysis, similar to that made to determine the critical behaviour of spin systems on a lattice, completes the determination of the gas-liquid phase transition by showing that, near critical point, the LJ system behaves as expected for a system belonging to the Ising universality class. It also gives a very accurate value of the critical temperature. This successful application of the finite size scaling analysis to the gas-liquid transition in the continuous systems has motivated an attempt to contribute to the clarification of the nature of the RPM critical behaviour by using Monte-carlo (MC) simulations in the grand canonical (GC) ensemble [14, 15].

The determination of the universality class of a phase transition by numerical simulations supposes, as it has been demonstrated in the case of spin systems on lattice, that the sizes of the considered systems are large and the accuracy of the simulation data is excellent. However, the simulations performed for the LJ fluid show that size systems of thousand times the molecular volume and GC simulation runs of 10^8 - 10^{10} MC elementary trial configurations seem sufficient to establish the conclusion that the simulation data are compatible with the expected critical behaviour of finite systems belonging to a given universality class. It is such a conclusion that we have tried to obtain for the 3 and 2 dimensional (2d, 3d) RPM.

2. RPM and MC simulations

The RPM depends on two parameters, namely the reduced density ρ^* of the charged hard spheres defined by $\rho^* = N\sigma^3/V$ (σ diameter of the hard spheres, N total number of ions, V volume), and the reduced coupling constant β^* equal to $q^2/\epsilon\,\sigma k_B T$ (ϵ dielectric constant of the dielectric continuum, k_B Boltzmann's constant and T temperature). The interaction of ions i and j having charges q_i and q_j separated by a distance r is

$$v_{ij}(r)\,/\,k_B T = \begin{cases} +\infty & r \leq \sigma \\ \dfrac{q_i q_j}{\epsilon k_B T \sigma}\dfrac{\sigma}{r} & r > \sigma \end{cases} \tag{1}$$

The GC simulations are performed in finite volumes with given boundary conditions. Then the long range of the coulombic interaction must be taken into account in a way such as the potential between charges satisfies the Poisson's equation and is compatible with the boundary conditions of the simulation cell volume. In our GC simulations of the RPM, two different ways to treat this problem have been used. The traditional way consists in using a cubic simulation cell with periodic boundary conditions. In this case the electrostatic potential energy is expressed as that of a Wigner crystal made of a periodic array of identical point charges embedded in a neutralizing background. The electrostatic energy is rewritten by the Ewald's method in a convenient manner for the simulations, i.e., as the sum of two fastly convergent series in real space and in reciprocal space on the sets of the positions of the charges and of their periodic replicas. This procedure is well documented in the literature.

We will give more details on a second way used to define the electrostatic interaction in a finite space by confining the charged particles on the surface of a hypersphere in a 4d space [16]. In this finite space S_3, characterizing a point of the hypersphere surface by a four dimensional vector $\mathbf{OM}$ (O center of the hypersphere), the Poisson's equation, which should give the potential in a point M due to a positive unit charge located at M_o,

$$\Delta_M^{S_3} v(M) = -4\pi\delta^{S_3}(M_o, M), \tag{2}$$

has no solution. This difficulty can be overcome, as in the case of a charge in 3d volume with periodic boundary conditions, by associating to this charge a uniform neutralizing backgroung of opposite charge. An other possibility, used in our simulations, consists to consider charged objects, called bicharges, made of two unit charges of opposite sign located at M_o^+ and at the antipodal position M_o^- such that $\mathbf{OM_0^+} = -\mathbf{OM_0^-}$. The potential in M due to a bicharge is given by the Poisson's equation

$$\Delta_M^{S_3} v(M) = -4\pi(\delta^{S_3}(M_o^+, M) - \delta^{S_3}(M_o^-, M)), \tag{3}$$

and its solution is

$$v(M) = \frac{1}{R} \cot \psi \tag{4}$$

where

$$\psi = \cos^{-1}\left(\frac{\mathbf{OM}}{R}\frac{\mathbf{OM_o^+}}{R}\right) \tag{5}$$

and R is the radius of the hypersphere. A complete analogy between an ionic fluid of bicharges in S_3 and the RPM in a 3d Euclidean space is obtained by adding between bicharges, a repulsive hard-core potential $v_{hs}(\psi)$, at short distance, centered on each charged end and defined by

$$v_{hs}(\psi)\,/\,kT \;=\; +\infty \quad \psi \leq \sigma/R \tag{6}$$

$$=\; 0 \quad \psi > \sigma/R. \tag{7}$$

In a configuration of a system of N bicharge, there are N_+ bicharges with their positive ends and $N_- = N - N_+$ with their negative ends in the northern hemisphere of S_3. Considering that the potential $v(M)$ of a bicharge acts on each charged end of another bicharge it is convenient to attribute to each end of the bicharges a charge $\pm q/\sqrt{2}$. Then the total energy of the system U can be written

$$U = \sum_{i<j}^{N} v_{hs}(\psi_{ij}) + N\,e_{self} + \sum_{i<j}^{N_+} q^2\,v(\psi_{ij}) + \sum_{i<j}^{N_-} q^2\,v(\psi_{ij}) - \sum_{i=1}^{N_+}\sum_{j=1}^{N_-} q^2\,v(\psi_{ij}). \tag{8}$$

e_{self}, equal to $-q^2 \tan(\sigma/4R)/R$, is the self energy term which fixes the zero energy and is chosen such that, in 3d space and S_3, the Onsager bounds coincide.

The two boundary conditions, hyperspherical and periodic, can be used to realize simulations of the 3d RPM confined to a surface. In the first case, the centers of the charged hard spheres are located on the surface of a 3d sphere. In the second case, the centers stay in a plane and the simulation cell is a square with periodic boundary conditions. It is important to notice that, in these simulations, the interaction between the charges conserves the same analytical form in a 3d space and on a 2d surface.

The GC sampling of the configurations of the RPM is done by inserting and deleting pairs of charged hard spheres in volume with periodic boundary conditions and pairs of bicharges in S_3 and, also, by moving charged hard spheres or bicharges. However, for the thermodynamic states corresponding to the gas - liquid transition, the configurations of the RPM consist mostly of neutral pairs of charged hard spheres of opposite sign almost at contact and, in S_3, of pairs of bicharges having their ends of opposite charges almost at contact. Then it is not possible to insert a neutral pair of hard spheres or a pair of bicharges by selecting their positions at random in the volume, because the acceptance probability of the insertion of such a pair will be very low. It is also not possible to delete an existing pair by selecting at random two hard spheres or bicharges, because, most of choices will correspond to break pairs of strongly associated spheres or bicharges and to increase largely the energy of the configuration and the deletion will be refused.

The use of a biased scheme for the insertion or deletion of pairs is needed in order to sample efficiently the configuration space of the GC ensemble [4, 5]. This scheme, in the case of the insertion, corresponds to insert the first hard sphere or bicharge at random in the volume and to select the position of the second with a probability which decreases exponentially with the distance separating the two spheres or bicharges. For the deletion, the neutral pair of hard spheres or bicharges is selected with a probability taking into account its potential energy. Obviously, due to this bias in the a priori sampling, the usual acceptance criterion of an insertion or deletion in the GCMC simulations is modified in order that the detailed-balance condition is satisfied.

As mentioned above, in the gas phase near the critical point or the gas-liquid coexistence line, the ions (charged hard spheres or bicharges) form pairs and also clusters. To increase the efficiency of the sampling of the GC phase space at constant N, global trial displacement moves of the ions associated in clusters are attempted and accepted according to the detailed-balance condition.

3. Critical Point and Finite size scaling analysis

From computer simulations and theory [3, 4, 5], it is established that the values of the temperature T_c^* and the density ρ_c^* of the RPM critical point are in the ranges 0.050-0.056 and 0.02-0.05, respectively. The uncertainties on the values determined by simulations are the consequences of the difficulties to sample efficiently the configurations of the GC or Gibbs ensemble phase space, and of the fact that, in these numerical works, no finite size scaling analysis of the simulations results has been performed.

We summarize the finite size scaling procedure used to analyze simulation data of the RPM which is identical to that proposed in ref. [11] for the study of the critical behaviour of the LJ system.

Fluid systems, as the LJ system or the RPM, do not possess near critical point the particle-hole symmetry characterizing Ising models or lattice gas. This leads to consider that the relevant scaling fields driving the phase transition are linear combinations of the variations $\mu^* - \mu_c^*$ and $\beta_c^* - \beta^*$ of the chemical potential $\mu^* = \mu/kT$ and the inverse β^* of the temperature near their critical values μ_c^* and β_c^* [17] and are defined by

$$h = \mu^* - \mu_c^* + r(\beta_c^* - \beta^*) \tag{9}$$

$$\tau = \beta_c^* - \beta^* + s(\mu^* - \mu_c^*) \tag{10}$$

where s and r are coefficients defined in ref. [11], in particular r is the derivative $\partial\mu^*/\partial\beta^*$ at $\beta^* = \beta_c^*$ along the coexistence line.

The conjugated operators $\mathcal{E}$ and $\mathcal{M}$ associated to these scaling fields are obtained from the GC partition function Ξ by

$$< \mathcal{E} > = \frac{1}{V}\frac{\delta}{\delta\tau} ln\ \Xi \tag{11}$$

$$< \mathcal{M} > = \frac{1}{V}\frac{\delta}{\delta h} ln\ \Xi \tag{12}$$

where

$$\mathcal{E} = \frac{1}{1 - sr}(u - r\rho) \tag{13}$$

$$\mathcal{M} = \frac{1}{1 - sr}(\rho - su) \tag{14}$$

and it is convenient to define

$$\delta\mathcal{M} = \mathcal{M} - < \mathcal{M} >_c \tag{15}$$

$$\delta\mathcal{E} = \mathcal{E} - < \mathcal{E} >_c \tag{16}$$

where $< \mathcal{M} >_c$ and $< \mathcal{E} >_c$ are average values at criticality.

In the finite size scaling procedure used in the analysis of simulation data, the joint distribution function $p_L(\mathcal{M}, \mathcal{E})$ for the operators $\mathcal{M}$ and $\mathcal{E}$ is supposed, in the immediate vicinity of the critical point ($T \to T_c$, $L \to \infty$), to have the following form

$$p_L(\mathcal{M}, \mathcal{E}) \simeq a_{\mathcal{M}}^{-1} a_{\mathcal{E}}^{-1} L^{d-y_\tau} L^{d-y_h} \times \tilde{p}_{\mathcal{M},\mathcal{E}}(a_{\mathcal{M}}^{-1} L^{d-y_h} \delta\mathcal{M}, a_{\mathcal{E}}^{-1} L^{d-y_\tau} \delta\mathcal{E}, a_{\mathcal{M}} L^{y_h} h, a_{\mathcal{E}} L^{y_\tau} \tau, a_i L^{y_i}) \tag{17}$$

where L (expressed in units of σ) is the linear dimension of the system (taken as $V^{1/3}$). The coefficients $a_{\mathcal{M}}$, $a_{\mathcal{E}}$, a_i are non-universal constants setting the scales of the variables while the scaling function $\tilde{p}_{\mathcal{M},\mathcal{E}}$ is universal. In Eq. (17) contribution from the least-irrelevant correction-to-scaling field has been incorporated. The renormalized exponents are given in terms of the usual exponents β, ν and θ (Wegner's correction-to-scaling exponent [18]), assuming hyperscaling, by $y_h = d - \beta/\nu$, $y_\tau = 1/\nu$ and $y_i = -\theta/\nu$.

By integrating over $\mathcal{E}$, one obtains the ordering distribution

$$p_L(\mathcal{M}) \simeq a_{\mathcal{M}}^{-1} L^{d-y_h} \tilde{p}_{\mathcal{M}}(a_{\mathcal{M}}^{-1} L^{d-y_h} \delta\mathcal{M}, a_{\mathcal{M}} L^{y_h} h, a_{\mathcal{E}} L^{y_\tau} \tau, a_i L^{y_i}) \quad (18)$$

where $\tilde{p}_{\mathcal{M}}$ is a universal function. Along the line $h \approx 0$ which corresponds, to a good approximation, at a chemical potential such that $p_L(\mathcal{M})$ is symmetric, one has

$$p_L(\mathcal{M}) \simeq a_{\mathcal{M}}^{-1} L^{d-y_h} \tilde{p}_{\mathcal{M}}(a_{\mathcal{M}}^{-1} L^{d-y_h} \delta\mathcal{M}, a_{\mathcal{E}} L^{y_\tau} \tau, a_i L^{y_i}) \quad (19)$$

where the zero argument in $\tilde{p}_{\mathcal{M}}$ has been suppressed.

The moments of the ordering distribution, by using the scaled variable $x = a_{\mathcal{M}}^{-1} L^{d-y_h} \delta\mathcal{M}$, are

$$\begin{aligned} < \delta\mathcal{M}^k >_L &= a_{\mathcal{M}}^k L^{-k(d-y_h)} \int x^k \tilde{p}_{\mathcal{M}}(x, a_{\mathcal{E}} L^{y_\tau} \tau, a_i L^{y_i})\, dx \quad (20) \\ &= a_{\mathcal{M}}^k L^{-k(d-y_h)} \tilde{f}_k(a_{\mathcal{E}} L^{y_\tau} \tau, a_i L^{y_i}) \quad (21) \end{aligned}$$

where $\tilde{f}_k$ is again a universal function. From these moments, a dimensionless ratio can be defined

$$Q_L = \frac{< \delta\mathcal{M}^2 >_L{}^2}{< \delta\mathcal{M}^4 >_L} = \frac{\tilde{f}_2^2(a_{\mathcal{E}} L^{y_\tau} \tau, a_i L^{y_i})}{\tilde{f}_4(a_{\mathcal{E}} L^{y_\tau} \tau, a_i L^{y_i})} \quad (22)$$

related to the fourth-order (Binder) cumulant by $U_L = 1 - \frac{1}{3} Q_L^{-1}$. In the limit $\tau \to 0$, at fixed L, but L sufficiently large so that the contribution from the irrelevant field has vanished, Q_L will tend to a universal finite constant Q^* independent of L. When the critical temperature is approached, at fixed L, the r.h.s. of Eq. (19) can be expand in the scaling fields (x defined above) [19]

$$\begin{aligned} p_L(\mathcal{M}) \simeq a_{\mathcal{M}}^{-1} L^{d-y_h} \quad & [\tilde{p}_{\mathcal{M}}^*(x) + a_{\mathcal{E}} L^{y_\tau} \tau p_1^*(x) + a_{\mathcal{E}}' L^{2y_\tau} \tau^2 p_2^*(x) + \ldots \\ & + a_i L^{y_i} p_3^*(x) + \ldots] \end{aligned} \quad (23)$$

where, we recall, along the coexistence line, $\tau \sim \beta_c^* - \beta^*$. Similarly $Q_L(\beta^*)$ can be expanded as [20],

$$\begin{aligned} Q_L(\beta^*) &= Q^* + q_1(\beta^* - \beta_c^*) L^{y_\tau} + q_2(\beta^* - \beta_c^*)^2 L^{2y_\tau} + q_3(\beta^* - \beta_c^*)^3 L^{3y_\tau} \\ & \quad + \ldots + b_1 L^{y_i} + \ldots \end{aligned} \quad (24)$$

where the q_i and b_i are non-universal constants.

For the Ising model, identifying $\mathcal{M}$ with the magnetization m, $\mathcal{E}$ with the energy density u and τ with $\beta^* - \beta_c^*$, $\tilde{p}_{\mathcal{M}}$ reduces to Binder's finite size scaling ansatz for the magnetization distribution. This possible identification of $\tilde{p}_{\mathcal{M}}$ with the distribution of the order parameter at criticality of a system of known universality class is the basis of the finite size scaling analysis proposed in ref. [11] for the continuous systems.

Clearly, if, for instance, a continuous fluid belongs to the universality class of the Ising model, in the vicinity of the critical point of the gas-liquid transition, for systems of sufficiently large sizes and with an adequate choice of the s parameter, the distribution functions $p_L(\mathcal{M})$ should behave as the magnetization distributions $p_L(m)$ of Ising systems of finite sizes near the critical temperature of the Ising model. Then, Eq. (23), supposing small the term of second order in τ, implies that it is possible to match the distributions $p_L(\mathcal{M})$ of fluid systems of different volumes on the distribution $\tilde{p}^*_{\mathcal{M}}(x)$ for values of $\beta^* = \beta^*_c(L)$ which vary with L as

$$\beta^*_c(\infty) - \beta^*_c(L) \propto L^{-(\theta+1)/\nu}. \tag{25}$$

The extrapolation of the values $\beta^*_c(L)$ in the limit $L \to \infty$ determines the value $\beta^*_c(\infty)$, i.e. the critical temperature of a fluid system in the limit of an infinite volume. This evaluation of the critical temperature needs to know, at criticality, the order parameter distribution $\tilde{p}^*_{\mathcal{M}}(x)$.

For the class of universality of the Ising model, this distribution being known at 2d and 3d the previous procedure has allowed to obtain a very accurate value of the critical temperatures for the 2d and 3d LJ systems. There is no doubt that the LJ systems belong to the Ising universality class, but in the case of the RPM the possibility to match the distribution $p_L(\mathcal{M})$ on the distribution $\tilde{p}^*_{\mathcal{M}}(x)$ of the Ising model is not so obvious. However this matching will be a good indication that the critical point of the RPM belongs to the Ising universality class, if it is realized at values of $\beta^*_c(L)$ satisfying the relation (25) where θ and ν are the exponents of the Ising model. In this case, it is possible to conclude for the RPM that the simulation data are compatible with a critical behaviour of Ising type.

Certainly, it would be more convincing to obtain the same conclusion from a finite size scaling procedure similar to that used to analyse the simulation data of lattice systems. This procedure is based on the determination of the unknown parameters (coefficients, exponents, values of $\beta^*_c(\infty)$ in the expression of $Q_L(\beta^*)$). For continuous systems, this determination must be made by computing, in critical region, $< \delta\mathcal{M}^2 >_L^2$ and $< \delta\mathcal{M}^4 >_L$ for different volumes at chemical potentials and values of s parameter such that $p_L(\mathcal{M})$ is symmetric. Taking into account the numbers of unknown parameters, very accurate values of $< \delta\mathcal{M}^2 >_L^2$ and $< \delta\mathcal{M}^4 >_L$ are needed to obtain a credible estimate of these parameters.

4. Results for the 3d and 2d RPM

To assess the critical behaviour of the RPM, we have determined on isotherms, near critical point, the values of μ corresponding to the liquid-gas coexistence. As mentioned above, at a given temperature, the value of the chemical potential on the coexistence line is obtained from the requirement that $p_L(\mathcal{M})$ is symmetric. This symmetry condition is fulfilled by adjusting both μ and the parameter s. This would be arduous without the use of the histogram reweighting technique, which enables to compute the joint distribution of N and $u = U/V$ $p(N,u)$ for any state (μ, β) knowing this distribution for neighboring states (from $p(N,u)$ it is easy to calculate $p_L(\mathcal{M},\mathcal{E})$ for given values of the parameters s and r). This technique is based on the relation:

$$p(N,u,\beta',\mu') = \frac{\Xi(\beta,\mu)}{\Xi(\beta',\mu')} e^{(\beta'\mu' - \beta\mu)N - (\beta' u' - \beta u)} p(N,u,\beta,\mu). \qquad (26)$$

where, in practice, the values of the ratios $\Xi(\beta',\mu') / \Xi(\beta,\mu)$ are computed by a method proposed in ref. [21].

In the case of the 3d RPM we have performed simulations for the volumes $V/\sigma^3 = 1500$, 2500, 3500, 5000 and 10000. For these volumes we have realized the collapse of $p_L(\mathcal{M})$ on the distribution $\tilde{p}^*_{\mathcal{M}}(x)$ of the 3d Ising model which has been determined by simulations in ref. [22]. Examples of these collapses are shown in Figure 1.

The values of $T^*_c(L)$, corresponding to the best matching, are equal to 0.05034, 0.04980, 0.04959, 0.04928 and 0.04915 for the five volumes from 1500 to 10000 with an error of the order of ± 0.00001. They vary linearly as a function of $L^{-(\theta+1)/\nu}$ where θ and ν are equal to their best estimates for the 3d Ising model : 0.54 and 0.629. The extrapolation to the infinite volume gives the critical temperature of the 3d RPM $T^*_c = 0.0488 \pm 0.0002$. $\mu^*_c(\infty) - \mu^*_c(L)$ scales similarly to $\beta^*_c(\infty) - \beta^*_c(L)$ and extrapolation of the values of $\mu^*_c(L)$ yields the infinite volume critical chemical potential -13.70±0.01.

The densities $\rho^*_c(L)$, obtained from $\int d\rho\, \rho p_L(\rho)$ calculated at $\beta^*_c(L)$ and $\mu^*_c(L)$ appear, within statistical error, nearly constant giving an estimate of the infinite volume critical density $\rho^*_c = 0.080 \pm 0.005$. The values of the parameter s and r corresponding to the collapse of the distributions $p_L(\mathcal{M})$ and $p_L(\mathcal{E})$ on the equivalent distributions of the Ising model at infinite volume, are equal to -1.46 and -0.62. They are large, but in agreement with the strong asymmetry of the gas-liquid coexistence curve.

The multiplicity of parameters and the statistical error on our simulation data (involving $\sim 10^8$ independent configurations selected from $2.5\,10^{10}$ elementary trial configurations), do not allow to obtain accurate estimates of the critical exponents from the fit of the computed values of Q_L. We can

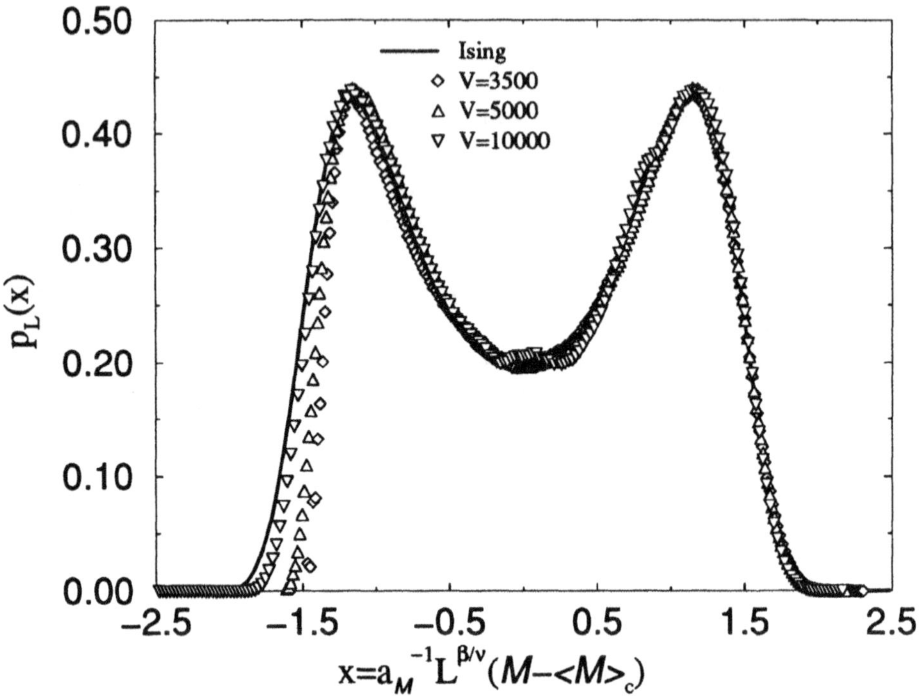

Figure 1. Collapse of the ordering operator distribution function $p_L(\mathcal{M})$ of the 3d RPM on the universal fixed point ordering operator distribution $\tilde{p}^*_{\mathcal{M}}(x)$ for $V/\sigma^3 = 3500$, $T^*_c(L) = 0.04959$, $V/\sigma^3 = 5000$, $T^*_c(L) = 0.04928$ and $V/\sigma^3 = 10000$, $T^*_c(L) = 0.04915$. The universal distribution $\tilde{p}^*_{\mathcal{M}}(x)$ (solid line) is the MC result of Wilding and Müller [22]. The scaling variable is $a^{-1}_{\mathcal{M}} L^{\beta/\nu}(\mathcal{M} - < \mathcal{M} >_c)$. Scale factors are chosen such that the distributions have unit variance.

establish only the compatibility between our simulations results for Q^*, y_T and y_i and their Ising values.

For the 2d RPM a similar finite size scaling analysis has been realized [15] by using the distribution $\tilde{p}^*_{\mathcal{M}}(x)$ of the 2d Ising model and the values of the critical exponents, in particular θ and ν respectively equal to 4/3 and 1. The matching of $p_L(\mathcal{M})$ and $\tilde{p}^*_{\mathcal{M}}(x)$ has been realized and gives the values of $T^*_c(L) = 0.0401$, 0.0403 and 0.04037 for the volumes $V/\sigma^2 = 1000$, 2000 and 4000 respectively. The linear extrapolation as a function of $L^{-7/3}$ of these values to infinite volume, shown in Figure 2, gives $T^*_c = 0.04045 \pm 0.000005$. From the density distributions having a very large extension from $\rho^* = 0.01$ to 0.45, at best, very rough lower 0.1 and upper 0.3 bounds can be given for ρ^*_c.

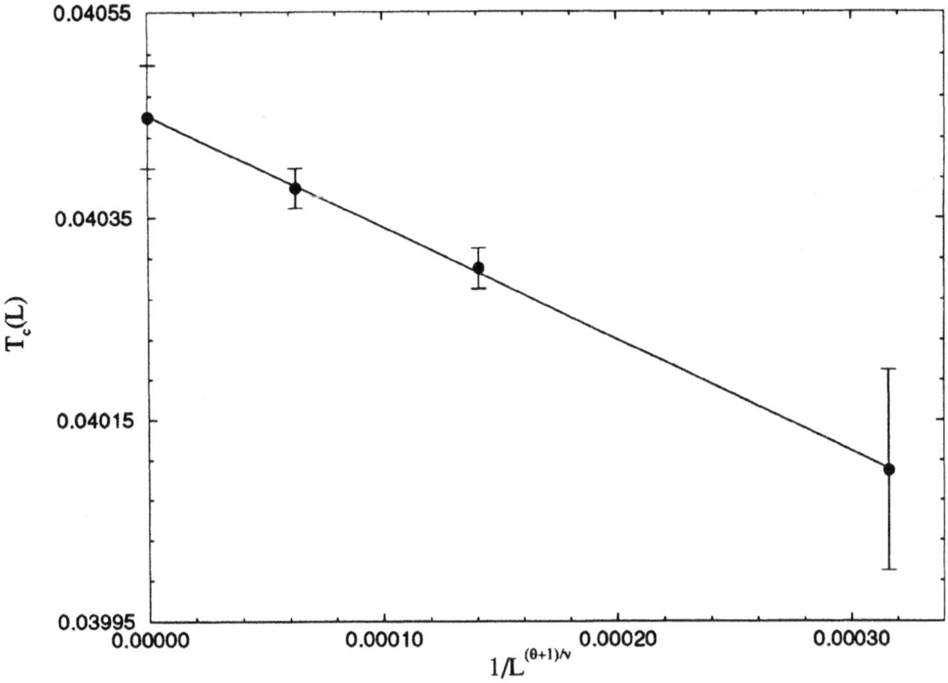

Figure 2. The apparent critical temperature $T_c^*(L)$ of the 2d RPM as function of $L^{-(\theta+1)/\nu}$, with $\theta = 4/3$ and $\nu = 1$. Extrapolation of the least-square fit gives the infinite volume temperature estimate $T_c^* = 0.04045 \pm 0.00005$.

5. Conclusion

By performing GCMC simulations in the vicinity of the gas-liquid critical point and applying a finite size scaling method to analyze the results, we have attempted to contribute to the determination of the nature of the criticality of the 2d or 3d RPM. However, opposed to the case of lattice systems, the system sizes which can be considered for these continuous systems hardly exceed a few thousand particles due the long range of the interactions. This means that if no assumption is made concerning the universality class of the critical point, the number of variables to be determined will be too large to expect an unambiguous estimate within the uncertainty of the simulations results.

The limitation on the system size precludes a satisfactory sampling of the very low density configurations with the consequence that the ordering operator distribution $p_L(\mathcal{M})$ cannot be very accurately symmetrized. This limitation also implies that the correction-to-scaling contributions stay important, especially for the 3d RPM .

The question arises whether the correspondence betwen $p_L(\mathcal{M})$ and $\tilde{p}^*_{\mathcal{M}}(x)$ is sufficient to decide that the RPM belongs to the Ising universality class. In particular: is the mean-field type critical behaviour excluded without ambiguity ? The absence of a distribution $\tilde{p}^*_{\mathcal{M}}(x)$ of precision similar to that of the Ising model precludes to show the incompatibility of our simulations data with the mean field universality class.

The accuracy of our results for the RPM is comparable to that of the results obtained twenty years ago for the lattice models. It allows us to conclude in favor of the compatibility of the critical behaviour of the RPM, determined by simulation, with the Ising universality class. But, it is necessary to increase this precision by an order of magnitude and to investigate larger system size to obtain an unambiguous confirmation of this conclusion.

References

1. Levelt Sengers, J.M.H. and Given, J.A. (1993) Critical behaviour of ionic fluids, *Mol. Phys.* **80**, 899-913.
2. Fisher, M.E. (1994) The story of coulombic criticality, *J. Stat. Phys.***75**, 1-35.
3. Zuckerman, D.M., Fisher, M.E. and Lee, B.P. (1997) Critique of primitive model electrolyte theories *Phys. Rev. E* **56**, 6569-6580.
4. Orkoulas, G. and Panagiotopoulos, A. Z. (1994) Free energy and phase equilibria for the restricted primitive model of ionic fluids from Monte Carlo simulations, *J. Chem. Phys.* **101**, 1452-1459.
5. Caillol, J.M. (1994) A Monte Carlo study of the liquid-vapor coexistence of charged hard spheres, *J. Chem. Phys.* **100**, 2161-2169.
6. Caillol, J.M. and Weis, J.J. (1995) Free energy and cluster structure in the coexistence region of the restricted primitive model, *J. Chem. Phys.* **102**, 7610-7621.
7. Zhang K.C., Briggs M. E., Gammom, R.W. and Levelt Sengers, J.M.H. (1992) The susceptibility critical exponent for non aqueous ionic binary mixture near a consolute point, *J. Chem. Phys.* **97**, 8692-8697.
8. Schroër, W., Kleemeier, M., Plikat, M., Weiss, V. and Wiegand, S. (1996) Critical behaviour of ionic solution in non-polar solvents with a liquid-liquid phase transition, *J. Phys.: Condens. Matter* **8**, 9321-9327.
9. Stell, G. (1993) Criticality and Phase Transitions in ionic fluids, *J. Stat. Phys.* **78**, 197-238.
10. Fisher, M.E. and Lee, B.P. (1996) Ginzburg criterion for coulombic criticality *Phys. Rev. Lett.* **77**, 3561-3564.
11. Wilding, N.B. and Bruce, A.D. (1992) Density fluctuations and field mixing in the critical fluid, *J. Phys.: Condens. Matter* **4**, 3087-3108.
12. Wilding, N.B. (1995) Critical point and coexistence curve properties of the Lennard-Jones: a finite-size-scaling study, *Phys. Rev. E* **52**, 602-611.
13. Caillol, J.M. (1998) Critical-point of the Lennard-Jones fluid: a finite-size scaling study, *J. Chem. Phys.* (to be published)
14. Caillol, J.M., Levesque, D. and Weis, J.J. (1997) A Monte Carlo finite size scaling study of charged hard sphere criticality, *J. Chem. Phys.* **107**, 1565-1575.
15. Weis, J.J., Levesque, D. and Caillol, J.M. (1998) Restricted primitive model of an ionic solution confined to a plane. *J. Chem. Phys.* (to be published)
16. Caillol, J.M. (1993) A new potential for the numerical simulations of electrolyte

solutions on a hypersphere, *J. Chem. Phys.* **99**, 8953-8963.

17. Rehr, J.J. and Mermin, N.D. (1973) Revised scaling equation of state at liquid vapour critical point, *Phys. Rev. A.* **8**, 472-480.
18. Wegner, F.G. (1972) Corrections to scaling law, *Phys. Rev. B.* **5**, 4529-4536.
19. Nicolaides, D. and Bruce, A.D. (1998) Universal configurationnal structure in two dimensional scalar models, *J. Phys. A: Math. Gen.* **21**, 233-244.
20. Blöte, H.W., Luijten, E. and Heringa, J.R. (1995) Ising universality in the three dimensions: a Monte Carlo study, *J. Phys. A: Math. Gen.* **28**, 6289-6313.
21. Ferrenberg, A.M. and Landau, D.P. (1991) Critical behavior of the three-dimensional Ising model: A high-resolution Monte Carlo study, *Phys. Rev. B.* **44**, 5081-5091.
22. Wilding, W. B. and Müller, M. (1995) Liquid-vapor asymmetry in pure fluids: a Monte-Carlo study, *J. Chem. Phys.* **102**, 2562-2573.

PHASE DIAGRAM OF DIPOLAR HARD-CORE MOLECULE SYSTEMS

D. LEVESQUE
Université Paris Sud
Laboratoire de Physique Théorique et Hautes Energies
Bâtiment 211, Université Paris-Sud
91405 Orsay Cedex, France

1. Introduction

The interaction between the permanent or induced dipoles of molecules has two important characteristics: its range and its anisotropy. Between two dipoles $\vec{\mu}_i$ and $\vec{\mu}_j$ located at $\vec{r}_i$ and $\vec{r}_j$, this interaction is

$$v_{DD}(\vec{r}_{ij}, \vec{\mu}_i, \vec{\mu}_j) = -\frac{1}{r_{ij}^3} \left[3 \frac{\vec{\mu}_i \,.\, \vec{r}_{ij} \; \vec{\mu}_j \,.\, \vec{r}_{ij}}{r_{ij}^2} - \vec{\mu}_i \,.\, \vec{\mu}_j \right]. \quad (1)$$

The above mentioned properties are obvious in eq. (1), in particular, if $\vec{\mu}_i$, $\vec{\mu}_j$ and $\vec{r}_{ij} = \vec{r}_j - \vec{r}_i$ are parallel and $|\vec{r}_{ij}| = 1$

$$v_{DD} = -2\,\mu^2 \quad (2)$$

or if, at the same distance, $\vec{\mu}_i$ and $\vec{\mu}_j$ are parallel and perpendicular to $\vec{r}_{ij}$

$$v_{DD} = \mu^2 \quad (3)$$

The long range of the dipolar interaction is at the origin of the dielectric properties of the molecular fluids or solids where the molecules have permanent dipoles. The properties of the dielectric systems have been studied and computed in numerous theoretical works and simulations. But, in this lecture, the main purpose is to present the consequences that the anisotropic character of v_{DD} seems to have on the phase diagram of simple systems of dipolar molecules [1, 2, 3, 4, 5].

It has been remarked that the anisotropy of v_{DD} makes the formation of linear clusters or "chains" of molecules possible at low temperatures.

C. Caccamo et al. (eds.), *New Approaches to Problems in Liquid State Theory*, 61–70.

However it was supposed that for temperatures T such that the coupling constant between dipoles is

$$\mu^2/k_BT\sigma^3 = \mu^{*2} \sim 1 \tag{4}$$

where σ is a typical dimension of the molecules, the rotation of the molecules should average the anisotropic effects. Then, it was believed that the phase diagram of the systems of molecules interacting by a short range repulsive interaction and a dipole-dipole interaction was similar to that of fluid interacting by a potential of van der Waals type. By using pertubative developments in power of $\mu^2/k_BT\sigma^3$ or mean field theories, theoretical works seemed to confirm this hypothesis and the critical point of the gas-liquid transition of the system of dipolar hard spheres was located at a temperature $T_c^* = k_BT_c\sigma^3/\mu^2 \sim 0.25 - 0.3$ and a density $\rho_c^* = \rho_c\sigma^3 \sim 0.1 - 0.2$ [6]. But this result was not well established due to the questionable validity of the used approximations in the low temperature domain.

Numerous works, using numerical simulations have been realized with the aim to test the theoretical results obtained, in particular, for the systems of dipolar hard spheres or molecules interacting *via* a Stockmayer potential. The studies of the phase diagrams of systems of dipolar molecules are interesting from a theoretical point view, but also in practice, because these simple fluids seem relevant models of real systems such as the ferrofluids and the electrorheologic fluids.

The simulations have been realized in the canonical (NVT), isobaric (NpT) and Gibbs ensemble. In these computations, the simulations are performed in cubic or parallelepipedal volumes with periodic boundary conditions. The long range of the interaction v_{DD} is generally taken into account by the method of the Ewald summation. The Hamiltonian of the studied systems has the form:

$$\begin{aligned} H &= -\frac{1}{2}\sum_{i\neq j} v_{ij} \\ &- \frac{1}{2}\sum_{i,j}\sum_{n}{}' (\vec{\mu}_i.\nabla_i)(\vec{\mu}_j.\nabla_j)\frac{\operatorname{erfc}(\alpha|\vec{r}_j - \vec{r}_i + \vec{n}L|)}{|\vec{r}_j - \vec{r}_i + \vec{n}L|} \\ &+ \frac{1}{2\pi V}\sum_{k,k\neq 0}\frac{4\pi}{k^2}e^{-k^2/4\alpha^2}\sum_i \vec{\mu}_i.\vec{k}\, e^{i\vec{k}.\vec{r}_i}\sum_j \vec{\mu}_j.\vec{k}\, e^{-i\vec{k}.\vec{r}_j} \\ &- \sum_i \frac{2\alpha^3\mu_i^2}{3\sqrt{\pi}} + \frac{2\pi}{(2\epsilon'+1)V}[\sum_i \vec{\mu}_i]^2. \end{aligned} \tag{5}$$

if the simulation cell is a cubic volume V of side L. The pair potential v_{ij} corresponds to the hard core or van der Waals interaction between molecules.

The vectors $\vec{n}$ have integer components, the vectors $\vec{k}$ have components of the form $2\pi m/L$ where m is an integer and the prime in the sum over $\vec{n}$ indicates that $i \neq j$ for $\vec{n} = 0$.

The last term takes into account the reaction field induced by the total moment $\sum_i \vec{\mu}_i$ of the molecules in the simulation cell in the dielectric continuum of dielectric constant ϵ' supposed surrounding the simulation cell and its periodic replica. Another way to take into account the long range dipolar interaction is the hypersphere method described above for the case of the charge-charge interaction.

It is important to stress the technical difficulties encountered in the simulations of strongly interacting dipolar hard-core molecules, in particular at low densities and temperatures. The main problem for the simulations performed in the NpT or Gibbs ensemble is the association of molecules in clusters or "chains" which precludes an efficient sampling of the volume V and, sometimes, a clear conclusion concerning the existence or localization of a phase equilibrium.

We discuss successively, the results obtained at low, liquid and solid densities for the dipolar hard spheres, and those found in the case of non-spherical molecules.

2. Dipolar hard spheres: phase diagram at low densities

The first studies, by numerical simulation, of the gas-liquid transition of a dielectric fluid have considered the fluid of Stockmayer molecules where the interaction between molecules is the sum of two potentials: the potential v_{DD} and the Lennard-Jones potential. They have allowed the localization of the critical point of this transition and have shown that the critical temperature is an increasing function of the value of the molecular dipole μ. The v_{DD} interaction seemed to favour the gas-liquid transition [7].

However, the numerical simulations in the canonical isobaric and Gibbs ensembles made in ref. [8] for the system of dipolar hard spheres at the temperatures $T^* = 0.222$ and 0.18, below the theoretically estimated critical temperature, do not show any evidence of a phase coexistence.

This negative conclusion was confirmed by simulations realized for a system of molecules where the potential v_{ij} has the form [9]

$$v_{ij}\,(r_{ij}) \;=\; 4\epsilon_{LJ}\left[\frac{\sigma^{12}}{r_{ij}^{12}} \;-\; \lambda\frac{\sigma^{6}}{r_{ij}^{6}}\right]. \tag{6}$$

The λ parameter, varied between 1 and 0, allows one to study the localization of the gas-liquid transition for van der Waals interactions of decreasing amplitude at a constant value of μ^*. The results obtained by MC simulations in the Gibbs ensemble show that the critical temperature T_c^* decreases linearly with λ. At the values of $\lambda < 0.3$ it is not possible to obtain any evidence of a phase transition.

Motivated by these works, a study [5] was made with the purpose to characterize the configurations at low density and temperature of the dipolar hard sphere system. From simulations realized along the isotherm $T^* = 0.0816$ ($\mu^* = 3.5$), it was established that in the density domain $\rho^* < 0.1$ the configurations of this system were mainly constituted by arrangements of "chain" type where each sphere is almost at contact with these two nearest neighbours. The distance vector between two neighbouring spheres is almost parallel with their two dipole moments. This last point is clearly confirmed by the value of the energy per molecule which, at these low densities, was almost independent of ρ^*, varying of less than 10% when the density ρ^* goes from 0.005 to 0.05.

The properties of these linear arrangements could be studied by using the following criterion to define the association between two spheres: two spheres are considered associated if their potential energy is smaller than $-\ 1.4\ \mu^{*2}$ (noticed that the minimal value is $-2\ \mu^{*2}$).

By using this criterion, it was possible to compute the mean length l of the "chain" arrangements and also the persistence length l_p for the arrangements of sufficient length, from the expressions

$$l = \sum_s s n_s \ / \ \sum_s n_s \tag{7}$$

and

$$l_p = \frac{1}{\sigma} \sum_{\text{all chains ,with } s>10} \ \sum_{k=1}^{s-2} \vec{e}_1 . \vec{e}_{1+k} \ / \ \sum_s n_s \tag{8}$$

where $\vec{e}_i = \vec{r}_{i+1} - \vec{r}_i$.

These "chain" arrangements in the considered domain of low density and temperature change considerably during a typical simulation run of $\sim 10^5$ trials move per molecule, but they keep their qualitative charactheristics. This evolution seems to allow the conclusion that the sampled configurations are representative of the equilibrium state of the system and the average quantities (energy, l, l_p, ...) are significant.

For the densities $\rho^* < 0.1 - 0.2$ when the temperature is increased, a fast increasing of the "monomer" number (isolated sphere following the above criterion) was found at $T^* \sim 0.25$, below this temperature the association of the dipolar hard sphere in chain of polymer type seems to prevent the gas-liquid transition.

The effect of the boundary conditions on the linear arrangements of the molecules has been studied by considering 2048 dipolar hard spheres enclosed in a spherical volume limited by an hard surface. The dimension of the system was sufficient to have a sizeable amount of bulk phase in the center of the volume and to show that the molecular arrangements were identical for similar thermodynamic conditions to those obtained in volumes with periodic boundary conditions.

At larger densities $0.2 < \rho^* < 0.6$ and low temperatures, it is not so easy to characterize the molecular arrangements due the small separation between molecules. There is no evidence of "chains" in the snapshots and the association criterion defined above does not seem relevant to these densities. The system of the dipolar hard spheres in this domain of thermodynamic states looks similar to a van der Waals dense gas.

3. Dipolar hard spheres: liquid and solid phases

By molecular dynamic simulations [4], it was demonstrated that the fluid of molecules in interaction by a short range repulsive potential v_{ij}

$$v_{ij}\,(r_{ij}) \;=\; 4\,\epsilon_{SS}\;\frac{\sigma^{12}}{r_{ij}^{12}} \tag{9}$$

and the dipole-dipole interaction v_{DD} at sufficiently low temperatures and densities larger than $\rho^* > 0.6$ has a finite polarization corresponding to a transition from a state with zero polarization to a ferroelectric state, in addition for $\rho^* \sim 0.9$, there is a transition from a ferroelectric liquid to a ferroelectric solid.

A systematic study [10] of the isotherm $T^* = 0.16$ shows that the dipolar hard sphere system has a similar phase diagram and, when the density increases, goes from a fluid state to a ferroelectric fluid state and a ferroelectric solid state. In these polarized states the modification of dielectric boundary conditions made by changing the values from $\epsilon' = \infty$ to $\epsilon' = 1$ in eq. (5) induces, as expected on the influence of the reaction field, a division of the system in sub-domains of polarization oriented such that the global polarization of the system is zero. For the system of size $N = 500$ the phase transition of isotropic phase to the ferroelectric phase occurs in the vicinity of $\rho^* = 0.7$. This phase is stable until $\rho^* = 1.0$.

At higher densities, in the solid phase, the mechanical stability of the fcc and bct (body centered tetragonal) phases has been studied [10]. For $\epsilon' = \infty$, the bct lattice is characterized by polarized alignments of spheres along the direction z. This phase seems stable on the domain of density 0.95 - 1.20. The fcc stucture is mechanically stable in the domain of density 1.02 -1.20. The polarization is oriented along the direction [001] and in this direction the orientation of the dipoles varies helically from a lattice plane to the other. At density below $\rho^* = 1.06$, a mechanical stable phase of type bco (body centered orthorhombic) is also observed. The fcc-bco transition results from a change in direction of the preferential ordering of the dipole moments from the [001] to the [110] direction which induces a contraction of the system along the [110] direction until near contact of the hard spheres is obtained.

For $\epsilon' = 1$, by using as initial configuration an equilibrium configuration obtained for $\epsilon' = \infty$, the fcc and bct lattices stay mechanically stable for densities larger than 1.06. The ferroelectric ordering is divided in sub-domains. In the bct lattice for 500-576 particle systems, it is found that the system splits into two domains and the fcc lattice in four domains. In the bct lattice, the two domains have polarizations parallel to a diagonal of the xy plane and perpendicular to the sphere alignments. In the fcc lattice, the four domains have polarizations parallel to the xy plane and pointing in different directions (roughly [110], [1,-1,0], [-1,-1,0] and [1,-1,0]). Obviously due to the system size, the bulk part of these domains is small and there are important boundary effects.

Recently the system of dipolar hard spheres with centers of mass and dipole moments confined to a plane has been studied by simulation [11]. In this system, the interaction between dipole moments is similar to v_{DD} where $\vec{r}_i$, $\vec{r}_j$, $\vec{\mu}_i$ and $\vec{\mu}_j$ are two component vectors. The simulation cell is a square with periodic boundary conditions and the long range of the dipolar interaction is handled by Ewald summation. Along the isotherm $T^* = 1/9$, at low density ($\rho^* = 0.05$) the molecules are organized in chains as in the 3d systems with the formation of ring structures. At higher density $\rho^* = 0.20$ the chains have a large extension with repeated branching. Upon further increasing the density, the chain structure span the simulation box. At the highest density, vortex-like structures are manifest as shown in Figure 3. The main result of this work is that, as in the 3d system, the formation of arrangements of "chain" type seems to preclude the gas-liquid transition and the vortex-liquid arrangement the formation of the ferroectric ordering of the dipoles. Similar conclusion are obtained in the case where the dipole-dipole interaction is derived from the 2d coulomb potential $ln\ r$.

4. Hard-cylinder and hard dumbbell systems

The hard-spherocylinder system has several stable phases: isotropic, nematic, smectic and solid. The influence of the dipolar interaction due to a permanent dipole moment located on the axis of the cylinder with an orientation parallel or perpendicular to this axis has been studied by numerical simulations [12]. As in the case of the dipolar hard spheres for most of the values of the aspect ratio L/σ (L cylinder length and σ diameter of the hemispherical cap on each end) in the isotropic phase the gas-liquid transition does not exist.

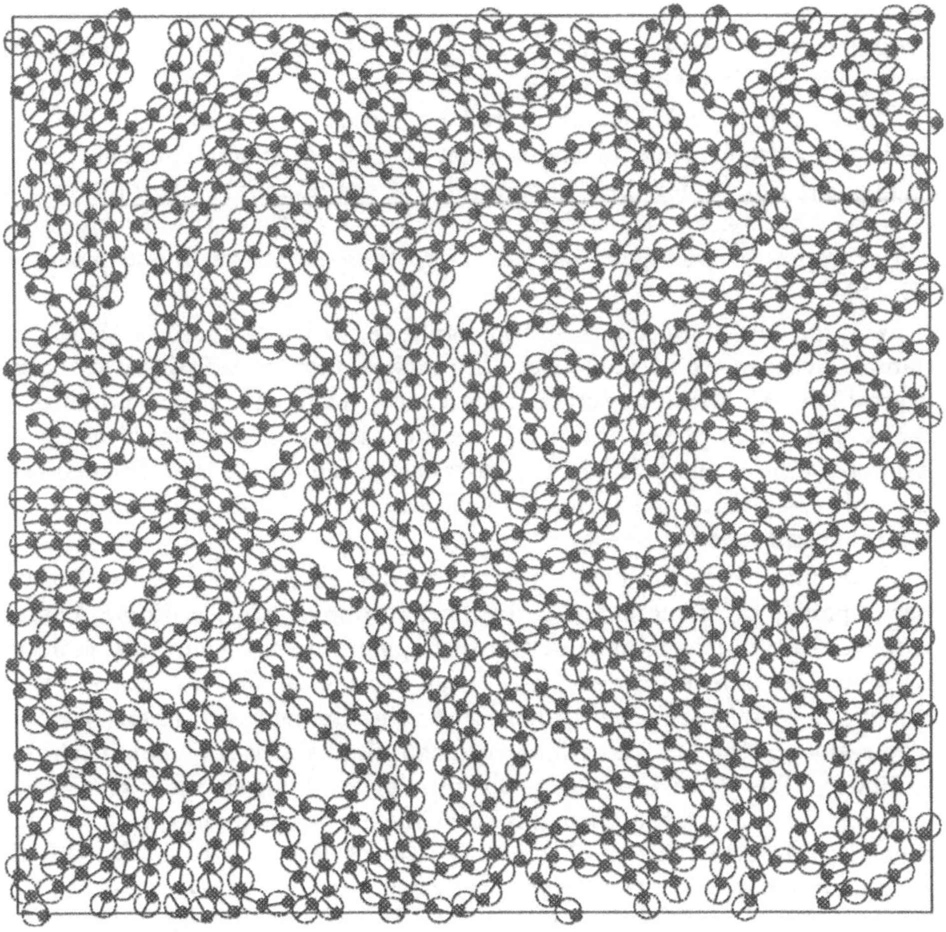

Figure 1. Snapshot of an instantaneous configuration of 1024 dipolar hard disks at $\rho^* = 0.8$ and $T^* = 1/9$. The box length is $L = 35.78\sigma$.

However, for the aspect ratios of values $\sim 1/4$, it has been shown that the gas-liquid transition can be recovered. At $L/\sigma = 1/4$ the critical temperature has been estimated equal to $T_c^* = 0.124$ and the critical density equal to $\rho_c^* = 0.105$. If, from the value 1/4, the aspect ratio is increased or decreased, the gas-liquid transition disappears. At the temperature $T^* = 0.12$, the gas- liquid coexistence is limited to the domain of L/σ values from 0.19 to 0.28. For the values of $L/\sigma \sim 0.$ - 0.1, the sphero-cylinders are arranged in almost linear chains similar to those found in the dipolar hard sphere system. For the case of $L/\sigma \sim 1$, there is formation of chains where the dipole moments are side by side in a antiparallel configuration. These two arrangements at $L/\sigma \sim 0$ and ~ 1 preclude the clustering that seems necessary for gas-liquid coexistence.

These results can be interpreted by noticing that, for $L/\sigma \sim 0$ and ~ 1, the arrangements of the nearest neighbour sphero-cylinders in the chains correspond, for these elongations, to the minimal energy of a pair of sphero-cylinders. For the aspect ratio $L/\sigma \sim 1/4$ these two arrangements have comparable energies; this fact seems to make the gas-liquid transition possible.

For the higher densities and elongations, in the domain of thermodynamic states where the hard core interaction induces transition to liquid-crystal phases, the interaction between dipole moments can stabilize or destabilize these phases following the localization of the dipole along the axis of the molecule [13, 14]. For instance, if the dipole is at the center of the sphero-cylinder the nematic phase can be destabilized relative to the isotropic and smectic phases.

The phase behaviour of dipolar hard dumbbells, which consist of two hard spheres of diameter σ, at fixed separation L, with a dipole moment μ located at equal distance of the two sphere centers along the molecule axis, has been investigated [15]. These molecules, for low values of the aspect ratio L/σ, are almost identical to the dipolar hard spherocylinders. The results of the GC and Gibbs ensemble simulations have confirmed this similarity. The phase diagram of the dipolar hard dumbbells is also almost identical to that of the hard sphero-cylinders. At temperature $T^* \sim 0.1$, a gas-liquid coexistence is obtained for aspect ratios $L/\sigma \sim 0.5$, with the possibility that this coexistence persists for values of $L/\sigma \sim 0.1 - 0.15$ smaller than those obtained for the dipolar hard spherocylinders.

The phase diagram of the hard dumbbells consisting of two spheres embedded with equal and opposite charges of magnitude q has been determined at the low densities and temperatures by GCMC simulations [16]. This model is a version of the model described above where the dumbbells have an extended dipole moment. In addition, since near the critical point the RPM is dominated by pairs and neutral clusters, it is interesting to see to what extent the phase diagram of the RPM is simply a system of charged hard dumbbells. The simulation results strongly suggests that the phase diagram of charged hard dumbbell fluid and that of the RPM are similar. The critical point of the gas-liquid transition for an aspect ratio equal to 1 is located, by using reduced temperature and density identical to those of the RPM, at $T^* \sim 0.05$ and $\rho^* \sim 0.05 - 0.1$. These results suggest that the RPM close to the critical point can behave as a fluid of neutral bounded pairs of ions perturbed by free ions.

The systems of dipolar hard discoidal particles have been studied. These systems present at high density several type of liquid crystal phases in particular columnar phase. The dense liquid crystal phases can be ferroelectric or antiferroelectric following the values of the aspect ratio [17, 18].

5. Conclusion

The results, discussed above, show the variety of the phase diagrams of dipolar hard-core molecules where there is a competition in the ordering of the molecules between the dipolar interaction and the steric effects. In the real systems, to these two interactions add the van der Waals interactions and the polarization effects. It may be difficult to determine the role of these different interactions in the localization and the symmetry of the stable phases.

For the systems of mesogenic molecules, an essential role had been attributed to the steric effects. This point of view was based on the results that most of the liquid crystal phases are obtained by considering only hard core interactions.

The simulations results, presented above, show that, in taking account the dipolar interactions in these hard core systems, it is possible to describe qualitatively the physical properties of systems having ferroelectric stable phases. Futhermore, if the dipole interaction can induce the formation of linear arrangements of the molecules, it seems to preclude the gas-liquid transition.

References

1. Groh, B. and Dietrich, S. (1996) Structural and thermal properties of orientationally dipolar fluids, *Phys. Rev. E.* **53**, 2509-2530.
2. Sear, R.P. (1996) Low-density fluid phase of dipolar hard spheres, *Phys. Rev. Lett.* **76**, 2310-2313.
3. Wei, D. W. and Patey, G.N. (1993) Orientational order in simple dipolar fluids: density-functionnal theory and absolute-stability conditions, *Phys. Rev. E.* **47**, 506-512.
4. Wei, D. W. and Patey, G.N. (1992) Orientational order in simple dipolar liquids: computer simulation of a ferroelectric nematic phase, *Phys. Rev. Lett.* **68**, 2043-2045.
5. Levesque, D. and Weis, J.J. (1994) Orientational and structural order in strongly interacting dipolar hard spheres, *Phys. Rev. E.* **49**, 5131-5140.
6. Ng, K.C., Valleau, J.P., Torrie, G.M. and Patey, G.N. (1979) Liquid-vapour coexistence of dipolar hard sphere, *Mol. Phys.* **38**, 781-788.
7. Smit, B. Williams, P.C., Hendricks, E. M. and De Leuw, S.W. (1989) Vapour -liquid equilibria for Stockmayer fluids, *Mol. Phys.* **68**, 765-769.
8. Caillol, J.M. (1993) Search of the gas-liquid transition of dipolar hard spheres, *J. Chem. Phys.* **98**, 9835-9849.
9. van Leeuwen, M.E. and Smit, B. (1993) What makes a polar liquid a liquid?, *Phys. Rev. Lett.* **71**, 3991-3994.
10. Weis, J.J. and Levesque, D. (1993) Ferroelectric phases of dipolar hard spheres, *Phys. Rev. E.* **48**, 3728-3740.

11. Weis, J.J. (1998) Orientational structure of quasi-two-dimensional dipolar hard spheres, *Mol. Phys.* **93**, 361-364.
12. McGrother, S.C. and Jackson, G. (1996) Island of vapour-liquid coexistence in dipolar hard-core systems, *Phys.Rev. Lett.* **76**, 4183-4186.
13. McGrother, S.C., Gil-Vilegas, A. and Jackson, G. (1996) The liquid crystalline behaviour of hard spherocylinders with terminal dipole, *J. Phys.: Condens. Matter* **8**, 9649-9655.
14. Gil-Vilegas, A., McGrother, S.C. and Jackson, G. (1997) Chain and ring structures in smectic phases of molecules with transverses dipoles, *Chem. Phys. Lett.* **269**, 441-447.
15. Shelley, J.C., Patey, G.N., Weis, J.J and Levesque, D. (1998) Liquid-vapour coexistence in fluids of dipolar hard dumbbells and spherocylinders (preprint).
16. Shelley, J.C. and Patey, G.N. (1995) A comparison of liquid-vapour coexistence in charged hard sphere and charged hard dumbbell fluids, *J. Chem. Phys.* **103**, 8299-8301.
17. Ayton, G., Wei, D.Q. and Patey, G.N. (1997) Liquid crystal phases of dipolar discoidal particles, *Phys. Rev. E.* **55**, 447-454.
18. Zarragoicoechea, G.J., Levesque, D. and Weis, J.J. (1993) Monte Carlo simulations of polar discoidal molecules, *Mol. Phys.* **78**, 1475-1492.

NEW RESULTS ON SOME IONIC-FLUID PROBLEMS

G. STELL
Department of Chemistry
State University of New York at Stony Brook
Stony Brook, New York 11794-3400, USA

1. Introduction

The ionic-fluid problems considered herein are all concerned with understanding and describing the thermodynamics and structural properties of strong electrolytes and molten salts. This general line of development has occupied the attention of theoreticians since the first quarter of the 20th century but the special focus of this seminar is on those thermodynamic states for which one finds criticality and phase separation (demixing) into two fluid states, either liquid-gas or liquid-liquid. This is a somewhat more recent concern among theoreticians that only began to receive serious attention over the past two decades or so. Well before that time ionic solutions that unmix had been observed and ionic melts were known to vaporize at sufficiently high temperature, despite the experimental inaccessibility of the critical regions of most of them. But whether a system of particles interacting via nothing but inverse-square Coulombic forces and repulsive core terms would have a critical point associated with liquid-gas phase separation was a question that had not been systematically studied previously. Stell, Larsen, and Wu made such a study that appeared in 1976 [1] for the system that has come to called to restricted primitive model (RPM) — a symmetric mixture of charged hard-sphere anions and cations of equal size and charge number. They concluded that such a system indeed has liquid-gas coexistence curve with a critical point, but also found that the curve's shape and location was very sensitive to the details of the approximations that inevitably must be introduced in such a theoretical study. The sensitivity was far more pronounced than that found in analogous model treatments of simple noble-gas fluids.

Phase separation and criticality in the RPM and related models has been the subject of a considerable number of theoretical studies as well as

C. Caccamo et al. (eds.), New Approaches to Problems in Liquid State Theory, 71–89.

Monte Carlo simulations since ref. [1] appeared. Moreover, several experimental studies have been made on demixing in ionic solutions selected for investigation because their interionic interactions appear to be reasonably well modeled by the RPM. The theoretical work includes an important study by Ebeling and Grigo [2], which extended the RPM results of the mean-spherical approximation (MSA) obtained in [1] by adding the effect of a mass action law (MAL) to describe ionic pairing of anions and cations. M.E. Fisher and his coworkers subsequently [3] chose as the starting point of a quantitative study of the RPM thermodynamics the linearized Poisson-Boltzmann equation of Debye and Hückel supplemented by Bjerrum's treatment of ion pairing. To this they added an ion-pair/free-ion contribution. Stell and his colleagues also continued their theoretical investigations, including a quantitative re-evaluation of the Ebeling-Grigo results that corrected a numerical error in the original Ebeling-Grigo work as well as an extension of the Ebeling-Grigo theory that added an ion-pair/free-ion contribution of a somewhat different form than that used by Fisher and his colleagues [4].

Although the above-mentioned theoretical studies, computer experiments, and experimental results have considerably deepened our understanding of various aspects of ionic-fluid phase coexistence and its modeling, a number of problematic features and apparent inconsistencies among the various sets of results have also emerged. The half-dozen or so Monte Carlo studies of the RPM made in the 1990's have yielded estimates of critical-point temperature and density ranging from $T_c^* = 0.07$ to 0.049 and $\rho_c^* = 0.025$ to 0.08 [5], with the two most recent published studies converging on $T_c^* = 0.049$ and $\rho_c^* = 0.07$ to 0.08. (Here $T^* = \sigma k_B T \epsilon / q^2$, and $\rho^* = \rho\sigma^3$, where T is absolute temperature, ρ is number density, σ is diameter, q is charge strength, ϵ is solvent dielectric constant, and k_B is Boltzmann's constant.) There are similar disparities among theoretical treatments. For example, the MSA-based treatment of [4] yields the critical-point values $T_c^* = 0.0716$, $\rho_c^* = 0.027$ while the Debye-Hückel-based treatment of [3] yields $T_c^* = 0.057$, $\rho_c^* = 0.028$. If one assumes that the most recent Monte Carlo estimates of T_c^* and ρ_c^* are the most reliable ones, one must conclude that these theoretical values are not very accurate, especially with respect to density.

In a monograph with Hafskjøld [6] on the properties of the RPM that appeared in 1982, Stell noted that one should expect the RPM critical behavior to be in the Ising-model universality class and susequently gave a detailed argument for this conclusion in a 1992 paper [7]. The conclusion was in apparent contradiction to the experimental results reported in 1990 by Singh and Pitzer [8] on an ionic fluid, $N_{2226}B_{2226}$ in diphenyl ether, that they felt should be well-modeled by the RPM. They found that it appeared to have mean-field criticality, although they noted that the data could be reconciled with a delayed crossover from mean-field to Ising-like behavior that occurs very close to the critical point. Their mean-field like results seemed to be supported by further experiments by Zhang et al. [9].

To help clarify this point, Narayanan and Pitzer [10] then studied a family of fluids consisting of a single solute, TBAB, in different solvents. In all cases, they found a crossover from mean-field to Ising-like behavior accuring at non-universal values of the reduced temperature $t = (T - T_c)/T_c$ that depended upon the solvent. Høye and Stell [11] pointed out that such non-universal crossover effects could be expected as a result of the presence of long-lived dipolar clusters such as ion pairs which will give rise to a charge-dipole term in the density-density correlation matrix of the fluid.

Although the Narayanan-Pitzer and Zhang et al. results seem to suggest that the Singh-Pitzer and Zhang et al. results for $N_{2226}B_{2226}$ are also best understood in terms of a mean-field to Ising crossover very close to critical, subsequent experiments on $N_{2226}B_{2226}$ by other groups [12] have revealed no detectable deviation from the Ising-like behavior expected in simple fluids, where deviations from mean-field behavior occur at reduced temperatures t much larger than that found in those systems investigated by Pitzer that showed a delayed crossover.

Ginzburg-criterion studies of the RPM by Leote de Carvalho and Evans [13] and by Fisher and Lee [14] yield a Ginzburg crossover temperature t_G that is no smaller for the RPM than for simple non-ionic fluids; in fact it tends to be somewhat larger. This is consistent with Stell's conclusion that the RPM is Ising-like. A finite-sized scaling analysis used in the Monte Carlo study of the RPM mode by Caillol et al. [5] also yields results consistent with Ising-like behavior. However a heat capacity analysis of Valleau's Monte Carlo data for the RPM has led Valleau and Torrie [15] to conclude that Valleau's results are fully consistent with mean-field criticality.

One can see from the foregoing remarks that neither the theoretical nor experimental picture is in fully satisfactory shape. Why are precise results for the criticality and coexistence behavior of the RPM proving to be so elusive? And why aren't reproducible experimental results that can be analyzed in terms of RPM theory more readily available? I believe a big part of the answer is that the RPM is not a simple fluid. It is a very complex fluid, despite the fact that its pair potentials can be written in a form that looks about as simple as a Lennard-Jones potential. And of the ionic fluids that have been investigated experimentally, even the most RPM-like among them are probably not very RPM-like after all, and may well require models considerably more complex than the RPM in order to be well described theoretically.

With respect to the RPM, there are two important reasons it is not simple.

> It is a mixture with very special symmetries and a very special potential range. (1)

> The interactions between particles are very strong compared to k_BT when the particles are close together at temperatures at which criticality occurs, so that highly nonlinear effects, due especially to the great attraction between anions and cations at their distance of closest approach, are important. In particular, this leads to ionic association involving anion/cation pairs as well as larger clusters, which has no counterpart in the structure of simple fluids. (2)

In order to discuss these two points precisely I shall first give in the next Section some theoretical background using the language of Ornstein-Zernike formalism. For a more detailed discussion as well as primary references to the various equations and approximations introduced in Section 2, see [6], [16] and [17].

2. Background

The pair potential between RPM particles of species i and j with centers a distance r apart has the form

$$\varphi_{ij}(r) = \varphi^{\rm HS}(r) + q_i q_j / r\epsilon \tag{3}$$

where $\varphi^{\rm HS}(r)$ is the hard-sphere potential

$$\varphi^{\rm HS}(r) = \begin{cases} \infty & \text{for } r < \sigma \\ 0 & \text{for } r > \sigma \end{cases} \tag{4}$$

and i can be either $+$ or $-$, as can j. One has $q_+ = -q_-$ and $q = |q_i|$. The ϵ in (3) represents the relative permitivity of the solvent, which is modeled as a uniform structureless medium. If the RPM is used as a molten-salt model the medium is taken to be a vacuum. In this context, the fluid-fluid coexistence and critical point that one finds is regarded as liquid-gas coexistence and criticality. When the RPM is used as an ionic solution model, the structureless solvent is typically thought of as being a liquid, and the coexistence is therefore thought of as liquid-liquid coexistence between solute-rich and solute-poor liquids.

For the binary mixture of RPM anions and cations, one has the usual set of Ornstein-Zernike equations relating the pair correlation functions h_{ij} and the direct correlated functions c_{ij},

$$h_{ij}(r) = c_{ij}(r) + \sum_{\ell} \rho_{\ell} \int h_{i\ell}(R) c_{\ell j}(|r - R|) d\mathbf{R} \tag{5}$$

with ℓ running over + and − and ρ_+ and ρ_- denoting the anion and cation density, respectively. Because of the high degree of symmetry in the RPM, $h_{++}(r) = h_{--}(r)$ and $h_{+-}(r) = h_{-+}(r)$, with corresponding equations for the $c_{ij}(r)$. Also $\rho_+ = \rho_-$ so there is only one independent density in the mixture, which can conveniently be expressed in terms of the total density $\rho = \rho_+ + \rho_-$. The h_{ij} are related to the familiar radial distribution functions g_{ij} and probability density functions ρ_{ij} through the usual relations $h_{ij}(r) = g_{ij}(r) - 1$ and $\rho_{ij}(r) = \rho_i \rho_j g_{ij}(r)$. For hard-core potentials of diameter σ_{ij} one has the core condition

$$h_{ij}(r) = -1 \text{ for } r < \sigma_{ij} \tag{6}$$

It is natural to introduce the density-density functions $h_{\rho\rho}(r)$ and $c_{\rho\rho}(r)$ and the charge-charge functions $h_{qq}(r)$ and $c_{qq}(r)$, which describe the density-density and charge-charge correlations, respectively. In the RPM these are just the sum and difference combinations for both the h_{ij} and the c_{ij}. For example $h_{\rho\rho}(r) = (h_{++}(r) + h_{+-}(r))/2$ and $h_{qq}(r) = (h_{++}(r) - h_{+-}(r))/2$, while the cross-correlation functions $h_{\rho q}$ and $c_{\rho q}$ are idendically zero in the RPM by symmetry. The $h_{\rho\rho}$ and $c_{\rho\rho}$ obey an OZ equation as do h_{qq} and c_{qq}, and these two equations are completely uncoupled,

$$h_{\rho\rho}(r) = c_{\rho\rho}(r) + \rho \int h_{\rho\rho}(R) c_{\rho\rho}(|r - R|) d\mathbf{R} \tag{7}$$

$$h_{qq}(r) = c_{qq}(r) + \rho \int h_{qq}(R) c_{qq}(|r - R|) d\mathbf{R} \tag{8}$$

with the coupling between density-density and charge-charge correlation coming entirely through a second independent set of closure relations between the $h_{ij}(r)$ and $c_{ij}(r)$. This can be expressed in terms of the well-known relation involving the pair potentials φ_{ij} and bridge functions b_{ij},

$$c_{ij}(r) = -\beta\varphi_{ij}(r) + h_{ij}(r) - \ln g_{ij}(r) + b_{ij}(r) \tag{9}$$

where $b_{ij}(r)$ have no simple closed form but instead are functionals of the h_{ij} and ρ_i that are highly nonlocal (i.e., $b_{ij}(r)$ for a particular r depends on $h_{ij}(r)$ for all r).

Neglecting the $b_{ij}(r)$ in (9) yields the well-known hypernetted chain approximation. Because the $g_{ij}(r)$ are identically zero for $r < \sigma$ in the RPM, one has, exactly

$$h_{\rho\rho}(r) = -1 \text{ for } r < \sigma \tag{10}$$

$$h_{qq}(r) = 0 \text{ for } r < \sigma \tag{11}$$

and the c_{ij} (or equivalently the $c_{\rho\rho}$ and c_{qq}) need only be given for $r > \sigma$ to yield a closed set of equations when used with (7,8) and (10,11).

For a simple fluid both the isothermal compressibility and the excess internal energy per particle are expressible in terms of the density-density correlation

$$\begin{aligned} \beta(\partial\rho/\partial p)_\beta &= 1 + \rho\int h_{\rho\rho}(r)\mathbf{dr} \\ &= [1 - \rho\int c_{\rho\rho}(r)\mathbf{dr}]^{-1} \end{aligned} \tag{12}$$

$$u^{\text{EX}} = (U/N)^{\text{EX}} = \frac{1}{2}\rho\int g_{\rho\rho}(r)\varphi(r)\mathbf{dr} \tag{13}$$

For the RPM, (12) remains, but instead of (13) one has

$$u^{\text{EX}} = \frac{1}{2}\rho\int g_{qq}(r)\varphi_{qq}(r)\mathbf{dr} \tag{14}$$

where $\varphi_{qq}(r)$ is the difference function $[\varphi_{++}(r) - \varphi_{+-}(r)]/2$ responsible for the charge-charge correlation,

$$\varphi_{qq}(r) = \begin{cases} 0 & \text{for } r < \sigma \\ q^2/\epsilon r & \text{for } r > \sigma \end{cases} \tag{15}$$

Conceptually, the difference between (13) and (14) is a consequence of statement (1). It is a reflection of the relative complexity of the RPM despite the fact that it has no more degrees of freedom than a pure simple fluid. Computationally, this difference proves to be of major significance, as we shall see.

For the φ_{ij} with hard cores typically used to model simple fluids, an adequate approximation to (9) is often obtained by replacing the $h_{ij} - \ln g_{ij} + b_{ij}$ in (9) outside the cores by a hard-sphere reference-system c_{ij}

$$c_{ij}(r) = c_{ij}^{\text{HS}}(r) - \beta\varphi_{ij}(r)\ , \quad r > \sigma_{ij} \tag{16}$$

Used with (5) and (6) this yields the useful result known as both gamma-ordered approximation (LOGA), and the optimized random phase approximation (ORPA). Furthermore, the $c_{ij}^{\text{HS}}(r)$ is small enough outside to core is that it can be entirely neglected there, simplifying (16) to

$$c_{ij}(r) = -\beta\varphi_{ij}(r)\ , \quad r > \sigma_{ij} \tag{17}$$

This yields the mean-spherical approximation (MSA). In the case of the RPM, the LOGA/ORPA result becomes

$$c_{\rho\rho}(r) = c^{\mathrm{HS}}(r) \ , \ r > \sigma \tag{18}$$

$$c_{qq}(r) = -\beta q^2/\epsilon r \ , \ r > \sigma \tag{19}$$

with the $c^{\mathrm{HS}}(r)$ set equal to zero for $r > \sigma$ in the MSA. When the $c_{qq}(r)$ of (19) is used with (8) and (11) to find the $g_{qq}(r)$ used in (14), the resulting "energy-route" thermodynamics (identical in the MSA and the LOGA/ORPA) is qualitatively adequate but quantitatively disappointing (it yields $T_c^* = 0.079, \rho_c^* = 0.014$). The MSA and LOGA/ORPA results for the RPM $h_{\rho\rho}(r)$ however are wholly inadequate; in these approximations $h_{\rho\rho}(r)$ is given by a hard-sphere contribution unperturbed by the presence of the Coulombic interaction, which has dropped out of (18). The exact RPM $c_{\rho\rho}(r)$ approaches $\frac{1}{2}h_{qq}^2(r)$ as $\rho \to 0$, as noted in refs. [7] and [17], and it is this contribution that is responsible for the Debye-Hückel limiting-law behavior of the free energy density, $\beta A/V \to \Gamma_{\mathrm{D}}^3/12\pi$, where Γ_{D} is the Debye-Hückel inverse charge-charge correlation length, $\Gamma_{\mathrm{D}}^2 = 4\pi\rho\beta q^2/\epsilon$.

We shall now use the results of a development in Section 2 of [17] that is based on a formally exact comparison between the thermodynamics of a pure non-ionic fluid and that of the RPM. For the pure non-ionic fluid we can drop the subscripts in (9) and rewrite it as

$$c(r) = -\beta\varphi(r) + T(\rho, h, r) \tag{20}$$

$$T(\rho, h, r) = h(r) + \ln g(r) + b(r) \tag{21}$$

The T is a functional, completely determined by ρ and $h(r)$, that goes to zero as $\rho \to 0$. One can think of the RPM thermodynamics obtained via (12) as the thermodynamics of a hypothetical pure non-ionic fluid with a pair potential defined by writing the RPM $c_{\rho\rho}(r)$ in terms of T as

$$c_{\rho\rho}(r) = -\beta\varphi(r)^{\mathrm{EQUIV}} + T(\rho, h_{\rho\rho}, r)\,. \tag{22}$$

From our remarks concerning $c_{\rho\rho}$ in the Debye-Hückel regime one sees that $-\beta\varphi(r)^{\mathrm{EQUIV}} \to \frac{1}{2}h_{qq}^2(r)$ as $\rho \to 0$. More generally, using the results of detailed analysis of the primitive model b_{ij} given by Kjellander and Mitchell [18], we find for $r > \sigma$

$$-\beta\varphi(r)^{\mathrm{EQUIV}} = A\,\mathcal{F}(r)\,h_{qq}^2(r) + Be^{-\mu r}/r + \text{higher order terms}\,, \tag{23}$$

where A and B are functions of ρ^* and β^*, $\mathcal{F}(r) \sim (\ln r)^{-2}$ as $r \to \infty$, and $A \to 1/2$, $B \to 0$, and $\mu \to 2\Gamma_D$ as $\rho \to 0$, such that for all $r > \sigma$ and a considerable range of ρ^* beyond the Debye-Hückel limiting-law regime, $-\beta\varphi(r)^{\mathrm{EQUIV}}$ can be well approximated by

$$-\beta\varphi(r)^{\mathrm{EQUIV}} = Ah_{qq}^2(r) \tag{24}$$

Although negligible in the Debye-Hückel regime, $T(\rho, h_{pp}, r)$ will not be negligible at higher densities. At any density, when $\beta^* \to 0$ it will reduce to the hard-sphere term in $c_{pp}(r)$ that remains when the RPM is discharged. This observation, together with (24), suggests the approximation

$$c_{pp}(r) = c_{pp}^{\mathrm{HS}}(r) + Ah_{qq}^2(r)\,,\; r > \sigma \tag{25}$$

as the appropriate RPM analog of the LOGA/ORPA result given by (16) for a non-ionic fluid.

To exploit (24) - (26) we must know something about $h_{qq}(r)$, which we shall write for convenience as

$$h_{qq}(r) = \sinh[-\beta W(r)] \tag{26}$$

As $\Gamma_{\mathrm{D}} \to 0$, the limiting form of $W(r)$ becomes

$$-\beta W(r) = -\beta q^2\,(\exp -\Gamma_{\mathrm{D}} r)/\epsilon r\,, \tag{27}$$

So for small Γ_{D} and large r, $h_{qq}(r) \approx -\beta W(r)$. As Γ_{D} increases, one expects the large-r form of $h_{qq}(r)$ to develop an inverse range Γ that departs from Γ_{D} and an amplitude that departs from $-\beta q^2/\epsilon$ as well as an oscillatory envelope. This can best be described in terms of $\tilde{W}(k)$, the three dimension Fourier transform of $W(r)$ in complex k space. One has at small Γ_{D}

$$\tilde{W}(\tilde{k}) \approx \tilde{C}\frac{q^2/\epsilon}{k^2 + \Gamma^2} \tag{28}$$

with $\tilde{C} \approx 1$ and $\Gamma \approx \Gamma_{\mathrm{D}}$. This $\tilde{W}(k)$ has poles at $k = \pm i\Gamma$; we need only follow the positive one. As Γ_{D} increases from zero, Γ begins to depart appreciably from Γ_D. Moreover the full structure of $\tilde{W}(k)$ is such that a second pole comes down the imaginary axis as Γ_{D} increases and collides with the pole of (22) at a particular value of Γ_{D} at this point, the pole that represents the dominant singularity of $\tilde{W}(k)$ given by (22) splits and moves off the imaginary axis as two poles located at $k = i\Gamma + \omega$ and $i\Gamma - \omega$. (See Fig. 1, where only the first quadrant is shown.) In simple approximations, this takes place at a value of Γ_D (and therefore at a value of the product ρ^*T^*) that is free of any further dependence on ρ^* (or equivalently, on T^*) so that it occurs along a locus in the T^*, ρ^* plane that is a straight line. But there is no reason to expect such a linear locus in an exact theory.

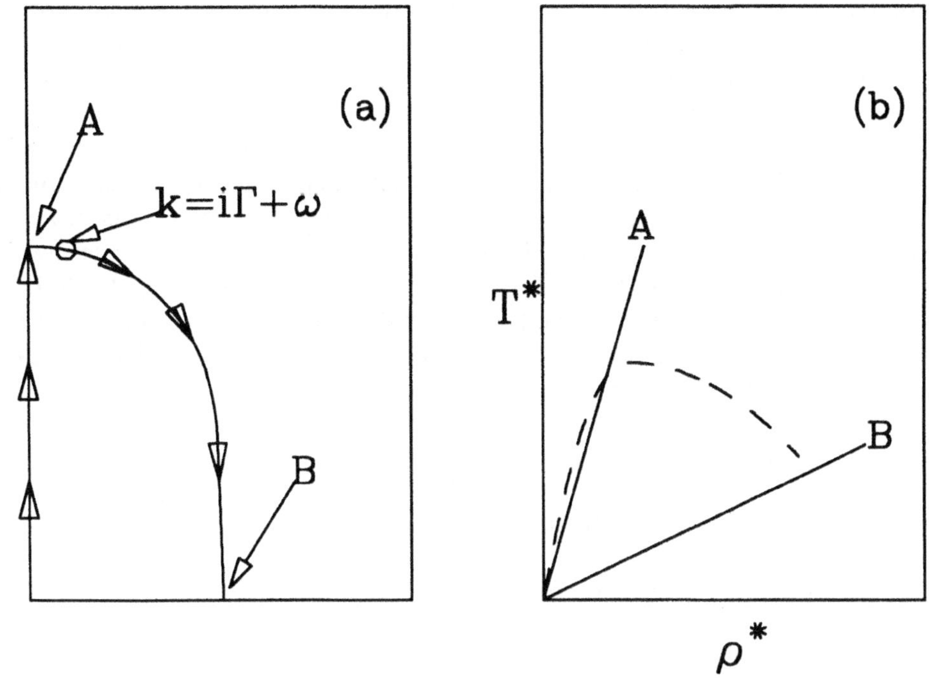

Figure 1. (a) Schematic of the trajectory of the dominant pole of the function $\tilde{W}(k)$ in complex k-space as Γ_D increases from zero to a value Γ_D^A at which the pole leaves the imaginary axis and is located at $k = i\Gamma + \omega$. At some higher value Γ_D^B one expects the pole to meet the real k-axis. (Only the first quadrant is shown here; there are images of the pole in all four quadrants.) (b) Schematic of the RPM ρ^*, T^* plane showing part of the coexistence curve (dashed) and the two lines $T^* = 4\pi\rho^*/(\Gamma_D^A)^2$ and $T^* = 4\pi\rho^*/(\Gamma_D^B)^2$ corresponding to the points A and B in Fig. 1 (a). As noted in the text, one expects Γ_D^A and Γ_D^B themselves to depend upon ρ^* (or equivalently, upon T^*) in an exact theory, so that these two loci will not necessarily be straight lines.

In real space the pole leaving the imaginary axis corresponds to $W(r)$ developing the form

$$W(r) \approx D\,e^{-\Gamma r}\cos(\omega r + \theta)/r \tag{29}$$

where $D(\rho, \beta)$ and $\theta(\rho, \beta)$ can be given in terms of the residue of the pole at $i\Gamma + \omega$.

It seems likely that when it reaches the real axis, the dominant pole will be joined by a confluence of other poles having larger imaginary values to produce a branch cut, so that when $\Gamma = 0$, $\omega \neq 0$, the power of r in the denominator of (29) is not unity but a critical exponent $1 + \eta$. This expectation follows from the observation that the RPM can be regarded as an antiferromagnetic spin fluid of spin dimensionality one, so that the point at which $\Gamma = 0$, $\omega \neq 0$ represents a kind of continuum-model analog of a Néel point for which one expects a nonzero η in 3 spatial dimensions.

The scenario for the expected form of $h_{qq}(r)$ and its related potential of mean forces as $\Gamma_D\sigma$ increases was first discussed by Kirkwood [19] and subsequently developed by Stillinger and his coworkers [20], Stell and Lebowitz [21], Stell [17], Kjellander and coworkers [18, 22], and others [23]. Among their results, Stell and Lebowitz developed systematic expansions of Γ/Γ_D in powers of $\Gamma_D\sigma$ (or equivalently $\rho^*\beta^*$) with ρ^*-dependent coefficients; see especially their Eqs. (4.24), (4.25), (5.1) - (5.13). More recently Kjellander and his colleagues have obtained HNC results for the pole trajectories that define the leading singularities of h_{qq} as well as exact expansion of Γ/Γ_D in powers of the plasma parameter Λ (our $\Gamma_D\sigma\beta^*$).

3. Implications

a. Charge-charge correlations

(i) From Eq. (14) we can write $u^{\rm EX}$ in terms of a scalar integral

$$u^{\rm EX} = (2\pi q^2 \rho/\epsilon) \int_\sigma^\infty r\, g_{qq}(r) dr \tag{30}$$

We would expect the behavior of $u^{\rm EX}$ to reflect the behavior of the integral obtained by substituting for the $g_{qq}(r)$ in (30) its large-r form that follows from (26) and (29)

$$g_{qq}(r) \approx -\beta D e^{-\Gamma r} \cos(\omega r + \theta)/r \tag{31}$$

to yield

$$u^{\rm EX} \approx -2\pi\beta D \rho\, I[\infty] \tag{32}$$

where

$$I[R] = \int_\sigma^R e^{-\Gamma r} \cos(\omega r + \theta) dr \tag{33}$$

We find that $I[\infty]$ and $\partial I[\infty]/\partial T$ will not be well approximated by $I[R]$ and $\partial I[R]/\partial T$ for $R = 2\sigma$ or 3σ when $\Gamma\sigma$, ω, θ, and D take values that are consistent with current simulation results in the neighborhood of the RPM critical point. The conclusion we draw from this is that the RPM internal energy and specific heat (at constant ρ) in the critical region cannot be reliably assessed by any method that involves a direct assessment of internal energy with a cell size given by such R. Using $I[R]$ instead of $I[\infty]$ can yield a fraction of the total $I[\infty]$ contribution that is rather small and quite sensitively R dependent, especially if the $\Gamma\sigma$ of the simulation is artificially small as a result of finite-size polarization effects, as it tends to be in simulations. We note that there is no analogous problem in using (13) to assess the internal energy of a simple fluid model such as truncated Lennard-Jonesium, since $\varphi(r) \approx 0$ for such R. (However, even for a simple-fluid model, gathering good enough fluctuation statistics over such a small cell to reliably estimate c_v is a formidably task.)

(ii) The RPM can be thought of as a spin system with a long-ranged antiferromagnetic exchange interaction, $J(\mathbf{r}) = -q^2/r\epsilon$, and one-dimensional spins $\mathbf{s}_i$ with $|\mathbf{s}_i| = 1$;

$$\varphi_{ij}(r) = -\mathbf{s}_i \cdot \mathbf{s}_j \, J(\mathbf{r}) \tag{34}$$

(Since the "spins" are charges, we should more properly refer to the interaction as antiferroelectric rather than antiferromagnetic.) If $J(r)$ were short ranged — say of Yukawa form $J(r) = -q^2 e^{-\alpha r}/r\epsilon$, rather than Coulomb form — we would still expect a similar scenario as that associated with Fig. 1, except that as $\rho\beta \to 0$, the dominant pole would stop at $k = i\alpha$ rather than go all the way down to zero. The lattice-gas version of such a spin fluid would simply be the spin-1 Ising model with $|\mathbf{s}| = 1, 0, -1$ where $|\mathbf{s}| = 0$ represents holes. For the short-ranged $J(r)$, one continues to expect a locus of Néel points at which

$$W(r) \approx D \cos(\omega r + \theta)/r^{1+\eta} \tag{35}$$

In the nearest-neighbor square-lattice lattice-gas case at close packing, this is a much-studied spin-1/2 Ising model that is exactly solvable in 2 dimensions. There one *knows* that (31) is true at Néel-point criticality, with $\eta = 1/4$. Moreover, although exact results are lacking in 3 dimensions, one expects for bipartite lattices such as simple cubic that as one lowers ρ^* along the λ-line of Néel points, in the ρ^*, T^* plane, one encounters a tricritical point below which the lattice gas phase-separates into paramagnetic and antiferromagnetic phases. The question is whether this still happens to the lattice gas when $J(\mathbf{r})$ becomes Coulombic, and further, whether it continues to happen as one goes from the bipartite lattice-gas version to the continuum-space version of the RPM. A host of related questions follow. In the lattice-gas cases, one does not expect coexistence curves that show both tricritical points and liquid-gas critical points, if one restricts oneself to simple nearest neighbor potentials or those Green's-function potentials that are the lattice analogs of the Yukawa potentials. But in the continuum-space RPM one already has strong evidence that there is a low-density critical point. It is rather the existence and likely location of a λ-line of points at which $\Gamma = 0$, $\omega \neq 0$ and that remains largely unexplored. Because the loss of charge-charge screening signals the onset of long-range charge ordering, it is natural to regard this line as a signature of crystallization, and it could well be that it cannot be realized in the RPM fluid but only in the RPM solid, perhaps along a phase boundary separating two solid phases of different symmetry. We note that from the Stillinger-Lovett [20] second moment condition, $\epsilon/\epsilon(k) = 1 - \Gamma_D^2[1 + \rho h_{qq}(k)]/k^2$, one has $\epsilon(k) = 0$ for k at which $h_{qq}(k) = \infty$, so $\epsilon(\omega) = 0$ along a λ-line.

(iii) On the basis of the observations we have made above, we might expect a good representation of the RPM $g_{qq}(r)$ in the vicinity of the λ-line to be given for large r by

$$g_{qq}(r) \approx -\beta E f(\Gamma r) \cos(\omega r + \theta)/r^{1+\eta} \qquad (36)$$

with $E = E(\rho^*, T^*)$ and $f(x) \to 1$ for $x \to 0$, $f(x) \to x^{\eta} e^{-x}$ for $x \to \infty$.

b. Density-density correlations

A second set of conclusions follows from contemplating the properties of the hypothetical simple fluid with the pair potential given by Eq. (24).

(i) The potential $\varphi^{\mathrm{EQUIV}}(r)$ is intrinsically negative and, in an overall sense, attractive, since $h_{qq}(r)$ is real and decaying as $r \to \infty$ so $h_{qq}^2(r)$ is non-negative with a decaying envelope. Thus one expects the density fluctuations to drive the system to liquid-gas criticality as Γ_D increases.

(ii) If the criticality occurs in the vicinity of a thermodynamic state in which Γ_D is reasonably close to the Γ_D^A of Fig. 1 (as it appears to be on the basis of simulation results and approximate theories) then the $\varphi^{\mathrm{EQUIV}}(r)$ will be an exponentially screened and relatively slowly varying function of ρ^* and T^*. Thus one would expect the criticality of the RPM to be in the same universality class as that of Lennard-Jonesium and the Ising model. (This is the argument of ref. [7] in somewhat different language from that used there.)

(iii) The fact that $\varphi^{\mathrm{EQUIV}}(r)$ is not a true mechanical potential energy but is rather a free energy that owes its existence to charge-charge fluctuations means that one cannot expect correction-to-scaling or crossover amplitudes and other correction-to-scaling effects to be system independent, with Ising-model magnitudes. Although the criticality can be expected to be Ising-like, all the detailed features of the criticality that would follow if $\varphi^{\mathrm{EQUIV}}(r)$ were a true potential cannot be assumed to hold *a priori*.

4. The spin-1 Ising model with Coulombic potential

In order to better understand the lattice-gas RPM Dickman and Stell have initiated Monte Carlo simulations on a simple cubic lattice for that system, using $J(\mathbf{r}) = -r^{-1}$. Our preliminary results are fully consistent with the presence of a λ-line of Néel points terminating in a tricritical point, below which (in T^*) there is coexistence of two lattice-gas phases (see Fig. 2). We have supplemented our simulation results with a mean-field analysis which is summarized here by Dickman in the Appendix. Høye and Stell [24] have also made a more general study of the spin-1 Ising model that included the solution of the MSA for a range-parameterized $J(\mathbf{r})$. As discussed by Høye and Stell, the MSA is not an appropriate approximation in which to study criticality and phase separation in the spin-1 model with Coulombic $J(\mathbf{r})$ (although it yields the correct Debye-Hückel limiting law, as Høye and Stell show). The Høye-Stell study goes beyond the MSA with a number of general observations that strongly support the conclusion that there will be tricriticality for a Coulombic $J(\mathbf{r})$.

5. Closing comments

The fact that one has two paths to the RPM thermodynamics, one involving only charge-charge fluctuation and one involving only density-density fluctuation, enables one to use the necessary consistency between those two paths in various special ways to devise improved treatments of the RPM thermodynamics, and also to learn more about both sorts of correlations. For example, using (7), (8), (10), (11), and (19) with (25) in which A is determined so that (12) and (14) give the same thermodynamics gives rise to an approximation that has the same thermodynamics as that of the MSA obtained through (14). It represents a variant of the Generalized MSA [25] that will yield the exact limiting behavior of the density-density correlation length $\rho \to 0$ that is discussed by Lee and Fisher [3]. Using the further information contained in (23) along with additional virial-equation consistency enables one to formulate approximations that incorporate an even higher degree of self-consistency.

We have not as yet explicitly discussed statement (2) of our Introduction. In a sense it follows directly from statement (1), because the fact that the anion/cation contact potential is so large compared to kT in the vicinity of the critical point can be regarded as a direct consequence of the special way in which the charge-charge and density-density correlations are coupled in the RPM. The coupling results in an $\varphi^{\mathrm{EQUIV}}(r)$ with an integrated strength great enough to produce criticality via Eq. (12) at the point at which $\varphi_{+-}(\sigma)$ is about $20kT$, a high value at which one finds a great deal of ionic clustering. As a consequence the RPM can profitably be regarded as a

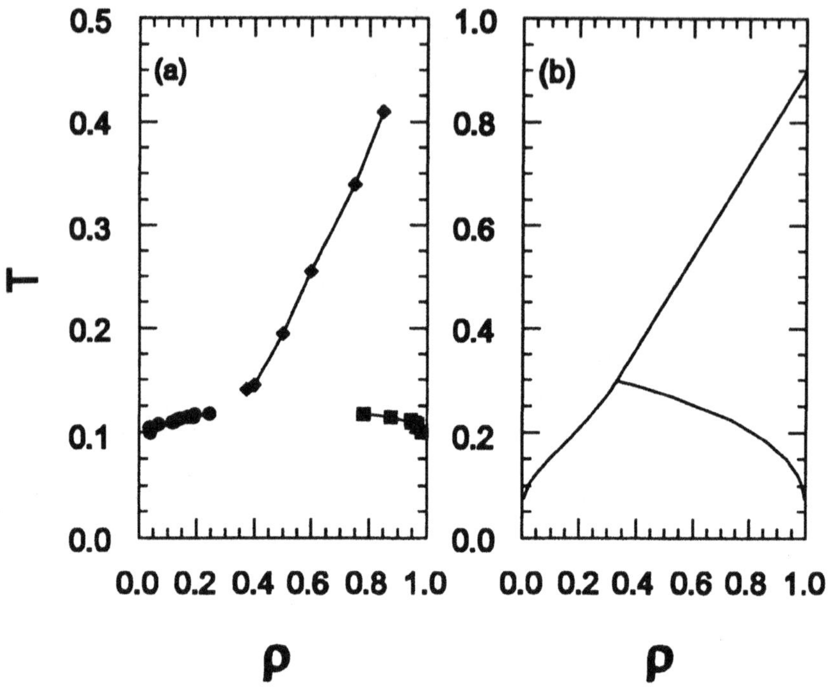

Figure 2. (a) Néel line and coexistence curve from cell histograms taken from Monte Carlo simulation results for the lattice RPM described in the text. (b) Néel line and coexistence curve from the simple mean-field theory described in the Appendix for the same lattice RPM. (Note the difference in the temperature scales.) The temperature scale shown here for the lattice RPM mean-field results differs by a factor of 0.1495 from the scale that describes the mean-field results for a nearest-neighbor potential, as discussed in the Appendix.

mixture of free ions and dipolar clusters, using an appropriate mass-action law, with the free ions treated by means of a theory (such as the MSA or the self-consistent theory discussed in the last paragraph) that does not itself incorporate the effects of the ionic association.

Acknowledgments

G.S. gratefully acknowledges the support of the Division of Chemical Sciences, Office of Basic Energy Sciences, Office of Energy Research, U.S. Department of Energy. The results reported by Professor Dickman in the Appendix were obtained while Prof. Dickman's contribution to our work was under the support of the National Science Foundation.

A. Mean-Field Theory for the Lattice Restricted Primitive Model; notes prepared by Ron Dickman

The energy of the lattice restricted primitive mode (LRPM) is

$$U = \frac{1}{2} \sum_{ij} \frac{s_i s_j}{\mathbf{r}_{ij}} , \tag{37}$$

where s_i is the charge (± 1) of particle i, residing at $\mathbf{r}_i$, $\mathbf{r}_{ij} = \mathbf{r}_i - \mathbf{r}_j$ and it is understood that particles occupy distinct lattice sites or, equivalently, occupy distinct cells that are centered at the vectors $\mathbf{r}_i$. We want to consider the LRPM with independent variables T and $\rho \equiv N/V$, where N is the number of particles and V the number of lattice sites or cells. We assume strict charge neutrality so that $N_+ = N_- = N/2$. It is helpful to think of this lattice gas system as a three-state (i.e., spin-1) antiferromagnetic Ising model with long-range interactions (i.e., $|J_{ij}| \sim 1/|\mathbf{r}_{ij}|$). In a simple mean-field theory (MFT) of this type of system (on a simple cubic or BCC lattice), we have only a single variable to characterize ordering, namely a sublattice charge disparity or staggered magnetization. In the context of MFT, this allows us to replace Eq. (37) by the corresponding expression for a *nearest neighbor* (NN) lattice gas, multiplied by a suitable factor.

Consider first the disordered system. Since the neighbors of any given particle are equally likely to bear the same or opposite charges, the MF estimate for the energy is zero, just as for the NN system. Next we suppose there is a sublattice ordering. Let ρ_+^{A} be the fraction of sites in sublattice A occupied by positive particles, etc., and let

$$\rho_+^{\mathrm{A}} = \rho(1+x)/2 \qquad \rho_+^{\mathrm{B}} = \rho(1-x)/2 \tag{38}$$

$$\rho_-^{\mathrm{A}} = \rho(1-x)/2 \qquad \rho_-^{\mathrm{B}} = \rho(1+x)/2 \tag{39}$$

so that

$$x = (\rho_+^{\mathrm{A}} - \rho_-^{\mathrm{A}})/\rho \tag{40}$$

and

$$\rho_+^{\mathrm{A}} + \rho_-^{\mathrm{A}} = \rho_+^{\mathrm{B}} + \rho_-^{\mathrm{B}} = \rho \tag{41}$$

Using these, the MF estimate for the energy of a single NN pair of sites is

$$\begin{aligned} u_{\mathrm{NN}} &= \rho_+^{\mathrm{A}}\rho_+^{\mathrm{B}} + \rho_-^{\mathrm{A}}\rho_-^{\mathrm{B}} - \rho_+^{\mathrm{A}}\rho_-^{\mathrm{B}} - \rho_-^{\mathrm{A}}\rho_+^{\mathrm{B}} \\ &= -\rho^2 x^2 \end{aligned} \tag{42}$$

Similarly, the MF estimate for the energy of a pair of second-neighbor sites (which must belong to the same sublattice) is $+\rho^2 x^2$, and so on. Thus the MF estimate for energy of the LRPM is

$$U_{\rm MF} = \frac{\rho^2 x^2}{2} \sum_{ij} \frac{(-1)^S}{r_{ij}} \tag{43}$$

Where $S = +1(-1)$ if sites i and j belong to the same (different) sublattices. The sum (including the factor of 1/2) is simply the electrostatic energy of an ionic crystal. For the simple cubic lattice, for instance

$$\frac{1}{2} \sum_{ij} \frac{(-1)^S}{r_{ij}} = -3V\frac{\alpha}{6} = -\frac{V\alpha}{2}\ , \tag{44}$$

as $V \to \infty$, where $\alpha = 1.74757$ is the Madelung constant for this lattice. Thus the MF energy per site is $u_{\rm MF} = -\alpha\rho^2 x^2/2$, while the corresponding expression for a NN system is $u_{\rm MF:NN} = -3\rho^2 x^2$. So for the purposes of MFT, we can replace the $1/r$ potential with a NN interaction.

¿From the above comparison, it might seem that we can treat the LRPM, in MFT, as if it were a NN lattice gas, but with an energy (i.e, temperature) scale $\alpha/6 \simeq 0.2913$ smaller. But a more meaningful way of relating temperature scales is to compare the energy change associated with an elementary excitation, in this case, the interchange of a NN pair in an otherwise perfectly ordered latice. For the simple cubic lattice gas such an exchange raises the energy by $\Delta U = 20$. (Ten bonds go from "good" to "bad"). For the ionic crystal, it's not hard to show that on the simple cubic lattice the corresponding energy change is $4(\alpha - 1) \simeq 2.9903$ so that $\Delta U_{\rm LRPM} \simeq 0.1495 \Delta U_{\rm NN}$. As is well known, the MF value for the critical temperature of the fully occupied NN lattice gas is 6, so the corresponding estimate for the LRPM is about 0.9. If we use the best numerical estimate of $T_{c,\rm NN} = 4.5115$, the corresponding LRPM value is 0.674. This is not too far from the value 0.52 that we obtain by extrapolating our MC data.

We can obtain the phase diagram for the LRPM by developing the MFT — for the NN system — in the usual way. We already have the MF energy, and now derive the entropy estimate. The number of allowed configurations for a sublattice ($(V/2)$ sites) is

$$\frac{(V/2)!}{N_+! N_-! [(V/2) - N_+ - N_-]!} \tag{45}$$

and using $N_+ = (V/2)\rho(1+x)/2$, etc., and Stirling's formula, we obtain the entropy per site:

$$s = -\{\rho \ln \rho + (1-\rho)\ln(1-\rho) - \rho \ln 2 + \frac{\rho}{2}[(1+x)\ln(1+x) + (1-x)\ln(1-x)]\} \tag{46}$$

Let $f = u - Ts$ be the (Helmholtz) free energy per site. To investigate the possibility of a free energy minimum with nonzero x, we note

$$\frac{\partial f}{\partial x}\Big|_{\rho} = \rho\Big[-6px + \frac{T}{2}\ln\frac{1+x}{1-x}\Big] \tag{47}$$

A free energy minimum therefore implies that

$$\frac{12\rho x}{T} = \ln\frac{1+x}{1-x} \tag{48}$$

Expanding the r.h.s. as $2x + 2x^3/3 + \ldots$, we see that solutions with nonzero x exist for $T < T_c = 6$ and $\rho > \rho_c = T/T_c$, and that for $\rho \geq \rho_c$,

$$x \simeq \pm\sqrt{3\Big(\frac{\rho}{\rho_c} - 1\Big)} \tag{49}$$

Thus $T = 6\rho$ is a line of second-order phase transitions (a Néel or lambda line), separating the high temperature phase ($x = 0$) from one with sublattice ordering. Next we need to see whether a uniform-density state is stable, or whether the system separates into high- and low-density phases.

To study phase coexistence we introduce the free energy per particle, $f_p = f/\rho$ and note that the pressure p and chemical potential μ are given by

$$p = \rho^2\left(\frac{\partial f_p}{\partial \rho}\right)_T = -T\ln(1-\rho) - 3\rho^2 x^2 \tag{50}$$

and

$$\mu = f_p + \frac{p}{\rho} \tag{51}$$

(In evaluating the derivative in Eq. (50) we use Eq. (48) to eliminate the term $\propto (\partial x/\partial \rho)$. As expected, the pressure in the high-temperture phase is just that of the simple lattice gas. But for $T < T_c$, $(\partial p/\partial \rho)_T$ suffers a discontinuity at ρ_c. Using Eq. (49) we have that for $\rho \geq \rho_c$,

$$\left(\frac{\partial p}{\partial \rho}\right)_T = \frac{T}{1-\rho} - 9\rho\left[3\left(\frac{\rho}{\rho_c}\right) - 2\right] \tag{52}$$

so that

$$\left(\frac{\partial p}{\partial \rho}\right)_{T;\rho=\rho_c} = T\left[\frac{1}{1-\rho_c} - \frac{3}{2}\right] \tag{53}$$

Thus the inverse compressibility vanishes when $\rho_c = 1/3$, which implies $T = 2$. The point ($\rho = 1/3$, $T = 2$) is a tricritical point, at which the Néel line meets the coexistence curve. (Note that for $T < 2$, the Néel line becomes a spinodal.)

To construct the coexistence curve we need to find densities ρ_1 and ρ_2 for which the pressure and chemical potential are equal. This is done numerically, by a routine which first chooses a density ρ_2 between ρ_s (the high-density spinodal) and $\rho_m > \rho_s$ (the density at which $p = p(\rho_c)$), evaluates the pressure, locates the corresponding point on the low-density portion of the isotherm via the relation $\rho_1 = 1 - e^{-p/T}$, and then compares the chemical potentials. If $\mu(\rho_2) > \mu(\rho_1)$, ρ_2 is too large, and vice-versa, and by adjusting the estimate for ρ_2 accordingly, the routine quickly locates the coexisting densities. At low temperatures ($T \leq 0.5$) one may use the fact that $x = 1 - 2e^{-12p/T} \approx 1$ to show that $\rho_2 \simeq 1 - e^{-3/T}$ and $\rho_1 \simeq 2\epsilon^{-3\rho_2/T}$. The phase diagram is shown in Fig. 2, with the T scale appropriate to the NN case renormalized by the 0.1495 factor discussed above Eq. (45) to facilitate comparison with our Monte Carlo results, which are also shown.

References

1. Stell, G., Wu, K.C. and Larsen, B. (1976) *Phys. Rev. Lett.* **37**, p. 1369.
2. Ebeling, W. and Grigo, M. (1980) *Ann. Phys. (Leipzig)* **37**, p. 21.
3. Fisher, M.E. and Levin, Y. (1993) *Phys. Rev. Lett.* **71**, p. 3826; Lee, B.D. and Fisher, M.E. (1996) *Phys. Rev. Lett.* **76**, p. 2906.
4. Yeh, S., Zhou, Y. and Stell, G. (1996) *J. Phys. Chem.* **100**, p. 1415.
5. Valleau, J.P. (1991) *J. Chem. Phys.* **95**, p. 584; (1993) *J. Chem Phys.* **99**, p. 4718; Caillol, J.M. (1994) *J. Chem. Phys.* **100**, p. 2161; Orkoulas, G. and Panagiotopoulos, A.Z. (1996) *J. Chem. Phys.* **101**, p. 145, *J. Chem. Phys.* **104**, p. 720; *J. Chem. Phys.* (to appear); Caillol, J.M., Levesque, D. and Weis, J.J. (1996) *Phys. Rev. Lett.* **77**, p. 4039; (1997) *J. Chem. Phys.* **107**, p. 1565.
6. Hafskjøld, B. and Stell, G. (1982) *The Equilibrium Statistical Mechanics of Simple Ionic Liquids* in *Studies in Statistical Mechanics* **Vol. III**. *The Liquid State of Matter: Fluids, Simple and Complex* Montroll, E.W. and Lebowitz, J.L., ed. North Holland, Amsterdam.
7. Stell, G. (1992) *Phys. Rev. A* **45**, p. 7628.
8. Singh, R.R. and Pitzer, K.S. (1990) *J. Chem. Phys.* **92**, p. 6775.
9. Zhang, K.C. et al. (1992) *J. Chem. Phys.* **97**, p. 8692.
10. Narayanan, T. and Pitzer, K.S. (1994) *Phys. Rev. Lett.* **73**, p. 3002; (1995) *J. Chem. Phys.* **102**, p. 8118.
11. Høye, J.S. and Stell, G. (1997) *J. of Molecular Liquids* **73,74**, p. 453.
12. Wiegand, S. et al. (1997) *J. Chem. Phys.* **106**, p. 2777; (1998) *J. Chem. Phys.* **109**, p. 4533.
13. Leote de Carvalho, R.J.F. and Evans, R. (1995) *J. Phys. Condens. Matter* **7**, p. L575.
14. Fisher, M.E. and Lee, B.P. (1996) *Phys. Rev. Lett.* **77**, p. 3561.
15. Valleau, J.P. and Torrie, G. (1998) *J. Chem. Phys.* **108**, p. 5169.
16. Stell, G. (1976) *Correlation Functions and Their Generating Functionals* in *Phase Transitions and Critical Phenomena* **Vol. 5B**, Domb, C. and Green, S., ed. Academic Press, London.
17. Stell, G. (1995) *J. Stat. Phys.* **78**, p. 197.
18. Kjellander, R.K. and Mitchell, D.J. (1994) *J. Chem. Phys.* **101**, p. 603.
19. Kirkwood, J.G. (1934) *J. Chem. Phys.* **2**, p. 351; (1936) *Chem. Rev.* **19**, p. 275.

20. Stillinger, F.H., Kirkwood, J.G. and Wojtowicz, P. (1960) *J. Chem. Phys.* **32**, p. 1837; Stillinger, F.H. and Lovett, R. (1968) *J. Chem. Phys.* **48**, p. 3858; Lovett, R. and Stillinger, F.H. (1968) *J. Chem. Phys.* **48**, p. 8869.
21. Stell, G. and Lebowitz, J.L. (1968) *J. Chem. Phys.* **49**, p. 3706.
22. Ennis, J., Kjellander, R. and Mitchell, D.J. (1995) *J. Chem. Phys.* **102**, p. 975.
23. Outhwaite, C.W. (1975) *Statistical Mechanics*, **Vol. 2**, p. 188. *Specialist Periodical Report*. The Chemical Society, London.
24. Høye, J.S. and Stell, G (1997) *J. Stat. Phys.* **89**, p. 177.
25. Stell, G. and Sun, S.F. (1975) *J. Chem. Phys.* **63**, p. 333.

FLUIDS WITH INTERNAL DEGREES OF FREEDOM

F. LADO
Department of Physics
North Carolina State University
Raleigh, NC, 27695-8202, USA

1. Introduction

The essential core of the modern integral equation approach to liquid state theory was arrived at nearly simultaneously in 1960 by a remarkable number of authors working independently [1]. They found that the density expansion of the pair distribution function $g(r)$ of a simple fluid with interatomic potential $\phi(r)$ could be grouped into infinite subsets of diagrams such that

$$g(r)e^{\beta\phi(r)} = 1 + S(r) + P(r) + B(r), \tag{1}$$

where, in the pictorial electrical language of M. S. Green [2], the diagrams of the *series* set $S(r)$ resemble series circuits and those of the *parallel* set $P(r)$ resemble parallel circuits; the remaining *bridge* set $B(r)$ begins with a diagram that looks like a Wheatstone bridge. Further, the diagrams of $P(r)$ could be summed in direct space to give $P(r) = g(r)\exp(\beta\phi(r)) - 1 - \ln[g(r)\exp(\beta\phi(r))]$, while those of $S(r)$ could be summed in Fourier space to yield $\tilde{S}(k) = \tilde{c}^2(k)/[1-\rho\tilde{c}(k)]$, which is the Ornstein-Zernike (OZ) equation in Fourier transform representation. Here, $c(r) = h(r) - S(r)$ is the sum of non-nodal graphs, or *direct correlation function*, while $h(r) = g(r) - 1$ is the *total correlation function*. In brief, the upshot of these analyses was that $g(r)$, the key to all fluid properties, could be obtained by the iteration of a simple, four-step recipe for the unknown $S(r)$, now more often denoted $\gamma(r)$ and called the *indirect correlation function*:

1. *Closure:* $\gamma(r) \to c(r)$.

$$c(r) = \exp[-\beta\phi(r) + \gamma(r) + B(r)] - 1 - \gamma(r) \tag{2}$$

2. *Transform:* $c(r) \to \tilde{c}(k)$.

$$\tilde{c}(k) = \frac{4\pi}{k}\int_0^\infty dr\, rc(r)\sin(kr) \tag{3}$$

C. Caccamo et al. (eds.), New Approaches to Problems in Liquid State Theory, 91–105.

3. *OZ equation:* $\tilde{c}(k) \to \tilde{\gamma}(k)$.

$$\tilde{\gamma}(k) = \rho\tilde{c}^2(k)/[1 - \rho\tilde{c}(k)] \tag{4}$$

4. *Inverse transform:* $\tilde{\gamma}(k) \to \gamma(r)$.

$$\gamma(r) = \frac{1}{2\pi^2 r}\int_0^\infty dk\, k\tilde{\gamma}(k)\sin(kr) \tag{5}$$

Upon convergence, meaning the output $\gamma(r)$ from (5) matches the input $\gamma(r)$ in (2), the pair distribution function is found as $g(r) = \exp[-\beta\phi(r) + \gamma(r) + B(r)]$. In these equations, ρ is the number density and $\beta = 1/k_BT$.

This recipe of course requires knowing $B(r)$. But the bridge function diagrams could not be summed and the original authors all hopefully proposed $B(r) \approx 0$ as a first approximation. The result, known as the hypernetted chain (HNC) equation, in fact was predated by the Percus-Yevick equation, obtained in a completely different fashion [3] but equivalent to $B(r) \approx -P(r)$. These familiar closures have since been supplemented with numerous other approximations for $B(r)$ [1] but remain still popular today.

Equations (2)-(5) apply to "simple" fluids in three dimensions with spherically symmetric potentials $\phi(r)$. They can be generalized for "molecular" fluids [4], with potentials $\phi(\mathbf{r}, \omega_1, \omega_2)$ that depend on orientational degrees of freedom ω in addition to the interparticle separation $\mathbf{r}$. The generalization is based on orthogonal expansions in the angular variables. In the present work, I am concerned with a further generalization to fluids whose molecules have *internal* degrees of freedom, such as size σ with distribution $f(\sigma)$ or polarization $\mathbf{p}$ with distribution $f_0(p)$ for an isolated molecule. In the latter case, the dipole-dipole interactions will rearrange the distribution $f_0(p)$ in dense fluids into a new shape $f(p)$ that is an additional unknown. The new generalization will similarly be based on expansions in the additional variables, σ, $\mathbf{p}$, *etc.*, but now the basis functions will be explicitly constructed to be orthogonal with weight function $f(\sigma)$, $f(p)$, *etc.*

2. Polydisperse colloidal fluids

Colloidal particles are stable conglomerates of atoms whose length scale is on the boundary between the microscopic and macroscopic worlds. The description of the structure of these liquid-like suspensions using the tools of simple atomic liquids is complicated primarily by the fact that the particles of any given suspension display a continuous distribution of sizes; *i.e.*, they are inherently *polydisperse.*

Let $f(\sigma)$ be the distribution of particle diameters σ in a homogeneous assembly of N polydisperse colloidal particles contained in volume V at temperature T. The density of particles at point $\mathbf{r}$ with diameter σ is then

$$\rho^{(1)}(\mathbf{r},\sigma) = < \sum_{i=1}^{N} \delta(\mathbf{r}-\mathbf{r}_i)\delta(\sigma-\sigma_i) > = \rho f(\sigma), \tag{6}$$

where $\rho = N/V$, while the two-body density is

$$\begin{aligned} \rho^{(2)}(\mathbf{r},\sigma,\mathbf{r}',\sigma') &= < \sum_{i\neq j} \delta(\mathbf{r}-\mathbf{r}_i)\delta(\sigma-\sigma_i)\delta(\mathbf{r}'-\mathbf{r}_j)\delta(\sigma'-\sigma_j) > \\ &= \rho^2 f(\sigma) f(\sigma') g(|\mathbf{r}-\mathbf{r}'|,\sigma,\sigma'), \end{aligned} \tag{7}$$

which defines the pair distribution function $g(r,\sigma,\sigma')$. In these expressions, $\mathbf{r}_i$ is the location of particle i and σ_i its diameter; the brackets denote a canonical ensemble average.

The generalized OZ equation between $\gamma(r,\sigma_1,\sigma_2)$ and $c(r,\sigma_1,\sigma_2)$ can again be deconvoluted using Fourier transforms and now yields

$$\tilde{\gamma}(k,\sigma_1,\sigma_2) = \rho \int_0^\infty d\sigma_3 f(\sigma_3)[\tilde{c}(k,\sigma_1,\sigma_3) + \tilde{\gamma}(k,\sigma_1,\sigma_3)]\tilde{c}(k,\sigma_3,\sigma_2), \tag{8}$$

with one remaining integral. The final integration in Eq. (8) can also be eliminated and the evaluation of the OZ equation reduced to algebra by now expanding all σ-dependent functions in orthogonal polynomials $p_j(\sigma)$, $j = 0,1,2,\ldots$, defined such that

$$\int_0^\infty d\sigma\, f(\sigma) p_i(\sigma) p_j(\sigma) = \delta_{ij}, \tag{9}$$

where δ_{ij} is the Kronecker delta. Then with

$$\tilde{\gamma}(k,\sigma_1,\sigma_2) = \sum_{i,j=0}^{\infty} \tilde{\gamma}_{ij}(k) p_i(\sigma_1) p_j(\sigma_2) \tag{10}$$

and a similar expansion of $\tilde{c}(k,\sigma_1,\sigma_2)$, Eq. (8) becomes

$$\tilde{\gamma}_{ij}(k) = \rho \sum_l [\tilde{c}_{il}(k) + \tilde{\gamma}_{il}(k)]\tilde{c}_{lj}(k), \tag{11}$$

which are now matrix equations. Orthonormality allows the easy inversion of Eq. (10); similar expansions and inversions hold for functions in r space. The inversion of $c(r,\sigma_1,\sigma_2)$, for example, is shown below in Eq. (14). The four-step recipe now becomes

1. *Closure:* $\gamma_{ij}(r) \to c_{ij}(r)$. Form

$$\begin{aligned} \gamma(r,\sigma_1,\sigma_2) &= \sum_{i,j} \gamma_{ij}(r) p_i(\sigma_1) p_j(\sigma_2), && (12) \\ c(r,\sigma_1,\sigma_2) &= \exp[-\beta\phi(r,\sigma_1,\sigma_2) + \gamma(r,\sigma_1,\sigma_2) + B(r,\sigma_1,\sigma_2)] \\ &\quad -1 - \gamma(r,\sigma_1,\sigma_2), && (13) \end{aligned}$$

   and numerically evaluate the coefficients

$$c_{ij}(r) = \int d\sigma_1 d\sigma_2 f(\sigma_1) f(\sigma_2) c(r,\sigma_1,\sigma_2) p_i(\sigma_1) p_j(\sigma_2) \qquad (14)$$

   using n-point Gaussian quadrature based on the zeroes of $p_n(\sigma)$.
2. *Transforms:* $c_{ij}(r) \to \tilde{c}_{ij}(k)$, as in Eq. (3).
3. *OZ equation:* $\tilde{c}_{ij}(k) \to \tilde{\gamma}_{ij}(k)$.

$$\tilde{\Gamma}(k) = \rho\tilde{C}(k)\tilde{C}(k)[I - \rho\tilde{C}(k)]^{-1} \qquad (15)$$

   Here $\tilde{\Gamma}(k)$ and $\tilde{C}(k)$ are symmetric matrices with elements $\tilde{\gamma}_{ij}(k)$ and $\tilde{c}_{ij}(k)$, respectively, and I is the unit matrix.
4. *Inverse transforms:* $\tilde{\gamma}_{ij}(k) \to \gamma_{ij}(r)$, as in Eq. (5).

With a converged solution in hand, the thermodynamic properties of the colloidal suspension, such as internal energy U, pressure p, and isothermal compressibility K_T, are readily obtained by quadratures. Further, the Helmholtz free energy A can also be obtained [5] from the coefficients of the pair functions in, for example, HNC approximation.

Such calculations have been carried out [5] for charge-stabilized, polydisperse colloidal suspensions whose particle diameters σ are distributed about a mean diameter $\bar{\sigma}$ according to the Schulz distribution,

$$f(\sigma) = \left(\frac{\alpha+1}{\bar{\sigma}}\right)^{\alpha+1} \frac{\sigma^\alpha e^{-(\alpha+1)\sigma/\bar{\sigma}}}{\Gamma(\alpha+1)}, \qquad (16)$$

where $\Gamma(z)$ is the gamma function. The relative standard deviation of this distribution is $s_\sigma \equiv (<\sigma^2> - <\sigma>^2)^{1/2}/<\sigma> = (\alpha+1)^{-1/2}$, which is used to characterize the suspensions and define the parameter α. The orthonormal polynomials for the Schulz distribution are

$$p_j(\sigma) = \left[\frac{j!\Gamma(\alpha+1)}{\Gamma(j+\alpha+1)}\right]^{1/2} L_j^\alpha\left((\alpha+1)\frac{\sigma}{\bar{\sigma}}\right), \qquad (17)$$

where the $L_j^\alpha(t)$ are the associated Laguerre polynomials.

The charged mesospheres are assumed to have a constant surface charge density, so that their charge polydispersity is mapped onto the size polydispersity according to $Z(\sigma) = Z_{\bar{\sigma}}(\sigma/\bar{\sigma})^2$, where $Z_{\bar{\sigma}}$ is the charge on a particle

TABLE 1. Thermodynamics of a polydisperse colloidal suspension with screened Coulomb potential and Schulz distribution of relative standard deviation s_σ; MC data are from D'Aguanno and Klein [6].

	s_σ	$\beta U/N$	$\beta p/\rho$	$\rho k_B T K_T$	$\beta A/N$
MC		28.00	48.50		
RY	0	28.30	48.78	0.00851	
HNC		28.71	49.20	0.00999	31.91
MC		27.66	48.14		
RY	0.1	27.94	48.45	0.00854	
HNC		28.35	48.87	0.01005	31.55
MC		26.64	47.16		
RY	0.2	26.89	47.48	0.00863	
HNC		27.31	47.90	0.01025	30.49
MC		24.97	45.50		
RY	0.3	25.24	45.86	0.00879	
HNC		25.67	46.30	0.01059	28.83

of mean diameter $\bar{\sigma}$, and to interact, beyond hard sphere contact, through a screened Coulomb potential,

$$\beta\phi(r,\sigma_1,\sigma_2) = \begin{cases} \infty, & \text{if } r < \sigma_{12} \\ A(\sigma_1)A(\sigma_2)\exp(-\kappa r)/r, & \text{if } r > \sigma_{12} \end{cases}, \tag{18}$$

where $\sigma_{12} = (\sigma_1 + \sigma_2)/2$ and $A(\sigma) = Z(\sigma)L_B^{1/2}\exp(\kappa\sigma/2)/(1+\kappa\sigma/2)$. Here, $L_B = e^2/4\pi\epsilon_0\epsilon k_B T$ is the Bjerrum length and $\kappa = (4\pi L_B \rho \bar{Z})^{1/2}$ is the inverse Debye-Hückel screening length, with $\bar{Z} = Z_{\bar{\sigma}}(1+s_\sigma^2)$.

This is the same model studied by D'Aguanno and Klein [6], who judiciously converted the polydisperse system to a finite mixture of discrete diameters. To illustrate the polynomial expansion method, I have recalculated their results for the state with $\bar{\sigma} = 250$ Å, $Z_{\bar{\sigma}} = 200$, and $L_B = 7.01$ Å, at the reduced density $\rho\bar{\sigma}^3 = 0.005$, using the HNC and Rogers-Young (RY) [7] closures,

The computed thermodynamic values are shown in Table 1 for three Schulz distributions of increasing width, $s_\sigma = 0.1$, 0.2, and 0.3, as well as the monodisperse case, $s_\sigma = 0$, for reference. The present integral equation results are in good agreement with those found earlier [6] using the "mixture" method.

Pair functions of interest can be expressed directly in terms of the computed coefficients $g_{ij}(r)$. In particular, the number-number pair distribution function $g_{NN}(r)$ [6] is precisely $g_{00}(r)$. Further, local fluctuations in density

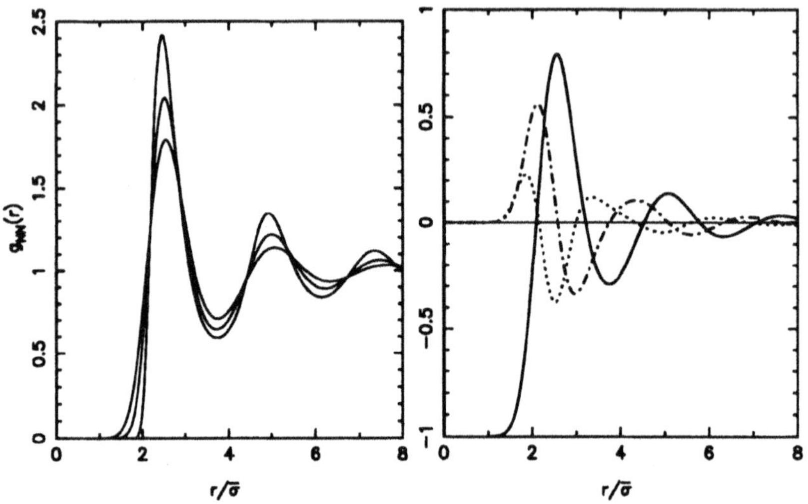

Figure 1. Pair distribution function coefficients for a polydisperse colloidal *monolayer* with screened Coulomb potential at density $\rho\bar{\sigma}^2 = 0.15$, from the RY equation. (a) *Left figure.* $g_{NN}(r) = g_{00}(r)$ for size dispersion of the spheres, in descending order of the first peak, of $s_\sigma = 0$, 0.2, and 0.3. (b) *Right figure.* $g_{00}(r) - 1$ (solid line), $g_{10}(r)$ (dash-dot line), and $g_{11}(r)$ (dotted line) for size dispersion $s_\sigma = 0.3$.

and size may be expressed in normalized form as

$$\delta\rho_0(\mathbf{r}) = \frac{1}{\rho}\left(\sum_{j=1}^{N}\delta(\mathbf{r}-\mathbf{r}_j)-\rho\right), \tag{19}$$

$$\delta\rho_1(\mathbf{r}) = \frac{1}{\bar{\sigma}\rho s_\sigma}\left(\sum_{j=1}^{N}\sigma_j\delta(\mathbf{r}-\mathbf{r}_j)-\bar{\sigma}\rho\right). \tag{20}$$

Their spatial correlations are then given by

$$< \delta\rho_0(\mathbf{r})\delta\rho_0(\mathbf{0}) > = \frac{\delta(\mathbf{r})}{\rho} + [g_{00}(r) - 1], \tag{21}$$

$$< \delta\rho_1(\mathbf{r})\delta\rho_0(\mathbf{0}) > = -g_{10}(r), \tag{22}$$

$$< \delta\rho_1(\mathbf{r})\delta\rho_1(\mathbf{0}) > = \frac{\delta(\mathbf{r})}{\rho} + g_{11}(r). \tag{23}$$

Rather than show these correlation functions for the three-dimensional calculation described above, I display instead the results for a similar quasi-two-dimensional polydisperse colloidal *monolayer* [8]. [The only change needed in the calculation algorithm for the 2D version is the replacement of the Fourier transforms, with kernel $\sin(kr)$, by Hankel transforms, with kernel $J_0(kr)$, in Steps 2 and 4, where $J_0(x)$ is the Bessel function of order zero.] The effect of growing polydispersity on the number-number distribution function $g_{NN}(r)$ is seen in Fig. 1(a) for $\rho\bar{\sigma}^2 = 0.15$. The general pattern of decreasing structure with increasing dispersion reproduces the behavior of polydisperse colloids in three dimensions [6]. The functions $g_{00}(r) - 1$, $g_{10}(r)$, and $g_{11}(r)$ for correlations of density-density, size-density, and size-size fluctuations, respectively, Eqs. (21)-(23), are shown in Fig. 1(b) for density $\rho\bar{\sigma}^2 = 0.15$ and dispersion $s_\sigma = 0.3$. We see, for example, that at the first peak of $g_{00}(r)$, which occurs at $r = 2.54\bar{\sigma}$, the size-size fluctuations are *negatively* correlated, while the cross correlation of size and density fluctuations passes through zero. This means that large spheres (large charges) are generally surrounded by small spheres (small charges) and vice versa.

These results are obtained for the same size-distribution, Eq. (16), and potential, Eq. (18), as the three-dimensional model. The potential parameters used are the experimental values of Crocker and Grier [9]: $\bar{\sigma} = 650$ nm, $L_B = 0.715$ nm, $\kappa\bar{\sigma} = 4.0$, and $Z_{\bar{\sigma}} = 1990$. Under these conditions, electrostatic repulsion keeps the hard spheres from touching, so that only charge polydispersity is relevant in practice.

3. Polarizable nonpolar fluids

Molecules lacking a permanent dipole moment will nonetheless display instantaneous moments $\mathbf{p}$ that fluctuate randomly within some overall constraints and so will interact through dipole-dipole forces with other such molecules. I will follow Høye and Stell [10] and Pratt [11] in assuming an intrinsic thermal distribution of Gaussian form,

$$f_0(p) = \frac{1}{(2\pi\alpha_0/\beta)^{3/2}} \exp\left(-\frac{\beta p^2}{2\alpha_0}\right), \tag{24}$$

for the spontaneous dipole moment $\mathbf{p}$ of an isolated molecule, where α_0 is its polarizability. These authors have independently solved the mean spherical approximation (MSA) for this model (with inclusion of a permanent dipole in the case of the Høye-Stell work) and find that the fluctuating dipole distribution $f(p)$ in the dense liquid remains Gaussian, an inherent restriction of the MSA, with a larger polarizability α. The present approach [12] removes any such prior restriction.

We consider then a system of N molecules, each with intrinsic polarizability α_0 and fluctuating polarity but no permanent dipole moment, interacting pairwise through through a Lennard-Jones (LJ) potential,

$$u_0(r) = 4\epsilon[(\sigma/r)^{12} - (\sigma/r)^6], \tag{25}$$

and the usual dipole-dipole potential,

$$u_{dd}(\mathbf{r}, \mathbf{p}_1, \mathbf{p}_2) = -\frac{1}{r^3}[3(\hat{\mathbf{r}} \cdot \mathbf{p}_1)(\hat{\mathbf{r}} \cdot \mathbf{p}_2) - \mathbf{p}_1 \cdot \mathbf{p}_2]. \tag{26}$$

The one-body and two-body densities are defined as in Eqs. (6) and (7) with $\mathbf{p}$ replacing σ, so we get $\rho^{(1)}(\mathbf{r},\mathbf{p}) = \rho f(p)$ and $\rho^{(2)}(\mathbf{r},\mathbf{p},\mathbf{r}',\mathbf{p}') = \rho^2 f(p) f(p') g(\mathbf{r}-\mathbf{r}',\mathbf{p},\mathbf{p}')$, where now both f and g are to be determined. The two distribution functions are of course coupled by the intermolecular potential u_{dd}. A direct expression of this coupling is obtained by differentiating $f(p)$ to yield

$$\frac{d}{dp}\ln\left[\frac{f(p)}{f_0(p)}\right] = -\rho\int d\mathbf{r}d\mathbf{p}'\, f(p')g(\mathbf{r},\mathbf{p},\mathbf{p}')\beta u_{dd}(\mathbf{r},\hat{\mathbf{p}},\mathbf{p}'), \qquad (27)$$

the first member of a Kirkwood-Born-Green-Yvon hierarchy. The plan now will be to calculate $g(\mathbf{r},\mathbf{p},\mathbf{p}')$ for a given $f(p)$ [starting with $f_0(p)$] using a procedure like that for polydisperse systems and then return to Eq. (27) to update $f(p)$. This will begin a new round for an updated g, and so the process continues until both f and g are self-consistently determined.

The conventional first step in solving molecular equations is to explicitly break out the angular dependence of all functions in the form of expansions in spherical harmonics [4]; *e.g.*,

$$\begin{aligned}\gamma(\mathbf{r},\mathbf{p}_1,\mathbf{p}_2) &= \gamma(r,p_1,p_2,\omega_1,\omega_2)\\ &= 4\pi\sum_{l_1,l_2,m}\gamma_{l_1l_2m}(r,p_1,p_2)Y_{l_1m}(\omega_1)Y_{l_2\bar{m}}(\omega_2),\end{aligned} \qquad (28)$$

where $\bar{m} = -m$. In this and similar expressions, the vector $\mathbf{r}$ has been implicitly chosen as the z direction in the specification of the Euler angles $\omega = (\theta,\phi)$. What makes this expansion particularly useful is of course the orthogonality of the spherical harmonics, so that the coefficients of the expansion are immediately obtainable by quadratures.

We now similarly break out the fluctuating-dipole dependence in the form of expansions in polynomials of p,

$$\gamma_{l_1l_2m}(r,p_1,p_2) = \sum_{n_1,n_2}\gamma_{l_1l_2m}^{n_1n_2}(r)Q_{n_1l_1}(p_1)Q_{n_2l_2}(p_2), \qquad (29)$$

which are constructed to be orthogonal with weight function $f(p)$,

$$4\pi\int_0^\infty dp\,p^2 f(p)Q_{nl}(p)Q_{n'l}(p) = \delta_{nn'}, \qquad (30)$$

so that coefficients of the expansion are again obtainable by quadratures.

Since the distribution function $f(p)$ will evolve during the course of an iterative solution, the associated polynomials $Q_{nl}(p)$ will also change. In principle, for a general $f(p)$ one may have to determine the $Q_{nl}(p)$ by elementary methods, such as Gram-Schmidt orthogonalization [13], starting

TABLE 2. Computed thermodynamic and electrostatic properties of a polarizable nonpolar Lennard-Jones liquid for states characterized by ρ, T, and α_0 values; molecular dynamics values of the dielectric constant ε are from Pollock and Alder [16].

$\rho\sigma^3$	k_BT/ϵ	α_0/σ^3	$\beta U/N$	$\beta p/\rho$	ξ_2	ξ_4	α/σ^3	ε RHNC	ε MD
0.84	0.70	0.02	-8.709	-0.213	-0.007	0.000	0.0201	1.227	1.23
0.84	0.72	0.05	-8.470	-0.023	-0.041	0.000	0.0514	1.646	1.64
0.20	2.01	0.10	-0.689	0.808	-0.037	0.001	0.1025	1.280	1.27
0.70	1.97	0.10	-2.332	2.028	-0.144	0.001	0.1096	2.284	2.24
0.84	0.67	0.10	-9.304	-0.686	-0.167	0.001	0.1112	2.688	2.64
0.84	0.80	0.14	-7.825	0.486	-0.362	0.003	0.1738	4.150	3.94

from $Q_{00}(p) = 1$. At least initially, however, the distribution $f(p)$ in the present calculation will be Gaussian and the needed polynomials are immediately found in the form of the eigenfunctions of the quantum-mechanical harmonic oscillator in three dimensions using spherical coordinates. For a Gaussian distribution of variance α/β, they are [14]

$$Q_{nl}(p) = \left[\frac{\Gamma\left(\frac{1}{2}(n-l)+1\right)\Gamma\left(\frac{3}{2}\right)}{\Gamma\left(\frac{1}{2}(n+l)+\frac{3}{2}\right)}\right]^{1/2} \left(\frac{\beta p^2}{2\alpha}\right)^{l/2} L^{l+\frac{1}{2}}_{\frac{1}{2}(n-l)}\left(\frac{\beta p^2}{2\alpha}\right). \quad (31)$$

For the cases studied in Ref. [12], it turns out that in fact the computed deviations of $f(p)$ from a Gaussian are very small, permitting retention of these polynomials (with changing α) for the entire calculation.

With the internal degrees of freedom included in the expansion bases, the solution of the (OZ+closure) equations for a polarizable fluid follows along a path already taken for purely molecular fluids [15] and polydisperse colloids [5]. Thus, upon introducing the full expansions for $\tilde{\gamma}$ and $\tilde{c}$ and exploiting the orthonormality of the basis functions, one finds that the OZ equation again goes over into a set of matrix equations for the respective coefficients,

$$\tilde{\gamma}^{n_1n_2}_{l_1l_2m}(k) = (-1)^m\rho\sum_{n_3,l_3}\left[\tilde{\gamma}^{n_1n_3}_{l_1l_3m}(k) + \tilde{c}^{n_1n_3}_{l_1l_3m}(k)\right]\tilde{c}^{n_3n_2}_{l_3l_2m}(k), \quad (32)$$

which has the solution, in matrix notation,

$$\tilde{\Gamma}_m(k) = (-1)^m\rho\tilde{C}_m(k)\tilde{C}_m(k)\left[I - (-1)^m\rho\tilde{C}_m(k)\right]^{-1}. \quad (33)$$

Here $\tilde{\Gamma}_m(k)$ and $\tilde{C}_m(k)$ are symmetric matrices with elements $\tilde{\gamma}^{n_1n_2}_{l_1l_2m}(k)$ and $\tilde{c}^{n_1n_2}_{l_1l_2m}(k)$, respectively, $n, l \geq m$, and I is the unit matrix.

The simplification these expansions effect is immediately reflected in the evaluation of Eq. (27) for $f(p)$, which can now be integrated to give the explicit solution [12]

$$\ln\left[\frac{f(p)}{f_0(p)}\right] = \sum_{n=2,4,6,\ldots} \left(\frac{4}{3n}\right)^{1/2} \xi_n Q_{n0}(p) = \sum_{j=1}^{\infty} \left(\frac{2}{3j}\right)^{1/2} \xi_{2j} \hat{L}_j^{1/2}\left(\frac{\beta p^2}{2\alpha}\right), \tag{34}$$

$$\xi_n = \frac{1}{2}\rho \int d\mathbf{r} \sum_{m=-1,0,1} g_{11m}^{n-1,1}(r)\beta u_{11\bar{m}}^{11}(r). \tag{35}$$

Here $\hat{L}_j^{1/2}(t)$ is the associated Laguerre polynomial renormalized to unity.

The computed coefficients of g will now yield not only the usual thermodynamic properties as before but also the electrostatic properties of the model; in particular, the dielectric constant is given by [12]

$$\varepsilon = \frac{1 - \rho\tilde{h}_{111}^{11}(0)}{1 + \rho\tilde{h}_{110}^{11}(0)}. \tag{36}$$

The calculated properties of a system of nonpolar polarizable Lennard-Jones molecules are summarized in Table 2 for six states for which molecular dynamics (MD) data of Pollock and Alder [16] are available. The specific closure used is the optimized reference-HNC (RHNC) approximation [17, 18, 19], with the bridge function B represented by just a spherically symmetric hard-sphere term, $B(\mathbf{r}, \mathbf{p}_1, \mathbf{p}_2) \approx B_{\mathrm{HS}}(r; \sigma_{\mathrm{HS}})$. A better approximation would include at least the same coefficients of B as are present in the total potential, but little is known about the higher coefficients.

4. Polarizable polar fluids

Assigning to each atom a permanent dipole moment $\boldsymbol{\mu}_0$ in addition to its fluctuating moment $\mathbf{p}$ introduces for each atom two new degrees of freedom, the Euler angles ω_0 for the direction of $\boldsymbol{\mu}_0$. The dipolar interactions now depend on the total moments $\mathbf{m}_j = \boldsymbol{\mu}_{0j} + \mathbf{p}_j$ of each particle j, with the fluctuating moments $\mathbf{p}_j$ again distributed according to the $f_0(p)$ of Eq. (24). However, the directions ω_0 can be immediately integrated out, so that we get a new distribution of just the relevant *total* moment m,

$$\begin{aligned} F_0(m) &\equiv \frac{1}{4\pi}\int d\omega_0 f_0(|\mathbf{m} - \boldsymbol{\mu}_0|) \\ &= \frac{\exp[-\beta\left(m^2 + \mu_0^2\right)/2\alpha_0]}{(2\pi\alpha_0/\beta)^{3/2}} \frac{\sinh(\beta\mu_0 m/\alpha_0)}{\beta\mu_0 m/\alpha_0}. \end{aligned} \tag{37}$$

It is then clear that with the substitution $\mathbf{p}, f_0(p) \to \mathbf{m}, F_0(m)$, this is the *same calculation* as that of nonpolar polarizable molecules and can

TABLE 3. Computed thermodynamic and electrostatic properties of a polarizable polar LJ fluid for $\rho\sigma^3 = 0.80$ and $k_BT/\epsilon = 1.35$; MD data are from Pollock *et al.* [24]. The simulation uncertainties in the last digit of ε are given in parenthesis. $\mu^* \equiv \mu/(\epsilon\sigma^3)^{1/2}$.

μ_0^*	α_0/σ^3		$U/N\epsilon$	$\beta p/\rho$	ξ_2	ξ_4	μ^*	α/σ^3	ε
0.50	0	MD	-5.28				0.50		1.78(4)
		RHNC	-5.326	2.347	0	0	0.50	0	1.75
		HNC	-5.030	3.846	0	0	0.50	0	1.75
		SSC	-5.034	3.829	0	0	0.50	0	1.75
0.50	0.050	MD	-5.41				0.53		2.79(4)
		RHNC	-5.504	2.247	-0.195	0.001	0.528	0.0528	2.76
		HNC	-5.217	3.744	-0.203	0.001	0.529	0.0529	2.76
		SSC	-5.231	3.683	-0.205	0	0.530	0.0530	2.78
0.50	0.075	MD	-5.45				0.57		3.41(7)
		RHNC	-5.648	2.177	-0.307	0.002	0.554	0.0830	3.52
		HNC	-5.371	3.661	-0.323	0.003	0.556	0.0834	3.53
		SSC	-5.394	3.563	-0.326	0	0.557	0.0835	3.58
1.50	0	MD	-7.96				1.50		21(1)
		RHNC	-7.914	1.125	0	0	1.50	0	20.5
		HNC	-7.720	2.537	0	0	1.50	0	20.0
		SSC	-7.971	1.654	0	0	1.50	0	26.1
1.50	0.025	MD	-8.43				1.62		31(2)
		RHNC	-8.753	0.854	-2.790		1.612	0.0269	31.5
		HNC	-6.907	2.220	-2.923		1.617	0.0270	32.7
		SSC	-8.897	0.968	-2.920	0	1.617	0.0270	44.1

be solved with the same algorithm. Only the numerical coefficients in the polynomials $Q_{nl}(m)$ will be different. Since $F_0(m)$ with finite μ_0 is not a classical weight function, the $Q_{nl}(m)$ are constructed explicitly using Gram-Schmidt orthogonalization [13, 20]. For this one needs the moments $\langle m^{2j} \rangle$ of $F_0(m)$, which are readily calculated. As in the nonpolar case, it is found in practice [21] that the functional form of the distribution $F_0(m)$ remains essentially unchanged in a calculation, but the polarizability and permanent moment take on new effective values, $\alpha_0 \to \alpha$ and $\mu_0 \to \mu$.

The thermodynamic, dielectric, and structural properties of a system of polar polarizable LJ atoms have been calculated for a large number of states [21]. A sample of results is shown in Table 3 [22]. These calculations follow the same integral equation procedures described earlier, using three specific closures: optimized RHNC (again with hard-sphere bridge function), HNC, and single superchain (SSC) [23], also known as linearized-HNC. The simulation data used for comparisons are from Pollock *et al.* [24]. The three

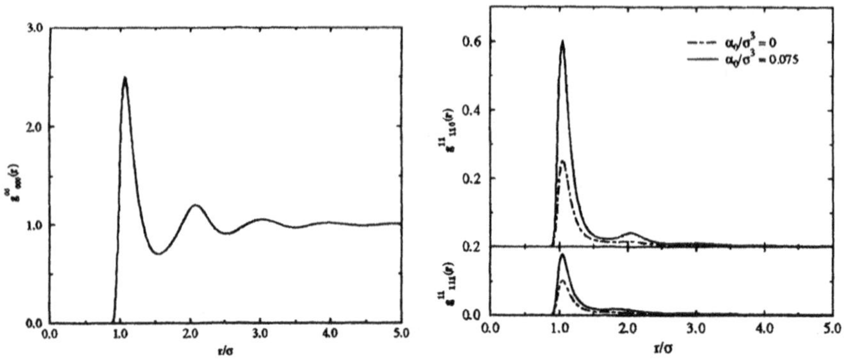

Figure 2. Distribution function coefficients of a polarizable polar Lennard-Jones liquid at density $\rho\sigma^3 = 0.80$ and temperature $k_BT/\epsilon = 1.35$. Computed results using the RHNC equation with permanent dipole moment $\mu_0 = 0.50(\epsilon\sigma^3)^{1/2}$ and two polarizabilities, $\alpha_0 = 0$ and $\alpha_0 = 0.075\sigma^3$. (a) *Left figure.* $g^{00}_{000}(r)$; the two curves are indistinguishable on this scale. (b) *Right figure.* $g^{11}_{110}(r)$ and $g^{11}_{111}(r)$.

closures give generally similar results. Where the disparities are greatest, the RHNC value gives the closest agreement with simulation.

For the smaller dipole moment in Table 3, the electrostatic effect of increasing the polarizability from 0 to $0.075\sigma^3$ is to double the dielectric constant ε. In Figs. 2(a) and 2(b), I show the corresponding structural effects. One sees that the geometrical packing of the atoms, reflected in the radial distribution functions of Fig. 2(a), is hardly affected by the increase in polarizability; the two curves are not distinguishable on this scale. However, the principal electrostatic coefficients of $g(\mathbf{r}_{12}, \mathbf{m}_1, \mathbf{m}_2)$ (which determine ε), shown in Fig. 2(b), are greatly increased in amplitude. This behavior agrees qualitatively with simulation findings [24] and is representative in general of the computed results.

It will be noted that, compared to earlier studies of fluids of polar non-polarizable molecules [25], the cost of polarizabilty in the previous section was an additional two degrees of freedom in the pair functions and so two additional indices in the expansion coefficients, a manageable increase in numerical complexity. In this section, we find that extending the technique to polar polarizable molecules entails *no further increase in complexity.* The orthogonal polynomial method makes polar polarizable fluids as easy to study as nonpolar polarizable systems. In programming terms, one merely replaces the subroutine that calculates the orthogonal polynomials.

5. Charged particle in a polarizable fluid

The strength $E(r)$ of the electric field created by a point charge q immersed in a dielectric fluid is conventionally described as that of the charge in vacuum reduced by a constant factor ε, the dielectric constant of the medium:

$$\mathbf{E}(\mathbf{r}) = \frac{q}{\varepsilon r^2}\hat{\mathbf{r}}, \tag{38}$$

where $\hat{\mathbf{r}}$ is a unit vector in the radial direction $\mathbf{r}$. At the molecular level, this relation will still hold for r large, but for distances r out to a few atomic diameters from the particle with point charge q the Maxwell field $E(r)$ will reflect the local structure of the dielectric medium, so that ε is no longer constant. More generally, we may define a screening function $F(r)$ such that

$$E(r) = \frac{q}{r^2}\left[1 - F(r)\right]. \tag{39}$$

Then for large r we should expect $F(r) \sim (\varepsilon - 1)/\varepsilon$, while for small r the finite size of the charged particle should lead to $F(r) \approx 0$ inside its first near-neighbor shell, or core region. With the polarizable fluid modeled as in Section 3, it is possible to express $F(r)$ in terms of a *computable* pair distribution function coefficient [26].

The model consists of $N+1$ particles, labeled 0 through N, all of which interact pairwise with each other through a common LJ potential $u_0(r)$. Particle 0 carries a point charge q and its location defines the origin of coordinates. It is not polarizable. The remaining N particles are polarizable nonpolar molecules with molecular polarizability α_0. Their instantaneous point dipoles lead to additional interactions, through dipole-dipole forces with each other and through charge-dipole forces with particle 0. Thus the total potential energy of an instantaneous configuration of positions $\mathbf{r}_j$ and dipole moments $\mathbf{p}_j$ is

$$\mathcal{U} = \sum_{0\le i<j\le N} u_0(r_{ij}) + \sum_{1\le i<j\le N} u_{dd}(\mathbf{r}_{ij}, \mathbf{p}_i, \mathbf{p}_j) + \sum_{1\le j\le N} u_{cd}(\mathbf{r}_{0j}, \mathbf{p}_j), \tag{40}$$

where $u_{cd}(\mathbf{r}, \mathbf{p}) = -q(\hat{\mathbf{r}}\cdot\mathbf{p})/r^2$.

With the contribution of the dipoles added to the central field made by charge q at the origin, the mean electric field at position $\mathbf{r}$ is

$$\mathbf{E}(\mathbf{r}) = \frac{q}{r^2}\hat{\mathbf{r}} + \int d\mathbf{r}'d\mathbf{p}'\frac{\rho f(p')g(\mathbf{r}', \mathbf{p}')}{|\mathbf{r}-\mathbf{r}'|^3}\left[3\frac{\mathbf{p}'\cdot(\mathbf{r}-\mathbf{r}')}{|\mathbf{r}-\mathbf{r}'|^2}(\mathbf{r}-\mathbf{r}') - \mathbf{p}'\right], \tag{41}$$

where $g(\mathbf{r}, \mathbf{p})$ is the pair distribution function of particle 0 with the polarizable molecules of instantaneous polarization $\mathbf{p}$. To evaluate the integrals

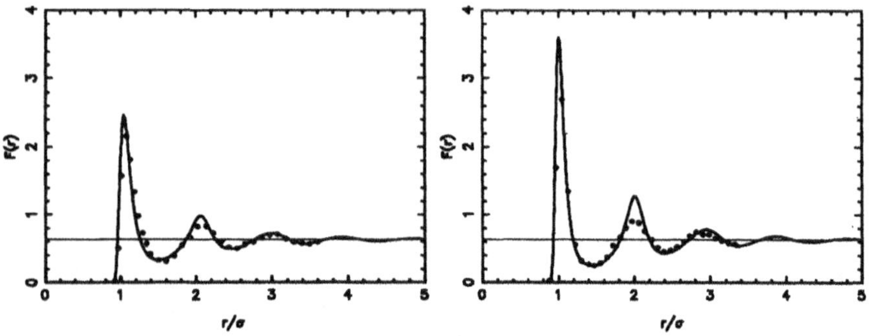

Figure 3. Screening functions $F(r)$ for a charge q in a liquid of polarizable nonpolar molecules: solid lines – HNC; points – MD [28]. The light horizontal lines are the asymptotic limit $(\varepsilon - 1)/\varepsilon = 0.63$. (a) *Left figure.* $q = 0.01(\epsilon\sigma)^{1/2}$. (b) *Right figure.* $q = 10(\epsilon\sigma)^{1/2}$.

in Eq. (41), we make use of the polynomial basis sets of Section 3 to break out the $\mathbf{p}$ dependence in $g(\mathbf{r}, \mathbf{p})$,

$$g(\mathbf{r}, \mathbf{p}) = (4\pi)^{1/2} \sum_{n,l} g_l^n(r) Q_{nl}(p) Y_{l0}(\omega). \tag{42}$$

The integrations can now be done explicitly; it is straightforward to show that the total radial field becomes simply [26]

$$E(r) = \frac{q}{r^2} - \left(\frac{\alpha}{\beta}\right)^{1/2} 4\pi\rho g_1^1(r) \tag{43}$$

and so the screening function is $F(r) = 4\pi\rho\alpha r^2 g_1^1(r)/q(\alpha\beta)^{1/2}$.

The system composed of charge q immersed in a fluid of N polarizable molecules constitutes a two-component mixture with one component at infinite dilution. We can thus adapt the usual equations for liquid state mixtures [1] to calculate the charge-dipole and dipole-dipole distribution functions, $g(\mathbf{r}_{01}, \mathbf{p}_1)$ and $g(\mathbf{r}_{12}, \mathbf{p}_1, \mathbf{p}_2)$, respectively. The latter calculation was described in Section 3; $g(\mathbf{r}_{01}, \mathbf{p}_1)$ is found by a similar, but simpler, procedure which incidently ensures the correct asymptotic limit, $F(r) \sim (\varepsilon - 1)/\varepsilon$ [26].

The only available simulation data for a model of a charged particle in a polarizable nonpolar fluid appear to be those of Pollock and Alder [27, 28] for a liquid state of Lennard-Jones molecules near the triple point: $\rho\sigma^3 = 0.844$, $k_BT/\epsilon = 0.85$, $\alpha_0/\sigma^3 = 0.10$. Thus, the results given here are for this same state. (The computed values of α and ε for this state are $\alpha/\sigma^3 = 0.1115$ and $\varepsilon = 2.699$.)

Figures 3(a) and 3(b) show the screening functions $F(r)$ obtained using the HNC equation for two different charges, $q/(\epsilon\sigma)^{1/2} = 0.01$ and 10. The second of these corresponds to about half an elementary charge, $q \approx 0.5e$, if argon values are used for the LJ parameters: $\sigma = 3.4$ Å and $\epsilon = 120k_B$, so $e/(\epsilon\sigma)^{1/2} = 20.2$. The figures are drawn to the same scale so the effect of increasing the charge q is more readily apparent. Comparison of the HNC results with the simulation data shows reasonably good agreement.

References

1. J. P. Hansen and I. R. McDonald, *Theory of Simple Liquids* (Academic, London, 1986).
2. M. S. Green, J. Chem. Phys. **43**, 1403 (1960). The parallel set $P(r)$ of this paper is not the same as that of Eq. (1).
3. J. K. Percus and G. J. Yevick, Phys. Rev. **110**, 1 (1958).
4. C. G. Gray and K. E. Gubbins, *Theory of Molecular Fluids* (Clarendon, Oxford, 1984), Vol. 1.
5. F. Lado, Phys. Rev. E **54**, 4411 (1996).
6. B. D'Aguanno and R. Klein, Phys. Rev. A **46**, 7652 (1992) and private communication.
7. F. J. Rogers and D. A. Young, Phys. Rev. **30**A, 999 (1984).
8. F. Lado, J. Chem. Phys. **108**, 6441 (1998).
9. J. C. Crocker and D. G. Grier, Phys. Rev. Lett. **73**, 352 (1994).
10. J. S. Høye and G. Stell, J. Chem. Phys. **73**, 461 (1980).
11. L. R. Pratt, Mol. Phys. **40**, 347 (1980).
12. F. Lado, Phys. Rev. E **55**, 426 (1997).
13. See, for example, G. Arfken, *Mathematical Methods for Physicists* (Academic, Orlando, 1985), Chap. 9.
14. P. M. Morse and H. Feshbach, *Methods of Theoretical Physics* (McGraw-Hill, New York, 1953), p. 1662ff.
15. F. Lado, Mol. Phys. **47**, 283 (1982).
16. E. L. Pollock and B. J. Alder, Phys. Rev. Lett. **39**, 299 (1977).
17. F. Lado, Phys. Rev. A **8**, 2548 (1973).
18. Y. Rosenfeld and N. W. Ashcroft, Phys. Rev. A **20**, 1208 (1979).
19. F. Lado, Phys. Lett. **89A**, 196 (1982).
20. N. I. Akhiezer, *The Classical Moment Problem* (Hafner, New York, 1965), Chap. 1.
21. F. Lado, E. Lomba, and M. Lombardero, J. Chem. Phys. **108**, 4530 (1998).
22. The computed results for $\alpha_0 = 0$ in Table 3 are obtained using the simpler algorithm for nonpolarizable dipolar systems described some time ago [25]. They are included here for completeness.
23. M. S. Wertheim, Mol. Phys. **26**, 1425 (1973).
24. E. L. Pollock, B. J. Alder, and G. N. Patey, Physica **108A**, 14 (1981).
25. F. Lado, M. Lombardero, E. Enciso, S. Lago, and J. L. F. Abascal, J. Chem. Phys. **85**, 2916 (1986).
26. F. Lado, J. Chem. Phys. **106**, 4707 (1997).
27. E. L. Pollock and B. J. Alder, Phys. Rev. Lett. **41**, 903 (1978).
28. E. L. Pollock, B. J. Alder, and L. R. Pratt, Proc. Natl. Acad. Sci. USA **77**, 49 (1980).

REPLICA AND GLASSES

G. PARISI
Dipartimento di Fisica and INFN,
Università di Roma La Sapienza
P. A. Moro 2, 00185 Roma (Italy)

1. Introduction

Many progresses have been done in the last years in the study of glass transitions using the replica approach. The basic ideas are those of Kauzmann, Adams, Di Marzio and Gibbs, however the use of modern theoretical techniques gives a much better insight.

The basic hypothesis in this approach is that at low temperature the energy (or better the free energy) landscape contains an exponential number of minima. The system may live in one of this mimima, but the entropy of the system will get contribution also from the so called configurational entropy or complexity.

I will concentrate on three points in my lectures.

- The study of soluble models in which this scenario is exact. The advantage of studying these model are the following:
  - You are able to obtain a precise formulation and define in a clear and precise manner the various concepts involved.
  - The resulting picture is much more complex than what you can be naively expect and a much better insight is obtained,
  - You also obtain new qualitative predictions.
  - It is possible positioning the mode coupling computations inside this scenario. This is crucial in order to understand the utility and the limits of mode coupling approach.
- Model independent predictions for the generalization of the fluctuation dissipation theorem in off-equilibrium aging dynamics. The fluctuation dissipation theorem does not hold when the system is reaching equilibrium and new relations are present, which are model independent: they

C. Caccamo et al. (eds.), New Approaches to Problems in Liquid State Theory, 107–138.

have been tested in numerical simulation and we are eagerly waiting for an experimental test.

– The construction of first principle analytic computations which allow the computation of the property of glass forming quantity at all temperatures (i.e. above and below the glass transition) by writing down and solving the appropriate integral equation for the correlation function.

The tools used are mostly the replica theory and generalized two times mode coupling equations for the dynamics.

In these lectures I will not present the detailed proof of the various result: I am trying to distill the conclusions of about one hundred papers. I will mostly state the main results, in some cases sketch the proof and refer to the original literature for more details.

2. The basic scenario

In the nutshell many of the ideas I am going to present are not new: they are already in the original papers of Gibbs and Di Marzio. However the comparison of glasses and generalized spin glasses, introduced in ref. [1], allows us to put these ideas in a sharper form and to test them in numerical (and eventually real) experiments. In this talk I will not discuss the theoretical basis under which this scenario has been derived (i.e. the mathematical tool needed to derive the results stated for the generalized spin glasses) but I will concentrate the attention of the physical picture.

The basic ideas are quite simple [2, 3, 4, 5, 6, 7, 8]. Let us consider a system of N particles and let us assume that we can introduce a free energy functional $F[\rho]$ which depends on the density $\rho(x)$ and on the temperature. We suppose that at sufficiently low temperature this functional has many minima (i.e. the number of minima goes to infinity with the number (N) of particles). Exactly at zero temperature these minima coincide with the mimima of the potential energy as function of the coordinates of the particles. Let us label them by an index α. To each of them we can associate a free energy F_α and a free energy density $f_\alpha = F_\alpha/N$.

In this low temperature region we suppose that the total free energy of the system can be well approximated by the sum of the contributions to the free energy of each particular minimum:

$$Z \equiv \exp(-\beta N f_S) = \sum_\alpha \exp(-\beta N f_\alpha). \tag{1}$$

When the number of minima is very high, it is convenient to introduce the function $\mathcal{N}(f,T,N)$ which is the density of minima whose free energy is equal to f. With this notation we can write the previous formula as

$$Z = \int df \exp(-\beta N f)\mathcal{N}(f,T,N). \tag{2}$$

In the region where $\mathcal{N}$ is exponentially large we can write

$$\mathcal{N}(f,T,N) \approx \exp(N\Sigma(f,T)), \tag{3}$$

where the function Σ is called the complexity or the configurational entropy (it is the contribution to the entropy coming from the existence of an exponentially large number of locally stable configurations).

The relation (3) is valid in the region $f_m(T) < f < f_M(T)$. The minimum possible value of the free energy is given by $f_m(T)$. Outside this region we have that $\mathcal{N}(f,T) = 0$. It all cases known $\Sigma(f_m(T),T) = 0$, and the function Σ is continuous at f_m.

For large values of N we can write

$$\exp(-N\beta f_S) \approx \int_{f_m}^{f_M} df \exp(-N(\beta f - \Sigma(f,T)). \tag{4}$$

We can thus use the saddle point method and approximate the integral with the integrand evaluated at its maximum. We find that

$$\beta f_S = \min_f \Phi(f) \equiv \beta f^* - \Sigma(f^*,T), \tag{5}$$

where

$$\Phi(f) \equiv \beta f - \Sigma(f,T). \tag{6}$$

(This formula is quite similar to the well known homologous formula for the free energy ,i.e. $\beta f = \min_E(\beta E - S(E))$, where $S(E)$ is the entropy density as function of the energy density.)

If we call f^* the value of f which minimize $\Phi(f)$. we have two possibilities:

- The minimum is inside the interval and it can be found as solution of the equation $\beta = \partial\Sigma/\partial f$. In this case we have

$$\beta\Phi = \beta f^* - \Sigma^*, \quad \Sigma^* = \Sigma(f^*,T). \tag{7}$$

  The system may stay in one of the many possible minima. The number of minima at which is convenient for the system to stay is $\exp(N\Sigma^*)$. The entropy of the system is thus the sum of the entropy of a typical minimum and of Σ^*, which is the contribution to the entropy coming from the exponential large number of microscopical configurations.

- The minimum is at the extreme value of the range of variability of f. We have that $f^* = f_m$ and $\Phi = f_m$. In this case the contribution of the complexity to the free energy is zero. The different states which contribute to the free energy have a difference in free energy density which is of order N^{-1} (a difference in total free energy of order 1). Sometimes we indicate the fact that the free energy is dominated by a few different minima by saying that the replica symmetry is spontaneously broken [9, 10].

Form this point of view the behaviour of the system will crucially depend on the free energy landscape [11], i.d. the function $\Sigma(f,T)$, the distance among the minima, the height of the barriers among them... Although our final task should be to compute the properties of the free energy landscape from the microscopic form of the Hamiltonian, we can tentatively assume that the landscape of fragile glasses is similar to that of some soluble long range models in presence of quenched disorder [1, 11].

I cannot discuss here in details the rationale for this hypothesis; it is also clear that it cannot be exact and some differences should be present among the predictions of the mean field approximation and the real world. Here I will present the scenario coming from mean field, stressing some of the predictions that should have a wider range of validity and comparing them with numerical simulations for fragile glasses.

We can distinguish a few temperature regions.

- For $T > T_f$ the only minimum of the free energy functional is given by a constant density. The system is obviously in the fluid phase.
- For $T_f > T > T_D$ there is an exponentially large number of minima with a non-constant density $\rho(x)$ [5, 12]. There are values of the free energy density such that the complexity Σ is different from zero, however the contribution coming from these minima is higher that the one coming from the liquid solution with constant $\rho(x)$.
- The most interesting situation happens in the region where $T_D > T > T_K$. In this region the free energy is still given by the fluid solution with constant ρ and at the same time the free energy is also given by the sum over the non trivial minima [7, 8].
  Although the thermodynamics is still given by the usual expressions of the liquid phase, and the final free energy is analytic at T_D, below this temperature the liquid phase corresponds to a system which at each given moment may stay in one of the exponentially large number of minima. It is extremely surprising that the free energy of the liquid can be written in this region as the sum of the contribution of the minima, according to formula (4).
  The time to jump from one minimum to another minimum is quite large: it is an activated process which is controlled by the height of

the barriers which separate the different minima. The correlation time will become very large below T_D and for this region T_D is called the dynamical transition point. The correlation time (which should be proportional to the viscosity) should diverge at T_K. The precise form of this divergence is not well understood. It is natural to suppose that we should get divergence of the form $\exp(A/(T-T_K)^\nu)$ for an appropriate value of ν [13], whose reliable analytic computation is lacking [1, 14]. The equilibrium complexity is different from zero (and it is a number of order 1) when the temperature is equal to T_D, it decreases when the temperature decreases and vanishes linearly at $T = T_K$. At this temperature (the so called Kauzmann temperature) the entropy of a single minimum becomes equal to the total entropy and the contribution of the complexity to the total entropy vanishes.

– In the region where $T < T_K$ the free energy is dominated by the contribution of a few minima of the free energy having the lowest possible value. Here the free energy is no more the analytic continuation of the free energy in the fluid phase. A phase transition is present at T_K and the specific heat is discontinuous here.

The free energy landscape is quite usual; we can try to present the following pictorial interpretation, which is a rough simplification [15]. At temperature higher than T_D the system stays in a region of phase space which is quite flat and corresponds to a minimum of the total free energy. On the contrary below T_D the phase space is similar to the one shown pictorially in fig. 1. The region of maxima and minima is separated by the region without barriers by a large nearly flat region. The minima in the region at the left are still present also when $T_f > T > T_D$, but they do not correspond to a global minimum.

At temperatures higher than T_D the system at thermal equilibrium states in the right region. When the temperature reaches T_D the system arrives in the flat region. Here the potential is flat and this causes a more or less conventional Van Hove critical slowing down which is well described by the well known mode coupling theory [16] (which is exact in the mean field approximation). The mode coupling theory describes the critical slowing down which happens near T_D [17].

In the mean field approximation the height of the barriers separating the different minima is infinite and the temperature T_D is sharply defined as the point where the correlation time diverges. In the real world activated process (which are neglected in the mean field approximation and consequently in the mode coupling theory) have the effect of producing a finite (but large) correlation time also at T_D. The precise meaning of the dynamical temperature beyond mean field approximation is discussed in details in [3]

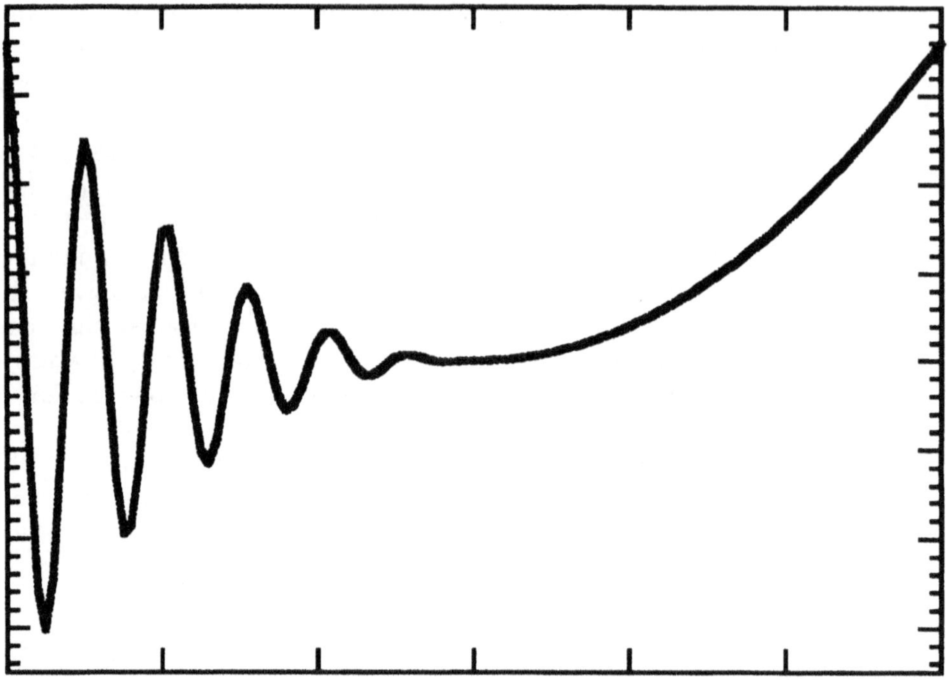

Figure 1. The qualitative dependence of the free energy as function of the configuration space in the region relevant for the dynamical transition, i.e. for $T < T_D$.

When the temperature is smaller that T_D we must be more precise in describing the dynamics of the system. Let us start from a very large system (of N particles) at high temperature and let us gradually cool it. We find that it should go at equilibrium in the region with many minima. However coming from high free energy (from the right) it cannot enter in the region where are many maxima and minima if we wait a finite amount of time (the time to cross the barriers diverges as $\exp(AN)$. If we do not wait an exponentially large amount of time the system remains confined in the flat region. In this case [2] the so called dynamical energy,

$$E_D = \lim_{t \to \infty} \lim_{N \to \infty} E(t, N), \tag{8}$$

is higher that the equilibrium free energy. The situation is described in fig. 2.

Below T_D the system is trapped in metastable states when cooled. The time needed to escape from these states diverges when N goes to infinity in the mean field approach where activated processes are forbidden. Of course the difference of the static and dynamic energy is an artifact of the mean

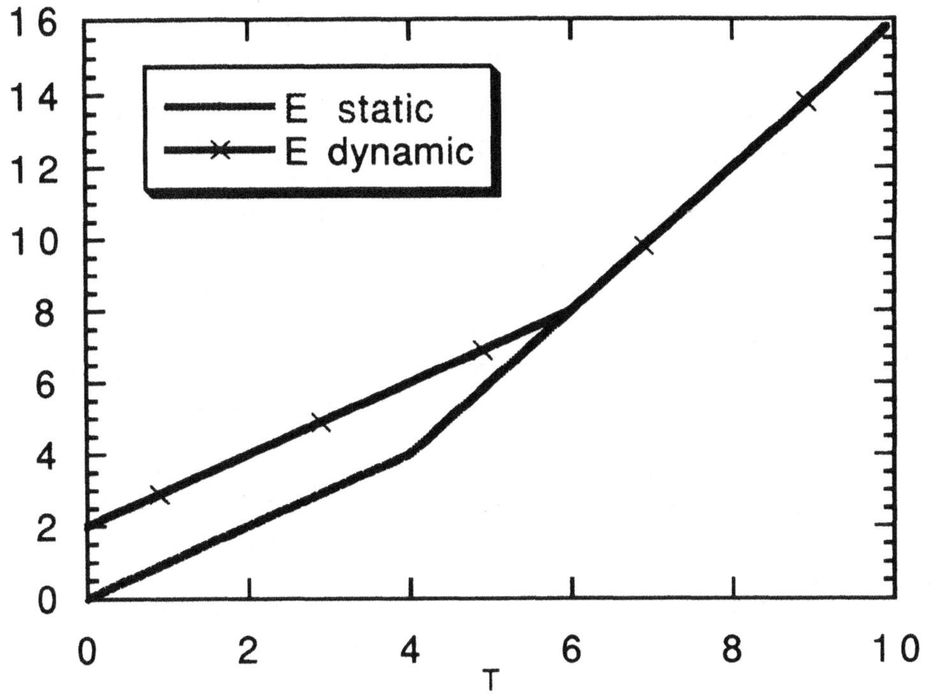

Figure 2. The qualitative behaviour of the equilibrium and of the dynamical energy as function of the temperature.

field approximation if we take literally the limit $t \to \infty$ in the previous equation because, as matter of fact, there are no metastable states with strictly infinite mean life. However it correctly describes the situation on laboratory times, where metastable states are observed.

In real systems, beyond the mean field approximation, the height of the barriers is finite also below T_D and the mean life of the metastable states is finite, albeit very large. Some mechanisms have been described which imply a divergence of the correlation time in real systems at the Kauzmann temperature [1, 3, 14]. The conventional glass temperature, i.e. the temperature at which the microscopic correlation time becomes macroscopic (e.g. of order of the minute) is between the two temperatures (i.e. $T_D > T_G > T_K$).

It should be clear that in this framework the dynamical temperature T_D is not so well defined and it corresponds to a crossover region below which the dynamics is dominated by activated processes [3].

In the mean field approximation there are very interesting phenomena that happen below T_D when the system is cooled from the high temperature phase. These phenomena are related to the fact that the system does not really go to an equilibrium configuration but wanders in the phase space never reaching equilibrium. The phenomena are the following:

- The energy approaches equilibrium slowly when the system is cooled from an high energy configuration. In other words for large times we have
$$E(t,T) = E_D(T) + B(T)t^{-\lambda(T)}, \qquad (9)$$
where the exponent $\lambda(T)$ does not vanish linearly at zero temperature as it should happen for an activated process.
- Aging is present, i.e. the correlation functions and the response functions in the region of large time do depend on the story of the system [18, 19, 20].
- In the region where aging is present the hypothesis at the basis of the fluctuation-dissipation theorem are no more valid. New generalized relations are satisfied [21, 22], which replace the fluctuation dissipation theorem.
- In the region where the diffusion constant is zero, a new phenomenon, logarithmic diffusion, is present.

All these phenomena are well known also in the case of spinodal decomposition but have a more general validity.

3. The Random Energy Model

3.1. THE DEFINITION OF THE MODEL

The Random Energy Model [23] is the simplest model for glassy systems. It has various advantages: it is rather simple (its properties may be well understood with intuitive arguments, which may become fully rigorous) and display very interesting and new phenomena.

The Random Energy Model is defined as following. There are N Ising spins (σ_i, $i = 1, N$) which may take values ± 1; the total number of configurations is equal to $M \equiv 2^N$ and they can be identified by a label k in the interval $1 - M$.

Generally speaking the Hamiltonian of the system is given when we know the values of the energies E_k for each of the M configurations of the system. Usually one writes explicit expression for the energies as function of the configuration; on the contrary here we assume that the values of the E_k are random, with a probability distribution $p(E)$ which is Gaussian:

$$p(E) \propto \exp(-\frac{E^2}{2N}). \qquad (10)$$

The partition function is simply given by

$$Z_{\mathcal{E}} = \sum_{k=1,M} \exp(-\beta E_k) = \int \rho(E) \exp(-\beta E), \tag{11}$$

$$\rho(E) \equiv \sum_{k=1,M} \delta(E - E_k).$$

The value of the partition function and of the free energy density ($f_{\mathcal{E}} = -\ln(Z_{\mathcal{E}})/(N\beta)$) depends on $\mathcal{E}$, i.e. all the values of the energies E_k. We would like to prove that when $N \to \infty$ the dependence on $\mathcal{E}$ of $f_{\mathcal{E}}$ disappears with probability 1. If this happens, the most likely value of $f_{\mathcal{E}}$ coincides with the average of $f_{\mathcal{E}}$, where the average is done with respect to all the possible values of the energy extracted with the probability distribution given by eq. (10) .

The model is enough simple to be studied in great details; exact expressions can be derived also for finite N. Here we sketch the results giving a plausibility argument.

3.2. EQUILIBRIUM PROPERTIES OF THE MODEL

The crucial observation is the following. The probability of finding a configuration of energy E is

$$\mathcal{N}_0(E) \equiv 2^N \exp(-\frac{E^2}{2N}) = \exp(N(\ln(2) - \frac{1}{2}e^2)), \tag{12}$$

where $e \equiv E/N$ is the energy density. It is reasonable to assume (and it is confirmed by a detailed computation) that in the case of a generic system (with probability 1 when $N \to \infty$) no configurations are present in the region where $\mathcal{N}_0(E) << 1$, i.e. for

$$e^2 < e_c^2 \equiv 2\ln(2). \tag{13}$$

We can thus write for the generic choice of the energies ($\mathcal{E}$):

$$\rho(E) \approx \mathcal{N}(E) \equiv \mathcal{N}_0(E)\theta(E_c^2 - E^2). \tag{14}$$

The partition function can be written as

$$\int \mathcal{N}(E) \exp(-\beta E). \tag{15}$$

Evaluating the integral with the saddle point method one finds that in the high temperature region, i.e.

$$\beta < \beta_c \equiv e_c^{-1}, \tag{16}$$

the internal energy density is simply given by $-\beta$. This behaviour must end somewhere because we know that the energy is bounded form below also when $\beta \to \infty$. Indeed in the low temperature region one finds that the integral is dominated by the boundary region $E \approx E_c$ and the energy density is exactly given by $-e_c$.

The entropy density is positive in the high temperature region, vanishes at β_c and remains zero in the low temperature region. It follows that in the high temperature region an exponentially large number of configurations contributes to the partition function, while in the low temperature region it is possible that the probability is concentrated on a finite number of configurations.

3.3. PROPERTIES OF THE LOW TEMPERATURE PHASE

It is worthwhile to study in more details the structure of the configurations which mostly contribute to the partition function in the lower temperature phase. At this end it is useful to sort the configurations with ascending energy. We rename the configurations and we introduce new labels such that ($E_k < E_i$ for $k < i$).

It is convenient to introduce the following quantity:

$$w_k \equiv \frac{\exp(-\beta E_k)}{Z}. \tag{17}$$

We have obviously that

$$\sum_{k=1,2^N} w_k = 1 \tag{18}$$

A detailed computation shows [23, 9, 10] that in the low temperature region (i.e. $\beta > \beta_c$) the previous sum is dominated by the first terms. Indeed

$$\sum_{k=1,L} w_k = 1 - O(L^{-\lambda}), \quad \lambda = \frac{1-m}{m}, \tag{19}$$

where

$$m = \frac{T}{T_c}. \tag{20}$$

In the same region the sum in the following equation is convergent and its average value is given by

$$\sum_{k=1,2^N} w_k^2 = 1 - m. \tag{21}$$

Generally speaking one finds that one can introduce the quantities F_k such that

$$w_k \propto \exp(-\beta F_k) \tag{22}$$

Here the variables F_k coincide with the total energy (not the energy density!) apart form an additive constant. Their probability distribution at the lower end (which is the relevant region for thermodynamics in the low temperature region) can be approximated as

$$P(F) \approx \exp(\beta m F). \tag{23}$$

In this model everything is clear: in the high temperature region the number of relevant configurations is infinite (as usual) and there is a transition to a low temperature region where only few configurations dominate.

This phenomenon can be seen also in the following way. We introduce a distance among two configurations α and γ as

$$d^2(\alpha,\gamma) \equiv \frac{\sum_{i=1,N}(\sigma_i^\alpha - \sigma_i^\gamma)^2}{2n}. \tag{24}$$

Sometimes it is convenient to introduce also the overlap q defined as

$$q(\alpha,\gamma) \equiv \frac{\sum_{i=1,N}\sigma_i^\alpha \sigma_i^\gamma}{2N} = 1 - d^2(\alpha,\gamma). \tag{25}$$

The distance squared is normalized in such a way that it spans the interval $0-2$. It is equal to

- 0, if the two configuration are equal ($q = 1$).
- 1, if the configuration are orthogonal ($q = 0$).
- 2, if $\sigma_i^\alpha = -\sigma_i^\gamma$ ($q = -1$).

It is convenient to introduce the function $Q(d)$ and $P(q)$, i.e. the probability that two equilibrium configurations are at distance d, or overlap q, respectively. We find

- For $T > T_c$

$$Q(d) = \delta(d-1), \quad P(d) = \delta(q). \tag{26}$$

- For $T < T_c$

$$Q(d) = (1-A)\delta(d-1) + A\delta(d), \quad P(d) = (1-A)\delta(d) + A\delta(q-1). \tag{27}$$

  where A is equal to $\sum_{k=1,2^N} w_k^2$. The average of A over the different realizations of system is equal to $1-m$.

As soon as we enter in the low temperature region, the probability of finding two equal configurations is not zero. The transition is quite strange from the thermodynamic point of view.

- It looks like a *second* order transition because there is no latent heat. It is characterized by a jump in the specific heat (which decreases going toward low temperatures).

– It looks like a *first* order transition. There are no divergent susceptibilities coming form above of below (which beyond mean field theory should imply no divergent correlation length). Moreover the minimum value of d jumps discontinuously (from 1 to 0).
– If we consider a system composed by two replicas (σ^1 and σ^2) [5] and we write the Hamiltonian

$$H(\sigma^1, \sigma^2) = H(\sigma^1) + H(\sigma^2) + N\epsilon d^2(\sigma_1, \sigma^2) \tag{28}$$

the thermodynamics is equal to that of the previous model (apart a factor 2) for $\epsilon = 0$, but we find a real first order thermodynamic transition, with a discontinuity in the internal energy, as soon as $\epsilon > 0$. The case $\epsilon = 0$ is thus the limiting case of real first order transitions.

These strange characteristics can be summarized by saying that the transition is of order one and half, because it share some characteristics with both the first order and the second order transitions.

It is impressive to note that the thermodynamic behaviour of real glasses near T_c is very similar to the order one and half transition of REM. We will see later that this behaviour is typical of the mean field approximation to glassy systems.

3.4. DYNAMICAL PROPERTIES OF THE MODEL

The dynamical properties of the model can be easily investigated in a qualitative way. Interesting behavior is present in the region where the value of N is large with respect to the time. Different results will be obtained for different definition of the dynamics if we consider some rather artificial form of the dynamics.

Let us first consider a single spin flip dynamics. In other words we assume that in a microscopic time scale, which for simplicity we consider of order unit, the system explores all the configurations which differ from the original one by a single spin flip and goes to one of them (or remains in the original one) with probability which is proportional to $\exp(-\beta H)$. This behaviour is typical of many dynamical process, like Glauber dynamics, monte Carlo, heath bath.

In this dynamical process each configuration C has N nearby configurations C' to explore. The energies of the configurations C' are uncorrelated to the energy of C, so that they would be of order one in most of the case. The lowest energy of the configurations C' would be of order $-(N\ln(N))^{1/2}$, the corresponding energy density $(-(\ln(N)/N)^{1/2})$ vanishes in the large N limit.

If the configuration C has an energy density e less that zero, but greater than the equilibrium energy, the time needed to do a transition to a configuration of lower energy will be, with probability one, exponentially large.

For short times a configuration of energy e will be completely frozen. Only at larger times it may jump to a typical configuration of energy zero. At later times different scenarios are possible: the configuration comes back to the original configuration of of energy e or, after some wandering in the region of configurations of energy density ≈ 0, it falls in another deep configuration of energy e'. A computation of the probabilities for these different possibilities has not yet been done, although it should not be too difficult.

The conclusions of this analysis are quite simple.

- If we start from a random configuration, after a time which is finite when $N \to \infty$, the system goes to a configuration whose energy is of order $-\ln(N)^{1/2}$ and stops there.
- If we start from a random configuration and we study the system at exponentially large times the system will reach an energy density which may be different from the original one.

4. Models with partially correlated energy

4.1. THE DEFINITION OF THE MODELS

The random energy model (REM) is rather unrealistic in that it predicts that the energy is completely upset by a single spin flip. This feature can be eliminated by considering a more refined model, the so called p-spins models [24, 25] , in which the energies of nearby configurations are also nearby. We could say that energy density (as function of the configurations) is not a continuous function in the REM, while it is continuous in the p-spins models, in the topology induced by the distance (eq. (24)). In this new case some of the essential properties of the REM are valid, but new features are present.

The Hamiltonian we consider depends on some control variables J, which have a Gaussian distribution and play the same role of the random energies of the REM and by the spin variable σ. For $p = 1, 2, 3$ the Hamiltonian is respectively

$$H_J^1(\sigma) = \sum_{i=1,N} J_i \sigma_i \tag{29}$$

$$H_J^2(\sigma) = \sum_{i,k=1,N}' J_{i,k} \sigma_i \sigma_k \tag{30}$$

$$H_J^3(\sigma) = \sum_{i,k,l=1,N}' J_{i,k,l}\sigma_i\sigma_k\sigma_l$$

where the primed sum indicates that all the indices are different. The variables J must have a variance of $O(N^{(1-p)/2})$ in order to have a non trivial thermodynamical limit.

It is possible to prove by an explicit computation that if we send first $N \to \infty$ and later $p \to \infty$, one recover the REM. Indeed the energy differences corresponding to one spin flip are of order p for large p (they ar order N in the REM, so that in the limit $p \to \infty$ the energies in nearby configurations become uncorrelated and the REM is recovered.

4.2. EQUILIBRIUM PROPERTIES OF THE MODELS

The main new property of the model is the correlation of energies. This fact implies that if C is a typical equilibrium configuration, all the configurations which differ from it by a finite number of spin flips will also have a finite energy. The equilibrium configurations are no more isolated (as in REM), but they belong to valleys, such that the entropy restricted to a single valley is proportional to N and it is and extensive quantity.

The thermodynamical properties at equilibrium can be computed using the replica method [9, 10] . Let us discuss firstly what happen for $p > 2$ In the simplest version of this method [24, 25] one introduces the typical overlap of two configurations inside the same valley (sometimes denoted by q_{EA}). Something must be said about the distribution of the valleys. Only those which have minimum free energy are relevant for the thermodynamics. One finds that these valleys have zero overlap and have the same distribution of free energy as in the REM

$$P(F) \propto \exp(\beta m(F - F_0)). \tag{31}$$

Indeed the average value of the free energy can be written in a self consistent way as function of m and q ($f(q,m)$) and the value of these two parameters can be found as the solution of the stationarity equations:

$$\frac{\partial f}{\partial m} = \frac{\partial f}{\partial q} = 0. \tag{32}$$

The quantity q (which would be 1 in the REM) is here of order $1 - \exp(-A\beta p)$ for large p, while the parameter m has the same dependence of the temperature as in the REM, i.e. 1 at the critical temperature, and a linear behaviour al low temperature. The only difference is that m is no more strictly linear as a function of the temperature.

The thermodynamical properties of the model are the same as is the REM: a discontinuity in the specific heat, with no divergent susceptibilities. Let us first recall the description of these glassy systems at equilibrium according to the predictions of the replica theory. We consider a systems and we denote by C a generic configuration of the system. For simplicity we will assume that there are no symmetry in the Hamiltonian, in presence of symmetries the arguments must be slightly modified. It is useful to introduce an overlap $q(C, C')$. There are many ways in which an overlap can be defined; for example in spin system we could define

$$q = \frac{\sum_{i=1,N} \sigma_i \tau_i}{N}, \tag{33}$$

N being the total number of spins or particles and σ and τ are the two spin configurations. In a liquid a possibility is given by

$$q = \frac{\sum_{i=1,N} \sum_{k=1,N} f(x(i) - y(k))}{N}, \tag{34}$$

where f is a function which decays in a fast way at large distances and is substantially different from zero only at distances smaller that the interatomic distance (x and y are the two configurations of the system).

In the high temperature phase for very large values of N the probability distribution of the overlap ($P_N(q)$) is given by

$$P_N(q) \approx \delta(q - q^*). \tag{35}$$

In the low temperature phase $P_N(q)$ depends on N (and on the quenched disorder, if it is present). When we average over N we find a function $P(q)$ which is not a simple delta function. In all known case one finds that

$$P(q) = a_m \delta(q - q_m) + a_M \delta(q - q_M) + p(q), \tag{36}$$

where the function $p(q)$ does not contain delta function and its support is in the interval $[q_m, q_M]$.

The non triviality of the function $P(q)$ (i.e. the fact that $P(q)$ is not a single delta function and consequently q is an intensive fluctuating quantity) is related to the existence of many different equilibrium states. Moreover the function $P_N(q)$ changes with N and its statistical properties (i.e the probability of getting a given function $P_N(q)$) can be analytically computed [9, 26].

In this equilibrium description a crucial role is given by the function $x(q)$ defined as

$$x(q) = \int_{q_m}^{q} P(q')dq' \,. \tag{37}$$

In the simplest case the function $p(q)$ is equal to zero. i.e the function $P(q)$ has only two delta functions without the smooth part. In this case, which corresponds to one step replica symmetry breaking, there are many equilibrium states, labeled by α, and the overlaps among two generic configurations of the same state and of two different states are respectively q_M and q_m. The probability of finding a state with total free energy f is proportional to

$$\exp\left(m\beta(f - f_R)\right), \tag{38}$$

where f_R is a reference free energy and m is the value of $x(q)$ in the interval $[q_m, q_M]$.

In the more complicated situation where the function $p(q)$ is non zero, couples of different states may have different values of the overlaps. The joint probability distribution of the states and of the overlaps can be described by formulae similar to eq. (38), but more complex [9].

Although these predictions are quite clear, it is not so simple to test them for many reasons:

- They are valid at thermal equilibrium, a condition that is very difficult to reach for this kind of systems.
- Experimentally is extremely difficult to measure the values of the microscopic variables, i.e all the spins of the system at a given moment. These measurements can be done only in numerical simulations, where the observation time cannot be very large.

A very important progress has been done when it was discovered [2] that the function X, which describes the violations of the fluctuation-dissipation theorem, is equal to the function x which is relevant for the statics. This equality is very interesting because function $X(C)$ can be measured relatively easily in off-equilibrium simulations [21].

The temperature dependence of the function $X(C)$ (or equivalently $x(q)$) is interesting also because rather different systems can be classified in the same universality class according to the behaviour of this function. It has been conjectured long time ago that the equilibrium properties of glasses are in the same universality class of some simple generalized spin glass models [1, 27].

4.3. THE FREE ENERGY LANDSCAPE

It would be interesting to characterize better the free energy landscape of the model, especially in order to understand the dynamics. Indeed we have already seen that in the REM the system could be trapped in metastable configurations. Here the situation is more complicated. Although the word *valley* has a strong intuitive appeal, we must first define what a valley is in a more precise way.

There are two different (hopefully equivalent) definitions of a valley:

- A valley is a region of configuration space separated by the rest of the configuration space by free energy barriers which diverge when $N \to \infty$. More precisely the system, in order to go outside a valley by moving one spin at once, must cross a region where the free energy is higher that of the valley by a factor which goes to infinity with N.
- A valley is a region of configuration space in which the system remains for time which goes to infinity with N.

The rationale for assuming that the two definitions are equivalent is the following. We expect that for any reasonable dynamics in which the systems evolves in a continuous way (i.e. one spin flip at time), when it goes from a valley to another valley, the system must cross a configuration of higher free energy and therefore the time for escape from a valley is given by

$$\tau \simeq \tau_0 \exp(\beta \Delta F) \tag{39}$$

where ΔF is the free energy barrier.

It is crucial to realize that in infinite range models there can be valleys which have an energy density higher that that of equilibrium states. This phenomenon is definitely not present in short range models. No metastable states with infinite mean life do exist in nature.

Indeed let us suppose that the system may stay in phase (or valleys) which we denote as A and B. If the free energy density of B is higher than that of A, the system can go from B to A in a progressive way, by forming a bubble of radius R of phase A inside phase B. If the surface tension among phase A and B is finite, as it happens in any short range model, for large R the volume term will dominate the free energy difference among the pure phase B and phase B with a bubble of A of radius R. This difference is thus negative at large R, its maximum will thus be finite. In the nutshell a finite amount of free energy in needed in order to form a seed of phase A starting from which the spontaneous formation of phase A will start. For example, if we take a mixture of H_2 and O_2 at room temperature, the probability of a spontaneous temperature fluctuation in a small region of the sample, which lead to later ignition and eventually to the explosion of the whole sample, is greater than zero (albeit quite a small number), and obviously it does not go to zero when the volume goes to infinity.

We have two possibilities open in positioning this mean field theory prediction of existence of real metastable states:

- We consider the presence of these metastable state with *infinite* mean life an artefact of the mean field approximation and we do not pay attention to them.
- We notice that in the real systems there are metastable states with very large (e.g. much greater than one year) mean life. We consider

the *infinite* time metastable states of the mean field approximation as precursors of these *finite* states. We hope (with reasons) that the corrections to the mean field approximation will give a finite (but large) mean life to these states (how this can happen will be discussed in the next section).

Here we suppose that the second possibility is the most interesting and we proceed with the study of the system in the mean field approximation. The strategy for investigating the properties of these metastable states consists in considering systems with R replicas (two or more) of the same system with Hamiltonian given by

$$\beta H = \sum_{r=1,R} \beta_r H(\sigma^r) + \sum_{r,s=1,R} \epsilon_{r,s} q_{r,s}. \tag{40}$$

Different replicas may stay at different temperature. The quantities ϵ are just Legendre multipliers needed to enforce specific value of the the overlaps q. In this way (let us consider for simplicity the case where all temperature are zero) we find (after a Legendre transform) a free energy density as function of the q.

Let us consider for simplicity the case where we set

$$q_{1,r} = q \quad \text{for} \quad r = 2, R \tag{41}$$

($q_{r,r}$ is identical equal to 1) and the others q are left free [5] . A simple computation shows the final free energy density that we obtain (let us call $f_R(q)$ is given by

$$f_R(q) - Rf = -\lim_{N\to\infty} \frac{\ln\left(\sum_\sigma \left(\sum_\tau \delta(q(\sigma,\tau) - q)\right)^{(R-1)}\right)}{\beta N} \equiv \\ -\lim_{N\to\infty} \frac{\ln < P_\sigma(q)^{(R-1)} >}{\beta N}, \tag{42}$$

where f is the unconstrained free energy.

A particular case, which is very interesting, is given by the limit $R \to 1$ [14, 28] :

$$W(q) \equiv \lim_{N\to\infty} \frac{\ln(P_\sigma(q))}{\beta N} = \frac{\partial(f_R(q) - Rf)}{\partial R}|_{R=1} \tag{43}$$

The potential $W(q)$ has usually a minimum at $q = 0$, where $W(0) = 0$. It may have a secondary minimum at $q = q_D$. We can have three quite different situations

- $W(q_D) = 0$. This happens in the low temperature region, below T_c, where we can put two replicas both at overlap 0 and at overlap q_{EA} without paying any prize in free energy. In this case $q_D = q_{EA}$.

- $W(q_D) > 0$. This happens in an intermediate temperature region, above T_c, but below T_D, where we can put one replica σ at equilibrium and have the second replica τ in a valley near it. It happens that the internal energy of both the σ configuration (by construction) and of the τ configuration are equal to the equilibrium one. However the number of valley is exponentially large so that the free energy a single valley will be smaller. One finds in this way that $W(q_D) > 0$ is given by

$$W(q_D) = \frac{\ln \mathcal{N}_e}{N} \tag{44}$$

where $\mathcal{N}_e$ is the average number of the valleys having the equilibrium energy [8, 29] .
- At $T > T_D$ the potential $W(q)$ has only the minimum at $q = 0$. The quantity q_D cannot be defined and no valley with the equilibrium energy are present. This is more or less the definition of the dynamical transition temperature T_D. A more careful analysis shows that for $T_D < T < T_V$ there are still valleys with energy *less* than the equilibrium one, but these valleys cover so small a region of phase space that they are not relevant for equilibrium physics.

It is also possible to study the properties of the free energy for $R \neq 1$; we can force the σ configuration to be not an equilibrium one and in this way we control the properties of the valleys having an energy different than the equilibrium one. This method can give rather detailed information on the free energy landscape, which I do not have time to discuss in details and which have not yet fully studied.

The most interesting result is that for $T < T_D$ the entropy of the system can be written as

$$S = S_V + W \tag{45}$$

where S_V is the entropy inside a valley and W is the configurational entropy, or complexity, i.e. the term due to the existence of an exponentially large number of states. The W contribution vanishes at T_c and becomes exactly equal to zero for $T < T_c$ [1] .

In the REM limit ($p \to \infty$) the temperature T_D goes to infinity. In this limit the third region does not exist. Therefore the dynamical transition is a new feature which is not present in the REM.

5. Static and dynamic properties

The simplest way to study the dynamics of the problem is to consider a system which evolves according to Langevin equation or to some sort of

Glauber dynamics. For example we can suppose that:

$$\frac{d\sigma_i}{dt} = -\frac{\delta H}{\sigma_i} + \eta_i(t), \tag{46}$$

where η is an appropriate white noise.

In this model is convenient to introduce the single site correlation function (C) and the response (G) function of the times. One finds that they can be defined as

$$C(t_1, t_2) = < \sigma_i(t_1)\sigma_i(t_2) > \tag{47}$$

$$G(t_1, t_2) = < \frac{\delta\sigma_i(t_1)}{h_i(t_2)} > \tag{48}$$

where $h(t)$ is an external magnetic field.

If the systems is at equilibrium (or in a metastable state), the correlation and the response functions will depend only on the time difference. If the system is out of equilibrium these functions will depend in a non trivial way from both the arguments.

In both cases one can write down closed equations for the correlation functions in the case where the size N goes to infinity at fixed times [2]. These equations have a rather complex structure. We could discuss two different regimes:

1. We start at time zero from an equilibrium configuration.
2. We start at time zero from a random configuration.

In the first case we find that, if we approach the dynamical temperature from above, the correlation time diverges as a power of $T - T_D$, and the usual analysis of the mode coupling theory can be done in this region. Mode coupling theory is essentially correct in this region.

In the second case one finds that below the dynamical transition, the energy does not go anymore to the equilibrium value, but it goes to an higher value, which in some cases can be computed analytically.

The phenomelogy is rather complex, the aging properties of the systems are particularly interesting, but we cannot discuss them for lack of space. It is interesting to note that the mode coupling theory become exact in the mean field theory and describes what happens nearby the dynamical phase transition at T_D, which is a temperature higher that the equilibrium transition temperature at T_c.

Here we would like to understand the dynamical properties of the model starting from results which we have already derived on the free energy landscape without having to compute explicitly the solution of the equation of motion for the correlations C and G. This can be done by assuming the this very slow dynamics is an activated process dominated by barrier crossing and that the height of the barriers can be computed using the method of the previous sections.

Some of the question we ask are the following:

- How long does a system remain in the same valley?
- If the system is out of equilibrium at time 0, which is the asymptotic value of the energy at large times?

The results are similar to those obtained by the explicit dynamical analysis of the previous section. We obtain also more detailed information on the region where the time scale is exponentially large. This region can be studied only with very great difficulties using the dynamical equations.

In the first case (i.e. the system starts from a thermalized configuration) below T_D, the system is confined to a valley. A first estimate of the time needed to escape from a valley can be obtained as follows. We introduce the parameter

$$q(t) \equiv \frac{\sum_{i=1,N} \sigma_i(0)\sigma_i(t)}{N}. \tag{49}$$

We assume that on the large time scale the evolution of q is more or less the same of a system with only one degree of freedom with potential $W(q)$. This assumption is not completely correct, but it is likely to be enough to give a first estimate (to be refined later) of the escape time. Therefore it is reasonable to assume that in a first approximation the time needed to escape from the valley is given by

$$\tau = \tau_0 \exp(N\beta\Delta W), \tag{50}$$

where ΔW is the difference in free energy among the minimum at q_D and the maximum al lower values of q.

For large time (but not exponentially divergent with N) we have that

$$\lim_{t\to\infty} q(t) = q_D. \tag{51}$$

Eventually the system will escape from a valley for times greater that τ.

In the other situation (the system is originally out equilibrium) things are more complicated. Metastable valleys of high energies do exist, however it is not clear that a system cooled from an high temperature region must be trapped in the valley with high energy. Indeed it will be trapped in the valley with largest attraction domain, but the properties of the valley cannot be found if we do not use the dynamics in an explicit way. Some educated conjectures can be done, but the question has never been investigated in details. The problem is not easy, because the system is strongly out of equilibrium.

A better understanding comes if we consider the case in which the system is slowly cooled, from high to low temperature. We consider here the case of slow, not ultra slow, cooling, i.e. the time scale if fixed when N goes to infinity. In this case it is reasonable to assume that the system is frozen in a valley at the dynamical temperature, and we have to follow the energy of that valley when we cool. That can be done by using the formulation with different replicas at different temperatures. In this way one finds that the system is frozen in a configuration which is near to the configuration at the dynamical temperature.

A detailed comparison of the results for the energy of the metastable states obtained by doing different assumption is still lacking, but it should be not too difficult.

6. Systems without quenched disorder

Apparently the previous discussions are restricted to systems where quenched disorder is present. The requirement of quenched disorder would limit ourselves in the applications of the replica method and one would cut all those systems, like glasses, which have a translational invariant Hamiltonian and where no quenched disorder is present.

This prejudice (on the need of quenched disorder(was so strong that it took a few years to realize that the replica method can also applied to system without quenched disorder.

There are many facts that clearly indicate the possibility of applying the replica methods to non-random systems.

- In the infinite range case there are pairs of systems with Hamiltonian respectively H_Q and H, where H_Q contains quenched disorder and no disorder is present in H, such that the high temperature expansion for the two systems coincide [30, 31]. It is natural to suppose that the free energies of the two systems are identical at all temperatures, so that replica symmetry breaking can be applied to both.
- It is possible to look for replica symmetry breaking in the expression for the free energy systems without disorder (e.g. soft spheres) inside a given approximation (e.g. hypernetted chain) and find out that replica symmetry breaks at low temperature [32].
- In the replica method we can introduce coupled replica potentials [33] in order to characterize the phase space of the system and these potentials can also be computed for non-random systems, obtaining the same results as for random systems [28]. This may be done analytically for soft spheres using the same approximation as before [34].
- It is now clear that the replica method may be applied to any stochastic stable system. Indeed stochastic stable systems are the limit of disorder systems where the replica method can be applied without problems. Systems without quenched disorder may be stochastically stable if the

free energy is computed using the Cesareo limit (i.e. averaging over N).

This new perspective allows us to use the replica method in systems quite different from the usual one, e.g. structural glasses, where no quenched disorder is present.

7. Fluctuation and dissipation relations in aging dynamics

7.1. THEORETICAL CONSIDERATIONS

When we suddenly decrease the temperature in an Hamiltonian system, many interesting phenomena happen if the initial and the final temperatures correspond to different phases. When the low temperature phase can be characterized by a simple order parameter (e.g. the magnetization for ferromagnets) we find the familiar phenomenon of spinodal decomposition characterized by growing clusters of different phases. There is a dynamical correlation length (i.e. the size of the clusters), which increases as a power of the time t after the quench, and the energy approaches equilibrium with power like corrections (e.g. $E(t) \approx E_\infty + At^{-1/2}$). In this region aging phenomena are also present [18].

The situation is more intriguing in the case of structural glasses and spin glasses where the low temperature phase cannot be characterized in term of a simple order parameter. Remarkable progresses in understanding the off-equilibrium dynamics and its relations to the equilibrium properties have been done by noticing that a crucial off-equilibrium feature is the presence of deviations from the well known *equilibrium* fluctuation-dissipation relations. On the basis of analytic results for soluble models it has been conjectured that we can define a function $X(C)$, C being an autocorrelation function at different times [2, 20, 17]. This function characterizes the violations of the fluctuation-dissipation theorem (which is correct only at equilibrium). It is remarkable that (at least in the case of spin glasses) the function $X(C)$ is equal to the function $x(q)$ (q being the overlap of two spin configurations) which plays a central role in the equilibrium computation of the free energy [9].

Let us be more precise. We concentrate our attention on a quantity $A(t)$. We suppose that the system starts at time $t = 0$ from an initial condition and subsequently it remains at a fixed temperature T. If the initial configuration is at equilibrium at a temperature $T' > T$, we observe an off-equilibrium behaviour. We can define a correlation function

$$C(t, t_w) \equiv \langle A(t_w) A(t + t_w) \rangle \tag{52}$$

and the response function

$$G(t,t_w) \equiv \left.\frac{\delta\langle A(t+t_w)\rangle}{\delta\epsilon(t_w)}\right|_{\epsilon=0}, \tag{53}$$

where we are considering the evolution in presence of a time dependent Hamiltonian in which we have added the term $\int dt\epsilon(t)A(t)$.

The usual equilibrium fluctuation-dissipation theorem (FDT) tells us that

$$G^{eq}(t) = -\beta\frac{dC^{eq}(t)}{dt}, \tag{54}$$

where

$$G^{eq}(t) = \lim_{t_w\to\infty} G(t,t_w), \quad C^{eq}(t) = \lim_{t_w\to\infty} C(t,t_w). \tag{55}$$

It is convenient to define the integrated response:

$$R(t,t_w) = \int_0^t d\tau G(t-\tau, t_w+\tau), \quad R^{eq}(t) = \lim_{t_w\to\infty} R(t,t_w), \tag{56}$$

$R(t,t_w)$ is the response of the system at time $t+t_w$ to a field acting for a time t starting at t_w. The usual FDT relation becomes

$$R^{eq}(t) = \beta(C^{eq}(t) - C^{eq}(0)). \tag{57}$$

The off-equilibrium fluctuation-dissipation relation [2] states that the response function and the correlation function satisfy the following relation for large t_w:

$$R(t,t_w) \approx \beta\int_{C(t,t_w)}^{C(0,t_w)} X(C)dC. \tag{58}$$

If we plot R versus βC for large t_w the data collapse on the same universal curve and the slope of that curve is $-X(C)$. The function $X(C)$ is system dependent and its form tells us interesting information. in the case of three dimensional spin glasses.

We must distinguish two regions:

- A short time region where $X(C) = 1$ (the so called FDT region) and C belongs to the interval I (i.e. $C_1 < C < C_2$.).
- A large time region (usually $t = O(t_w)$) where $C \notin I$ and $X(C) < 1$. In the same region the correlation function often satisfies an aging relation i.e $C(t,t_w)$ depends only on the ration $s \equiv t/t_w$ in the region where both t and t_w are large: $C(t,t_w) \approx C^a(t/t_w)$.

In the simplest non trivial case, i.e. one step replica symmetry breaking [9, 27] , the function $X(C)$ is piecewise constant, i.e.

$$X(C) = m \quad \text{for } C \in I, \quad X(C) = 1 \quad \text{for } C \notin I. \tag{59}$$

One step replica symmetry breaking for glasses has been conjectured in ref. [1].

In all known cases in which one step replica symmetry holds, the quantity m vanishes linearly with the temperature at small temperatures. It often happens that $m = 1$ at $T = T_c$ and $m(T)$ is roughly linear in the whole temperature range.

7.2. STOCHASTIC STABILITY

Stochastic stability is a property which is valid in the mean field approximation; it is however possible to conjecture that is valid in general also for short range models. It has been introduced quite recently [**35, 36, 37, 38**] and strong progresses have been done on the study of its consequences.

In order to decide if a system with Hamiltonian H is stochastically stable, we have to consider the free energy of an auxiliary system having the following Hamiltonian:

$$H + \epsilon^{1/2} H_R. \tag{60}$$

If the average (with respect to H_R) free energy is a differentiable function of ϵ (and the limit volume going to infinity commutes with the derivative with respect to ϵ), for a generic choice of the random perturbation H_R inside a given class and ϵ near to zero, the system is stochastically stable.

The definition of stochastic stability may depend on the class of random perturbations we consider. Quite often it is convenient to chose as a random perturbation an infinite range Hamiltonian, e.g.

$$H_R = \sum_{i,k,l} J_{i,k,l} \sigma_i \sigma_k \sigma_l \tag{61}$$

where the sum runs over all the N points of the system and the J's are random variables with variance $1/N$.

In the nutshell stochastic stability tell us that the Hamiltonian H does not have any special features and that it properties are quite similar to those of similar random systems.

Although it seems quite natural, stochastic stability has quite deep consequences. For example we could consider a system in which there are many equilibrium states, labeled by α, and the overlaps among two generic configurations of the same state and of two different states are respectively q_M and q_m, the free energies of the different states are uncorrelated... The situation would be quite similar to the one described by one step replica symmetry breaking. However we may not specify the form of the probability distribution of the free energies which is characterized by a function $\mathcal{P}(f)$ which a priori may have an arbitrary shape.

It is a simple computation to verify that stochastic stability implies that the probability distribution of the free energies ($\mathcal{P}(f)$) must have the form given in eq. (38) with an appropriate choice of m. The most dramatic effect of stochastic stability is to link the behaviour of the function $\mathcal{P}(f)$ in the region of large f (where a large number of states do contribute) to the low f behaviour, which controls the distribution of the states which are dominant in the partition function.

We have seen that stochastic stability strongly constraints the properties of the systems and many of the qualitative results of the replica approach can be derived as mere consequences of stochastic stability. Stochastic stability apparently does not imply ultrametricity, which seems to be an independent property [37]. This independence problem is still open as far as the only explicitly constructed probabilities distribution of the free energies of the states are ultrametric.

7.3. NUMERICAL SIMULATIONS

Let us consider the case of spin glasses at zero magnetic field (in this case the replica symmetry is fully broken [40]). The natural variable to consider is a single spin ($A = \sigma_i$)). In this case the correlation $C(t, t_w)$ is equal to the overlap among two configurations at time t and t_w:

$$C(t, t_w) = \frac{\sum_{i=N} \sigma_i(t)\sigma_i(t_w)}{N}. \tag{62}$$

The response function is just the magnetization in presence of an infinitesimal magnetic field. In this case the situation is quite good because there are reliable simulations for the system at equilibrium [40].

In fig. 3 (taken from [22]) we plot the prediction for the function R versus C, obtained at equilibrium (i.e. using the equilibrium probability distribution of the overlaps, $P(q)$) by means of a simulation of a 16^3 lattice using parallel tempering [39, 40]. The simulation involves the study of 900 samples of a $L = 16$ lattice.

During the off-equilibrium simulations [22] in a first run without magnetic field the autocorrelation function has been computed. In a second second run from $t = 0$ until $t = t_w$ the magnetic field is zero and then (for $t \geq t_w$) there is an uniform magnetic field of small strength h_0. The starting configurations were always chosen at random (i.e. the system is suddenly quenched from $T = \infty$ to the simulation temperature T).

In fig. 3 there are the results of the off-equilibrium simulations [22] where $t_w = 10^5$ and $t_w = 10^4$, with a maximum time of $5 \cdot 10^6$ Monte Carlo sweeps. The lattice size in was 64, and $T = 0.7$ (well inside the spin glass phase, the critical temperature is close to 1.0). We plot the response function R times T (in this case R is equal to m/h_0) against $C(t, t_w)$. We have plotted

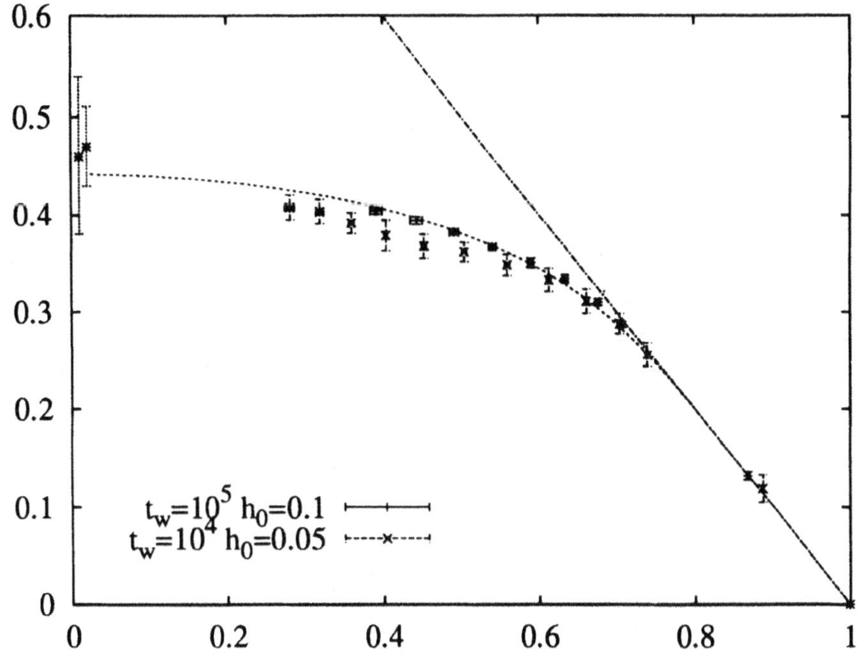

Figure 3. **The response R times T versus C at $T = 0.7$ for the three dimensional Ising spin glass [10]. The curve is the prediction for the function $R(C)$ obtained from the equilibrium data. The straight line is the FDT prediction. We have plotted the data of the two runs: $t_w = 10^5$, and $t_w = 10^4$.**

also a straight line with slope -1 in order to control where the FDT is satisfied. Finally we have plotted two points, in the left of the figure, that are obtained with the infinite time extrapolation of the magnetization.

The agreement among the absolute theoretical predictions (no free parameters) coming from the statics and the dynamical numerical data is quite remarkable. These data show the correctness of the identification of the functions x of the statics and X of the dynamics.

Let us now go to the case of glass forming materials. I will present the data for binary mixture of soft spheres [41]. Theoretically there have been many speculations whether the glass transition is described by one step replica symmetry breaking [1, 27, 32, 3]. Here the equilibrium properties are no so well known as in spin glasses, although there are some evidence that the homologous of the function $P(q)$ is non trivial [26]. On the other side, as we shall see, off-equilibrium simulations [42] show that the function $X(C)$ seems to be given by the one step formula (59) with an approximate linear dependence of m on the temperature.

We consider a mixture of soft particles of different sizes. Half of the particles are of type A, half of type B and the interaction among the particles is given by the Hamiltonian:

$$H = \sum_{i<k} \left(\frac{(\sigma(i) + \sigma(k))}{|\mathbf{x}_i - \mathbf{x}_k|} \right)^{12}, \tag{63}$$

where the radius (σ) depends on the type of particles. This model has been carefully studied in the past [41, 42]. The choice $\sigma_B/\sigma_A = 1.2$ strongly inhibits crystallisation and the system goes into a glassy phase when it is cooled. Using the same conventions of the previous investigators we consider particles of average radius 1 at unit density. It is usual to introduce the quantity $\Gamma \equiv \beta^4$. For quenching from $T = \infty$ the glass transition is known to happen around $\Gamma_c = 1.45$ [41].

The best quantity we can measure to evidenziate off-equilibrium effects is the diffusion of the particles:

$$\Delta(t, t_w) \equiv \frac{\sum_{i=1,N} \langle |\mathbf{x}_i(t_w) - \mathbf{x}_i(t_w + t)|^2 \rangle}{N}. \tag{64}$$

The usual diffusion constant is given by $D = \lim_{t\to\infty} \Delta(t, t_w)/t$.

The other quantity we measure is the response to a force. At time t_w we add to the Hamiltonian the term $\epsilon\ \mathbf{f} \cdot \mathbf{x}_k$, where f is vector of squared length equal to $d = 3$ and we measure the response

$$R(t_w, t) = \left. \frac{\partial \langle \mathbf{f} \cdot \mathbf{x}_k(t_w + t) \rangle_\epsilon}{\partial \epsilon} \right|_{\epsilon=0} \approx \frac{\langle \mathbf{f} \cdot \mathbf{x}_k(t_w + t) \rangle_\epsilon}{\epsilon} \tag{65}$$

for sufficiently small ϵ. The usual fluctuation theorem tells that at equilibrium $\beta \Delta^{eq}(t) = R^{eq}(t)$.

In the following we will look for the validity in the low temperature region of the generalized relation $\beta X(\Delta) = \partial R/\partial \Delta$. This relation (with $X \neq 1$) can be valid only in the region where the diffusion constant D is equal to zero. Strictly speaking also in the glassy region $D \neq 0$, because diffusion may always happen by interchanging two nearby particles (D is different from zero also in a crystal); however if the times are not too large the value of D is so small in the glassy phase that this process may be neglected in a first approximation.

The simulations we present are done using a Monte Carlo algorithm, which is a discretized form of a Langevin dynamics. In fig. 4 we show R versus $\beta\Delta$ at $t_w = 2048$ and $t_w = 8192$ for $\Gamma = 1.6$ and $t \leq 4t_w$ at $N = 66$. We also show the data for $t_w = 2048$ at $N = 130$. We do not observe any significant systematic shift in this plot among three data sets. We distinguish two linear regions with different slope as expected from one

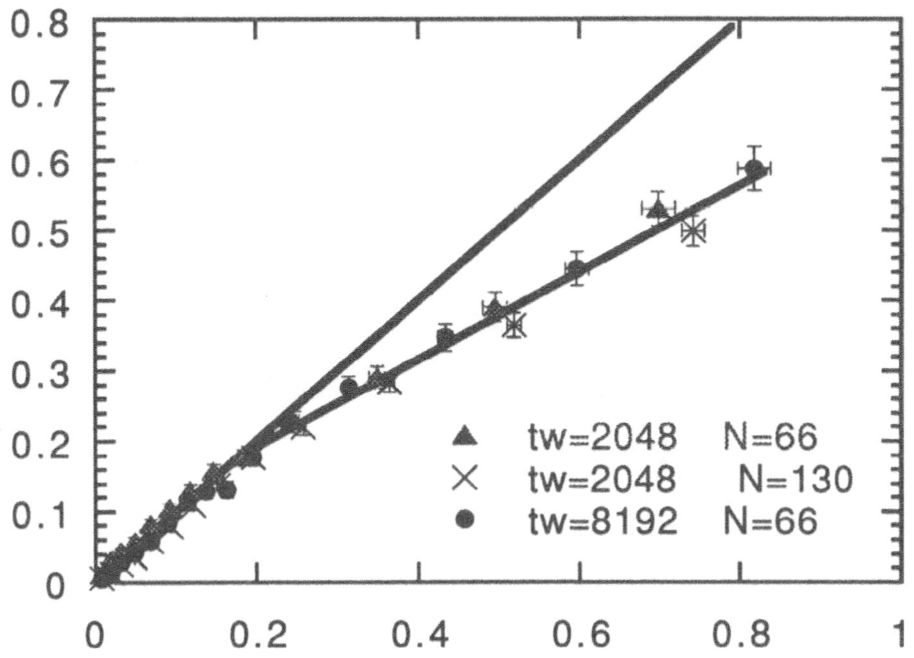

Figure 4. R versus $\beta\Delta$ at $\Gamma = 1.6$ for $t_w = 8192$ and $t_w = 2048$ at $N = 66$ and for $t_w = 2048$ at $N = 130$. The two straight lines have slope 1 and .62, respectively.

step replica symmetry breaking. The slope in the first region is compatible with 1, as expected from the FDT theorem, while the slope in the second region is near 0.62. Also the data at different temperatures for all values of $\Gamma \geq 1.5$ show a similar behaviour. The value of R, in the region where the FDT relation does not hold, can be very well fitted by a linear function of Δ as can be seen in fig. 4. The region where a linear fit (with $m < 1$) is quite good corresponds to $t/t_w > 0.2$.

The fitted value of $m \equiv \partial R/\partial(\beta\Delta)$ is displayed in fig. 5. When m becomes equal to 1, the fluctuation-dissipation theorem holds in the whole region and this is what happens at higher temperatures. The straight line is the prediction of the approximation $m(T) = T/T_c$, using $\Gamma_c = 1.45$.

All the results are in very good agreement with the theoretical expectations based on our knowledge extracted from the mean field theory for generalized spin glass models. The approximation $m(T) = T/T_c$ seems to work with an embarrassing precision. We can conclude that the ideas developed for generalized spin glasses have a much wider range of application than the models from which they have been extracted. It is likely that they reflect quite general properties of the phase space and therefore they can be applied in cases which are very different from the original ones.

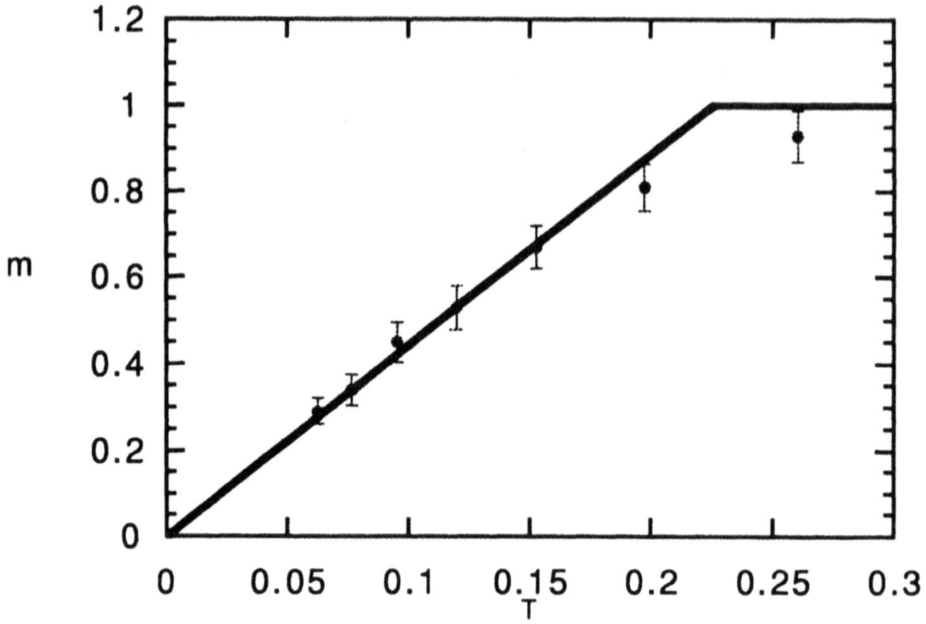

Figure 5. The quantity $m \equiv \frac{\partial R}{\partial \beta \Delta}$ for $t_w = 2048$ as a function of the temperature. The straight line is the prediction of the approximation $m(T) = T/T_c$.

7.4. SOME SUGGESTIONS FOR PLANNING AN EXPERIMENT

The most interesting development would be to measure experimentally the function X both in spin glasses and in structural glasses. Clearly the most difficult task is the measurement of the fluctuations. In spin glasses it is clear how it should be done: the measurement of the thermal fluctuations of the magnetization is a delicate, but feasible experiment. In the case of structural glasses some ingenuity is needed in planning the experiments. (An open interesting possibility would be to do the measurements in the case of rubber, where a transition with similar characteristics should take place.)

References

1. T.R. Kirkpatrick and D. Thirumalai, Phys. Rev. Lett. **58**, 2091 (1987); T.R. Kirkpatrick and D. Thirumalai, Phys. Rev. **B36**, 5388 (1987); T.R. Kirkpatrick, D. Thirumalai and P.G. Wolynes, Phys. Rev. **A40**, 1045 (1989).
2. L.Cugliandolo, J.Kurchan, Phys. Rev. Lett.**71** (1993) 173.
3. S.Franz and G. Parisi, Phys. Rev. Letters **79**, 2486 (1997) and *Effective potential in glassy systems: theory and simulations*, cond-mat/9711215.
4. For a review see also G. Parisi, *Proceedings of the ACS meeting*, Orlando (1996), cond-mat/9701068, *Lecture given at the Sitges conference, June 1996*

cond-mat/9701034 and *Lectures given at the Varenna summer school* 1996, cond-mat/9705312.
5. J. Kurchan, G. Parisi, and M. A. Virasoro, J. Phys. I France **3**, 1819 (1993).
6. A. Crisanti and H.-J. Sommers, J. Phys. I (France) **5**, 805 (1995).
7. S. Franz, G. Parisi, J. Physique I **5** (1995) 1401.
8. E. Monasson Phys. Rev. Lett. **75** (1995) 2847.
9. M.Mézard, G.Parisi and M.A.Virasoro, *Spin glass theory and beyond*, World Scientific (Singapore 1987).
10. G.Parisi, *Field Theory, Disorder and Simulations*, World Scientific, (Singapore 1992).
11. For a careful analysis of the free energy landscape see A. Cavagna, I. Giardina and G. Parisi, J. Phys. A: Math. Gen. 1997, **30**, 7021 and references therein.
12. A. Barrat, R. Burioni, and M. Mézard, J. Phys. A **29**, L81 (1996).
13. H. Vogel, Phys. Z, **22**, 645 (1921); G.S. Fulcher, J. Am. Ceram. Soc., **6**, 339 (1925).
14. G. Parisi, in *The Oskar Klein Centenary*, ed. by U. Lindström, World Scientific, (1995), Il nuovo cimento **16**, 939 (1994).
15. J.Kurchan and L. Laloux, J. Phys. A **29**, 1929 (1996).
16. For review see, W. Gotze, *Liquid, freezing and the Glass transition*, Les Houches (1989), J. P. Hansen, D. Levesque, J. Zinn-Justin editors, North Holland; C.A. Angell, Science, **267**, 1924 (1995)
17. J.-P. Bouchaud, L. Cugliandolo, J. Kurchan., M Mézard, Physica A **226**, 243 (1996).
18. J.-P. Bouchaud, J. Phys. France **2** 1705, (1992).
19. L. C. E. Struik; *Physical aging in amorphous polymers and other materials* (Elsevier, Houston 1978).
20. S. Franz and M. Mézard Europhys. Lett. **26**, 209 (1994).
21. S. Franz and H. Rieger Phys. J. Stat. Phys. **79** 749 (1995).
22. E. Marinari, G. Parisi, F. Ricci-Tersenghi, J.J. Ruiz-Lorenzo *Violation of the Fluctuation Dissipation Theorem in Finite Dimensional Spin Glasses*, cond-mat /9710120.
23. B.Derrida, Phys. Rev. **B24** (1981) 2613.
24. D. J. Gross and M. Mezard, *The Simplest Spin Glass*, Nucl. Phys. **B240** (1984) 431.
25. E. Gardner, Nucl. Phys. **B257** (1985) 747.
26. B. Coluzzi and G. Parisi *On the Approach to the Equilibrium and the Equilibrium Properties of a Glass-Forming Model*, cond-mat/9712261.
27. G. Parisi, *Gauge theories, spin glasses and real glasses*, cond-mat/9411115; *Slow Dynamics in Glasses*, cond-mat/9412034; *On the mean field approach to glassy systems*, cond-mat/9701034; *On the replica approach to glasses*, cond-mat/9701068; *Slow dynamics of glassy systems*, cond-mat/9705312; *New ideas in glass transitions*, cond-mat/9712079.
28. S.Franz and G.Parisi, J. Phys. **I** (France) **5**(1995) 1401.
29. G. Parisi and M. Potters, J. Phys. A, Math. Gen. **28** (1995) 5267, Europhys. Lett. **32** (1995) 13.
30. E.Marinari, G.Parisi and F.Ritort, J.Phys.A (Math.Gen.) **27** (1994), 7615; J.Phys.A (Math.Gen.) **27** (1994), 7647.
31. S. Franz, J.Hertz, Phys. Rev. Lett. **74**, (1995) 2114.
32. M. Mezard, G. Parisi, J. Phys. A **29**, (1996) 6515.
33. G. Parisi and M. A. Virasoro, J. Physique **50**, (1989) 3317.
34. M. Cardenas, S. Franz, G. Parisi, preprint cond-mat/9712099, submitted to J. Phys. A Lett. .
35. F. Guerra, Int. J. Phys. B, **10**, 1675 (1997).
36. M. Aizenman and P. Contucci, *On the stability of the quenched state in mean field spin glass models*, cond-mat 9712129.
37. G. Parisi, *On the probabilistic formulation of the replica approach to spin glasses*, cond-mat/9801081.
38. S. Franz, M. Mézard, G. Parisi and L. Peliti, *Measuring equilibrium properties in*

aging systems cond-mat/9803108.
39. K. Hukushima and K. Nemoto, J. Phys. Soc. Japan, **65**, 1604 (1996).
40. E. Marinari, G. Parisi and J. J. Ruiz-Lorenzo. *Numerical Simulations of Spin Glass Systems* in *Spin Glasses and Random Fields*, edited by P. Young. Word Scientific (Singapore 1997).
41. B.Bernu, J.P. Hansen, Y. Hitawari and G. Pastore, Phys. Rev. **A36** 4891 (1987), J.L. Barrat, J-N. Roux and J.P. Hansen, Chem. Phys. **149**, 197 (1990),J.P. Hansen and S. Yip, Trans. Theory and Stat. Phys. **24**, 1149 (1995).
42. G. Parisi, Phys. Rev. Lett. **79**, 3660 (1997).

NEMATIC TO SMECTIC: A "HARD" TRANSITION

Y. MARTÍNEZ-RATÓN AND J. A. CUESTA
Grupo Interdisciplinar de Sistemas Complicados (GISC), Departamento de Matemáticas, Universidad Carlos III de Madrid, c/ Butarque, 15, 28911-Leganés, Madrid, SPAIN

R. VAN ROIJ
H. H. Wills Physics Laboratory, Bristol University, Tyndall Avenue, Bristol BS8 1TL, UNITED KINGDOM

AND

B. MULDER
FOM Institute for Atomic and Molecular Physics, Kruislaan 407, 1098 SJ Amsterdam, THE NETHERLANDS

1. Introduction

The history of the Nematic-Smectic (N-S) transition in hard rod models is a colourful one. Until the mid 1980's, it was still widely believed that attractive forces where necessary to stabilize the smectic phase, in spite of preliminary theoretical work by Hosino, Nakano and Kimura [1]. Only after MC simulations by Stroobants, Lekkerkerker and Frenkel [2] was the occurrence of a Smectic phase purely on entropic grounds accepted. Subsequent density functional theories by several groups (Somoza and Tarazona [3], Poniewierski and Hołyst [4], Poniewierski and Sluckin [5]) mapped out the phase diagram of the hard rod system, predicting among others the existence of a tricritical point on the N-S transition line as a function of the aspect ratio L/D of the rods. Intuitively one expects the transition to be continuous for large aspect ratios. In that case the nematic order at the N-S transition is nearly saturated and the particles more or less fully aligned. For perfectly aligned rods a rather general Landau-type symmetry argument then predicts that the transition to the smectic phase should be second order. However, Poniewierski's later work [6] on the hard rod model in the limit $L/D \to \infty$ which included the contribution from the third virial coefficient, quite surprisingly predicted a first order transition even in this

C. Caccamo et al. (eds.), New Approaches to Problems in Liquid State Theory, 139–150.

limit. In a seemingly independent development van Roij et al. [7] showed that the N-S transition is, at least for moderately elongated rods, strongly influenced by the coupling between orientational and positional order, an aspect which the older Density Functional Theory (DFT) work had neglected. In order to disentangle the various factors at play here, we have studied the Zwanzig model of rectangular rods with restricted orientations for moderately elongated rods, a regime in which the discrete nature of the orientations is not yet too dominant. Employing bifurcation analysis we are able to fully characterize the nature of the transition. We have applied this technique to the "worst" and the "best" density functionals around: the Onsager functional and one based on Fundamental Measure Theory (FMT) respectively, enquiring separately into the role that orientation-position coupling plays. The results are: (i) the tricritical point predicted by the older theories is most probably an artifact of the decoupling of orientational and positional order and (ii) the FMT functional predicts a first order transition throughout the relevant range of aspect ratios in contrast to the Onsager functional for which the transition is continuous. The whole story is an object lesson in the subtlety of hard particle systems, as well as a showcase for some otherwise highly successful techniques that have been developed to deal with them.

2. Generalities

2.1. THE FREE ENERGY FUNCTIONAL

The description of choice for systems exhibiting spontaneous order is the density functional formalism [8] (conventionally abbreviated as DFT, although the T for Theory is probably a misnomer), in which the intrinsic free energy $\mathcal{F}$ is given as a functional of the one particle distribution function $\rho(\boldsymbol{r}, \hat{\omega})$, where $\boldsymbol{r}$ and $\hat{\omega}$ denote the position of the center of mass and the orientation of a rod, respectively. The normalization of $\rho(\boldsymbol{r}, \hat{\omega})$ is given by

$$\int d\boldsymbol{r} d\hat{\omega}\, \rho(\boldsymbol{r}, \hat{\omega}) = N = \rho V \tag{1}$$

where N is the total number of particles, V the total volume and ρ without arguments is the number density. The general form of $\mathcal{F}$ reads:

$$\beta\mathcal{F}[\rho(\boldsymbol{r}, \hat{\omega})] = \int d\boldsymbol{r} d\hat{\omega}\, \rho(\boldsymbol{r}, \hat{\omega})[\log \rho(\boldsymbol{r}, \hat{\omega})\mathcal{V}_{\mathrm{T}} - 1] + \beta\mathcal{F}^{ex}[\rho(\boldsymbol{r}, \hat{\omega})], \tag{2}$$

with $\mathcal{V}_{\mathrm{T}}$ the thermal volume i.e. the product of the de Broglie wavelengths associated with the degrees of freedom of the particle. Specific theories are obtained by formulating approximations for the, otherwise unknown, excess free energy functional $\mathcal{F}^{ex}$. The equilibrium distribution function at a fixed number density is found by requiring that $\mathcal{F}$ is minimized with respect to $\rho(\boldsymbol{r},\hat{\omega})$ under the constraint that the number of particles is fixed, which requires the following stationarity relation to hold

$$\frac{\delta}{\delta\rho(\boldsymbol{r},\hat{\omega})}\left\{\mathcal{F}[\rho(\boldsymbol{r},\hat{\omega})]-\mu\int d\boldsymbol{r}'d\hat{\omega}'\,\rho(\boldsymbol{r}',\hat{\omega}')\right\}=0 \tag{3}$$

As we are here concerned with the nematic phase (a homogeneous fluid with uniaxial orientational order) and the smectic phase (a fluid with a one dimensional periodic density modulation and uniaxial orientational order around the wavevector of the density modulation) we can parametrize our distribution function as

$$\rho(\boldsymbol{r},\hat{\omega})=\rho\left(\psi_0(\hat{\omega})+\sum_{j=1}^{M}\Delta_j\psi_j(\hat{\omega})\cos(jqz)\right) \tag{4}$$

where $\int d\hat{\omega}\psi_k(\hat{\omega})=1$ and the z-axis is along the wavevector of the density modulation. In the past several groups have tried to avoid the complexity of this expansion by assuming $\psi_j=\psi_0$, so that

$$\rho_{\rm dec}(\boldsymbol{r},\hat{\omega})=\rho\psi_0(\hat{\omega})\left(1+\sum_{j=1}^{M}\Delta_j\cos(jqz)\right)\equiv\psi_0(\hat{\omega})\rho(z) \tag{5}$$

which decouples the orientational order from the positional order. As we will see this has rather dramatic consequences for the nature of the N-S transition.

2.2. LANDAU EXPANSION

A first impression of the phase behaviour is obtained by inserting the form Eq. (4) into the functional Eq. (2) and expanding in terms of powers of the amplitudes Δ_j, which yields:

$$\begin{aligned}\frac{\beta\mathcal{F}}{N}\equiv f &= f_0+\frac{1}{2}A_1\Delta_1^2+\frac{1}{2}A_2\Delta_2^2-\frac{1}{2}B\Delta_1^2\Delta_2+\\ &\quad \frac{1}{4!}C_1\Delta_1^4+\frac{1}{4!}C_2\Delta_2^4+\frac{1}{4}C_{12}\Delta_1^2\Delta_2^2+\cdots,\end{aligned} \tag{6}$$

This expansion is symmetry dictated i.e. any functional will lead to the same form. The expansion coefficients $A_\Box, B, C_\Box$, however, are functionals of $\psi_j(\hat{\omega})$ and functions of ρ and q that depend on the specific choice of the functional. f_0 denotes the free energy of the reference nematic phase. One justifies this expansion, because the smectic order parameters are expected to be small in the vicinity of the N-Sm transition. Stationarity of $\mathcal{F}$ w.r.t. Δ_2 then gives

$$\frac{\partial f}{\partial \Delta_2} = 0 \Rightarrow \Delta_2 = \frac{B}{2A_2}\Delta_1^2 + \cdots \tag{7}$$

Hence f is to sixth order in Δ_1 given by

$$f = f_0 + \frac{1}{2}A_1\Delta_1^2 + \frac{1}{4!}\widetilde{C}_1\Delta_1^4 + D_1\Delta_1^6, \tag{8}$$

where we have defined the renormalized quartic coefficient

$$\widetilde{C}_1 \equiv C_1 - \frac{3B^2}{A_2}. \tag{9}$$

Conventional Landau theory predicts a N-SmA phase transition if $A_1 = 0$. This transition is second order if $\widetilde{C}_1 > 0$, first order if $\widetilde{C}_1 < 0$, and tricritical if $\widetilde{C}_1 = 0$. In the latter cases we must require that $D_1 > 0$. This simple Landau-picture is, as we will show, slightly too naive, because it neglects the question of the mechanical stability of the smectic phase.

2.3. BIFURCATION ANALYSIS

Similar in spirit to the Landau analysis, but closer to the "bone" of the density functional is the bifurcation analysis of the stationarity equation Eq. (3). Here we perurbatively construct a new solution bifurcating from the parent phase, in this case the nematic. This is done by introducing an arbitrary expansion parameter ε, expanding the independent and dependent variables of the problem in powers of ε, inserting these expansions into Eq. (3), and solving the resultant set of equations order by order in ε. In our case we get the expansions

$$\rho = \rho^* + \rho^{(1)}\varepsilon + \rho^{(2)}\varepsilon^2 + \ldots \tag{10}$$

$$\psi_k(\hat{\omega}) = \psi_k^*(\hat{\omega}) + \psi_k^{(1)}(\hat{\omega})\varepsilon + \psi_k^{(2)}(\hat{\omega})\varepsilon^2 + \ldots \qquad k = 0, 1, 2, \ldots \tag{11}$$

$$\Delta_j = \Delta_j^{(1)}\varepsilon + \Delta_j^{(2)}\varepsilon^2 + \ldots \qquad j = 1, 2, \ldots \tag{12}$$

$$q = q^* + q^{(1)}\varepsilon + q^{(2)}\varepsilon^2 + \ldots \tag{13}$$

The complexity of the case with freely rotating particles, however, precludes the actual program to be carried through using reasonable means. In the next section we therefore turn to a model, which keeps as much of the relevant physics as possible, but is sufficiently simple to allow the full bifurcation analysis to be carried out.

3. The Zwanzig model

3.1. DEFINITION

The Zwanzig model [9] consists of rectangular hard rods with dimensions $L \times \sigma \times \sigma$, hence aspect ratio $\lambda = L/\sigma$, which are free to move in space but restricted to point along three mutually orthogonal directions, which define the axes of our reference frame. For intermediate aspect ratios this model is expected to show both a nematic and a smectic phase. For very high aspect ratios the discreteness of the orientations will ruin the game as the degree of order of the nematic phase will be so high that there are effectively no "transverse" particles. In that case a simple spatial rescaling of the system along the nematic ordering axis will turn it into a system of parallel hard cubes. The system of parallelhard cubes obviously does not posses a liquid crystalline phase, but as Kirkpatrick first showed undergoes a continuous transition to a solid phase [18].

We will denote the orientation of a Zwanzig rod by $\mu = x, y, z$. The diameter $\sigma_{\mu i}$ of a rod in orientation μ along the i'th coordinate direction, can be read off from the matrix

$$\boldsymbol{\sigma} = \begin{pmatrix} L & \sigma & \sigma \\ \sigma & L & \sigma \\ \sigma & \sigma & L \end{pmatrix} \tag{14}$$

Normalized distributions of the discrete orientations are conveniently described by

$$\psi(\mu) = \frac{1}{3}\left(1 + 2s\cos\left(\frac{2\pi n_\mu}{3}\right)\right) \qquad n_x = 1, n_y = 2, n_z = 3 \tag{15}$$

which introduces the order parameter s, the discrete analogon of the Mayer-Saupe $S = \langle P_2 \rangle$ order parameter. An alternative point of view is to consider particles of different orientations as members of a different "species". The quantities $x_\mu = \rho_\mu/\rho$ can then be interpreted as the mole fraction of the species μ. The problem of finding the the equilibrium orientation distributions becomes equivalent to a semi-grand ensemble treatment [10], where the total number of particles is kept fixed and the chemical potential differences between the species are fixed to be zero.

3.2. TREATMENT

We will treat the Zwanzig model introduced above using two different density functional theories. First, the "good old" Onsager approach [11], formally only valid for infinitely elongated freely rotating particles in the isotropic phase, but nevertheless the work-horse of non-spherical hard-particle research. Both formally, and in a practical sense, it is to hard particles what mean-field theory is to thermally driven systems. Second, we apply Fundamental Measure Theory [12, 13], the "new kid on the block", which in the past couple of years has shown itself to provide a single functional able to deal with both the disordered and ordered phases of hard spheres, while its formulation seems rather firmly planted in general notions concerning the statistical mechanics of hard particles. For aligned rectangular particles its application turns out to be surprisingly simple [14, 16].

3.2.1. *The Onsager approximation*
The Onsager approximation consists of replacing the excess free energy functional by its lowest order contribution in a formal expansion in powers of the distribution functions.

$$\beta\mathcal{F}^{ex}_{\mathrm{ONS}}[\rho_\mu(\boldsymbol{r})] = -\frac{1}{2}\sum_{\mu,\mu'=\{x,y,z\}}\int d\boldsymbol{r}d\boldsymbol{r}'\rho_\mu(\boldsymbol{r})\rho_{\mu'}(\boldsymbol{r}')f_{\mu\mu'}(\boldsymbol{r}-\boldsymbol{r}') \tag{16}$$

The Mayer function f is simple for hard rectangular blocks and is just given by

$$f_{\mu\mu'}(\boldsymbol{r}) = -\theta\left(\frac{\sigma_{\mu x}+\sigma_{\mu' x}}{2}-|x|\right)\theta\left(\frac{\sigma_{\mu y}+\sigma_{\mu' y}}{2}-|y|\right)\theta\left(\frac{\sigma_{\mu z}+\sigma_{\mu' z}}{2}-|z|\right) \tag{17}$$

so that apart from the minus sign it is just a truth function for the question whether the two particles in question overlap or not.

3.2.2. *The Fundamental Measure Theory*
Unfortunately, this is not the place to delve deeply into the FMT. Let us suffice by giving the basic idea here. In FMT one assumes that the excess free energy is a local functional of a small set of weighted densities

$$\beta\mathcal{F}^{ex}_{\mathrm{FMT}}[\rho_\mu(\boldsymbol{r})] = \int d\boldsymbol{r}\Phi\left(\left\{n^{(\alpha)}(\boldsymbol{r})\right\}\right) \tag{18}$$

These weighted densities are defined through

$$n^{(\alpha)}(\boldsymbol{r}) = \sum_\mu\int d\boldsymbol{r}'\rho_\mu(\boldsymbol{r}')\omega^{(\alpha)}_\mu(\boldsymbol{r}-\boldsymbol{r}') \tag{19}$$

where the ω^α_μ are the "fundamental measures", which have either a scalar or a vector nature. The label α , which runs from 0 to D relates to the dimension of the measure through $\left[\omega^\alpha_\mu\right] = [\mathrm{Volume}]^{(\alpha-D)/D}$ where D is the dimensionality of our sytem. The measures are related to the geometrical properties of the particles involved, and one way of obtaining them is by expanding the Mayer function $f_{\mu\mu'}(\boldsymbol{r})$ (For the most general statement of this method see [15])

$$f_{\mu\mu'}(\boldsymbol{r}) = \sum_{\alpha+\gamma=D} \phi^{\alpha\gamma} \int d\boldsymbol{r}' \omega_{\mu}^{(\alpha)}(\boldsymbol{r}')\, \omega_{\mu'}^{(\gamma)}(\boldsymbol{r}-\boldsymbol{r}') \tag{20}$$

where the constraint $\alpha + \gamma = D$ serves to keep f dimensionless. If we Fourier transform Eq. (17) we see that the results factorise into functions relating each of the two particles separately, from which the measures can then be determined. We refer to the Appendix for the explicit form of the measures for Zwanzig rods. Recently some of us [16] have shown that the requirement of dimensional cross-over i.e. the fact that a D-dimensional functional should correctly reduce to a $D-1$-dimensional one when it is evaluated for a distribution which is confined to a $D-1$-dimensional subspace, is sufficient to deduce the FMT in any dimensionality from the 0-dimensional limit i.e. a system in a volume which allows at most a single particle inside it. The 0-dimensional functional is given by [17]

$$\begin{aligned} \beta\mathcal{F}_0^{ex}[\rho_\mu(\boldsymbol{r})] &= \int d\boldsymbol{r}\Phi_0(\eta(\boldsymbol{r})) \qquad (21)\\ &= \int d\boldsymbol{r}\ \{\eta(\boldsymbol{r}) + (1-\eta(\boldsymbol{r}))\log(1-\eta(\boldsymbol{r}))\} \\ \eta(\boldsymbol{r}) &= n^{(3)}(\boldsymbol{r}) = \sum_{\mu}\int d\boldsymbol{r}'\rho_\mu(\boldsymbol{r})\omega_\mu^{(3)}(\boldsymbol{r}-\boldsymbol{r}') \qquad (22) \end{aligned}$$

If we define the differential operator

$$D_i = \sum_{\mu} \frac{\partial}{\partial \sigma_{\mu i}} \qquad i = x, y, z \tag{23}$$

we get the surprising result

$$\begin{aligned} &\Phi\left(\left\{n^{(\alpha)}(\boldsymbol{r})\right\}\right) = D_x D_y D_z \Phi_0(\eta(\boldsymbol{r})) = \qquad (24)\\ &-n^{(0)}(\boldsymbol{r})\log\left(1-n^{(3)}(\boldsymbol{r})\right) + \frac{\boldsymbol{n}^{(1)}(\boldsymbol{r})\cdot\boldsymbol{n}^{(2)}(\boldsymbol{r})}{1-n^{(3)}(\boldsymbol{r})} + \frac{n_x^{(2)}(\boldsymbol{r})\, n_y^{(2)}(\boldsymbol{r})\, n_z^{(2)}(\boldsymbol{r})}{\left(1-n^{(3)}(\boldsymbol{r})\right)^2} \end{aligned}$$

4. Results

Having formulated the density functionals for our two target theories, we can perform the bifurcation analysis as outlined in Section 2.3. Here we skip the rather tedious details. The variables we expand in are $\{\rho, q, s_0, s_1, \ldots, \Delta_1, \Delta_2, \ldots\}$. The main actors in our play close to the bifurcation will be $\{\rho, s_0, \Delta_1\}$. The remaining variables we will generically denote by ξ. The relevant lowest order results are the values of the bifurcation density ρ^* and the nematic order parmeter at the bifurcation s_0^*, the bifurcating wavenumber q^* and

$$\Delta_1^{(1)} = 1 \tag{25}$$

$$s_0^{(1)} = \rho^{(1)} = \xi^{(1)} = 0 \tag{26}$$

In the next order calculation we can determine the value of $s_0^{(2)}$ and $\rho^{(2)} = b^*/6c^*$. These ingredients are all we need to perform the required stability analysis, and the relevant results can all be expressed in the quantities b^* and c^*, known functions of the coefficients $A_\square, B$ and $C_\square$ and their derivatives with respect to ρ and s_0 taken at the bifurcation point ρ^*. We obtain the free energy difference of the smectic branch and the nematic branch *at the same density* $\rho^* + \varepsilon^2\rho^{(2)}$ as given by

$$f_{Sm} - f_N = \delta f_0 + \frac{1}{2}A_1\varepsilon^2 + \frac{1}{4!}a^*\varepsilon^4 = -\frac{1}{4!}b^*\varepsilon^4 \tag{27}$$

where we have omitted higher order terms in ε. Here f_{Sm} and f_N denote the free energy of the smectic and nematic branch, respectively. Note that f_{Sm} and f_N are no longer free energy *functionals*, but the (local) minimum of the functional f. The smectic branch corresponds to a lower free energy than the nematic branch if $b^* > 0$. Hence one would expect a second order N-S phase transition if $b^* > 0$ and a first order transition if $b^* < 0$. The condition for tricriticality is then $b^* = 0$. However, we have as yet neglected to consider the mechanical stability of the smectic branch. This stability is governed by the isothermal compressibility κ_T, or equivalently its inverse defined through

$$\kappa_T^{-1} = -V\left(\frac{\partial P}{\partial V}\right)_{N,T} = \rho^2\left(\frac{\partial^2\phi}{\partial\rho^2}\right)_{V,T} \tag{28}$$

where P is the pressure and $\phi = \rho f$ is the Helmholtz free energy per unit volume. In the vicinity of the N-S bifurcation we have from

$$\varepsilon^2 = \frac{\rho - \rho^*}{\rho^{(2)}} = \frac{6c^*(\rho - \rho^*)}{b^*} \tag{29}$$

and Eq. (27) that

$$f_{Sm} = f_N - \frac{3}{2}\frac{(c^*)^2}{b^*}(\rho - \rho^*)^2, \tag{30}$$

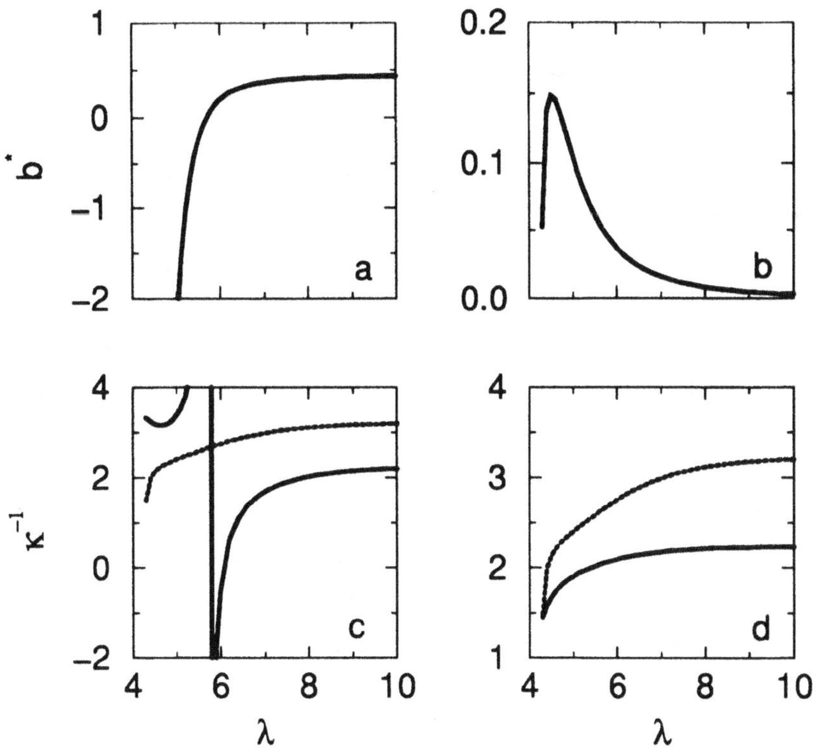

Figure 1. b^* and $\kappa_{T,Sm}^{-1}$ for the Onsager approximation. Dotted lines: inverse compressibility of the nematic at the transition. Left hand column: decoupling approximation. Right hand column: full theory

Inserting this into Eq. (28) yields the isothermal compressibility of the smectic phase $\kappa_{T,Sm}$ given by

$$\kappa_{T,Sm}^{-1} = \kappa_{T,N}^{-1} - \frac{3(c^*)^2}{b^*}(\rho^*)^3, \tag{31}$$

So we can have the situation that $b^* > 0$ but $\kappa_{T,Sm}^{-1} < 0$, which suppresses the continuous transition, turning into a first order one. This implies that the true criterion for tricriticality is $b^* > 0$ *and* $\kappa_{T,Sm}^{-1} = 0$. Of course when $b^* < 0$ the transition is first order regardless of the sign of $\kappa_{T,Sm}^{-1}$.

We have calculated b^* and $\kappa_{T,Sm}^{-1}$ both for the Onsager theory and the FMT. In both cases we have also considered the effect of the decoupling approximation Eq. (5) on the results. In Figure 1. we present the results for the Onsager approximation, while the result for the FMT are shown in Figure 2. From these results we must conclude that the decoupling approximation leads to the spurious prediction of a tricritical point ($\lambda_{tc} = 6.1$ for

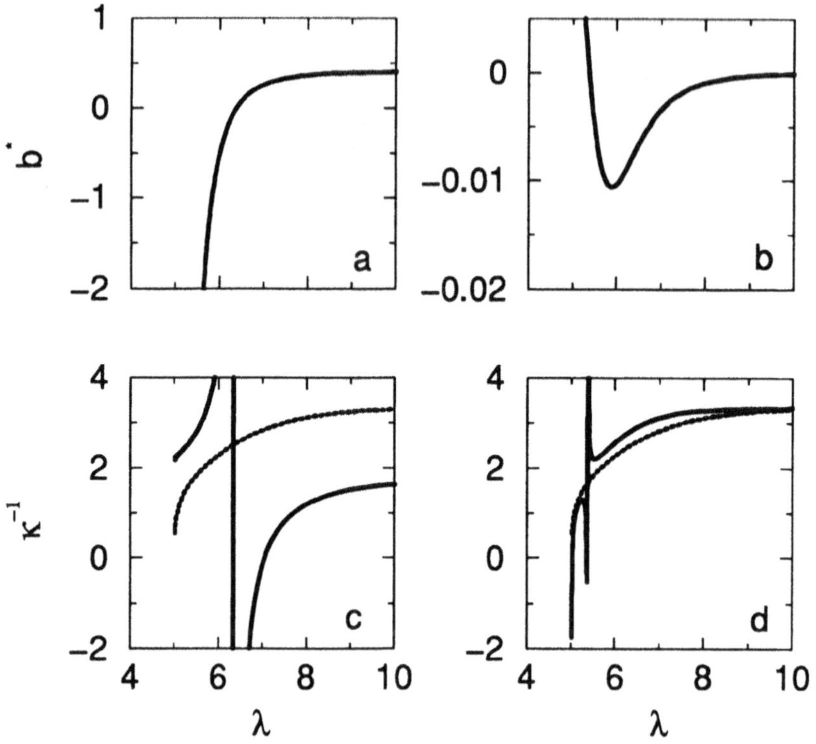

Figure 2. b^* and $\kappa_{T,Sm}^{-1}$ for the FMT. Dotted lines: inverse compressibility of the nematic at the transition. Left hand column: decoupling approximation. Right hand column: full theory

the Onsager theory and $\lambda_{tc} = 7.1$ for FMT) below which the transition is first order. Without this approximation the Onsager theory predicts a continuous transition throughout, whilst FMT predicts a first order transition throughout. We have to be a bit careful here because the N-S transition itself is bound to be pre-empted by the strong first order isotropic-(plastic) solid transition for smaller length-to-width ratios. Our current estimates put the value below which crystallization dominates at around $\lambda \sim 6$ for the FMT.

5. Conclusions

Our results show that the N-S transition in hard rods is indeed a "hard" one. First, a previously assumed to be innocuous approximation, the decoupling of orientational and positional order, leads to a tricritical point whose existence is most probably an artifact. Next, the order of the transition predicted by the Onsager theory differs from that of FMT. Of course, one might argue that the Onsager approximation is too naive. However, as far as the expected critical behaviour is concerned one would at first sight assume both theories to be on the same level: both are still essentially mean-field theories. Unfortunately, technical difficulties prevent us from performing the same in-depth analysis for the case where the particles are freely rotating. Nevertheless, we are confident that the FMT actually does capture the essential nature of the transition. This is corroborated by recent simulation results by Polson and Frenkel [19], who give evidence for a first order transition for rods in the limit $L/D \to \infty$. The remaining question is what actually causes the difference between FMT and the Onsager approximation. Poniewierski [6], who considered the problem for $L/D \to \infty$ in a theory including the third virial contribution to the the free energy, also predicted a first order transition throughout. This leads us to conjecture that it is indeed the influence of the third virial contribution, which by the way is treated exactly within the framework of the FMT, that triggers this remarkable change in behaviour.

Acknowledgements: The work of YM-R was supported by a travelling grant of the Universidad Carlos III de Madrid (Spain). JAC was supported by project no. PB96-0119 and the travelling grant PR95-558 of the Dirección General de Enseñanza Superior (Spain). They both want to thank AMOLF, Amsterdam for its hospitality. The work of RvR was supported by the European Community grant ERB4001GT963401. The work of BM is part of the research program of the "Stichting voor Fundamenteel Onderzoek der Materie (FOM)", which is financially supported by "Nederlandse Organisatie voor Wetenschappelijk Onderzoek (NWO)".

A. Fundamental measures of Zwanzig rods

Here we give the actual forms of the fundamental measures $\omega_\mu^{(\alpha)}$ for Zwanzig rods as encountered in the text.

$$\omega_\mu^{(0)}(\boldsymbol{r}) \equiv \frac{1}{8}\delta(\sigma_{\mu x}/2-|x|)\delta(\sigma_{\mu y}/2-|y|)\delta(\sigma_{\mu z}/2-|z|) \tag{32}$$

$$\begin{aligned}\boldsymbol{\omega}_\mu^{(1)}(\boldsymbol{r}) \equiv \frac{1}{4}\big(&\theta(\sigma_{\mu x}/2-|x|)\delta(\sigma_{\mu y}/2-|y|)\delta(\sigma_{\mu z}/2-|z|),\\ &\delta(\sigma_{\mu x}/2-|x|)\theta(\sigma_{\mu y}/2-|y|)\delta(\sigma_{\mu z}/2-|z|),\\ &\delta(\sigma_{\mu x}/2-|x|)\delta(\sigma_{\mu y}/2-|y|)\theta(\sigma_{\mu z}/2-|z|)\big)\end{aligned} \tag{33}$$

$$\begin{aligned}\boldsymbol{\omega}_\mu^{(2)}(\boldsymbol{r}) \equiv \frac{1}{2}\big(&\delta(\sigma_{\mu x}/2-|x|)\theta(\sigma_{\mu y}/2-|y|)\theta(\sigma_{\mu z}/2-|z|),\\ &\theta(\sigma_{\mu x}/2-|x|)\delta(\sigma_{\mu y}/2-|y|)\theta(\sigma_{\mu z}/2-|z|),\\ &\theta(\sigma_{\mu x}/2-|x|)\theta(\sigma_{\mu y}/2-|y|)\delta(\sigma_{\mu z}/2-|z|)\big)\end{aligned} \tag{34}$$

$$\omega_{\mu}^{(3)}(\boldsymbol{r}) \;=\; \theta\left(\frac{\sigma_{\mu x}}{2}-|x|\right)\theta\left(\frac{\sigma_{\mu y}}{2}-|y|\right)\theta\left(\frac{\sigma_{\mu z}}{2}-|z|\right) \qquad (35)$$

References

1. Hosino, M, Nakano, H. and Kimura, H. (1979) *J. Phys. Soc. Jap.* **46**, 1709
2. Stroobants, A., Lekkerkerker, H. N. W. and Frenkel, D. (1986) *Phys. Rev. Lett.* **86**, 1452
3. Somoza, A. M. and Tarazona, P. (1990) *Phys. Rev. A.* **41**, 965
4. Poniewierski, A. and Hołyst, R. (1990) *Phys. Rev. A.* **41**, 6871
5. Poniewierski, A. and Sluckin, T. J. (1991) *Phys. Rev. A.* **43**, 6837
6. Poniewierski, A. (1992) *Phys. Rev. A.* **45**, 5605
7. van Roij, R., Bolhuis, P., Mulder, B. and Frenkel, D. (1995) *Phys. Rev. E.* **52**, R1277
8. Evans, R. (1997) *Adv. Phys.* **28**, 143
9. Zwanzig, R. (1963) *J. Chem. Phys.* **39**, 1714
10. Kofke, D. A. and Glandt, E. D. (1987) *J. Chem. Phys.* **87**, 4881
11. Onsager, L. (1949) *Ann. N. Y. Acad. Sci.* **51**, 627
12. Rosenfeld, Y. (1988) *J. Chem. Phys.* **89**, 4272
13. Rosenfeld, Y. (1989) *Phys. Rev. Lett.* **63**, 980
14. Cuesta, J. A. (1996) *Phys. Rev. Lett.* **76**, 3742
15. Wertheim, M. S. (1994) *Mol. Phys.* **83**, 519
16. Cuesta, J. A. and Martinez-Raton, Y (1997) *Phys. Rev. Lett.* **78**, 3681; *ibid.* (1997) *J. Chem. Phys.* **107**, 6379
17. Rosenfeld, Y. , Schmidt, M.,Löwen, H. and Tarazona, P. (1996) *J. Phys. C : Cond. Matt.* **8**, L577
18. Kirkpatrick, T. R. (1986) *J. Chem. Phys.* **85**, 3515
19. Polson, J. and Frenkel, D. (1997) *Phys. Rev. E* **56**, R6260

INTERFACES AND CONFINED FLUIDS

FLUIDS IN MODEL PORES OR CAVITIES: THE INFLUENCE OF CONFINEMENT ON STRUCTURE AND PHASE BEHAVIOUR

R. EVANS
University of Bristol
H.H. Wills Physics Laboratory, Bristol BS8 1TL, U.K.

Abstract

When fluids are confined in narrow pores or capillaries, or in cavities their properties are altered drastically from those in bulk. The effects of wall-fluid forces and of finite-size modify the static structure, the dynamical properties and the phase behaviour of the fluid. These lecture notes summarise some recent investigations of confined fluids. The first lecture describes the use of density functional theory (DFT) and Monte Carlo simulation for determining the equilibrium structure (density profiles) of fluids confined in a small spherical cavity, where packing constraints lead to extreme inhomogeneities and where the results depend on the choice of the ensemble, i.e. whether the system is open (grand canonical) or closed (canonical). The second lecture focuses on simulation studies of the condensation and freezing transitions for simple (Lennard-Jones like) fluids confined in an ideal slit pore and describes the extent to which the location of the phase boundaries depends on the choice of wall-fluid potentials and on the size of the pore. The possible relevance of such studies of idealised model pores for understanding the phase behaviour of fluids confined in real mesoporous solids such as Vycor glass is discussed briefly.

1. DFT for Small Systems: Hard Spheres in a Spherical Cavity

1.1 INTRODUCTION

Understanding the behaviour of fluids inside micropores, i.e. pores in which the confining dimension is of the order of a nanometer or so, constitutes a fundamental problem in surface physics and chemistry [1,2]. Zeolites, clathrates, fullerenes and porous glasses are examples of materials containing micropores or cavities. Here we consider a very simple model for a fluid in a single micropore, namely the hard-sphere fluid confined in a hard spherical cavity. Despite its appealing simplicity, there have been relatively few studies of fluids in spherical cavities [3-6], most investigators preferring slit and cylindrical capillaries [7]. Earlier work for small cavities includes virial expansions of the grand potential [3,4], integral equation theories [5], and computer simulation [6]. Direct calculation of the partition functions can be made for cavities containing up to about six atoms [4]. One of the aims of the present study was to develop

C. Caccamo et al. (eds.), New Approaches to Problems in Liquid State Theory, 153–172.

a theoretical framework which could be used to determine the equilibrium properties for cavities which might contain large and small numbers of particles.

In most studies of adsorption it is natural to consider the fluid to be in contact with a reservoir so that the chemical potential μ and the temperature T are fixed and the grand canonical ensemble is the appropriate one. However, one is usually dealing with infinite substrate area and infinite available volume for the adsorbed fluid so results obtained from the ensembles are equivalent. By contrast, a fluid inside a single spherical cavity constitutes a finite system and in this case the statistical mechanical ensembles are not equivalent. If the spherical pore is closed, so that it cannot exchange particles with the external reservoir, then the number of particles N is a fixed integer and the canonical ensemble is appropriate. However, one can also allow exchange of particles with the reservoir and treat the system as open (this is equivalent to having an ensemble of independent cavities with an average number of particles $\langle N \rangle$ determined by μ). When N is very large, properties calculated for the closed pore should be close to those for the open pore but for small numbers of particles this is no longer the case and the choice of ensemble becomes significant. In the open cavity the structure of the confined fluid can be obtained conveniently by grand canonical ensemble Monte Carlo simulation (GCEMC) [2] or by DFT. For the closed cavity one can resort to molecular dynamics or canonical ensemble Monte Carlo (CEMC). However, there is, as yet, no canonical classical DFT approach which works with a fixed N constraint. Recall that classical DFT is formulated in terms of a minimum grand-potential principle with μ, T fixed by the reservoir [8]. If one wishes to obtain canonical results from DFT then one requires some procedure in addition to the standard (grand canonical) approach. Relating properties obtained in one ensemble to those obtained from another ensemble is a general problem in the statistical mechanics of small systems – another important example concerns the stability of small droplets or bubbles [9].

1.2 THE $1/\langle N \rangle$ EXPANSION

The procedure we have adopted consists of expanding the canonical one-body density profile (or other distribution functions) in powers of $1/\langle N \rangle$. The expansion uses as input grand canonical quantities which can be obtained from either DFT or GCEMC. Our procedure is based on an early analysis [10] for the canonical radial distribution function $g_N(r)$ of a bulk fluid in terms of its grand canonical counterpart – see also [11,12]. One can show [13] that the canonical density profile $\rho_{\langle N \rangle}(r)$ for a system with a fixed number of particles $N = \langle N \rangle$ is given by

$$\rho_{\langle N \rangle}(r) = \rho(\mathbf{r}) + f_1(\mathbf{r} < N >) + f_2(\mathbf{r}, < N >) + \mathrm{O}\left(\frac{1}{< N >^3}\right) \tag{1}$$

where $\rho(\mathbf{r})$ is the grand canonical profile,

$$f_1(\mathbf{r}, < N >= -\frac{1}{2} < (N - < N >)^2 \frac{\partial^2}{\partial < N >^2} \rho(\mathbf{r}) \tag{2}$$

and

$$f_2(\mathbf{r},<N>) = -\frac{1}{2}<(N-<N>)^2>\frac{\partial^2}{\partial<N>^2}f_1(\mathbf{r},<N>)$$
$$-\frac{1}{6}<(N-<N>)^3>\frac{\partial^3}{\partial<N>^3}\rho(\mathbf{r}) \quad (3)$$

are corrections of order $1/<N>$ and $1/<N>^2$, respectively. < > denotes the grand canonical ensemble average. Note that the integral of each correction term over the total volume must vanish [13]. Also, f_1 and f_2 can be further transformed depending on the procedure for obtaining the grand canonical ensemble averages. In particular, if one uses grand canonical ensemble DFT it is useful to express the fluctuations $<(N-<N>)^2>$ and $<(N-<N>)^3>$ in terms of derivatives of $<N>$ w.r.t. μ which can be performed numerically. Conversely, if one performs GCEMC simulations it is more convenient to express $f_1(\mathbf{r},<N>)$ and $f_2(\mathbf{r},<N>)$ in terms of fluctuations which can be readily measured in a simulation [13].

1.3 CHOICE OF DFT

Density functional theories have become a popular tool for investigating the structure and thermodynamic properties of inhomogeneous fluids and have provided a microscopic framework for understanding a variety of interfacial phenomena [14-16]. They have been applied to several different types of spatial inhomogeneity and have been successful in describing fluids confined in slits or in narrow cylindrical pores [7,14,15]. DFT is based on the idea that for a given interatomic potential the intrinsic Helmholtz free-energy F is a unique functional of the one-body density profile $\rho(\mathbf{r})$ [8]. For a fluid at fixed μ, T, subject to an external potential $V_{ext}(\mathbf{r})$, this density profile minimises the grand potential functional

$$\Omega[\rho] = F[\rho] + \int d\mathbf{r}\rho(\mathbf{r})[V_{ext}(\mathbf{r}) - \mu] \quad (4)$$

The exact form of $F[\rho]$ is, of course, not known and approximations must be used. For the particular case of the hard-sphere fluid there is a plethora of prescriptions [15] but we have chosen to employ the Fundamental Measure Theory (FMT) of Rosenfeld [17] which has proved successful [18] for several highly inhomogeneous situations. Moreover, recent refinements [19,20] of the FMT are specifically designed to account for the correct dimensional crossover from the bulk to situations where the effective dimensionality D of the fluid is reduced. These occur, for example, when a fluid is confined between parallel plates that are separated by a distance which permits only a single layer of particles so that D is reduced from 3 to 2. For a fluid confined in a narrow cylinder that allows only a single line of particles, D is reduced from 3 to 1. In the most extreme case one considers a very small spherical cavity which cannot hold more than one particle and refers to this as the zero dimension (0D) cavity [19,20]. Ascertaining the nature of dimensional crossover has already provided some important new insight into the analytical structure of free-energy functionals [19-22] and promises to yield much more [23]. In the present study we encounter quasi 0D situations where, as a result of packing constraints, the density profile develops a very pronounced peak in the centre of the cavity. Accounting accurately for the peak and its dependence on <N> provides a searching test of any DFT approximation for hard spheres. In our calculations [13] we have employed different prescriptions for the third contribution to the excess free energy density (denoted Φ_3) which enters FMT. These are ROS, the original Rosenfeld version [17], RSLT1 which uses the antisymmetrised form $\Phi_{3,\mathrm{asym}(q)}$ (with q = 2,3) in [19] and RSLT2 which uses the interpolated form $\Phi_{3,\mathrm{int}}$, also given in [19]. The remaining contributions Φ_1 and Φ_2 to the excess free energy density are the same as those in the original version.

The DFT calculations were performed by varying the chemical potential slowly in an iterative solution of the Euler-Lagrange equation resulting from minimising (4), so that the final equilibrium density profile integrates to the required average number of particles [13]. Such a procedure is equivalent to knowing in advance the correct μ_N that would give rise to the required average number $\langle N\rangle = N$. It is important to recognise that this constrained minimisation does *not* reduce the grand canonical DFT to a canonical one. Fixing $\langle N\rangle = N$ with N integer is simply equivalent to fixing the chemical potential μ_N; the resulting averages still refer to the grand canonical ensemble.

1.4 COMPARISON OF DFT RESULTS WITH THOSE OF SIMULATION

Both CEMC and GCEMC simulations were carried out for a range of cavity radii R_{cav} and particle numbers. Details can be found in [13]. The results from GCEMC can be compared directly with those from DFT. At low packings the different FMT prescriptions yield similar results and the density profiles $\rho(r)$ are close to those from simulation. For very small cavities $R_{\mathrm{cav}} \leq 1.5\sigma$, where σ is the hard-sphere diameter, there is a single shell of particles located near the cavity wall, giving high contact values of the local density, while the density in the centre is very small. Although the fluid is very inhomogeneous all the FMT prescriptions are rather accurate for these cases. The more interesting situation is that of larger cavities when the packing is such that a peak of $\rho(r)$ can arise in the centre of the cavity. This is illustrated in Figure 1, where we show results for $R_{\mathrm{cav}} = 1.85\sigma$ and $\langle N\rangle = 6,8,10$. The peak in $\rho(r)$ in the centre grows rapidly as $\langle N\rangle$ increases and for $\langle N\rangle = 10$, $\rho(0)$ is larger than the contact value. This constitutes a quasi-0D situation and it is evident that ROS grossly overestimates the height of the central peak whereas the modifications of FMT fare much better, with RSLT1 (q=2) and (q=3) bracketting the GCEMC results. Similar features are found for other quasi-0D situations and we conclude that the modifications of FMT do perform much better than the original ROS whenever a zone of strong particle localisation develops in the centre of the cavity. Although it is difficult to say which particular modification is the most accurate (the level of agreement with GCEMC depends somewhat on the choice of R_{cav} and $\langle N\rangle$) RSLT2 using the free energy density $\Phi_{3,\mathrm{int}}$, which interpolates between the original ROS and the exact result, $\Phi_{3,\mathrm{exact-0D}}$, for a 0D limiting situation [19], does provide accurate results for extreme cases such as $R_{\mathrm{cav}} = 1.85\sigma$ with $\langle N\rangle = 11$ where ROS appears to fail to converge [13].

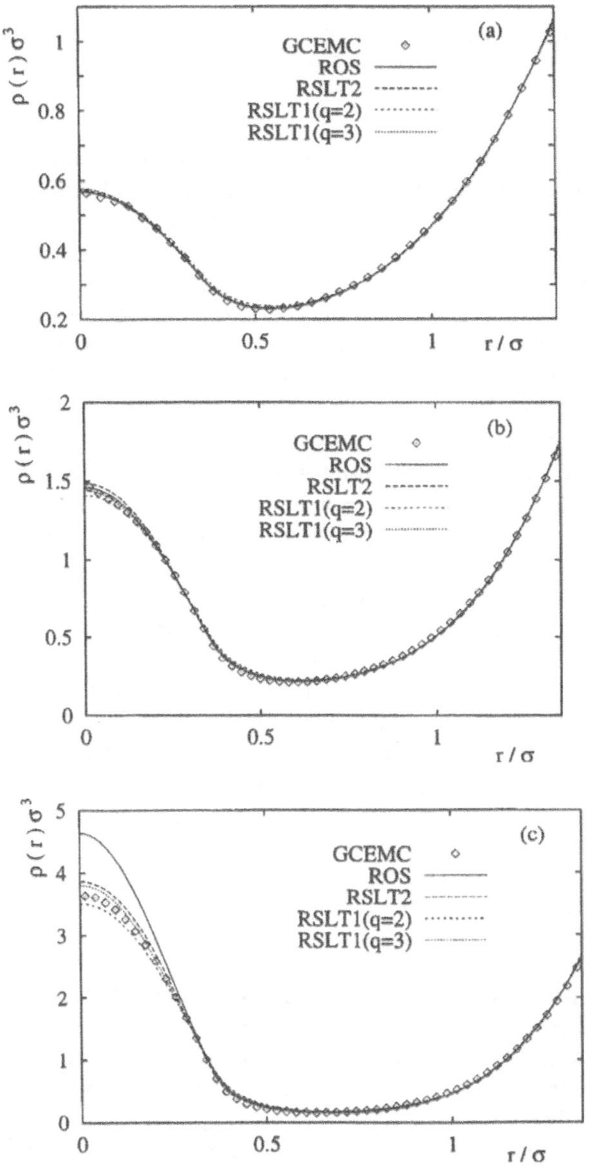

Figure 1. Grand canonical density profiles of hard spheres in a spherical cavity of radius R_{cav} = 1.85σ (maximum radius available to the centre of the hard sphere is $R_{cav} - 0.5\sigma$). (a) $\langle N \rangle = 6$, (b) $\langle N \rangle = 8$, and (c) $\langle N \rangle = 10$. The symbols are the results of GCEMC simulations and the lines are the results of various DFT prescriptions. Note the difference in vertical scales between the three figures.

One can also compare the DFT results for the quantities f_1 and f_2, defined in (1-3), with those obtained from GCEMC. For quasi-0D situations with small values of $\langle N\rangle$, f_1 and f_2 are sensitive to the choice of FMT prescription and ROS sometimes fails to yield physical values even though this prescription yields reasonable zeroth-order (grand canonical) profiles [13].

We turn now to the comparison of the CEMC simulation results for $\rho_{\langle N\rangle}(r)$ with those obtained from the expansion (1) truncated at second order. Not surprisingly, the main differences between the canonical and (zeroth order) grand canonical results are most evident when $\langle N\rangle$ is rather small. However, for $\langle N\rangle = 45$ and $R_{cav} = 2.8\sigma$, where a central peak develops in $\rho(r)$, $f_1(r,\langle N\rangle)$ is negative near the centre of the cavity and there is a substantial difference between GCEMC and CEMC results. $f_2(r,\langle N\rangle)$ is almost zero in this case so the first order correction is sufficient to bring the results into very close agreement. RSTL2 results also converge very rapidly but to a density profile which has a slightly higher central peak than the simulation result [13]. In Figure 2 we present simulation and DFT results for $R_{cav} = 1.85\sigma$ and $\langle N\rangle = 6,8,10$, the same system as in Figure 1. For clarity, and because it performs a little better than the other prescriptions, we show only RSLT2 results. For $\langle N\rangle = 6$ the canonical and grand canonical results are very different, especially in the centre of the cavity. Note that the grand canonical profile is higher in the centre as it is a sum of contributions from canonical profiles $\rho_N(r)$ weighted with $P(N)$, the probability that the system contains N particles. Contributions with $N < \langle N\rangle$ are substantial. Although the average number of particles is small, adding the first and second order corrections is sufficient to bring the two sets of simulation results into close agreement – even in the centre of the cavity where $f_1(r,\langle N\rangle)$ and $f_2(r,\langle N\rangle)$ are much larger in magnitude than for $r \geq 0.4\sigma$. In this case of moderate packing the results from GCEMC and RSLT2 are very similar at each order of the expansion. At a higher packing, $\langle N\rangle = 8$, the zeroth, first and second order results almost coincide for $r \geq 0.4\sigma$, whereas for $r < 0.4\sigma$ the zeroth-order (grand canonical) profile is much higher than the canonical result and the convergence towards the latter is not as rapid. This trend is reinforced for $\langle N\rangle = 10$, although the first and second order results from GCEMC continue to bracket those from CEMC. The RSTL2 results follow the same pattern as those of simulation which means that the theory provides a reasonable account of f_1 and f_2.

It is instructive to examine extreme cases where the $1/\langle N\rangle$ expansion might be expected to fail. For $\langle N\rangle = 1$ and 2 the canonical density profile can be calculated exactly and one can compare the results with those from DFT and from GCEMC data. When the average density of the fluid in the cavity is small the second order results are almost identical to the exact ones for $\langle N\rangle = 2$ [13]. On the other hand, for higher average densities, the second order densities are often negative in the central part of the cavity where volume exclusion forces the exact canonical profile for $N = 2$ to be zero for $r < 1.5\sigma - R_{cav}$, when $1.5\sigma > R_{cav} > 1\sigma$ [13].

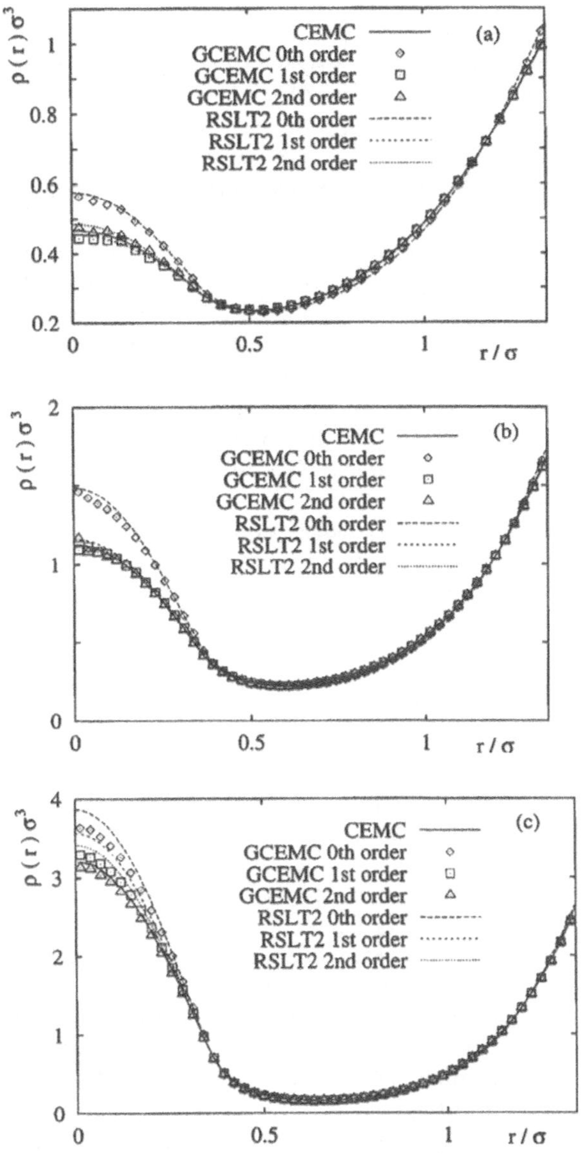

Figure 2. Density profiles of hard spheres in a spherical cavity of radius $R_{cav} = 1.85\sigma$. (a) $\langle N \rangle = 6$, (b) $\langle N \rangle = 8$, and (c) $\langle N \rangle = 10$. The symbols refer to GCEMC simulations while the solid lines are the results of CEMC simulations. The other lines are the zeroth, first and second order results of the RSLT2 DFT approximation. Note the difference in vertical scales between the three figures.

1.5 REMARKS

We conclude from our comparisons that (i) the truncated $1/\langle N \rangle$ expansion is a reliable means of obtaining canonical density profiles, even for cavities with a small average number of particles and (ii) the modifications of FMT, which are designed incorporate dimensional crossover and the 0D limit, perform significantly better than the original ROS prescription, especially as regards the calculation of the correction terms f_1 and f_2, in quasi-0D situations where the profile in the central part of the cavity exhibits slower convergence. Further investigations are required to ascertain what are the limitations of the (truncated) series expansion. From (2) and (3) we see that f_1 and f_2 depend on the fluctuations in particle number and on derivatives of $\rho(r)$ with respect to μ. Although it is clear that the values of f_1 and f_2 are sensitive indicators of packing constraints, it is not obvious how these quantities behave near maximum packing of the cavity when one might expect to find 'transitions' to frozen or glassy states. This issue warrants further attention. Note that one can derive expressions for the higher-order terms in the expansion but these are probably too complicated to evaluate either in simulation or in DFT. The present study has specialised to the structure of the hard-sphere fluid. However, it is clear that the $1/\langle N \rangle$ expansion for the density profile can be applied to any fluid and to any confining geometry. What is not so clear is how one should calculate thermodynamic functions for a canonical system using DFT methods.

2. Freezing and Condensation Transitions of Confined Fluids

2.1 INTRODUCTION

Confinement can have a profound influence on the phase behaviour of a confined fluid [7]. How the liquid-gas transition is shifted by confinement in narrow pores from that in bulk (capillary condensation) has been the subject of many theoretical, computer simulation and experimental studies and the factors which determine the shift are well-understood [7,14]. By contrast, the liquid-solid transition is less well-studied from a theoretical perspective and it is only recently that *systematic* simulation studies have been performed for fluids in ideal slit pores which attempt to answer basic questions such as how the freezing transition depends on the separation between the two walls and on the nature of the wall-fluid potential [24,25]. Consequently little is known about the overall phase diagram of a confined fluid. Much of the impetus for the present study stems from the (many) experimental investigations of freezing and melting phenomena of fluids in mesoporous solids. These are summarised in [24-26] and we simply state that experiments usually find a depression of the freezing temperature below the bulk value accompanied by pronounced hysteresis between freezing and melting, with melting still occurring below the bulk transition temperature. Duffy et al. [27] used a positronium annihilation technique to determine both the freezing and condensation phase boundaries for CO_2 in Vycor glass with a mean pore radius R_m ~20 Å. They found that confinement shifts the liquid-gas transition to a considerably higher and the liquid-solid transition to a considerably lower temperature than in bulk and that the triple point appears to be reduced in both pressure and temperature. Similar features have been found in a very different type of investigation using the surface force apparatus where it was shown that certain alcohols, confined between the mica surfaces, remain liquid down to 30K below the bulk freezing point [28]. However, other surface force experiments, albeit at very small mica separations, presented evidence for an increase in the freezing temperature of cyclohexane and other fluids [29].

Simple thermodynamic arguments are often invoked to explain the freezing point depression. These involve balancing volume contributions to the free energy of the confined fluid against interfacial (surface area) contributions. For a slit pore with planar walls separated by a distance ℓ, the change in freezing temperature $\Delta T_f = T_f - T_{fb}$ is given by [30]

$$\frac{\Delta T_f}{T_{fb}} = -2\frac{(\gamma_{ws} - \gamma_{wl})v}{\ell \Delta H_{fb}} \tag{5}$$

where T_{fb} is the bulk freezing temperature, ΔH_{fb} is the bulk latent heat of freezing and v is the molar volume. This formula is a direct analogue of the version of the Kelvin equation [7,30] used to describe the shift in condensation temperature. In the case of condensation (γ_{ws} - γ_{wl}) is replaced by (γ_{wl} - γ_{wg}), the difference in wall-liquid and wall-gas interfacial tensions and the latent heat is that of condensation. Both formulae should be valid in the limit of large ℓ [30]. Capillary condensation, the shift of the condensation line to higher temperatures (lower pressures), occurs for walls which favour liquid over gas, i.e. $\gamma_{wl} < \gamma_{wg}$. Capillary evaporation, a shift to lower temperatures (higher pressures), occurs when the walls are sufficiently repulsive that $\gamma_{wl} > \gamma_{wg}$. The Kelvin equation for condensation is known to remain reliable down to surprisingly small separations ℓ (10 molecular diameters or so). Although the regime of validity is not known for (5), this equation , with ℓ replaced by the mean pore radius R_m, provides the basis for the technique of thermoporometry, which is used to characterise pore size distributions of porous materials by measuring the freezing point depression of a liquid (usually water) [31]. On the basis of (5) depression (negative ΔT_f) should occur because the wall-solid interfacial tension $\gamma_{ws} > \gamma_{wl}$, the wall-liquid tension, and the magnitude of the depression should vary as $1/\ell$. If the walls preferred the solid phase, which could be the case when the wall-fluid attraction is much stronger than fluid-fluid attraction, then $\gamma_{ws} < \gamma_{wl}$ and (5) predicts $\Delta T_f > 0$, i.e. an increase in the freezing temperature.

Computer simulation is a natural technique to employ in investigations of freezing and other phase behaviour and early studies are described in [2]. Other simulations of freezing in slits have been carried out [32], but none of these calculated free energies so they were unable to locate the equilibrium transition. The most detailed studies are the GCEMC simulations of Miyahara and Gubbins [24] (for slits) and Maddox and Gubbins [26] (for cylinders) which investigated how the freezing of the confined Lennard-Jones fluid depends on pore size and on the choice of wall-fluid potential. Our present study addresses similar issues but unlike [24,26] we employ MC techniques which enable us to calculate the Gibbs free-energies of the confined phases and determine the equilibrium transitions, i.e. the phase boundaries. A full account of this work is given in [25] and these lecture notes merely summarise the main results.

2.2 THE MODEL AND THE SIMULATIONS

We consider a Lennard-Jones fluid (cut-off and shifted at $r_c = 2.5\sigma$) between two (structureless) walls which exert a confining potential $\phi_\ell(z) = \phi(z) + \phi(\ell - z)$ on the fluid, z is measured normal to the walls and the potential due to a single wall is the 10-4-3 potential:

$$\phi(z) = 2\rho_W \pi \sigma'^2 \varepsilon' \Delta \left[\frac{2}{5}\left(\frac{\sigma'}{z}\right)^{10} - \alpha\left(\frac{\sigma'}{z}\right)^{4} - \alpha \frac{\sigma'^4}{3\Delta(z + 0.61\Delta)^3} \right] \quad (6)$$

When $\alpha = 1$, (6) reduces to the well-known Steele potential used to model fluids at a graphite substrate with the lattice planes separated by a distance Δ. ρ_w is the number density of atoms in the substrate and σ' and ε' are substrate-fluid parameters. By varying the parameter α we could investigate purely repulsive walls ($\alpha = 0$) as well as different strengths of attraction.

MC simulations were performed in the NP_zT ensemble. The more usual GCEMC can fail for the very high density systems of interest because of the difficulty of transferring particles into the confined system from the reservoir. Such problems are avoided in the NP_zT ensemble where the pressure P_z normal to the walls is fixed externally and the system finds the average (equilibrium) value $<\ell>$ of the wall separation. Periodic boundary conditions were imposed in the x and y directions. A drawback is that for finite $<\ell>$, P_z is not related directly to any bulk pressure which means that it is not possible to make direct comparisons of the resulting phase diagrams, in the P_z,T plane for fixed N, with those from experiment (or GCEMC) which correspond to the P,T plane for fixed ℓ (pore volume), with P being the pressure of the reservoir. However, before embarking upon the determination of phase equilibria, the consistency of results from the NP_zT ensemble with those from GCEMC was confirmed for a particular fluid state [25].

The structure of the confined system was monitored by means of snapshots and by calculating the average density profile $\rho(z)$, the in-plane pair correlation function $g(r)$ and its Fourier transform, the two-dimensional structure factor. The Helmholtz free energy of the solid phase was calculated using the Frenkel-Ladd method [33] based on the classical Einstein solid as a reference system. For the free energy of the confined fluid we developed a new integration procedure which consists of slowly switching off the wall-fluid potential ϕ_ℓ by means of a parameter λ until the bulk fluid is reached, which then acts as the reference system [25]. Our procedure encounters difficulties if a phase transition occurs as the wall-fluid potential is turned on (or off). For intermediate and strongly attractive potentials solid-like layers develop near the walls while the main central portion of the system remains liquid-like, as ascertained from the structural data. In order to avoid the complications of these 'surface' transitions, which occur at pressures P_z lower than that for the freezing of the pore liquid as a whole, we restricted our calculations of free energies to weakly attractive ($\alpha = 0.5$ and small values of ε') and purely repulsive ($\alpha = 0$) walls.

2.3 RESULTS OF SIMULATIONS

In our first investigations of freezing the temperature T was fixed and the system was equilibrated at some normal pressure P_z in a liquid-like state. Subsequent simulations were carried out at a higher value of P_z, until the liquid appeared to freeze. Runs were also performed decreasing P_z until melting was observed. Very substantial hysteresis was found for all temperatures, potentials and system sizes. We carried out simulations for $N = 300$ and 600. For strongly attractive walls, i.e. $\alpha = 1$, $\varepsilon' = 7.237\,\varepsilon$, and the other parameters chosen to mimic argon on a very attractive graphite-type substrate, freezing was studied at a reduced temperature $T^* = k_B T/\varepsilon = 1.15$. ε and σ refer to the Lennard-Jones fluid parameters. The density profiles and in-plane pair correlation functions $g_i(r)$ in the inhomogeneous liquid and solid phases are quite different throughout the central layers of the system and the average wall separation $<\ell>$ is smaller in the solid phase, indicating an increase in the mean density $\rho = N/A<\ell>$ on solidification. The profiles for the first layer, i.e. those in contact with the wall, are nearly identical in the two phases and $g_1(r)$corresponds to the pair-correlation function of a two-dimensional hexagonal crystal. Snapshots and the $g_i(r)$ of the other layers show that these are also hexagonally ordered in the solid phase. Similar results were obtained for weaker-attraction, $\alpha = 1$, $\varepsilon' = 3.0\,\varepsilon$, and these are rather close to those given in [24] for Lennard-Jones methane confined between two 10-4-3 graphite walls. Only by reducing ε' below $1.0\,\varepsilon$, and keeping $\alpha = 1$, was it possible to 'melt' the solid-like contact layer. Thus, we carried out investigations of the phase equilibria for potentials with $\varepsilon' = 1.277\,\varepsilon$ and $\alpha = 0.5$ or 0. The other wall parameters are $\rho_w\sigma^3 = \sqrt{2}$, $\Delta = \sigma/\sqrt{2}$ and $\sigma' = 1.0946\sigma$, mimicking fluid argon at a weak substrate. Figure 3 shows the density profiles and in-plane pair correlation functions for $\alpha = 0$, calculated for $T^* = 1.15$ and $N = 300$ at the pressure corresponding to liquid-solid coexistence, i.e. when the two phases have equal Gibbs free energy.

The contact layer in the liquid phase remains liquid-like for $\alpha = 0$ and $\alpha = 0.5$. Indeed for the purely repulsive walls $g_1(r)$ is similar to the pair correlation functions in the central layers. For both values of α, and $N = 300$, the coexisting solid phase has 10 hexagonally ordered layers containing 30 atoms. The orientation and stacking of the layers is complex and there was some evidence of strain in the confined solid [25]. Simulations for $N = 600$, for the same temperature and wall area, yield similar profiles and $g_i(r)$ as those for $N = 300$ but with twice as many layers in the solid phase. The normal pressure at coexistence is $P_z^* = 7.48 \pm 0.08$, which is substantially lower than for $N = 300$, where $P_z^* = 8.52 \pm 0.09$. For $\alpha = 0$ the freezing pressure is increased by confining the liquid to narrower pores. In Figure 4 we plot the Gibbs free energy, calculated by the integration procedures we mentioned earlier, for the solid and liquid phases with $N = 600$ and $\alpha = 0$. The range of P_z^* over which liquid and solid branches exist simultaneously is substantial and, in this case, the equilibrium transition occurs roughly midway through the hysteresis region.

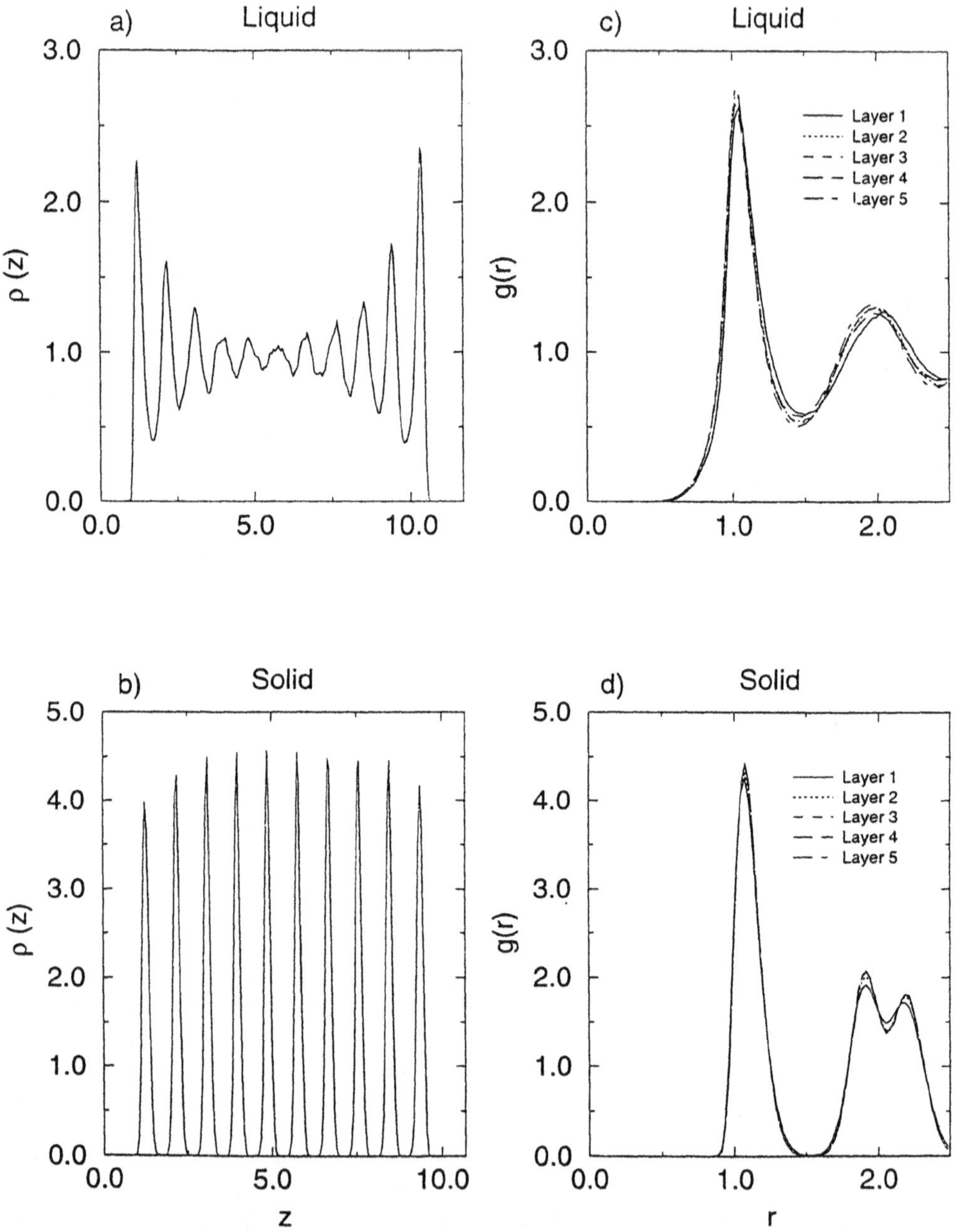

Figure 3. Density profiles $\rho(z)$ and in-plane pair correlation functions $g(r)$ for a Lennard-Jones fluid confined between purely repulsive walls: $\alpha = 0$, $N = 300$ and $T^* = 1.15$. The normal pressure $P_z^* = 8.52$ where liquid and solid coexist. The average wall separations are $<\ell>_{liquid} = 11.41$ and $<\ell>_{solid} = 10.68$. Layer 1 is in contact with the wall.

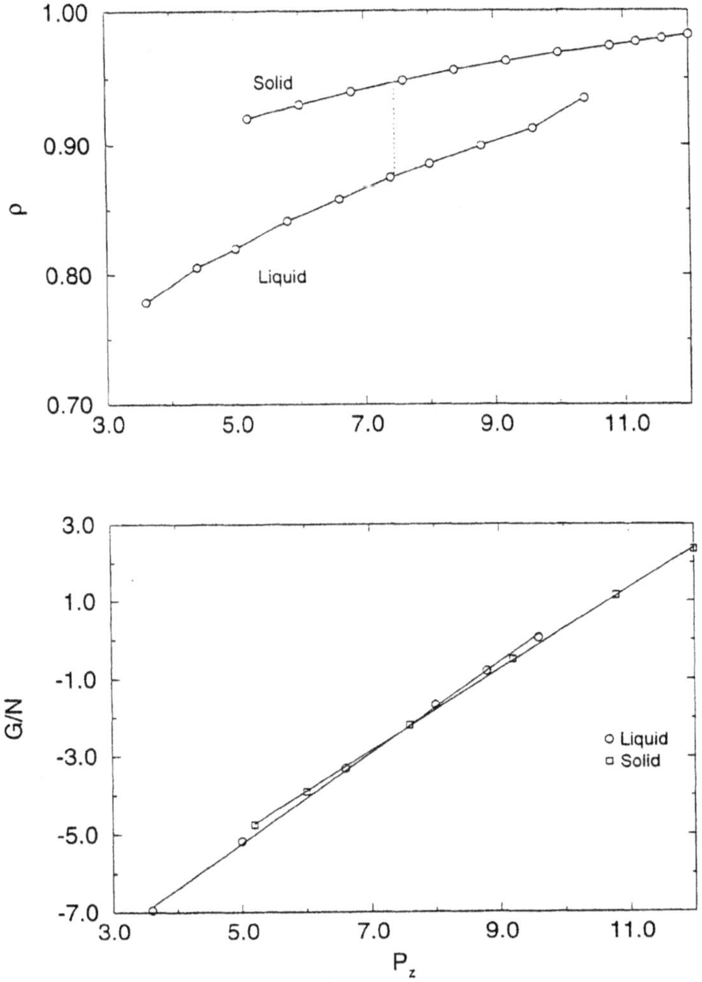

Figure 4. The mean density ρ and Gibbs free energy per particle G/N versus normal pressure P_z^* for a Lennard-Jones fluid confined between purely repulsive walls: $\alpha = 0$, $N = 600$ and $T^* = 1.15$. Error bars are of the same size as the symbols. The dashed vertical line in the top figure denotes the equilibrium liquid-solid transition which occurs at $P_z^* = 7.48$.

Calculating the Gibbs free energies of the two phases involves a large number of simulations and it is clear that an enormous amount of effort would be required to map out a freezing line or other phase boundaries using only this procedure. A more efficient approach is the 'Gibbs-Duhem' method used successfully by Kofke [34] for determining bulk coexistence lines. This approach was adapted to the confined system [25] and in Figure 5 we show results for the freezing lines, plotted on a logarithmic pressure scale, for $\alpha = 0$ with 300 and 600 particles and for $\alpha = 0.5$ with 300 particles. The low pressure portions are re-plotted in Figure 6 to display the triple-point region more clearly. These figures include our own results for the bulk freezing line.

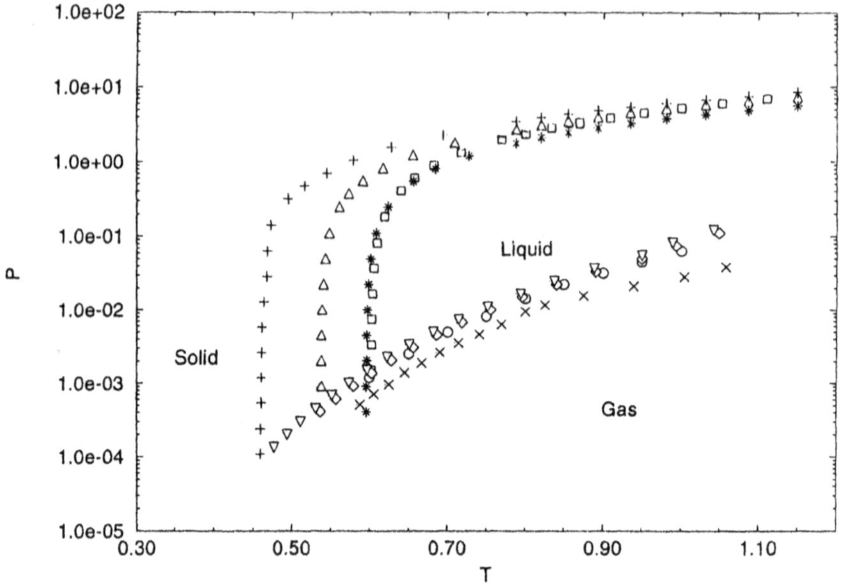

Figure 5. Coexistence curves of the confined and bulk Lennard-Jones fluid. Note that for the confined fluid P denotes the normal pressure P_z^* whereas for the bulk it is the bulk pressure. O, bulk liquid-gas; □ , bulk liquid-solid; ◇ , $N = 600$, $\alpha = 0$, liquid-gas; Δ, $N = 600$, $\alpha = 0$, liquid-solid; ∇ , $N = 300$, $\alpha = 0$, liquid-gas; +, $N = 300$, $\alpha = 0$, liquid-solid; x, $N = 300$, $\alpha = 0.5$, liquid-gas; *, $N = 300$, $\alpha = 0.5$, liquid-solid.

For purely repulsive walls ($\alpha = 0$) the freezing line is shifted to lower temperatures than in the bulk. At low pressures the shift for $N = 300$ is about twice that for $N = 600$ and the average wall separation is about half that of the large system. For weakly attractive walls ($\alpha = 0.5$) and $N = 300$ (see the asterisks in Figure 5) the freezing pressure P_z^* is lower than in bulk at high temperatures whereas at low temperatures, approaching T_{tr}, it is higher. Overall, the magnitude of the shift from bulk is small: at low pressures the (reduced) freezing temperature is lowered by <0.01, which should be contrasted with the large lowering ~0.14 found for $\alpha = 0$ and the same number of particles (Figure 6). It is evident that including attraction in the wall-fluid potential has a very significant effect on the freezing line at low pressures.

We also examined the liquid-gas transition so that we could map out a substantial fraction of the phase diagram for a given wall-fluid potential. Gibbs free energies of the confined liquid and gas phases were calculated at suitable starting temperatures and then the Gibbs-Duhem integration was performed in a similar manner to that for liquid-solid coexistence [25]. The resulting liquid-gas coexistence lines are plotted in Figures 5 and 6, along with that for the bulk, obtained using the parametrisation of the free energy for the bulk Lennard-Jones fluid given in [35]. The coexistence lines for $\alpha = 0$ lie above that of the bulk (circles) with the line for $N = 300$ (triangles) lying above the line for $N = 600$ (diamonds). These results illustrate the phenomenon of capillary evaporation, expected for a fluid confined between purely repulsive (hard or soft) walls [7,36]. As the pressure was reduced towards the value at coexistence the oscillations in the density profiles of the liquid phase disappeared and a

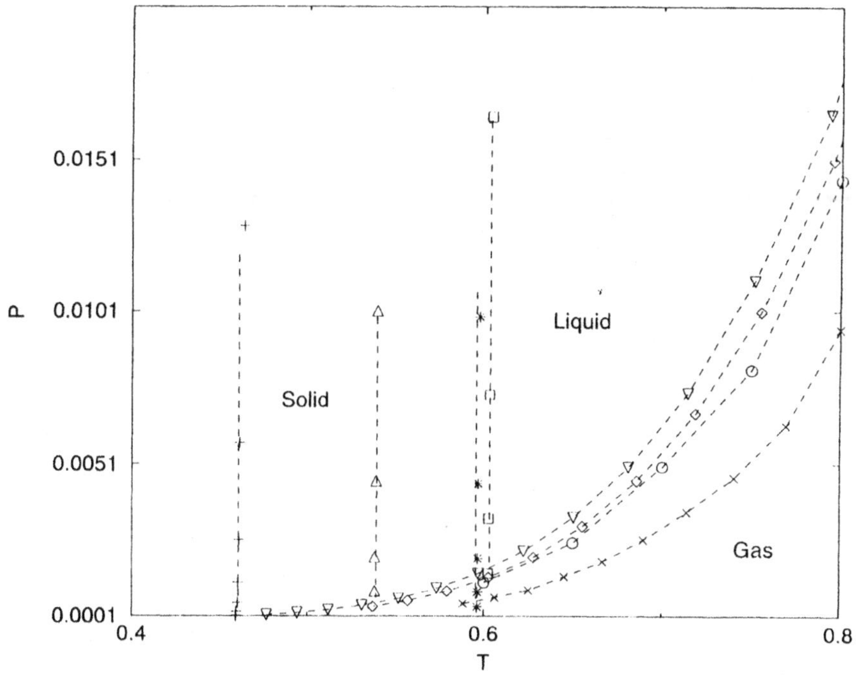

Figure 6. Low temperature-low pressure coexistence curves of the confined and bulk Lennard-Jones fluid. Symbols are as for figure 5; the dashed lines joining these are a guide to the eye. The freezing lines are almost vertical on this pressure scale.

film of low density gas began to intrude between the liquid and the walls [37]. For weakly attractive walls ($\alpha = 0.5$) the liquid-gas coexistence line (crosses) lies below that of the bulk. This is an example of capillary condensation since now the wall-fluid forces favour the liquid phase, leading to condensation at lower pressures than in bulk. In the system of 300 particles the shift is very pronounced for the whole range of temperatures. Finally, from Figure 6, we see that for $\alpha = 0$ and 300 particles the triple point is shifted to substantially lower temperatures and pressures than in bulk, whereas for $\alpha = 0.5$ and the same number of particles the shift is in the same direction but is very much smaller.

2.4 DISCUSSION

We have presented the results of NP_zT simulations for the freezing and condensation transitions of a Lennard-Jones fluid confined in a single slit pore whose walls exert different strengths of attraction on the atoms of the fluid. The main conclusions of this study are as follows: i) for walls that are purely repulsive the freezing line is shifted to temperatures well-below that of the bulk, the shift being larger for the system with a smaller number of particles, i.e. the narrower pore, and the condensation line is shifted to slightly higher pressures than in the bulk (capillary evaporation) and ii) for walls that exhibit weak attraction the freezing line is shifted only slightly from bulk and actually crosses the bulk line at $T^* = 0.65$, whereas the condensation line lies at lower pressures than in bulk (capillary condensation). Caution must be exercised in the interpretation of these results. As mentioned earlier, since P_z cannot be identified with the pressure of a bulk reservoir it is not straightforward to make direct comparisons of the results of Figures 5 and 6 with experimental studies such as those described in [27] or with simulation studies based on other ensembles. One obvious complication is that the average wall separation $<\ell>$ varies along the coexistence lines calculated at fixed N. Nevertheless, one can make qualitative comparisons with other studies.

The work of Miyahara and Gubbins [24] is the closest in spirit to the present. These authors carried out GCEMC simulations for a Lennard-Jones model of methane, taking a particular path in which μ first followed the bulk condensation line down to the triple point and then the bulk sublimation line; the pressure was not constant. For hard-walls they found a substantial depression of the apparent freezing temperature below that in bulk and this is consistent with our present results for purely repulsive walls. When their walls were made of methane, so that the 10-4-3 wall-fluid potential was weakly attractive, they found for separation $\ell = 7.5$ – see Fig. 18 of Ref. [24] – that the apparent freezing temperature was almost unchanged from that of bulk. (We use the terminology 'apparent' since in Ref. [24] the transition is identified with the jump of the mean density on cooling.) Once more this is consistent with our present results. Moreover these authors found that all the layers tended to freeze and melt in unison which is what we find. Miyahara and Gubbins [24] argued that the case of strongly attractive walls, mimicking graphite, is very different with the apparent freezing transition now occurring at temperatures above the bulk value and with the increase being larger for narrower pores. However, for graphite pores with $\ell \leq 5.3$ they found no evidence for a freezing transition. They also emphasised that the freezing of the contact layer occurred at a temperature higher than for the central layers. Although our technique did not allow us to investigate the equilibrium freezing transition for strongly attractive walls, our results for the structure of the confined liquid and solid are similar to those found in Ref. [24] and do suggest that the solidification of the contact layer is important in determining the freezing characteristics of the pore fluid. Clearly it would be valuable (see later) to develop some new technique which allows one to calculate the free energies of the liquid phase when the contact layers are solidified. It would also be useful to investigate further the influence of box shape on the freezing transition. In our study and in that of Ref. [24] the wall dimensions S_x and S_y were kept fixed during the simulations. Future studies might allow the shapes of walls to change, while keeping the area fixed, during the course of the simulations.

An issue we have not discussed but which is certainly relevant in the physics of confinement concerns the influence of the atomic corrugations which are present in real substrates. These have been ignored in our present model which considers two structureless substrates. One might expect atomic corrugations to influence the structure of the solid phases which form. We carried out some NP_zT simulations using a model of 'structured' walls in which each wall consisted of 4 crystalline layers each with 32 rigidly fixed atoms arranged in the fcc structure. The walls were separated by a distance ℓ and arranged so that the fluid was confined between two (100) faces which were perfectly out of registry. Such an arrangement favours the formation of a fcc solid with an even number of layers and our simulations for $N = 300$ and wall particle-fluid particle pair potential equal to that between fluid particles, i.e. with the same Lennard-Jones potential, did find ten well-ordered 'solid' layers in registry with the crystalline walls. The positions of the maxima of the in-plane pair correlation functions $g_i(r)$ and snapshots were all consistent with ordering characteristic of the (100) planes of the fcc lattice [37].

The propensity towards fcc ordering should depend on the surface density of atoms in the crystalline walls. Our own simulations were for a rather open lattice with a lattice parameter of 1.416σ so the corrugations were very pronounced. Miyahara and Gubbins [24] carried out simulations for a structured graphite wall in which the carbon-carbon distance was much smaller than the diameter of the adsorbed methane and found, in contrast to our result, that the confined solid phase ordered in the same way as for structureless walls, i.e. in hexagonal layers. Indeed the in-plane pair correlation functions and the apparent freezing temperature for $\ell = 7.5$ were almost identical for the two wall models they considered, leading them to conclude that corrugation effects should not be important for the freezing of most adsorbates in graphite pores. However, it is likely that corrugations play a much more significant role for crystalline walls whose lattice spacing is larger than the diameter of the adsorbates and further investigations are required.

Our present study has focused on the freezing and condensation transition in slit pores for which $\langle \ell \rangle \geq 10$. Such widths correspond to what is commonly termed the mesoporous regime. The emphasis was on ascertaining how the transitions are shifted from those of the bulk fluid. We have not investigated the nature of any transitions for fluids in very narrow pores where the details of the wall-fluid potential and the precise value of the wall separation (whether an odd or even number of layers can be accommodated) become crucially important [2,32,36]. The computational methods we have used here could be used for very narrow pores, but we should note that choice of ensemble is even more relevant in this case. Allowing the system to find its average separation for given NP_zT should correspond to a very different situation from confining the fluid between two walls at a fixed separation of 2σ or 3σ.

We conclude this part of the discussion by returning to the problem of calculating free-energies for strongly attractive walls. The method we have applied relies on finding a reversible thermodynamic path which does not intersect any phase transition in the fluid phase. This is not possible if the contact layers freeze. Very recently Radhakrishnan and Gubbins [38] have developed an alternative method which involves calculating the Landau free energy as a function of a suitable order parameter, using GCEMC. Their method *is* able to tackle strongly attractive walls and in the case of Lennard-Jones methane in a structureless graphite (10-4-3) slit of width $\ell = 7.5\sigma$ they find strong evidence for three distinct phases. In the high temperature phase A, all the layers have an isotropic liquid-like structure; the intermediate B phase is characterised by frozen contact layers, with the remainder having an isotropic structure; and in phase C, all the layers are frozen. There are two (weak) first order transitions, one corresponding to the freezing of the contact layers and the other at, a lower temperature, corresponding to the freezing of the central layers. The results for the angular structure function in the individual layers indicate that these transitions are from a two-dimensional liquid to a hexatic phase. Both transition temperatures are well-above that of the bulk liquid. This new method promises to find many applications in determining the location and the nature of phase transitions in models of confined fluids [38].

An infinitely long cylinder of finite radius is another popular model for a porous solid. In this case all phase transitions are rounded, as the system is effectively one-dimensional but the degree of rounding, in some appropriate thermodynamic variable, may be extremely small. We refer the reader to the recent simulation study of Maddox and Gubbins [26] for a valuable discussion of the effects of pore geometry, packing constraints and choice of wall-fluid potential on freezing. An important conclusion is that although strongly attractive walls tend to elevate the freezing temperature of a fluid in contact with the walls (as for planar walls) the geometrical constraints of the cylinder act to lower the apparent freezing temperature of a fluid in the central portion of the pore. Just how one defines the freezing temperature of such a system is not unambiguous, and a dynamic criterion is used in [26] to determine whether a given layer is solid or liquid.

What relevance might simulation studies for idealised pores have for experiments in real mesoporous solids? There is some consensus that the rapid increase in the amount of fluid adsorbed, as observed in adsorption isotherm measurements at temperatures above the bulk triple point, is associated with capillary condensation [7]. Although the detailed explanation of the observed hysteresis remains somewhat controversial, it is generally accepted that what is monitored is a shifted gas-liquid transition whose location in the phase diagram is determined by the 'average' pore radius [39]. The transition is smeared as a result of the distribution of pore sizes found in the real porous material. The interconnectivity of the pores certainly plays a role in determining the nature of the hysteresis [40]. The effects of pore size distributions, interconnectivity and geometry of the pores must also play an important part in determining the nature of freezing of fluids in porous solids such as Vycor and other silica glasses. The hysteresis which is observed [27] is complex and occurs over a wider temperature range than is found for capillary condensation. When crystallites form, their lateral extent, as estimated from the widths of diffraction peaks, is often much larger than a typical pore length, implying interconnection of crystallites in neighbouring pores. The picture which is inferred from diffraction and other studies is of crystallites forming in the central portion of the pores, leaving a fluid or amorphous region next to the walls. Whether crystallites form or not may depend on the degree of pore filling.

Although studies for single idealised pores cannot address the wealth of issues raised by the experiments one might hope to understand why freezing point depression is generally observed and what basic physical factors determine a phase diagram such as that reported in [27] for CO_2 in Vycor where we see that the ratio of the depression to the bulk freezing temperature, $(T_f - T_{fb}) / T_{fb}$, is about -0.05 for pressures near that of the bulk triple point. (The melting temperature is depressed by about half of this amount.)

On the basis of our present study we could argue that such shifts of the freezing line should be associated with wall-fluid interactions that are very weakly attractive. Very substantial freezing point depression is found for purely repulsive walls ($\alpha = 0$), see figures 5 and 6, with $(T_f - T_{fb})/ T_{fb} \sim -0.24$ for $N = 300$ ($<\ell> \sim 10$) and ~ -0.11 for $N = 600$ ($<\ell> \sim 20$). For the wall-fluid potential with $\alpha = 0.5$ we find a very small ratio of -0.01 near the bulk triple point for a system with 300 particles, and Miyahara and Gubbins [24] obtain a similar shift for methane between methane walls. It is feasible that for silica glasses the effective wall-fluid attraction is rather weak relative to the fluid-fluid attraction: certainly it is much weaker than for graphite substrates. However, we are now aware of explicit calculations of model potentials. Of course, there must be sufficient wall-fluid attraction to bring about the observed capillary condensation rather than capillary evaporation, i.e. the walls must prefer liquid to solid while also preferring liquid to gas. Finally, one cannot rule out the possibility that the effects of pore geometry and the inherent disorder of the porous network could bring about freezing point depression even when the wall-fluid interaction is strongly attractive [26].

3. Acknowledgements

The research reported in Section 1 was carried out in collaboration with A. González, J.A. White and F.L. Román (Salamanca) and that in Section 2 with H. Dominguez and M.P. Allen (Bristol). The author expresses his gratitude to both groups for stimulating interactions. He has benefitted from discussions with M.A. Alam, M. Dijkstra, H. Löwen, Y. Rosenfeld, M. Schoen and P. Tarazona and thanks K.E. Gubbins for helpful correspondence.

4. References

1. e.g. Klafter, J., and Drake, J.M. (eds.) (1989) *Molecular Dynamics in Restricted Geometry*, Wiley, New York.
2. Schoen, M. (1993) *Computer Simulation of Condensed Phases in Complex Geometries*, Springer, Heidelberg.
3. Rowlinson, J.S. (1986) *J. Chem. Soc. Faraday Trans.* 2 **82**, 1801; McQuarrie, D.A., and Rowlinson, J.S. (1987) *Mol. Phys.* **60**, 977.
4. Woods, G.B., Panagiotopoulos, A.Z., and Rowlinson, J.S. (1988) *Mol. Phys.* **63**, 49.
5. Zhou, Y., and Stell, G. (1989) *Mol. Phys.* **66**, 767.
6. Macpherson, A.K., Carignan, Y.P., and Vladimiroff, T. (1987) *J. Chem. Phys.* **87**, 1768.
7. e.g. Evans, R. (1990) *J. Phys: Condens. Matter* **2**, 8989.
8. Evans, R. (1979) *Adv. Phys.* **28**, 143; Mermin, N.D. (1965) *Phys. Rev. A* **137**, 1441.
9. Lee, D.J., Telo da Gama, M.M., and Gubbins, K.E. (1986) *J. Chem. Phys.* **85**, 490.
10. Lebowitz, J.L., and Percus, J.K. (1961) *Phys. Rev.* **122**, 1675.
11. Salacuse, J.J., Denton, A.R., and Egelstaff, P.A. (1996) *Phys. Rev. E.* **53**, 2382.
12. Román, F.L., White, J.A., and Velasco, S. (1997) *J. Chem. Phys.* **107**, 4635.
13. González, A., White, J.A., Román, F.L., and Evans, R. (1998) *J. Chem. Phys.* **109**, 3637; González, A., White, J.A., Román, F.L., Velasco, S., and Evans, R. (1997) *Phys. Rev. Lett.* **79**, 2466.
14. Evans, R. (1989) in J. Charvolin, J.F. Joanny and J. Zinn-Justin (eds.), *Liquids at Interfaces*, Les Houches Session 48, Elsevier, Amsterdam, p.1.
15. Evans, R. (1992) in D. Henderson (ed.), *Fundamentals of Inhomogeneous Fluids*, Dekker, New York, p.85.
16. Löwen, H. (1994) *Phys. Rep.* **237**, 249.
17. Rosenfeld, Y. (1989) *Phys. Rev. Lett.* **63**, 980.
18. Rosenfeld, Y. (1990) *Phys. Rev. A* **42**, 5978; Rosenfeld, Y., Levesque, D., and Weis, J.J. (1990) *J. Chem. Phys.* **92**, 6818; Rosenfeld, Y. (1993) *J. Chem. Phys.* **98**, 8126; see also Kierlik, E., and Rosinberg, M.L. (1990) *Phys. Rev. A* **42**, 3382.
19. Rosenfeld, Y., Schmidt, M., Löwen, H., and Tarazona, P. (1996) *J. Phys: Condens. Matter* **8**, 1577; (1997) *Phys. Rev. E* **55**, 4245.
20. Tarazona, P., and Rosenfeld, Y. (1997) *Phys. Rev. E* **55**, R4873.
21. González, A., White, J.A., and Evans, R. (1997) *J. Phys: Condens. Matter* **9**, 2375.
22. Cuesta, J.A., and Martínez-Ratón, Y. (1997) *Phys. Rev. Lett.* **78**, 3681; *J. Chem. Phys.* **107**, 6379.
23. See the lectures by P. Tarazona and Y. Rosenfeld in these ASI proceedings.
24. Miyahara, M., and Gubbins, K.E. (1997) *J. Chem. Phys.* **106** 2865.
25. Dominguez, H., Allen, M.P., and Evans, R. (1998) *Mol. Phys.* (to appear).
26. Maddox, M.W., and Gubbins, K.E. (1997) *J. Chem. Phys.* **107**, 9659.

27. Duffy, J.A., Wilkinson, N.J., Fretwell, H.M., Alam, M.A., and Evans, R. (1995) *J. Phys: Condens. Matter* **7**, L713.
28. Christenson, H.K. (1995) *Phys. Rev. Lett.* **74**, 4675.
29. Klein, J., and Kumacheva, E. (1995) *Science* **269**, 816.
30. Evans, R., and Marini Bettolo Marconi, U. (1987) *J. Chem. Phys.* **86**, 7138.
31. Strange, J.H., Rahman, M., and Smith, E.G. (1993) *Phys. Rev. Lett.* **71** 3589; Hansen, E.W., Stöker, M., and Schmidt, R. (1996) *J. Phys. Chem.* **100**, 2195.
32. e.g. Lupkowski, M., and Van Swol, F. (1990) *J. Chem. Phys.* **93**, 737; Hug, J.E., Van Swol, F., and Zukoski, C.F. (1995) *Langmuir* **11**, 111; Ma, W.J., Banavar, J.R., and Koplik, J.J. (1992) *J. Chem. Phys.* **97**, 485.
33. Frenkel, D., and Ladd, A.J.C. (1984) *J. Chem. Phys.* **81**, 3188.
34. Kofke, D.A. (1993) *J. Chem. Phys.* **98**, 4149; Agrawal, R., and Kofke, D.A. (1995) *Mol. Phys.* **85**, 43.
35. Johnston, J.K., Zollweg, J.A., and Gubbins, K.E. (1993) *Mol. Phys.* **78**, 591.
36. Tarazona, P., Marini Bettolo Marconi, U., and Evans, R. (1987) *Mol. Phys.* **60**, 573.
37. Dominguez, H. (1997) Ph.D. Thesis, University of Bristol.
38. Radhakrishnan, R., and Gubbins, K.E. (1998), preprint.
39. See the lectures by M.L. Rosinberg in these ASI proceedings for an alternative perspective.
40. Ball, P.C., and Evans, R. (1989) *Langmuir* **5**, 714 and references therein.

BULK AND SURFACE PROPERTIES OF DIPOLAR FLUIDS

B. GROH
FOM Institute for Atomic and Molecular Physics,
Kruislaan 407, 1098 SJ Amsterdam, The Netherlands

AND

S. DIETRICH
Fachbereich Physik, Bergische Universität Wuppertal,
D-42097 Wuppertal, Federal Republic of Germany

Abstract. Based on density-functional theory we analyze the full phase diagram, the occurrence of long-ranged orientational order, and the structural properties of dipolar fluids. As a model system we consider the Stockmayer fluid that consists of spherical particles interacting via a Lennard-Jones potential plus dipolar forces. For sufficiently strong dipole moments one finds a region where a fluid phase with long-ranged orientational order is stable. For all sample shapes with the exception of a long thin needle this phase exhibits a spatially inhomogeneous magnetization which depends on the actual shape. We determine the details of the magnetization structure in a cubic sample in the absence and in the presence of an external magnetic field. One obtains a vortexlike structure with an escape of the magnetization into the axis direction near the vortex axis and two point defects where the absolute value of the magnetization is strongly reduced. If the spherical cores of the particles are replaced by elongated or oblate shapes a nematic phase without spontaneous magnetization is also possible due to the anisotropic steric interactions. We study the interplay of this nematic ordering with ferromagnetism in fluids of dipolar hard ellipsoids. Orientational order arises locally in the isotropic fluid phases near the liquid-gas interface of the Stockmayer fluid. Density-functional theory allows us to determine density and orientational order profiles as well as the surface tension of this interface.

C. Caccamo et al. (eds.), New Approaches to Problems in Liquid State Theory, 173–196.

1. Introduction

Spontaneous symmetry breaking induced by phase transitions is one of the central themes of statistical physics. From a homogeneous and isotropic liquid phase transitions can occur to phases with positional or orientational long-ranged order of the individual particles. Here we are mainly interested in the case of pure orientational order. Well known examples of such systems are the nematic liquid crystals. Initially, it was suspected that dipolar interactions are responsible for the formation of the nematic state [1]. Later, it became clear that the short-ranged steric interactions between strongly anisotropic (elongated or disklike) nonpolar molecules are sufficient to generate a nematic phase. On the other hand, it is an interesting question whether dipolar interactions alone, i.e., in the absence of anisotropic steric interactions, are also capable of stabilizing an orientationally ordered liquid. That this is indeed possible has been shown for the first time some years ago in computer simulations of a dipolar soft sphere fluid by Wei and Patey [2, 3]. Similar observations were made for dipolar hard spheres [4, 5, 6] and the Stockmayer fluid [7], in which the spherically symmetric Lennard-Jones potential is added to the dipolar interactions. Depending on whether one deals with electric or magnetic dipole moments this new phase, in which the dipoles are aligned over macroscopic distances, is called a ferroelectric or ferromagnetic liquid. In the following we will adopt the magnetic language keeping in mind that all results are equally valid in the electric case. Theoretically this phase transition was predicted within mean-field theory for a dipolar lattice gas [8] and by a generalized van der Waals theory [9]. A density-functional theory using as input direct correlation functions of the isotropic liquid obtained from an integral equation theory qualitatively confirmed these findings [10]. However, more recent work along the same lines [11] predicts that the ferromagnetic liquid is metastable with respect to freezing. It might be possible to improve upon this approach by directly calculating the direct correlation function in the ordered phase using the methods of Ref. [12]. The present authors also developed a density-functional description [19, 21] using a relatively crude approximation for the pair correlation function, whose results will be reviewed in this paper. In contrast to Ref. [10] and to the simulations this approach enables one to map out whole phase diagrams which turn out to exhibit an interesting topology varying with the dipole strength. Other authors have focused on the isotropic-ferromagnetic spinodal which corresponds to a divergence of the dielectric constant, whose position can be estimated using correlation functions from different integral equation theories [13, 14]. Finally, we mention perturbation theory as an alternative approach to determine successfully liquid-vapor coexistence of Stockmayer fluids with rigid or polarizable dipoles [15, 16, 17, 18].

In all theoretical treatments particular care must be taken of the long range of the dipolar interaction [19, 20] which may give rise to diverging integrals. Several approaches to handle these problems have been suggested [10, 19, 20], the most consistent one being an explicit performance of the thermodynamic limit starting from a large but finite sample of a given shape [19]. For a homogeneously magnetized fluid one finds an energy contribution that depends on the sample shape. However, this shape dependence disappears if inhomogeneously magnetized configurations are taken into account [22, 21].

These complications do not arise in Heisenberg fluids governed by short-ranged exchange interactions for which similar phase transitions and phase diagrams have been found [23, 24, 25]. For this model as well as for the corresponding Ising fluid [28] there is an ongoing discussion concerning the values of the critical exponents describing the isotropic-ferromagnetic transition [26, 27, 28]. A Potts fluid with $q \geq 3$ states exhibits only first-order isotropic-ferromagnetic transitions [29]. For a review of the properties of dipolar and spin fluids as obtained from simulations see Ref. [24].

Experimentally a ferromagnetic liquid phase in dipolar fluids has not yet been found. The dipole moments of molecular fluids seem to be too low, though only by a small factor, compared to those that led to ferromagnetism in the simulations. In ferrofluids [30], i.e., colloidal suspensions of small single-domain ferromagnetic particles, chemical tailoring facilitates higher reduced dipole moments; the relevant dimensionless quantity is $k_B T m^2/\sigma^3$ where m is the dipole moment and σ the particle diameter. However, in the simulations the phase transition always occurs at high packing fractions that are presently not attainable in ferrofluids. Due to the combination of fluidity and strong paramagnetism ferrofluids already have a wide range of technical applications that would certainly be enlarged if a ferromagnetic phase could be established. Electro-rheological and magneto-rheological fluids [31] also have high technological relevance. In these systems dipole moments are induced in the colloidal particles by external electric or magnetic fields and large differences between the permitivities of the solute and the solvent which then give rise to strong dipolar particle interactions having a pronounced effect on their rheological properties. However, spontaneous orientational order in the absence of external fields is not possible in these fluids.

Ferromagnetic metals become paramagnetic already far below their melting temperature. Therefore it came as a surprise when recently ferromagnetism was detected for the first time in a liquid metal. In $Co_{80}Pd_{20}$ the melting temperature is sufficiently close to the Curie temperature so that ferromagnetism in the liquid state can be reached by undercooling

[32, 33, 34]. For the formation of the ferromagnetic order in these materials short-ranged exchange interactions play an important role. However, in addition dipolar interactions are also present and, as in solid ferromagnets, are essential in forming the domain structure. Therefore for these systems we expect strong similarities to the behavior of dipolar fluids as discussed here.

Two other closely related peculiarities of simple dipolar fluids have attracted much attention recently. Strongly dipolar hard or soft spheres form chains of head-to-tail aligned particles at very low densities [35, 36]. The chain formation hinders the condensation of liquid droplets. Thus a certain amount of spherically symmetric dispersion interaction is necessary in order to obtain liquid-gas coexistence in a dipolar fluid [37] (but see also Ref. [38]). Alternatively, liquid-gas coexistence of hard dipolar particles is promoted if they are slightly elongated, which renders the energies of the head-to-tail and the antiparallel side-to-side configuration more equal. It has been shown that for a narrow regime of aspect ratios a liquid phase condensates [39]. The relation between chaining and condensation has been studied in a number of theoretical papers [40, 41, 42, 38].

The solid phases of dipolar fluids are less well studied. In simulations a body-centered tetragonal structure, which has also been found experimentally in electrorheological fluids [43], as well as fcc structures with different types of orientational order were examined [5], but it was not possible to sort out the thermodynamically stable phase. Two early density-functional approaches [44, 45] did not find a ferromagnetic solid, but both suffer from algebraic errors [46] and an incorrect treatment of the long-ranged dipolar interaction [47]. Our own density-functional results for the freezing of the Stockmayer fluid are presented in Sec. 5. Ferromagnetism in amorphous dipolar solids is discussed in Refs. [48, 49, 50].

The ferromagnetic liquid should be more stable in fluids of oblate instead of spherical particles. Indeed a ferromagnetic phase is formed in a fluid of cut spheres while at lower temperatures a columnar phase with ferromagnetic order only within the columns is observed [51]. The relations between nematic and ferromagnetic order in fluids of dipolar hard ellipsoids are discussed in Sec. 6 within the framework of a simple density-functional theory.

Once the bulk phases and their coexistence are established one can study the interfacial properties which arise if different phases are brought into spatial contact with each other or with external walls. These cases can, in principle, be analyzed using the same density functionals as for the bulk. This is demonstrated in Sec. 7 for the interface between the isotropic gas and the isotropic liquid which exhibits local orientational order. For dipolar hard spheres and Stockmayer fluids other theoretical approaches to this problem can be found in Refs. [52, 53, 54]. There is also an extensive literature concerning more realistic models of specific substances such as water or methanol using lattice-gas theory [55, 56], integral equation theory [57], density-functional theory [58], and Monte Carlo simulation [59].

Non-planar interfaces arise in liquid drops and clusters. Metastable spherical liquid drops have been studied with density-functional theory by Talanquer and Oxtoby [60, 61] in connection with the analysis of liquid nucleation in supersaturated dipolar gases. In computer simulations of small dipolar clusters [62, 63] interestingly at low temperatures one finds vortex-like orientational order similar to the bulk structure of the ferromagnetic phase as discussed in Sec. 3.

2. Model and theoretical approach

We study the Stockmayer model of a dipolar fluid. It is defined by the interaction potential

$$w(\mathbf{r}_{12},\omega,\omega') = w_{LJ}(r_{12}) + w_{dip}(\mathbf{r}_{12},\omega,\omega') \tag{1}$$

where $\mathbf{r}_{12} = \mathbf{r} - \mathbf{r}'$ is the interparticle vector and the orientation of the dipole moments is denoted by $\omega = (\theta,\phi)$. The Lennard-Jones potential $w_{LJ}(r) = 4\epsilon[(\sigma/r)^{12} - (\sigma/r)^6]$ with parameters ϵ and σ describes the spherically symmetric dispersion interaction as well as the short-ranged repulsion. The dipolar potential due to point dipoles embedded in the center of the particles is given by

$$w_{dip}(\mathbf{r},\omega,\omega') = -\frac{m^2}{r^3}\left(3(\hat{\mathbf{m}}(\omega)\hat{\mathbf{r}})(\hat{\mathbf{m}}(\omega')\hat{\mathbf{r}}) - \hat{\mathbf{m}}(\omega)\hat{\mathbf{m}}(\omega')\right) \tag{2}$$

where carets denote unit vectors and m is the absolute value of the dipole moment. A natural extension of this model are mixtures of two polar or polarizable species which also are important in a number of technicological applications [64, 65].

For a start the Stockmayer fluid is treated within the framework of a relatively simple density-functional theory, which can be derived from an exact expression for the density functional using the low-density limit $\exp(-\beta w)$ for the pair distribution function [66]. (For more sophisticated choices see Ref. [11].) The resulting functional of the one-particle number density $\hat{\rho}(\mathbf{r},\omega)$ of particles at point $\mathbf{r}$ and with orientation ω is

$$\begin{aligned}\Omega[\{\hat{\rho}(\mathbf{r},\omega)\},T,\mu] &= \frac{1}{\beta}\int d^3r d\omega\, \hat{\rho}(\mathbf{r},\omega)\left[\ln\left(4\pi\Lambda^3\hat{\rho}(\mathbf{r},\omega)\right) - 1\right] \\ &+ \int d^3r\, f_{CS}(\rho(\mathbf{r})) - \mu\int d^3r d\omega \hat{\rho}(\mathbf{r},\omega)\end{aligned} \tag{3}$$

$$-\frac{1}{2\beta}\int d^3r d^3r' d\omega d\omega'\, \hat{\rho}(\mathbf{r},\omega)\hat{\rho}(\mathbf{r}',\omega')\Theta(r_{12}-\sigma)f(\mathbf{r}_{12},\omega,\omega').$$

Here Λ denotes the thermal de Broglie wavelength, $\rho(\mathbf{r}) = \int d\omega \hat{\rho}(\mathbf{r},\omega)$ is the orientationally averaged density profile, $f = \exp(-\beta w) - 1$ is the Mayer function, $\beta = 1/(k_B T)$ corresponds to the inverse temperature, and μ is the chemical potential. The first term in Eq. (3) corresponds to the ideal gas free energy while the second term represents the excess free energy of the hard sphere reference system in a local density approximation with the Carnahan-Starling free energy density of hard spheres of packing fraction $\eta = \frac{\pi}{6}\rho d^3$:

$$\beta f_{CS}(\rho) = \rho\frac{4\eta - 3\eta^3}{(1-\eta)^3}. \tag{4}$$

The temperature dependent diameter $d = \int_0^\sigma dr(1 - \exp(-\beta w_{LJ}(r)))$, as given by Barker and Henderson [67], accounts for the actually soft repulsion. The third term in Eq. (3) describes the contributions to the free energy due to the long-ranged part of the interactions in a mean-fieldlike manner. The equilibrium density distribution for a given thermodynamic state minimizes the above density functional.

For bulk fluid phases the orientationally averaged density ρ is taken to be constant and the orientational distribution to be axisymmetric, so that with

$$\hat{\rho}(\mathbf{r},\omega) = \rho(\mathbf{r})\alpha(\mathbf{r},\omega), \qquad \int d\omega\, \alpha(\mathbf{r},\omega) = 1, \tag{5}$$

in general, here one has

$$\hat{\rho}(\mathbf{r},\omega) = \rho\alpha(\cos\theta) = \frac{\rho}{2\pi}\sum_{l=0}^{\infty}\alpha_l P_l(\cos\theta). \tag{6}$$

In the isotropic phase $\alpha(\cos\theta) = 1/(4\pi)$, i.e., $\alpha_l = \delta_{l0}/2$, while in the ferromagnetic phase also the higher expansion coefficients $\alpha_{l\geq 1}$ take on non-zero values. If one inserts the expansion in Eq. (6) into the density functional and employs a suitable expansion of the Mayer function into rotational invariants of the three spatial angles ω, ω' and ω_{12}, one encounters a conditionally convergent integral that is proportional to α_1^2 [19]. This is a consequence of the slow decay of the dipolar potential (and therefore of the Mayer function). The value of this integral for fluid volume $V \to \infty$ depends on the shape of the sample adopted for this limiting process. This integral was determined for ellipsoidal shapes of arbitrary aspect ratio k and shown to be related to the demagnetization factor known from classical magnetostatics. On the other hand it has been shown by general arguments that the free energies and the phase diagrams of dipolar systems in zero field do not depend on the sample shape [22]. This apparent contradiction is reconciled by the observation that the true equilibrium state of the ferromagnetic liquid corresponds to such a shape-dependent inhomogeneous configuration that in the resulting free energy the shape dependence drops out. Using a general spatially inhomogeneous orientational distribution, within the present theory we have confirmed that the equilibrium free energy indeed does not depend on the sample shape and is the same as that of the homogeneous state in a needle-shaped volume ($k \to \infty$) [21]. Therefore the phase diagram determined for this special case is valid for all shapes keeping in mind that for a general shape the ferromagnetic phase exhibits a nontrivial spatial structure. Details of this structure are discussed in the next section. We remark that these inhomogeneities are suppressed by the so-called "tin-foil" boundary conditions that are usually applied in computer simulations, i.e., by surrounding the sample with an infinitely permeable material (a perfect conductor in the electric case, respectively). The free "magnetic charges" in the surrounding material cancel the demagnetization field which leads to shape independence also for homogeneous magnetization. In the following we only consider the more realistic open boundary conditions.

The properties of the Stockmayer fluid can be described in terms of the following dimensionless quantities:

$$T^* = k_B T/\epsilon, \qquad \rho^* = \rho\sigma^3, \qquad m^* = m/\sqrt{\sigma^3\epsilon}. \tag{7}$$

Figure 1 displays the phase diagrams in the temperature-density plane as obtained from the present theory for three different values of the reduced dipole moment m^*. At low dipole moments there is a large coexistence region of the isotropic gas and liquid phases and a second-order transition to the ferromagnetic liquid along a line of critical points. This line ends at the coexistence curve in a critical end point. Upon increasing the dipole moment the isotropic-ferromagnetic transition shifts to lower densities and higher temperatures and a tricritical point emerges below which the transition becomes first order. Moreover all three fluid phases coexist at a triple point. At even higher dipole moments the gas-liquid coexistence becomes metastable and is preempted by the isotropic gas-ferromagnetic liquid transition, so that one is left with only two fluid phases. A similar series of phase diagrams has been obtained from a phenomenological theory [9], while in simulations to date neither the order of the transition nor the topology of the phase diagram have been mapped out.

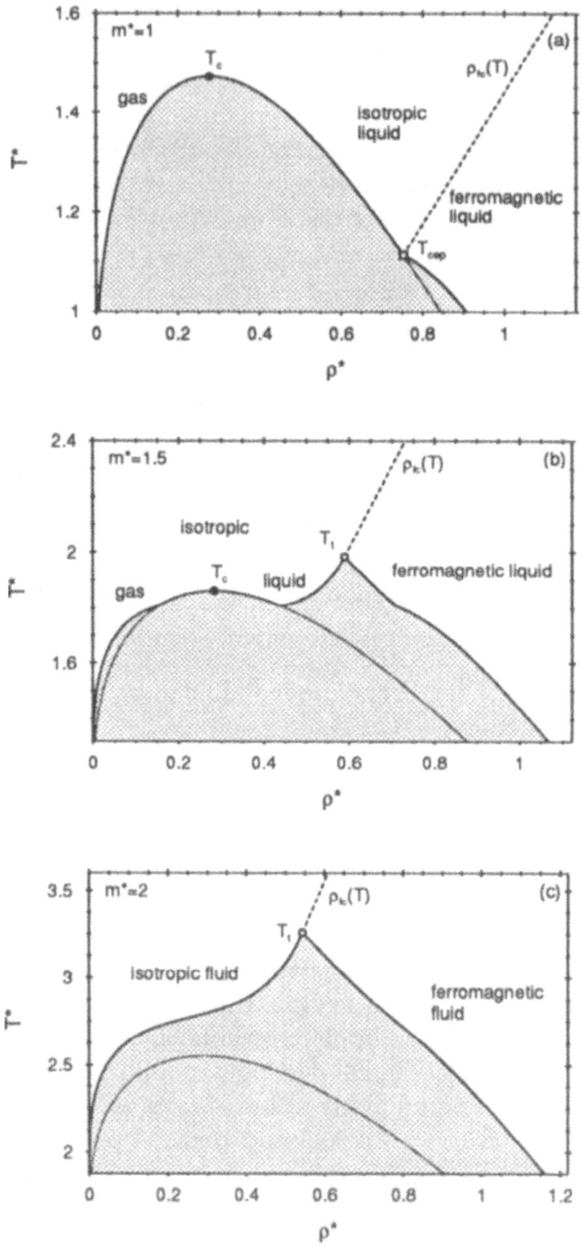

Figure 1. Phase diagrams of the Stockmayer fluid for three different values of the reduced dipole moment m^*, including the ferromagnetic liquid phase. The dashed line $\rho_{fc}(T)$ denotes a line of critical points, T_c is the gas-liquid critical point, and the shaded areas represent two-phase regions. In part (a) the critical line ends at the coexistence line in a critical end point T_{cep}, while in (b) and (c) a tricritical point T_t arises, below which the isotropic-ferromagnetic transition becomes first order. The dotted lines represent the metastable parts of the coexistence curve of the isotropic phases.

3. Inhomogeneous magnetization

The arguments leading to the proof of the shape independence mentioned above provide also a characterization of the actual inhomogeneous magnetization configuration for a given sample shape. Locally it should exhibit the same orientational distribution that has been determined for the needle-shaped sample, but with a spatially varying preferential direction. This variation should correspond to a magnetization field $\mathbf{M}(\mathbf{r})$ which on the scale of the system size has zero divergence and a vanishing normal component at the surface [21]. However, for a given shape these conditions do not determine $\mathbf{M}(\mathbf{r})$ uniquely and we are not aware of a systematic method to construct fields consistent with them. Therefore we pursued an alternative approach by computing this field via a quasi-free minimization of the density functional. A cubic discretization lattice with lattice constant a and lattice vectors $\mathbf{R}$ is superimposed on a cubic sample which we have chosen as an example. From Eq. (3) one derives the following approximation for the free energy difference ΔF between the isotropic and ferromagnetic phases [68]:

$$\begin{aligned}\beta\Delta F &= \rho a^3 \sum_{\mathbf{R}} \sum_{n=1}^{\infty} \frac{(\sqrt{12\pi}M(\mathbf{R}))^{2n}}{(2n-1)2n(2n+1)} \\ &\quad + \frac{1}{2}\rho^2 a^6 \sum_{\mathbf{R},\mathbf{R}'} \sum_{i,j} M_i(\mathbf{R}) w_{ij}(\mathbf{R}-\mathbf{R}') M_j(\mathbf{R}')\end{aligned} \tag{8}$$

with the interaction tensor

$$w_{ij}(\mathbf{R}) = \Theta(R-\sigma)\left[f_{110}(R)\delta_{ij} + f_{112}(R)(\delta_{ij} - 3\hat{R}_i\hat{R}_j)\right]. \tag{9}$$

The functions f_{110} and f_{112} are certain expansion coefficients of the Mayer function that are known analytically. To lowest order in m^2 one has $f_{110} \sim m^6$ and $f_{112}(r) \sim (1 - e^{-\beta w_{LJ}})m^2/r^3$ so that $w_{ij}(\mathbf{R}-\mathbf{R}')$ reproduces the dipolar interaction between the points $\mathbf{R}$ and $\mathbf{R}'$. This expression is minimized with respect to the values of $\mathbf{M}(\mathbf{R})$ at all lattice points using the simulated annealing algorithm [69]. Up to $N = 24$ points for each spatial dimension are used which amounts to $3 \times 24^3 = 41472$ variables.

Results for the thermodynamic parameters $m^* = 1.5$, $T^* = 2.25$, and $\rho^* = 0.94$ are shown in Fig. 2. One of the three perpendicular cubic axes is chosen spontaneously as a vortex axis around which the magnetization field circulates. As one can infer from the sections parallel to the vortex axis (i.e., the z axis) in Figs. 2(c) and (d) $\mathbf{M}(\mathbf{r})$ lies essentially in the vortex plane except close to the axis where it "escapes" into the third dimension in order to avoid a line singularity. Since the magnetization is approximately parallel

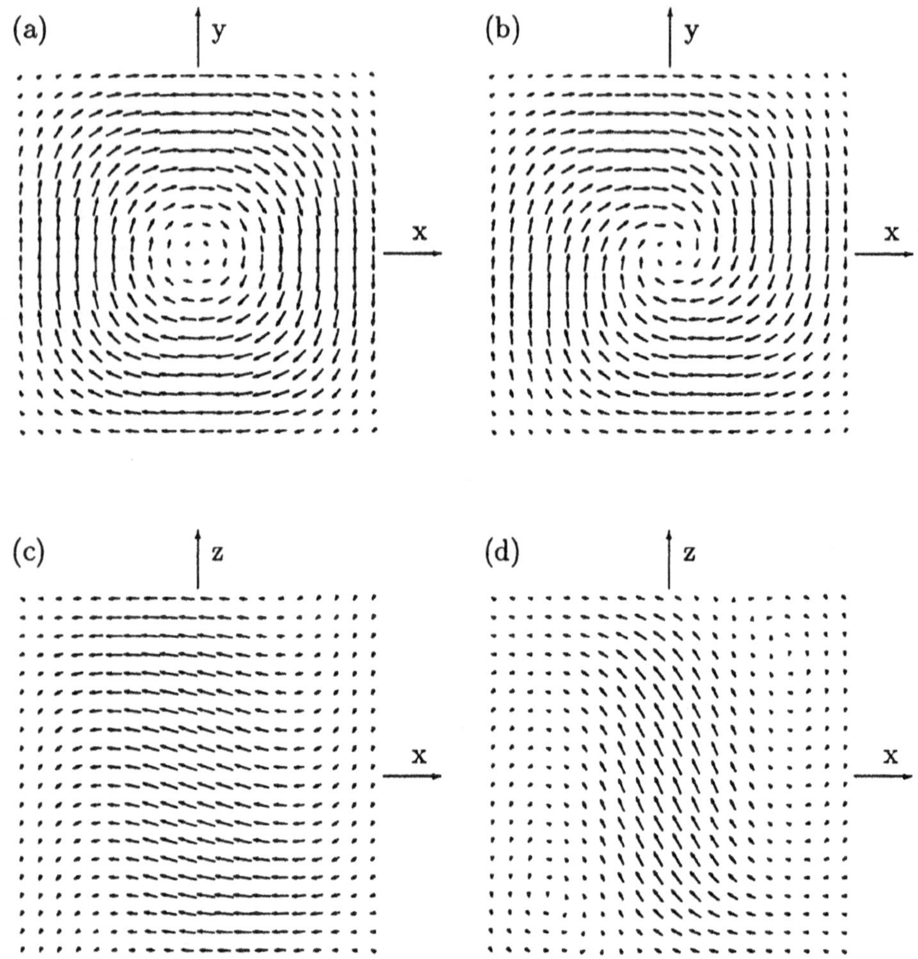

Figure 2. **Magnetization structure of a ferromagnetic liquid in a cubic volume with edge length $L/\sigma = 7.2$, 20^3 mesh points, and $T^* = 2.25$. The four parts of the figure represent sections (a) for fixed $z = 0.025L$, (b) fixed $z = -0.425L$, (c) fixed $y = -0.075L$, and (d) fixed $y = -0.225L$; the origin is at the center of the cube. The arrows represent projections of the local magnetization at their midpoint onto the section plane. The z axis is the vortex axis. Part (d) demonstrates that in the vicinity of this axis the magnetization "escapes" from the vortex plane.**

at the surface two point defects at the centers of the top and bottom face are inevitable. The absolute value $M(\mathbf{r})$, which in contrast to the common assumption in micromagnetic calculations [70] is not fixed at a constant value, is strongly reduced within the cores of the point defects whereas it is approximately constant throughout the rest of the volume with the exception of thin surface layers. At higher temperatures in the vicinity of the phase transition to the isotropic liquid a reduction of $M(\mathbf{r})$ occurs also along the whole vortex axis.

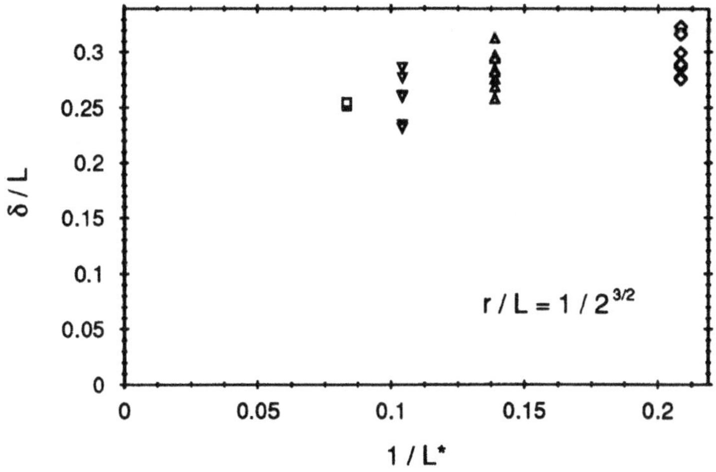

Figure 3. The domain wall thickness δ in a cubic sample measured within the plane $z = 0$ at distance $r = L/2^{3/2}$ from the center, i.e., at half distance between the center and the edges. The different results for the same value $L^* = L/\sigma$ correspond to different numbers of mesh points used for the minimization. The extrapolation to $1/L = 0$ suggests a finite value of δ/L in the thermodynamic limit.

The structure in Fig. 2 can also be considered to consist of four triangular domains with four broad 90° domain walls along the diagonals. These walls are clearly induced by the cube edges and would probably be absent in a cylindrical sample. We have computed a suitably defined wall thickness δ [21] for different sample sizes L. In Fig. 3 we show the results as function of $1/L$ up to the largest systems with $L/\sigma = 12$. The various points for the same L correspond to different values of N and thus give an indication of the error induced by the discretization. The data are consistent with the conjecture $\delta/L \to$ const for $L \to \infty$ which would mean that the wall thickness is determined by the system size [71]. This contrasts with the finite wall thickness in solid ferromagnets that is caused by the existence of easy axes induced by the lattice anisotropy. The whole magnetization configuration fulfills, to a good approximation, the scaling relation

$$\mathbf{M}(r/\sigma, L/\sigma) \approx \mathbf{M}^{(0)}(\mathbf{r}/L) \tag{10}$$

when different sizes within the accessible regime are compared. We conjecture that this holds also in the thermodynamic limit.

The average magnetization of the sample $\langle \mathbf{M} \rangle = V^{-1} \int d^3r \mathbf{M}(\mathbf{r})$ is approximately zero although $|\mathbf{M}(\mathbf{r})| \neq 0$. The positive contribution to $\langle M_z \rangle$ from the "escape" near the vortex axis is compensated by negative contributions from the regions close to the edges. However, a finite value of $\langle M_z \rangle$ is expected for cylindrical or spherical [72] samples.

4. External field

An external magnetic field tends to rotate all dipoles into the field direction and thus to destroy any inhomogeneity of $\mathbf{M}(\mathbf{r})$. Let us first consider sample shapes which induce a planar zero-field structure. As we have seen, this is approximately true for the cube, but most probably also for a torus [73]. We consider the case that the field is applied perpendicularly to the magnetization plane. If one assumes that the angle by which the magnetization is rotated out of plane is the same throughout the sample one can still treat the inhomogeneous problem quasi-analytically [21]. Upon increasing the external field the parallel component of $\mathbf{M}$ first increases linearly and finally saturates while the perpendicular components decrease and vanish with a square root singularity at a critical field strength H_c above which the sample is homogeneously magnetized. One can also determine phase diagrams at fixed external fields which now do depend on the sample shape via the demagnetization factor D [21]. The results for a cubic or spherical sample ($D = 1/3$) with $m^* = 1.5$ are shown in Fig. 4. The critical line between homogeneous isotropic and inhomogeneous ferromagnetic fluids in zero field turns into a critical line between homogeneously and inhomogeneously magnetized fluids. With increasing field strength this line together with the tricritical point shifts to higher densities and lower temperatures. (Only for a needle-shaped sample ($D = 0$) this line disappears and the tricritical point turns into a second critical point because all phases are homogeneous.) The small shift of the liquid-gas critical temperature is proportional to H^2 for small fields in accordance with the results of perturbation theory [75]. The influence of an external field on the liquid-gas coexistence has also been studied in Refs. [76, 7, 77].

If the external field is not perpendicular to the magnetization plane, for small fields the structure is still similar to the zero field result. On the other hand, for example in a cube there are six degenerate zero-field solutions. By using the free minimization procedure described in the previous section one can address the question which of them is selected by a small external field parallel to one of the cube axes. It turns out that a slightly perturbed vortex structure with its axis parallel to the field is favored [68]. The further evolution of the structure with increasing field strength takes place as described above. Again the azimuthal magnetization component around

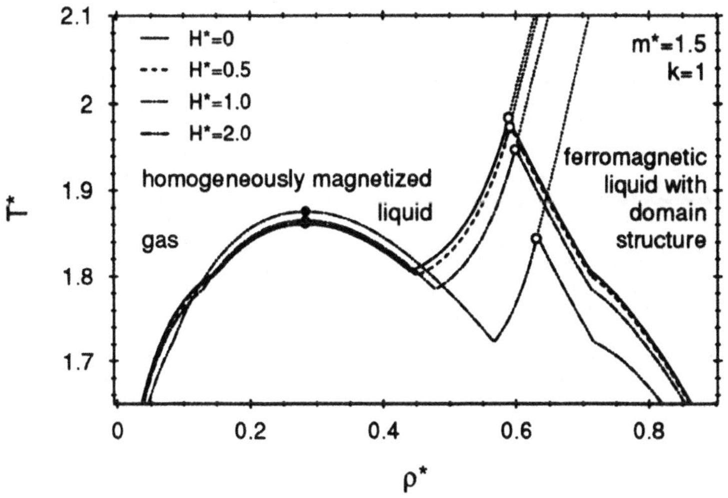

Figure 4. Phase diagrams for $m^* = 1.5$ and various fixed values of the reduced external field strength $H^* = H\sqrt{\sigma^3/\epsilon}$ for a sample with aspect ratio $k = 1$ (i.e., demagnetization factor $D = 1/3$). For nonzero fields the inhomogeneous domain structure is partly retained at high densities and low temperatures. The magnetic field shifts the transition line of critical points (dotted curve) and the tricritical point (open circle) towards higher densities and lower temperatures, whereas the gas-liquid critical temperature (full circle) is slightly increased.

the field direction vanishes at a critical field strength H_c, leaving an almost homogeneously magnetized sample for $H > H_c$. A comparison between the structures in zero field and in a finite field below H_c is given in Fig. 5. It is interesting to consider the corresponding behavior of $\mathbf{M}(\mathbf{r})$ for a *spherical* sample which possesses infinitely many equivalent directions. Fredkin and Koehler [74] have performed micromagnetic calculations for solid spherical particles. For low lattice anisotropy they find again a vortex state, but the vortex axis rotates away from the field direction as H vanishes.

5. Solid phases

It is important to obtain an estimate for the position of the freezing transition in the phase diagram in order to know whether the formation of the ferromagnetic liquid is preempted by freezing. In order to get a reasonable description of the solid phase we have chosen a more sophisticated density functional [47]. The hard sphere reference fluid is now treated within the modified weighted density approximation (MWDA) introduced by Denton and Ashcroft [78] which gives very accurate results for the freezing properties of hard spheres. The perturbative contribution is changed in a way that

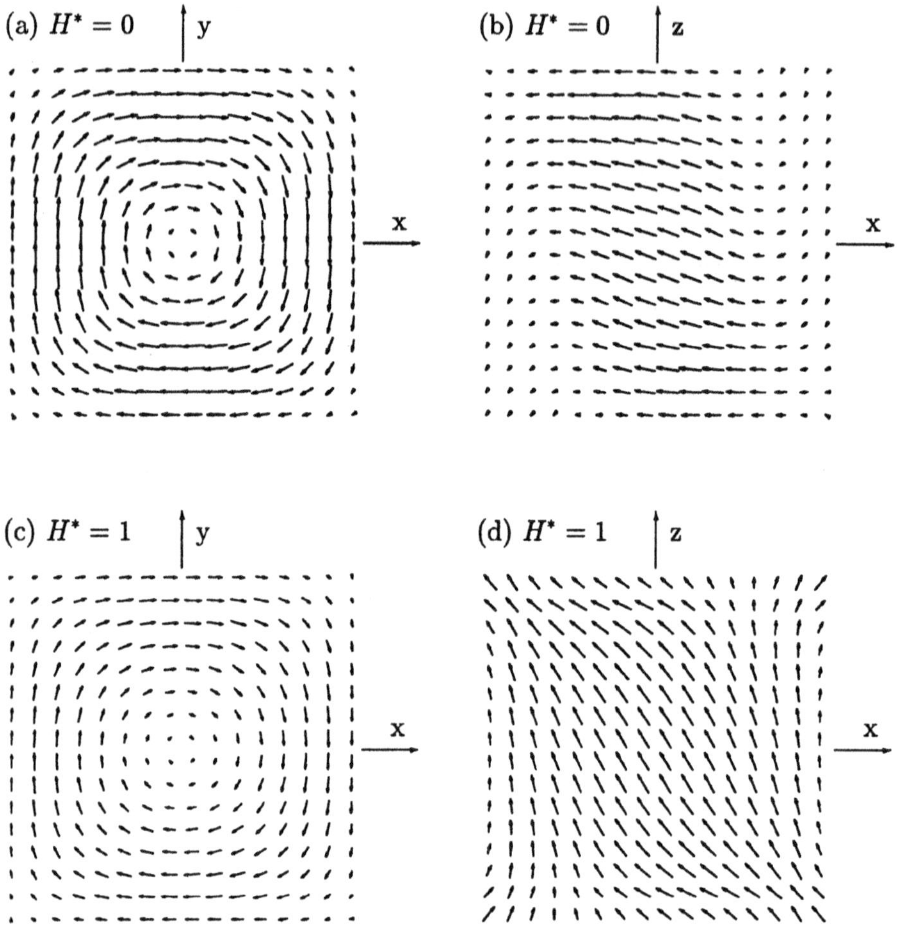

Figure 5. Sections of the magnetization structure for $L/\sigma = 9.6$ and $N = 16$ (a),(b) without and (c),(d) with an external field $H^* = H\sqrt{\sigma^3/\epsilon} = 1$ applied in the z direction. The scale factor which determines the lengths of the arrows is the same in all parts of this figure. Parts (a) and (c) are sections orthogonal to the field direction at $z/\sigma = -0.9$. The transversal components of the magnetization are reduced by the field but the overall features of the inhomogeneous structure are preserved. Parts (b) and (d) are sections perpendicular to the field for $y/\sigma = -2.1$. At this distance of the plane from the center the z component of the magnetization in zero field is only small, while the external field induces a large z component everywhere.

reproduces the results of Curtin and Ashcroft [79] in the limit of a Lennard-Jones fluid but also includes a contribution from the Mayer function of the dipolar potential as in Eq. (3). The number density is parametrized as

$$\hat{\rho}(\mathbf{r},\omega) = \rho(\mathbf{r})\alpha(\omega) = \frac{1}{2\pi}\left(\frac{\lambda}{\pi}\right)^{3/2} \sum_{\mathbf{R}} e^{-\lambda(\mathbf{r}-\mathbf{R})^2} \sum_{l=0}^{\infty} \alpha_l P_l(\cos\theta) \qquad (11)$$

where $1/\sqrt{\lambda}$ is the width of the density peaks at the lattice points $\mathbf{R}$. The shape dependent contribution to the density functional has been calculated also for a solid structure. For cubic lattices it has been shown to be the same as that of a liquid.

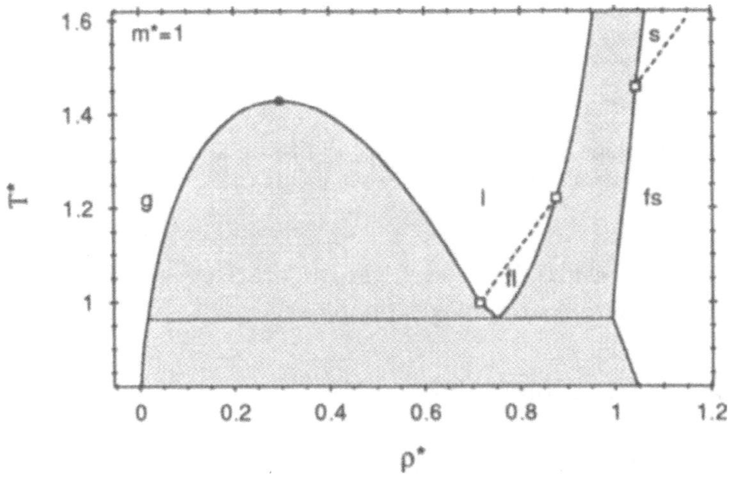

Figure 6. Phase diagram for $m^* = 1$ including the solid phase as obtained from the generalized density-functional theory. It encompasses five phases: the isotropic gas (g) and liquid (l), the ferromagnetic liquid (fl), the ferromagnetic solid (fs), and the solid without orientational order (s). Two-phase regions are shaded, the dashed lines indicate continuous phase transitions, and the dotted line denotes the triple point. Critical end points are denoted by squares.

For the considered parameter values ($m^* \leq 2$) the fcc lattice was found to be more stable than the bct (body centered tetragonal) structure observed in simulations [5]. The phase diagrams obtained by this approach (Figs. 6 and 7) comprise five different phases, among them a ferromagnetic solid (fs) magnetized in the (111) direction, and a plastic solid (s) without orientational order. The transition between the liquid phases with and without orientational order extends into the solid region. For dipole moments larger than $m^* \approx 1$ there is indeed a region with a stable ferromagnetic liquid (fl) phase. However, based on a different density-functional theory Klapp and Forstmann [11] have claimed recently that the latter phase is metastable with respect to the solid for all dipole moments, at variance with

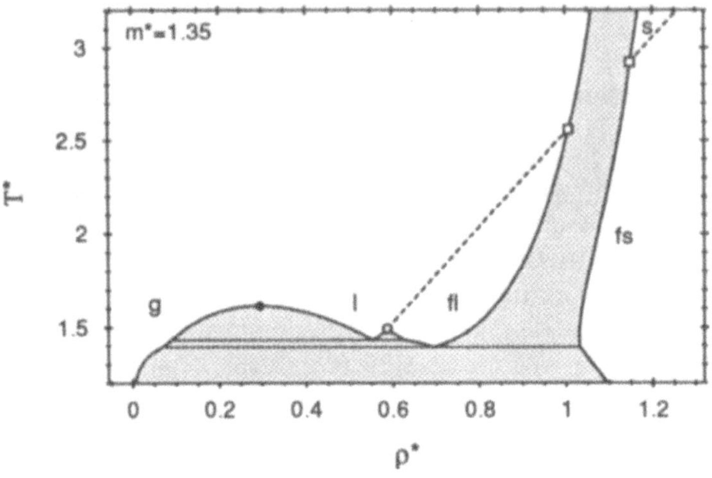

Figure 7. The same as Fig. 6, but for $m^* = 1.35$. In addition to the features of Fig. 6 a tricritical point (open circle) and two triple points (dotted lines) occur.

the simulation results. Thus even the qualitative features of the Stockmayer fluid phase diagram are not yet finally settled.

6. Nonspherical particles

In fluids consisting of nonspherical particles an orientationally ordered liquid phase may already occur in the absence of dipolar interactions. This nematic phase can be stabilized by short-ranged steric interactions between sufficiently elongated or disklike particles. It has a higher symmetry than a ferromagnetic liquid as particle orientations parallel and antiparallel to the preferential direction have identical probabilities. It is interesting to study the interplay of these two ordering mechanisms in a system of nonspherical dipolar particles. Specifically we have considered hard ellipsoids and hard spherocylinders with a central longitudinal dipole moment [80]. In this case, the reference part of the density functional coincides with the hard core part which is well approximated by the so-called decoupling approximation that amounts to replacing the direct correlation function by that of a hard-sphere fluid scaled by the distance of closest approach for given orientations of the particles and their joining vector. The dipolar interaction is again treated perturbatively within the low-density approximation that yields an expression bilinear in the densities with the Mayer function kernel as in Eq. (3).

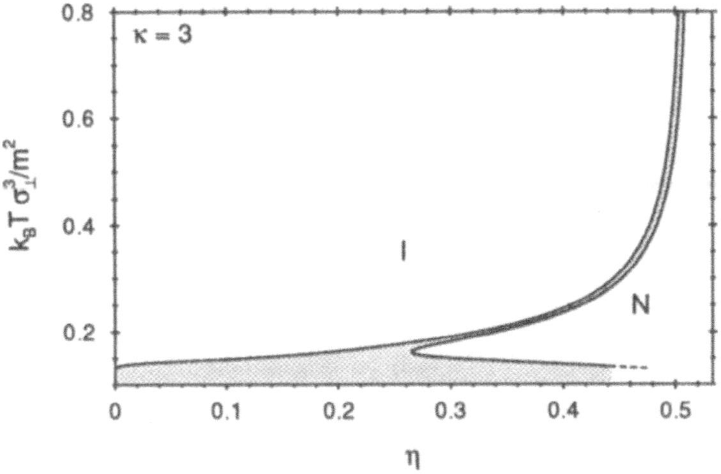

Figure 8. Phase diagram of hard ellipsoids of revolution with a longitudinal dipole moment and aspect ratio $\kappa = 3$. η is the packing fraction. At high temperatures or small dipole moments the transition between the isotropic (I) and the nematic (N) phases is weakly first order, but it widens into a gas-nematic coexistence at low temperatures. Possible smectic and solid phases cannot be described within the present theory.

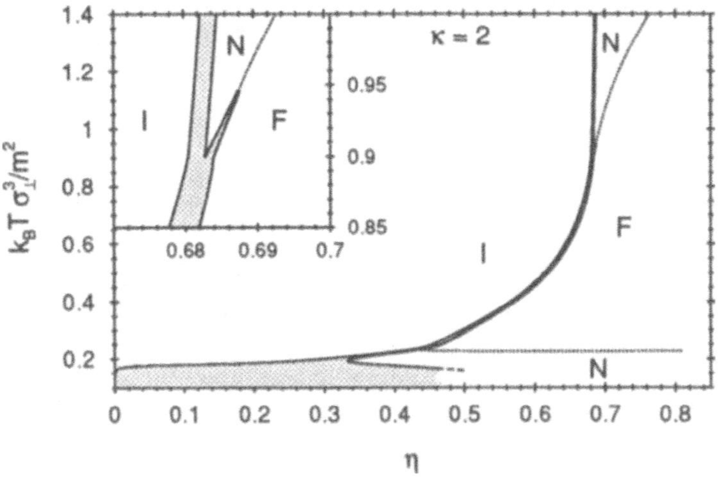

Figure 9. The same as Fig. 8 but for aspect ratio $\kappa = 2$. Besides the isotropic (I) and nematic (N) phases a ferromagnetically ordered liquid (F) occurs in the intermediate temperature range. The dotted and solid lines denote second- and first-order phase transitions, respectively. The density gap of the I-N transition at $\eta \simeq 0.68$ cannot be resolved on the present scale. The inset shows that in a narrow temperature range the high temperature continuous N-F transition is turned into a weakly first-order transition generating a tricritical point and an I-N-F triple point. The reentrant nematic phase at low temperatures is probably an artefact of the present theory.

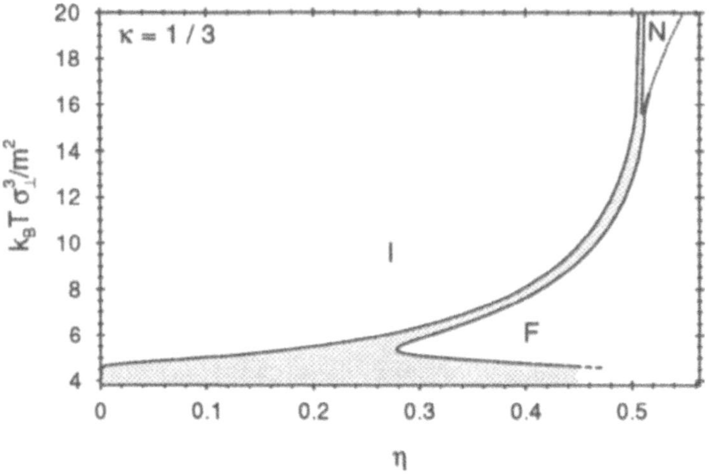

Figure 10. **Phase diagram for oblate ellipsoids with aspect ratio $\kappa = 1/3$. The I-N coexistence packing fractions for $T \to \infty$ are the same as those for $\kappa = 3$, but in contrast to Fig. 8 a ferromagnetic phase is stable at high densities.**

The properties of this model are determined by the packing fraction η, the dimensionless ratio $k_BT\sigma_\perp^3/m^2$, where $\sigma_\perp$ and $\sigma_\parallel$ are the particle diameters measured perpendicular and parallel to the dipole moment, respectively, and the aspect ratio $\kappa = \sigma_\parallel/\sigma_\perp$ of the particles. We find that for elongated particles the isotropic-nematic transition is shifted to lower densities by the dipole moment. This trend is in agreement with earlier results from the Onsager virial theory [81], integral equation theory [82], and a different version of density-functional theory [83]. The two-phase region widens into gas-nematic coexistence at high dipole moments or low temperatures (Fig. 8). For ellipsoids of aspect ratio $\kappa \lesssim 2$ (Fig. 9) a ferromagnetic fluid (F) appears at high densities, as well as a reentrant nematic phase at high dipole moments that most probably is an artefact of the theoretical approximations. The ferromagnetic phase is reached from the nematic or isotropic states by continuous (dotted lines) or weakly first-order (solid lines) transitions, depending on the temperature and particle aspect ratio. Similar phase diagrams are found for oblate particles ($\kappa < 1$, see Fig. 10). If one compares fluids with aspect ratios κ and $1/\kappa$ the isotropic-nematic coexistence packing fractions are the same for $m = 0$ (or $T \to \infty$) but the ferromagnetic phase becomes stable at much higher values of k_BTV_0/m^2 (V_0 is the particle volume) for a given packing fraction if $\kappa < 1$, because in this case the dipole moments can approach each other closer in the most favorable head-to-tail configuration. For the same reason the ferromagnetic liquid should be easier to attain in fluids of oblate than of spherical particles. For dipolar hard spherocylinders one finds an analogous series of phase diagrams as for elongated ellipsoids [80].

7. Gas-liquid interface

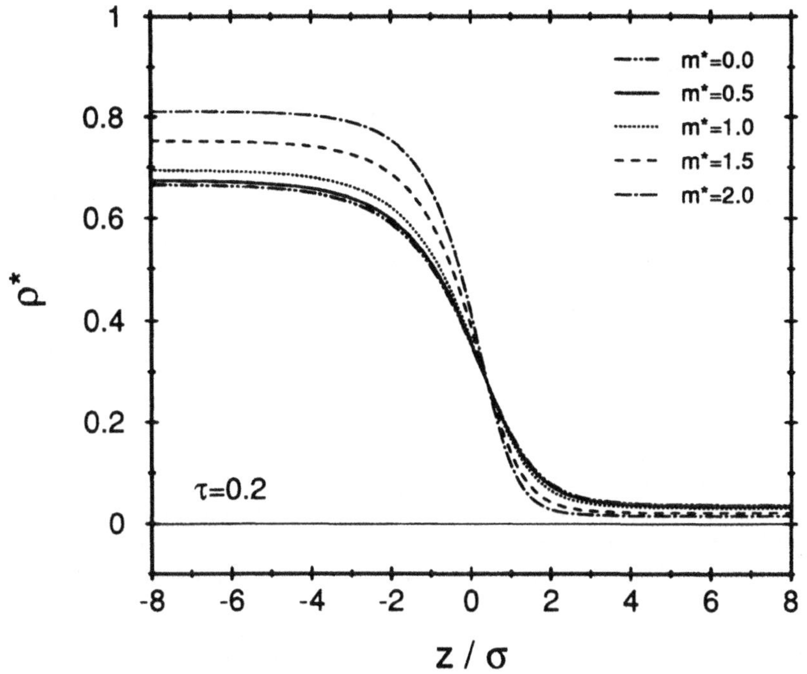

Figure 11. Density profiles for the gas-liquid interface of a Stockmayer fluid for various dipole moments at a fixed reduced temperature $\tau = 1 - T/T_c = 0.2$.

Up to this point we were only concerned with bulk properties of dipolar fluids and the influence of sample surfaces. If one allows for the presence of interfaces between coexisting bulk phases various new phenomena arise. By imposing appropriate boundary condition in principle one can examine interfaces between any two phases that can coexist according to the phase diagrams shown in the previous sections. These interfaces can be studied with the same density-functional theories, although the calculations will be much more involved due to the lower symmetry of $\hat{\rho}(\mathbf{r}, \omega)$. Here we restrain ourselves to the simplest case, i.e., the interface between the isotropic gas and the isotropic liquid [54, 66, 84]. For this system the number density depends on the coordinate z normal to the interface and the local orientational order near the interface is described by the first nontrivial coefficient

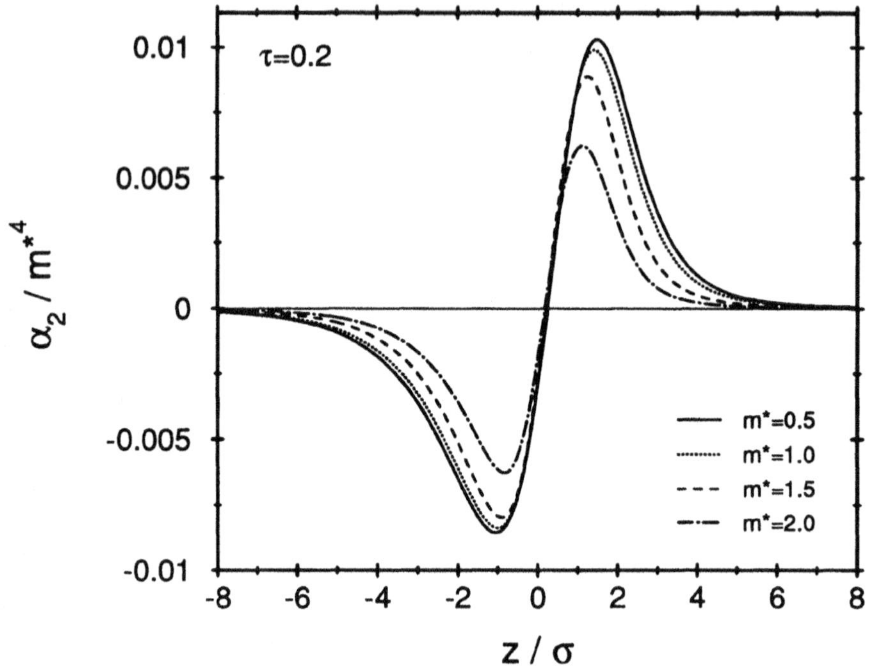

Figure 12. Orientational order profiles $\alpha_2(z)$ at the gas-liquid interface for the same parameters as in Fig. 11. Note that α_2 is scaled by m^{*4}. The dipoles are aligned preferentially parallel to the interface on the liquid side and perpendicular to the interface on the gas side. For $m^* \to 0$ the profiles approach a limiting form $\alpha_2(z/\sigma, m^*) = m^{*4}\alpha_2^{(0)}(z/\sigma)$.

$\alpha_2(z)$ in an expansion in terms of Legendre polynomials:

$$\hat{\rho}(\mathbf{r}, \omega) = \hat{\rho}(z, \cos\theta) = \frac{1}{2\pi}\rho(z)\left(\frac{1}{2} + \alpha_2(z)P_2(\cos\theta) + \dots\right). \tag{12}$$

(The higher coefficients $\alpha_{l>2}(z)$ are very small.)

The functions $\rho(z)$ and $\alpha_2(z)$ have been calculated numerically by solving the integral equation that results from the stationary condition for the density functional. The density profiles are monotonic and for small reduced dipole moment m^* they depend only weekly on m^* for fixed distance $\tau = 1 - T/T_c$ from the critical temperature T_c (see Fig. 11). Near the interface the particles are preferentially aligned parallel to the interface ($\alpha_2(z) < 0$) on the liquid side and perpendicular to the interface ($\alpha_2(z) > 0$) on the vapor side (Fig. 12). The amplitude of $\alpha_2(z)$ increases strongly ($\sim m^{*4}$) with the dipole moment. Near the critical point both the density profile

$$\rho^*(z, \tau \to 0) = \rho_c^* + A_\rho(m^*)\tau^\beta F_\rho(z/\xi) \tag{13}$$

and the orientational profile

$$\alpha_2(z, \tau \to 0) = A_{\alpha_2}(m^*)\tau^{\beta+2\nu} F_{\alpha_2}(z/\xi) \tag{14}$$

exhibit a scaling behavior. Here ξ is the bulk correlation length, F_ρ and F_{α_2} are scaling functions, and A_ρ and A_{α_2} are nonuniversal amplitudes. Within the density functional approach the critical exponents β and ν of the order parameter and of the correlation length take on their mean-field value $\frac{1}{2}$. Equation (14) has been used to interpret ellipsometric measurements of the liquid-liquid interface of the critical ionic mixture triethyl-n-hexylammonium triethyl-n-hexylboride dissolved in diphenyl ether [85].

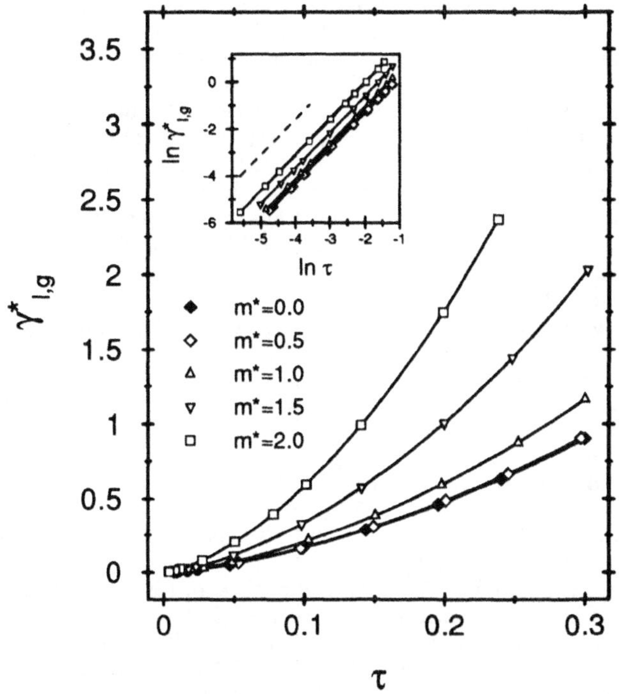

Figure 13. Temperature dependence of the liquid-gas surface tension $\gamma^*_{l,g} = \gamma_{l,g}\sigma^2/\epsilon$ for various dipole moments. The double logarithmic plot in the inset demonstrates that $\gamma_{l,g} \sim \tau^{3/2}$ for $\tau \to 0$.

The temperature dependence of the surface tension γ, which can be calculated from these profiles, is shown in Fig. 13. The limiting power law $\gamma(\tau \to 0) = \gamma_0 \tau^{3/2}$ is valid even quite far away from T_c. The amplitude γ_0 increases with increasing dipole strength. The values of γ are (for $m^* \leq 2$) practically unaltered if the orientational profile $\alpha_2(z)$ is replaced by $\alpha_2(z) \equiv 0$. This means that the main influence of the dipolar interaction on the surface tension is due to the change of the bulk phase diagram and of the number density profile but not due to preferential orientations at the interface. This result has been exploited for obtaining estimates of the surface tension of polar fluids from other, more simple analytical approaches and simulations [86].

References

1. M. Born, Sitz. Phys. Math. **25**, 614 (1916); Ann. Phys. **55**, 221 (1918).
2. D. Wei and G.N. Patey, Phys. Rev. Lett. **68**, 2043 (1992).
3. D. Wei and G.N. Patey, Phys. Rev. A **46**, 7783 (1992).
4. J.J. Weis, D. Levesque, and G.J. Zarragoicoechea, Phys. Rev. Lett. **69**, 913 (1992).
5. J.J. Weis and D. Levesque, Phys. Rev. E **48**, 3728 (1993).
6. M.J. Stevens and G.S. Grest, Phys. Rev. E **51**, 5962 (1995).
7. M.J. Stevens and G.S. Grest, Phys. Rev. E **51**, 5976 (1995).
8. K. Sano and M. Doi, J. Phys. Soc. Jpn. **52**, 2810 (1983).
9. H. Zhang and M. Widom, Phys. Rev. E **49**, R3591 (1994).
10. D. Wei, G.N. Patey, and A. Perera, Phys. Rev. E **47**, 506 (1993).
11. S. Klapp and F. Forstmann, Europhys. Lett. **38**, 663 (1997).
12. H. Zhong and R.G. Petschek, Phys. Rev. E **51**, 2263 (1995).
13. J.S. Høye and G. Stell, Mol. Phys. **86**, 707 (1995).
14. S. Klapp and F. Forstmann, J. Chem. Phys. **106**, 9742 (1997).
15. M.E. van Leeuwen, B. Smit, and E.M. Hendriks, Mol. Phys. **78**, 271 (1993).
16. G. Kronome, J. Liszi, and I. Szalai, J. Chem. Soc. Faraday Trans. **93**, 3053 (1997); and references therein.
17. I. Szalai, G. Kronome, and T. Lukács, J. Chem. Soc. Faraday Trans. **93**, 3737 (1997).
18. C. Kriebel and J. Winkelmann, Mol. Phys. **90**, 297 (1997); and references therein.
19. B. Groh and S. Dietrich, Phys. Rev. Lett. **72**, 2422 (1994); ibid **74**, 2617 (1995); Phys. Rev. E **50**, 3814 (1994).
20. M.A. Osipov, P.I.C. Teixeira, and M.M. Telo da Gama, J. Phys. A. **30**, 1953 (1997).
21. B. Groh and S. Dietrich, Phys. Rev. E **53**, 2509 (1996).
22. R.B. Griffiths, Phys. Rev. **176**, 655 (1968).
23. J.M. Tavares, M.M. Telo da Gama, P.I.C. Teixeira, J.J. Weis, and M.J.P. Nijmeijer, Phys. Rev. E **52**, 1915 (1995).
24. M.J.P. Nijmeijer and J.J. Weis, Ann. Rev. Comp. Phys. **IV**, 1 (1996).
25. A. Oukouiss and M. Baus, Phys. Rev. E **55**, 7242 (1997).
26. M.J.P. Nijmeijer and J.J. Weis, Phys. Rev. Lett. **75**, 2887 (1995).
27. M.J.P. Nijmeijer and J.J. Weis, Phys. Rev. E **53**, 591 (1996).
28. M.J.P. Nijmeijer, A. Parola, and L. Reatto, Phys. Rev. E **57**, 465 (1998).
29. M.A. Załuska-Kotur and L.A. Turski, Phys. Rev. A **41**, 3066 (1990).
30. R.E. Rosensweig, *Ferrohydrodynamics* (Cambridge University Press, Cambridge, 1985); E. Blums, A. Cebers, and M.M. Maiorov, *Magnetic Fluids* (de Gruyter, Berlin, 1997).
31. Proceedings of the 5th International Conference on Electro-Rheological Fluids, Magneto-Rheological Suspensions and Associated Technology, Int. J. Mod. Phys. B **10**, Numbers 23 & 24 (1996).
32. D. Platzek, C. Notthoff, D.M. Herlach, G. Jacobs, D. Herlach, and K. Maier, Appl. Phys. Lett. **65**, 1723 (1994).
33. J. Reske, D.M. Herlach, F. Keuser, K. Maier, and D. Platzek, Phys. Rev. Lett. **75**, 737 (1995).
34. T. Albrecht, C. Bührer, M. Fähnle, K. Maier, D. Platzek, and J. Reske, Appl. Phys. A **65**, 215 (1997).

35. J.J. Weis and D. Levesque, Phys. Rev. Lett. **71**, 2729 (1993).
36. D. Levesque and J.J. Weis, Phys. Rev. E **49**, 5131 (1994).
37. M.E. van Leeuwen and B. Smit, Phys. Rev. Lett. **71**, 3991 (1993).
38. J.M. Tavares, M.M. Telo da Gama, and M.A. Osipov, Phys. Rev. E **56**, R6252 (1997).
39. S.C. McGrother and G. Jackson, Phys. Rev. Lett. **76**, 4183 (1996).
40. R.P. Sear, Phys. Rev. Lett. **76**, 2310 (1996).
41. R. van Roij, Phys. Rev. Lett. **76**, 3348 (1996).
42. M.A. Osipov, P.I.C. Teixeira, and M.M. Telo da Gama, Phys. Rev. E **54**, 2597 (1996).
43. T. Chen, R.N. Zitter, and R. Tao, Phys. Rev. Lett. **68**, 2555 (1992).
44. W.E. McMullen and D.W. Oxtoby, J. Chem. Phys. **86**, 4146 (1987).
45. S.J. Smithline, S.W. Rick, and A.D.J. Haymet, J. Chem. Phys. **88**, 2004 (1988).
46. W.E. McMullen and D.W. Oxtoby, J. Chem. Phys. **88**, 1476 (1988).
47. B. Groh and S. Dietrich, Phys. Rev. E **54**, 1687 (1996).
48. B.E. Vugmeister and M.D. Glinchuk, Rev. Mod. Phys. **62**, 993 (1990).
49. H. Zhang and M. Widom, J. Magn. Magn. Mat. **122**, 119 (1993); Phys. Rev. B **51**, 8951 (1995).
50. G. Ayton, M.J.P. Gingras, and G.N. Patey, Phys. Rev. Lett. **75**, 2360 (1995).
51. G. Ayton and G.N. Patey, Phys. Rev. Lett. **76**, 239 (1996).
52. J. Eggebrecht, K.E. Gubbins, and S.M. Thompson, J. Chem. Phys. **86**, 2286 (1987).
53. M. Kasch and F. Forstmann, J. Chem. Phys. **99**, 3037 (1993).
54. P.I. Teixeira and M.M. da Gama, J. Phys. Condens. Matter **3**, 111 (1991).
55. M. Matsumoto, H. Mizukuchi, and Y. Kataoka, J. Chem. Phys. **98**, 1473 (1993).
56. N.A.M. Besseling and J. Lyklema, J. Phys. Chem. **98**, 11610 (1994).
57. M.J. Booth, D.-M. Duh, and A.D.J. Haymet, J. Chem. Phys. **101**, 7925 (1994).
58. B. Yang, D.E. Sullivan, and C.G. Gray, J. Phys. Condens. Matter **6**, 4823 (1994).
59. V.P. Sokhan and D.J. Tildesley, Mol. Phys. **92**, 625 (1997); and references therein.
60. V. Talanquer and D.W. Oxtoby, J. Chem. Phys. **99**, 4670 (1993).
61. V. Talanquer and D.W. Oxtoby, J. Chem. Phys. **103**, 3686 (1995).
62. H.B. Lavender, K.A. Iyer, and S.J. Singer, J. Chem. Phys. **101**, 7856 (1994).
63. D. Lu and S.J. Singer, J. Chem. Phys. **103**, 1913 (1995).
64. D.J. Henderson and W. Schmickler, J. Chem. Soc. Faraday Trans. **92**, 3839 (1992).
65. C. Kriebel and J. Winkelmann, Mol. Phys. **90**, 297 (1997); and references therein.
66. P. Frodl and S. Dietrich, Phys. Rev. A **45**, 7330 (1992); Phys. Rev. E **48**, 3203 (1993).
67. J.A. Barker and D. Henderson, J. Chem. Phys. **47**, 4714 (1967).
68. B. Groh and S. Dietrich, Phys. Rev. Lett. **79**, 749 (1997); Phys. Rev. E, **57**, 4535 (1998).
69. W.H. Press, B.P. Flannery, S.A. Teukolsky, and W.T. Vetterling, *Numerical Recipes* (Cambridge University Press, Cambridge, 1989).
70. W.F. Brown jr., *Micromagnetics* (Krieger, Huntington, 1978).
71. P.G. de Gennes and P.A. Pincus, Solid State Communications **7**, 339 (1969).
72. A. Aharoni and J.P. Jakubovics, J. Magn. Magn. Mat. **83**, 451 (1990).
73. P.J. Wojtowicz and M. Rayl, Phys. Rev. Lett. **20**, 1489 (1968).
74. D.R. Fredkin and T.R. Koehler, IEEE Trans. Magn. **MAG-24**, 2362 (1988); J. Appl. Phys. **67**, 5544 (1990).
75. D. Boda, I. Szalai, and J. Liszi, J. Chem. Soc. Faraday Trans. **91**, 889 (1995).
76. C.E. Woodward and S. Nordholm, J. Phys. Chem. **92**, 501 (1988).
77. D. Boda, J. Winkelmann, J. Liszi, and I. Szalai, Mol. Phys. **87**, 601 (1996).
78. A.R. Denton and N.W. Ashcroft, Phys. Rev. A **39**, 4701 (1989).
79. W.A. Curtin and N.W. Ashcroft, Phys. Rev. Lett. **56**, 2775 (1986).
80. B. Groh and S. Dietrich, Phys. Rev. E **55**, 2892 (1997).
81. A.G. Vanakaras and D.J. Photinos, Mol. Phys. **85**, 1089 (1995).
82. A. Perera and G.N. Patey, J. Chem. Phys. **91**, 3045 (1989).

83. C. Vega and S. Lago, J. Chem. Phys. **100**, 6727 (1994).
84. P. Frodl and S. Dietrich, Phys. Rev. E **48**, 3741 (1993).
85. C.L. Caylor, B.M. Law, P. Senanayake, V.L. Kuzmin, V.P. Romanov, and S. Wiegand, Phys. Rev. E **56**, 4441 (1997).
86. S. Abbas, P. Ahlströmega, and S. Nordholm, Langmuir **14**, 396 (1998).

FLUIDS IN CONTACT WITH STRUCTURED SUBSTRATES

S. DIETRICH
Fachbereich Physik, Bergische Universität Wuppertal
D-42097 Wuppertal, Germany

Abstract. The contact between fluids and microstructured surfaces opens a wide range of interesting applications. The detailed knowledge of the strongly inhomogeneous density distribution of simple liquids on geometrically or chemically structured surfaces is a prerequisite for the specific design of man-made soft structures. By using density functional theory, Landau theory, and simulations the equilibrium fluid structures near the three-phase contact line between a solid, liquid and vapor, in a wedge, across a chemical step, and on curved substrates, are studied with a particular emphasis on wetting properties. These examples serve as elementary paradigms for more complicated structures.

1. Introduction

Modern technology allows one to endow solid surfaces with stable geometrical or chemical structures which display a well-defined design on the scale of nanometers or micrometers. Lateral geometric structures can be formed by using various lithographic techniques such as, e. g., holographic [1], X-ray (LIGA) [2], soft [3], and nanosphere lithography [4]. These microfabrication techniques provide routes to high-quality patterns and structures with lateral dimensions down to tens of nm. These structures are either periodic in one lateral direction consisting of grooves with various shapes of the cross section or they display periodicity in both lateral dimensions. These structures provide a rich topography based on a single homogeneous material.

On the other hand by using microcontact printing [5] it has become feasible to generate surface patterns which consist of regular monolayer patches composed of different chemical species attached to a homogeneous substrate such that the resulting decorated surface remains flat on an atomic scale.

C. Caccamo et al. (eds.), New Approaches to Problems in Liquid State Theory, 197–244.

These self-assembled monolayers with a thickness of about 2nm form a smooth carpet which is firmly anchored onto the underlying solid substrate and which can be endowed with rich and specificly designed lateral patterns. These patterns can generate, e. g., a highly connected and well-defined network of hydrophilic areas separated by hydrophobic ones. Such a carpet can also be wrapped around small cylinders [6] with matching angular periodicity.

Spherical colloidal particle with a diameter in the μm range can be turned into "Janus beads" by chemical treatment of their surfaces such that they are half-hydrophilic, half-hydrophobic generating a chemical step at the equator [7]. The Janus beads serve as artificial amphiphiles which can stabilize oil-water emulsions. Due to their size, even if they are densely packed at an oil-water interface, they leave sufficiently wide pores between them such that ion transport between the two separated phases remains possible. For various systems it is advantageous to have such a porous but nonetheless stable amphiphilic sheet. Molecular amphiphiles are so densely packed at the interface that they block the exchange of ions.

As the last example demonstrates these man-made surfaces offer a wide range of possible applications if they are exposed to fluids. In the context of microfluidics (see, e. g., Ref. [8]) these geometrical or chemical structures, or combinations thereof, can be used as guiding systems in order to deliver tiny amounts (nanoliters) of very valuable liquids, which may contain rare biological products, to designated analysis centers on a solid surface [9]. These networks of chemical lanes along which liquids can spread have the potential to serve as chemical chips. Different fluids can be guided along different paths leading to possible chemical reactions at designated crossings. Chemically treated grooves may act as microfilters for multi-component fluids flowing through them. A major issue in constructing such miniaturized chemical factories concerns the precision with which liquids can be forced to stay on their designated lanes without spillover to neighboring lanes. By using statistical physics as a tool one might hope to explore – similar as quantum mechanics does for semiconductor chips – the extent to which miniaturization of spreading fluids on surfaces can be achieved. Moreover, by studying the dependence of these highly inhomogeneous fluid structures on the molecular interaction potentials between the fluid particles themselves as well as on the substrate potential one is put into the position to be able to optimize the performance of these systems systematically.

It will take many years of research to achieve these ambitious theoretical, experimental, and technological goals. In the present early stage it is useful to focus on elementary building blocks of these invisaged integrated systems in order to gain quantitative confidence. Although the study of fluids exposed to elementary geometrical and chemical structures is already interesting in its own right, the above discussion lends these studies an additional and interesting perspective. Many of the applications mentioned above are intricately linked to the dynamic properties of fluids under the influence of external forces which vary on microscopic scales. As a first step, in the present contribution the corresponding static properties are discussed as they prevail in the limit of thermodynamic equilibrium.

The standard situation to be considered is a structured surface exposed to the vapor phase of a volatile fluid which is brought close to the liquid-vapor coexistence curve of its bulk phase diagram so that a thin wetting film forms on the substrate. (Similarly one can consider a binary liquid mixture, consisting of A and B particles, close to two-phase coexistence between the A-rich liquid and B-rich liquid phase.) The theoretical aim is to monitor the morphology of such wetting films as function of pressure and temperature of the vapor reservoir and as functional of the interparticle potential $\phi(r)$ and of the substrate potential $V(\mathbf{r})$ which defines the inhomogeneous environment. The corresponding wetting phenomena on homogeneous and planar substrates have been studied intensively over the last twenty years, still remaining a very active field of research [10, 11]. This has been documented in several review articles [12, 13, 14, 15, 16, 17] so that the reader is referred to them concerning both the conceptual framework and the detailed experimental and theoretical results for wetting transitions. As a first step, in Sec. II the intrinsic structure of the vapor-liquid-substrate three-phase contact line is studied, followed by a discussion of wetting of curved substrates in Sec. III. The morphology of wetting films across a chemical step and the fluid sturctures in wedges are described in Sec. IV and V, respectively.

2. Three-phase contact line

2.1. BOUNDARY CONDITIONS

The lateral confinement of a liquid on a substrate necessarily requires the formation of a three-phase contact line where the liquid-vapor interface of a liquid film intersects the supporting substrate. In view of the strategic goals formulated in the Introduction it is therefore rather natural to start with the analysis of this structural component. In a certain sense the three-phase contact line can also be considered as the result of the simplest available structuring mechanism, i. e., imposing competing lateral boundary conditions on an otherwise homogeneous and planar substrate.

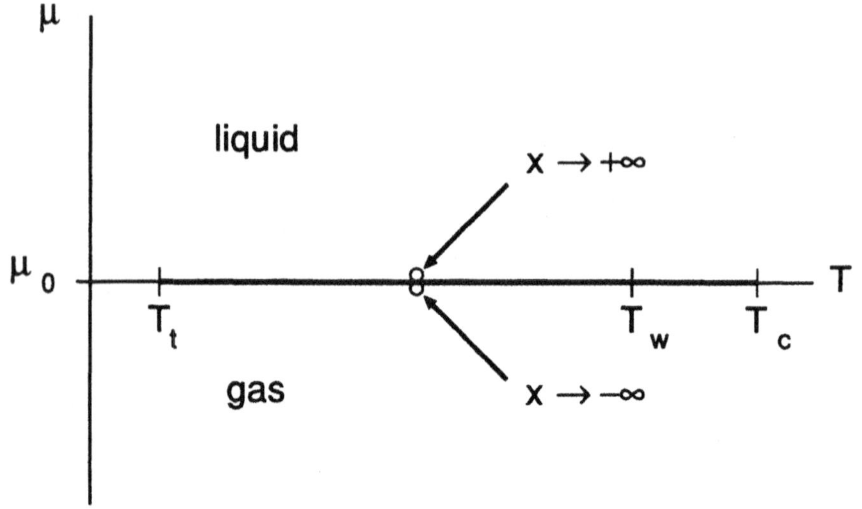

Figure 1. Schematic phase diagram of liquid-gas coexistence at μ_0 between the triple point temperature T_t and the critical temperature T_c. For simplicity μ_0 is taken to be temperature independent. Below the wetting transition temperature T_w the substrate remains nonwet if gas-liquid coexistence is approached from the gas side. If μ_0 is approached from the liquid side the whole sample is filled with liquid. If for $x \to +\infty$ the latter boundary condition is applied and for $x \to -\infty$ the former one, one obtains a density distribution as shown in Fig. 2.

To be specific a simple one-component fluid in a grand canonical ensemble is considered which is characterized by its chemical potential μ and temperature T. This fluid is confined by a homogeneous substrate and taken to be at liquid-vapor coexistence $\mu_0(T)$ such that it is possible to impose as lateral boundary conditions along the x axis so that for $x \to -\infty$ the substrate is in contact with the bulk vapor phase whereas for $x \to \infty$ it is in contact with the bulk liquid phase (Fig. 1). The system is regarded to be translationally invariant in the second lateral dimension y so that the number density $\rho(\mathbf{r} = (x, y, z))$ depends only on x and the distance z normal to the surface (Fig. 2). Below the wetting transition temperature T_w the boundary condition $\rho(x = -\infty, z \to \infty) = \rho_g$ leads to the formation of a thin wetting film of thickness $l_0(T)$. For $T \nearrow T_w$ $l_0(T)$ either diverges continuously (critical wetting transition) or jumps to a macroscopic value (first-order wetting transition). In both cases the contact angle θ (Fig. 2) vanishes continuously for $\tau = (T_w - T)/T_w \to 0$: $\theta(\tau \to 0) \sim \tau^{1/2}$ for first-order wetting and $\theta(\tau \to 0) \sim \tau^{3/2}$ for critical wetting. With the present

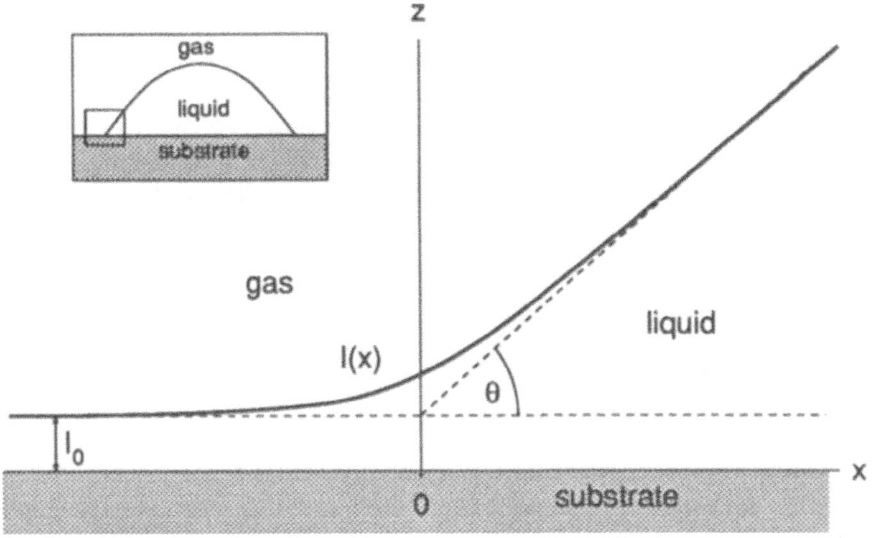

Figure 2. Equilibrium density distribution resulting from the boundary conditions $\rho(x = +\infty, z \to \infty) = \rho_l$ and $\rho(x = -\infty, z \to \infty) = \rho_g$ at liquid-vapor coexistence $\mu_0(T)$. $l(x)$ is the position of the suitably defined emerging liquid-gas interface. $l_0 = l(x \to -\infty)$ is the equilibrium thickness of the wetting film at $T < T_w$ and at coexistence $\mu_0(T)$. $l(x \to +\infty)$ diverges linearly such that the asymptote forms the contact angle θ with the substrate. Under suitable conditions $l(x)$ should resemble the density distribution of the three-phase contact region of a sessile liquid ridge. The actual shape of $l(x)$ depends on the order of the wetting transition at T_w. The position $x = 0$ is defined as the point at which the asymptotes of $l(x)$ intersect. The density distribution is assumed to be translationally invariant in the y direction normal to the xz plane.

choice of boundary conditions one obtains $\rho(x \to \pm\infty, z) = \rho_\pm(z)$ where $\rho_+(z)$ corresponds to the substrate-liquid interfacial profile and $\rho_-(z)$ corresponds to the substrate-vapor interface structure containing a liquidlike film of thickness l_0. The introduction of the local position of the liquid-vapor interface by, e. g., $\rho(x, z = l(x)) = \frac{1}{2}(\rho_l + \rho_g)$, renders a smooth lateral interface profile $l(x)$ which reduces to the value $l_0(T)$ for $x \to -\infty$ and which diverges for $x \to \infty$ (see Fig. 2). It can be shown [18] that the above boundary conditions lead to a linear divergence of $l(x \to +\infty)$ such that the asymptote of $l(x \to +\infty)$ forms an angle θ with the horizontal substrate which is the contact angle given by Young's law (see Fig. 2 and Refs. [12, 13, 14, 15, 16, 17]).

2.2. DENSITY FUNCTIONAL THEORY

Density functional theory offers a successful approach to determine the equilibrium density distribution $\rho(\mathbf{r})$ of spatially inhomogeneous fluids as function of T and μ and as functionals of $\phi(r)$ and $V(\mathbf{r})$ [19]. Within this approach the equilibrium number density distribution minimizes the grand canonical free energy functional $\Omega[\{\rho(\mathbf{r})\};T,\mu]$ which reads in a simple but nonetheless very efficient and successful version

$$\Omega[\{\rho(\mathbf{r})\};T,\mu] = \int_V d^3r\, f_{HS}(\rho(\mathbf{r}),T) + \tag{1}$$
$$\frac{1}{2}\int_V d^3r \int_V d^3r'\, w(|\mathbf{r}-\mathbf{r}'|)\rho(\mathbf{r})\rho(\mathbf{r}') + \int_V d^3r\, [\rho_w V(\mathbf{r}) - \mu]\rho(\mathbf{r}).$$

V is the volume accessible to the fluid, $\rho_w V(\mathbf{r})$ is the substrate potential, which for convenience contains an explicit factor ρ_w denoting the average number density of the substrate particles of diameter σ_w. For a homogeneous substrate one has $V(\mathbf{r}) = V(z) = -\sum_{i\geq 3} u_i z^{-i}$, neglecting atomic corrugation effects, which diverges for $z \to 0$ and decays $\sim z^{-3}$ for $z \to \infty$, disregarding retardation effects. The pair potential $\phi(r)$ between the fluid particles can be reasonably represented by the Lennard-Jones potential $\phi(r) = 4\epsilon[(\sigma/r)^{12} - (\sigma/r)^6]$. Following the Barker-Henderson scheme [20] $\phi(r)$ is split into an attractive part $w(r) = \phi(r)\Theta(r-\sigma)$, where Θ is the Heaviside function, and into a repulsive part $\phi(r \leq \sigma)$. Whereas the former is treated perturbatively in Eq. (1) the latter part gives rise to a reference free energy density $f_{HS}(\rho,T)$ of a hard sphere fluid with an effective hard sphere diameter $d(T) = \int_0^\sigma dr\, \{1 - \exp[-\phi(r)/(k_B T)]\}$. For the reference free energy the Carnahan-Starling expression

$$f_{HS}(\rho,T) = k_B T\rho\left(\ln\eta - 1 + \frac{4\eta - 3\eta^2}{(1-\eta)^2}\right), \qquad \eta = \frac{\pi}{6}\rho(d(t))^3, \tag{2}$$

is a good approximation [21].

One of the virtues of density functional theory is that for the class of density distributions which are compatible with the imposed boundary conditions it allows one to decompose the variational functional (Eq. (1)) systematically into bulk, surface, and line contributions which scale with the volume V, the surface area A, and the length L of the (straight) three-phase contact line, respectively [18]:

$$\Omega[\{\rho(\mathbf{r})\};T,\mu] = V\Omega_b + A\Omega_s + L\Omega_l. \tag{3}$$

First, the bulk contribution Ω_b alone is minimized yielding the bulk densities ρ_l and ρ_g and the bulk phase diagram. These equilibrium values are inserted into the surface (Ω_s) and the line (Ω_l) contribution.

In the case of *homogeneous* lateral boundary conditions one has $\rho(\mathbf{r}) = \rho(z)$ and the surface contribution leads to various surface tensions $\sigma_{\alpha\beta}$ depending on the boundary condition at $z \to \infty$:

$$\sigma_{\alpha\beta}(T,\mu) = \min_{\rho(z)} \Omega_s[\{\rho(z)\};T,\mu] \tag{4}$$

where $(\alpha,\beta) = (w,l)$ corresponds to the boundary condition $\rho(z\to\infty) = \rho_l$, $(\alpha,\beta) = (w,g)$ to $\rho(z\to\infty) = \rho_g$, and $(\alpha,\beta) = (l,g)$ to $\rho(z\to\infty) = \rho_g$ and $\rho(z\to-\infty) = \rho_l$ for $V(z) \equiv 0$. σ_{wl} and σ_{wg} are the wall-liquid and wall-gas surface tensions associated with the equilibrium profiles $\rho_+(z)$ and $\rho_-(z)$, respectively, as introduced above. σ_{lg} is the surface tension of the free liquid-gas interface defined for $\mu = \mu_0(T)$.

2.3. EFFECTIVE INTERFACE POTENTIAL AND WETTING TRANSITIONS

It has turned out [22] that wetting properties are described surprisingly well by applying the so-called sharp-kink approximation within which the minimization in Eq. (4) is restricted to the subspace of piecewise constant density profiles. Inserting a trial function as shown in Fig. 3 with a variable film thickness l into Eqs. (1) and (4) leads to the so-called effective interface potential

$$\Omega_s(l;T,\mu) = l\Delta\Omega + \sigma_{wl} + \sigma_{lg} + \omega(l). \tag{5}$$

The linearly increasing term $l\Delta\Omega$ with $\Delta\Omega = \Omega_b(\rho_l,T,\mu_0) - \Omega_b(\rho_g,T,\mu) = \Delta\mu\Delta\rho > 0$ corresponds to the price of building up a liquidlike film of thickness l in contact with an undersaturated vapor phase in the bulk. At two-phase coexistence one has $\Delta\Omega = 0$. In Eq. (5) $\omega(l)$ denotes the cost in free energy to maintain a wetting film of prescribed thickness l as compared to a wet configuration with $l = \infty$:

$$\omega(l) = \Delta\rho\left(\rho_l \int_{l-d_w}^{\infty} dz\, t(z) - \rho_w \int_{l}^{\infty} dz\, V(z)\right) = \sum_{i\geq 2} \frac{a_i}{l^i}, \qquad l \gg d_w, \tag{6}$$

where $V(z) = -\sum_{i\geq 3} u_i z^i$ and

$$t(z) = \int_z^{\infty} dz' \int_{-\infty}^{\infty} dx' \int_{-\infty}^{\infty} dy'\, w(\sqrt{x'^2+y'^2+z'^2}) = -\sum_{i\geq 3} t_i z^i; \tag{7}$$

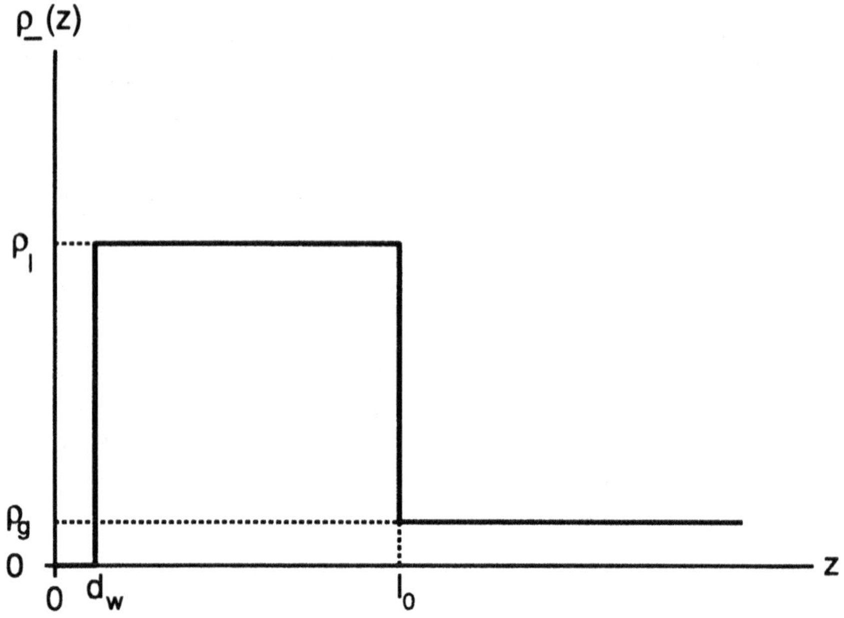

Figure 3. **Sharp-kink approximation of the number density profile $\rho_-(z)$ corresponding to an equilibrium wetting film thickness l_0. $d_w \simeq (\sigma_w+\sigma)/2$ describes the excluded volume due to the repulsive part of the substrate potential. σ_w and σ denote the diameter of the substrate atoms and of the fluid particles, respectively. The corresponding approximation for $\rho_+(z)$ yields a profile which equals ρ_l for $z > d_w$ and which is zero for $z < d_w$. ρ_l and ρ_g are taken at coexistence $\mu_0(T)$.**

a_i are expansion coefficients where a_2 is known as Hamaker constant. The equilibrium film thickness $l_0(T,\mu)$ minimizes $\Omega_s(l)$ and renders the wall-gas surface tension:

$$\sigma_{wg} = \min_l \Omega_s(l) = \Omega_s(l_0) \tag{8}$$

At coexistence, i. e., $\Delta\Omega = 0$, this implies $\sigma_{wg} = \sigma_{wl} + \sigma_{lg}$ so that the contact angle is given by

$$\theta = \arccos\left(1 + \frac{\omega(l_0)}{\sigma_{lg}}\right) \tag{9}$$

due to Young's law $\cos\theta = (\sigma_{wg} - \sigma_{wl})/\sigma_{lg}$. Within the present approximation one has

$$\sigma_{lg} = -\frac{1}{2}(\Delta\rho)^2 \int_0^\infty dz\, t(z) \tag{10}$$

and

$$\sigma_{wl} = -\frac{1}{2}\rho_l^2 \int_0^\infty dz\, t(z) - d_w \Omega_b(\rho_l) + \rho_l \rho_w \int_{d_w}^\infty dz\, V(z). \qquad (11)$$

In case of a first-order wetting transition, $\omega(l)$ has two minima separated by a free energy barrier. One of these minima is localized at a finite value l_0 and represents the global minimum for temperatures $T < T_w$, while the second minimum is given by $l = \infty$ where $\omega(l) = 0$. At $T = T_w$, the minimum at $l = \infty$ changes from a local to a global minimum, so that the film thickness on the planar substrate jumps from a microscopic value to infinity. Off coexistence this infinite jump is reduced to a finite jump occurring at the prewetting line which joins the gas-liquid coexistence curve tangentially at T_w and ends at a critical point where the size of the finite jump vanishes continuously. In the case of a second-order wetting transition, $\omega(l)$ has only one minimum, which is shifted continuously from a small value l_1 to $l = \infty$ as the temperature is raised from $T < T_w$ towards $T = T_w$. The necessary conditions for a continuous wetting transition, called critical wetting, are $a_2(T \lessgtr T_w) \lessgtr 0$, $a_2(T = T_w) = 0$, and $a_3(T_w) > 0$. Irrespective of the order of the transition one has $a_2(T > T_w) > 0$ so that l_0 diverges along an isotherm $T > T_w$ and $\Delta\mu \to 0$. This is called a complete wetting transition.

2.4. INTERFACIAL PROFILES

Equipped with this knowledge of film formation for a homogeneous boundary one can turn to the inhomogeneous case shown in Fig. 2. In this case in the thermodynamic limit the free energy proportional to the surface area A is determined by σ_{wg} and σ_{wl} (see Eqs. (8) and (11)). Thus with Ω_b and Ω_s fixed the shape $l(x)$ of the lateral interface profile is determined by minimizing the corresponding line contribution Ω_l in Eq. (3). Even for a simple density functional as given by Eq. (1) it is a very demanding numerical challenge to determine the full density distribution $\rho(x, z)$ associated with the three-phase contact line; in particular a temperature sweep close to the wetting transition becomes prohibitively difficult due to the lateral broadening of the structure. Therefore as a first step, which is also already very difficult to implement numerically, the local sharp-kink approximation

$$\hat{\rho}(x, z) = \Theta(z - d_w)\{\rho_l \Theta(l(x) - z) + \rho_g \Theta(z - l(x))\} \qquad (12)$$

with $l(x \to -\infty) = l_0$ and $l(x \to +\infty) = l_0 + x \tan\theta$ is applied. Inserting Eq. (12) into Eq. (1) yields a lengthy expression for Ω_l in terms of $l(x)$ (see Eq. (2.26) in Ref. [18]). The corresponding equation for the minimum $l(x)$ of Ω_l is given by

$$\Delta\Omega - \Delta\rho[\rho_l t(l(x)) - \rho_w V(l(x))] \qquad (13)$$

$$= -(\Delta\rho)^2 \int_{-\infty}^{\infty} dx' \int_{0}^{l(x')-l(x)} dz' \int_{-\infty}^{\infty} dy'\, w(\sqrt{(x-x')^2 + y'^2 + z'^2}).$$

Equation (13) is a nonlinear and *nonlocal* integral equation for the equilibrium interface profile. By enforcing a gradient expansion to the nonlocal expression on the right hand side of Eq. (13) one can approximate Eq. (13) by a *local* nonlinear differential equation:

$$\Delta\Omega - \Delta\rho[\rho_l t(l(x)) - \rho_w V(l(x))] \tag{14}$$
$$= \sigma_{lg} \frac{d^2 l(x)}{dx^2} \left[1 + \left(\frac{dl(x)}{dx}\right)^2\right]^{-3/2} + \mathcal{O}((l'')^2, l''').$$

The right hand side of Eq. (14) equals the negative functional derivative of $\sigma_{lg} \int dx(\sqrt{1 + [dl(x)/dx]^2} - 1)$ which is the phenomenological expression for the cost in free energy due to the increased surface area of the configuration $l(x)$ as compared with the flat one $l(x) =$ const, disregarding curvature contributions. For a detailed assessment of the reliability of this phenomenological and local approximation versus the full nonlocal theory see Refs. [23, 24, 25, 26]. It turns out that for long-ranged interparticle forces decaying like a power law the coefficients of the higher order terms on the right hand side of Eq. (14) become infinitely large so that the gradient expansion breaks down.

Figures 4 and 5 display the numerical solutions of Eq. (14) for a system undergoing first-order wetting and for one exhibiting critical wetting (see Ref. [18] for the choice of interaction potential parameters). $T^* = k_B T/\epsilon$ is the reduced temperature, σ is the diameter of the fluid particles, and $a(x) = l_0\Theta(-x) + (l_0 + x\tan\theta)\Theta(x)$ is the asymptote of the profile $l(x)$ so that the deviation $\delta l(x) = l(x) - a(x)$ vanishes for $|x| \to \infty$ and has a break in slope at $x = 0$ which is defined as the position of the intersection between the asymptotes of $l(x)$. For a first-order wetting transition $l(x \to \infty)$ approaches its asymptote from below so that $\delta l(x > 0) < 0$ whereas $l(x \to \infty)$ approaches its asymptote from above for critical wetting so that in this case $\delta l(x > 0) > 0$. In both cases $l(x \to -\infty)$ approaches its asymptote l_0 from above so that always $\delta l(x < 0) > 0$. For first-order wetting this positive part of $\delta l(x < 0)$ is barely visible on the scales of Fig. 4. For critical wetting $\delta l(x < 0)$ is substantially larger and comparable with $\delta l(x > 0)$ although $\delta l(x)$ vanishes more rapidly for $x \to -\infty$ than for $x \to \infty$. There is a huge difference of the magnitude and of the extent of the deviation of the profile from its asymptotes between first-order and critical wetting. At comparable temperatures the deviation is about an order of magnitude larger and

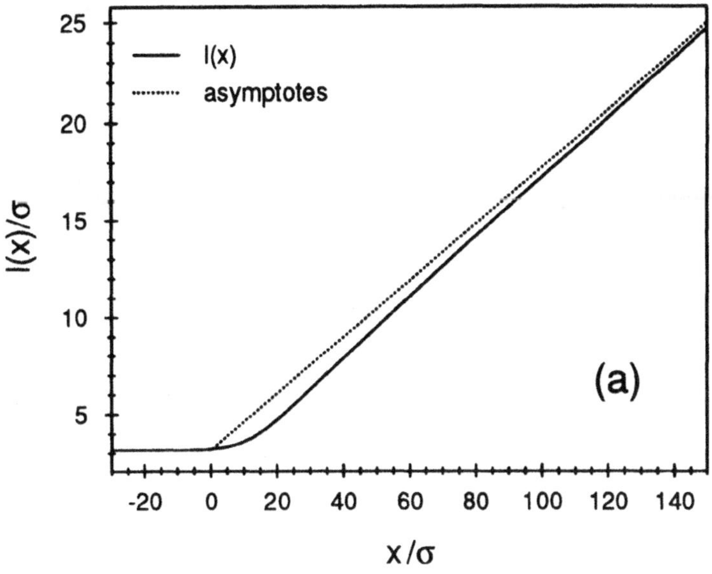

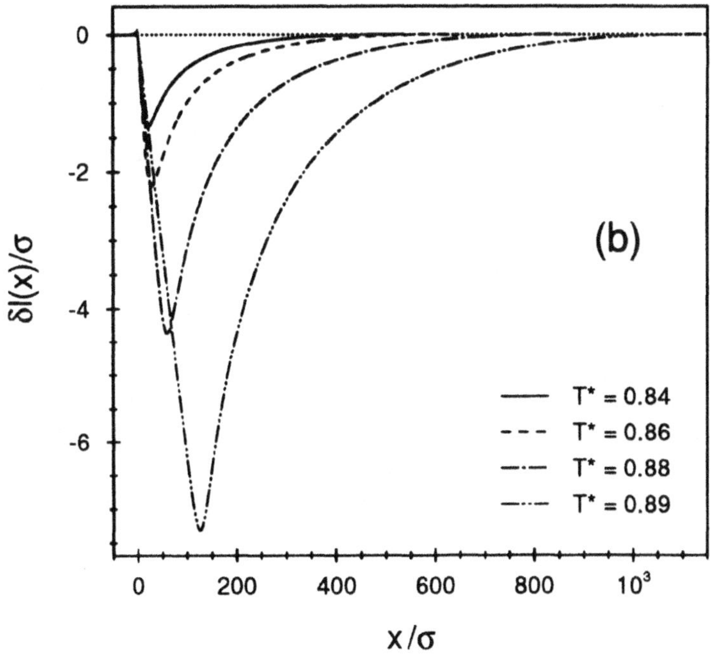

Figure 4. (a) Shape of the profile $l(x)$ of the liquid-gas interface at $T^* = 0.84$ for a first-order wetting transition within local approximation. The wetting temperature is $T^*_w = 0.9$. The dotted line is the asymptote $a(x)$. The angle given by the slope of the asymptote gives the contact angle θ at $T^* = 0.84$. (b) Deviation $\delta l(x) = l(x) - a(x)$ of the profile from its asymptotes for various temperatures.

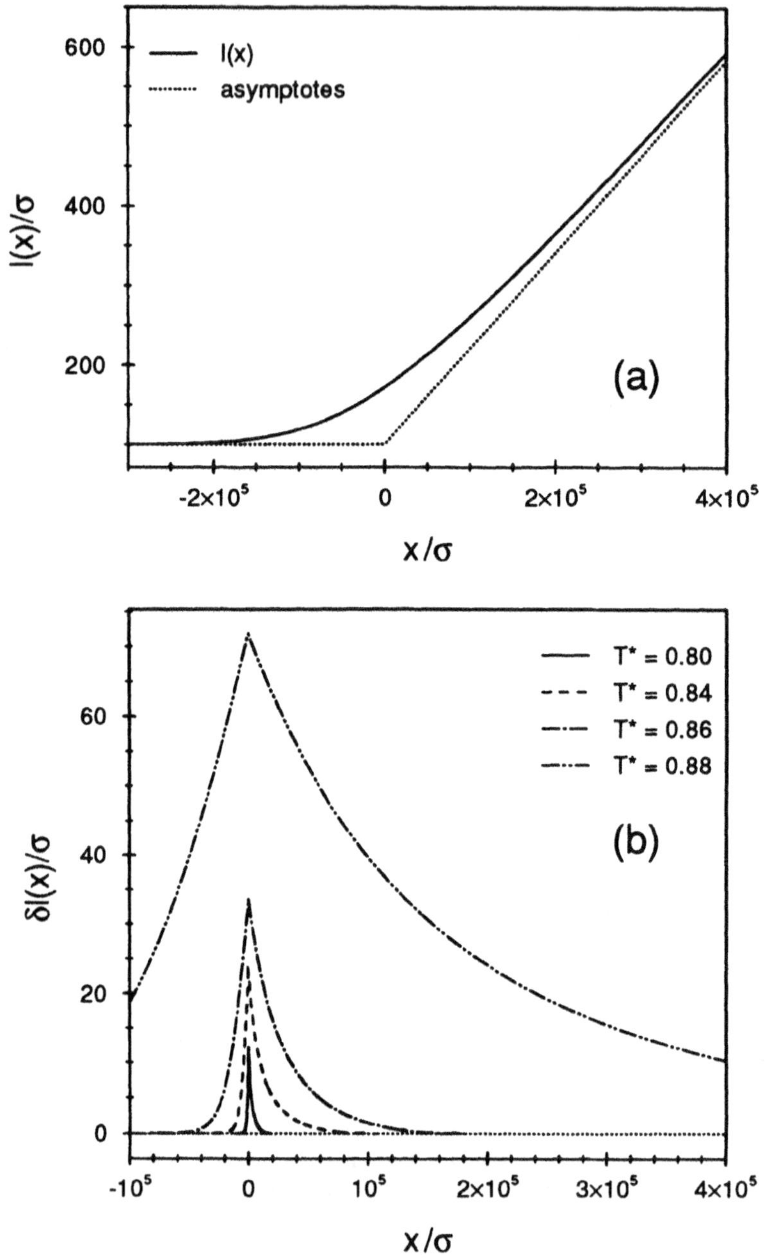

Figure 5. (a) Shape of the profile $l(x)$ of the liquid-gas interface at $T^* = 0.88$ for a critical wetting transition within local approximation. The wetting temperature is $T_w^* = 0.9$. The dotted line is the asymptote $a(x)$. The angle given by the slope of the asymptote gives the contact angle θ at $T^* = 0.88$. (b) Deviation $\delta l(x) = l(x) - a(x)$ of the profile from its asymptotes for various temperatures.

two orders of magnitude wider for critical wetting than for first-order wetting. Irrespective of the order of the transition the deviation $\delta l(x)$ increases upon approaching T_w and reaches a mesoscopic scale. By attributing a suitably defined width Δx to the profile $\delta l(x)$ the numerical data suggest a divergence $\Delta x \sim \theta^{-3/2}$ for $\theta \to 0$ irrespective of the order of the wetting transition [18]. With modern techniques such as reflection interference contrast microscopy (RICM) the local morphology of liquid-vapor interfaces can be probed quantitatively with a vertical resolution of 2Å and a lateral resolution of 2000Å [27] which roughly corresponds to 0.5 and 500 units of σ, respectively. Thus with a moderate improvement of the lateral resolution the structures displayed in Figs. 4 and 5 are experimentally accessible.

2.5. LINE TENSION

The careful separation of the grand canonical free energy into bulk, surface, and line contributions (Eq. (3)) as well as the identification of all artificial terms generated by the truncation of the finite sample volume V [18] leads to the following explicit expression for the line tension τ associated with the three-phase contact line:

$$\tau[l(x)] = \tau_a[a(x)] + \tau_\omega[l(x)] + \tau_i[l(x)]. \tag{15}$$

The first term is solely determined by the asymptotes of the profile, i. e., by θ and l_0:

$$\tau_a[a(x)] = \tilde{\tau}(\theta) + \tilde{\omega}(l_0, \theta) \tag{16}$$

where

$$\tilde{\tau}(\theta) = \frac{1}{2}(\Delta\rho)^2 \left[1 - \frac{\theta}{\tan\theta}\right] \int_0^\infty dx \int_0^\infty dz \int_{-\infty}^\infty dy\, xz\, w(\sqrt{x^2+y^2+z^2}) \tag{17}$$

and

$$\tilde{\omega}(l_0, \theta) = \frac{1}{\tan\theta} \int_{l_0}^\infty dl\, \omega(l). \tag{18}$$

The second term in Eq. (15) does depend on the full interface profile and resembles an integrated local effective interface potential:

$$\tau_\omega[l(x)] = \int_{-\infty}^\infty dx\, \{\omega(l = l(x)) - \omega(l = a(x))\}. \tag{19}$$

The form of the third term in Eq. (15) depends on whether the local or the nonlocal theory is applied (compare Eqs. (13) and (14)):

$$\tau_i^{nloc}[l(x)] = \frac{1}{2}(\Delta\rho)^2 \int_{-\infty}^{\infty} dx \int_{a(x)}^{l(x)} dz \int_{-\infty}^{\infty} dx' \int_{-\infty}^{\infty} dy' \tag{20}$$

$$\left\{ \int_{-\infty}^{a(x')} dz'\, w(\sqrt{(x-x')^2 + y'^2 + (z-z')^2}\,) - \right.$$

$$\left. - \int_{l(x')}^{\infty} dz'\, w(\sqrt{(x-x')^2 + y'^2 + (z-z')^2}\,) \right\}$$

The local approximation thereof is given by

$$\tau_i^{loc}[l(x)] = \sigma_{lg} \int_{-\infty}^{\infty} dx \left\{ \sqrt{1 + \left(\frac{dl(x)}{dx}\right)^2} - \sqrt{1 + \left(\frac{da(x)}{dx}\right)^2} \right\}. \tag{21}$$

Figure 6 displays the temperature dependence of the line tension within local approximation for the systems corresponding to Figs. 4 and 5. For first-order wetting the line tension diverges $\sim -\ln\theta$ for $T \to T_w$. For critical wetting the line tension is negative and vanishes $\sim \theta^{1/3}$ upon approaching T_w. This behavior agrees with the result obtained by Indekeu from a phenomenological local interface displacement model [28]. The present approach in the context of density functional theory enables one to derive quantitative predictions in terms of the interaction potential between the fluid particles and of the substrate potential. The explicit formulae in Eqs. (15)-(21) demonstrate how the line tension depends on the balance between these two interaction potentials. This can serve as a first step to understand the wide range of positive and negative values of line tensions ($10^{-5} - 3 \cdot 10^{-12}$J/m) reported experimentally for various systems [29]. Away from T_w, τ is of the order of ϵ/σ, i. e., about 10^{-11}J/m which is close to computer simulation results yielding values of about 10^{-12}J/m for a specific choice of interaction potentials [30]. As far as the comparison with experimental or simulation data is concerned one should keep in mind that the analysis presented above is based on a mean-field theory which neglects thermal fluctuations of the contact line along the y direction. Their contribution to the line tension as well as to the shape of the mean interface profile still awaits a thorough analysis. For first steps in that direction see Refs. [31] and [32]. Within mean-field theory it is an improvement to use

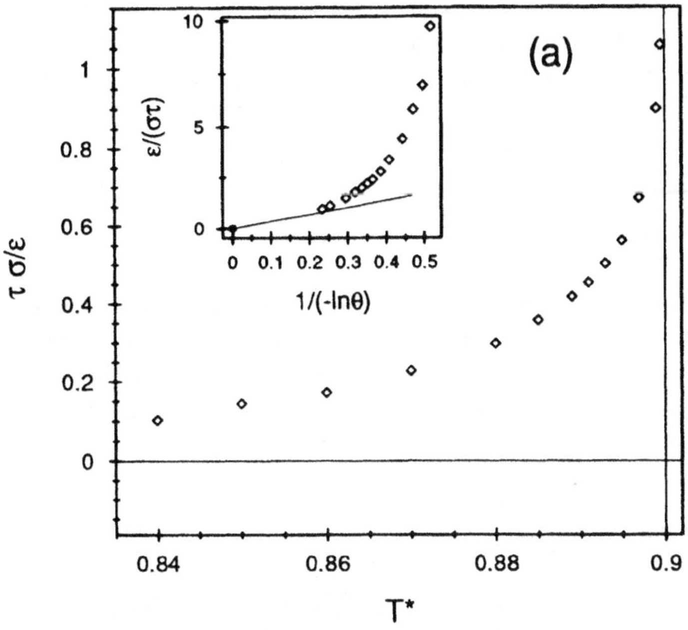

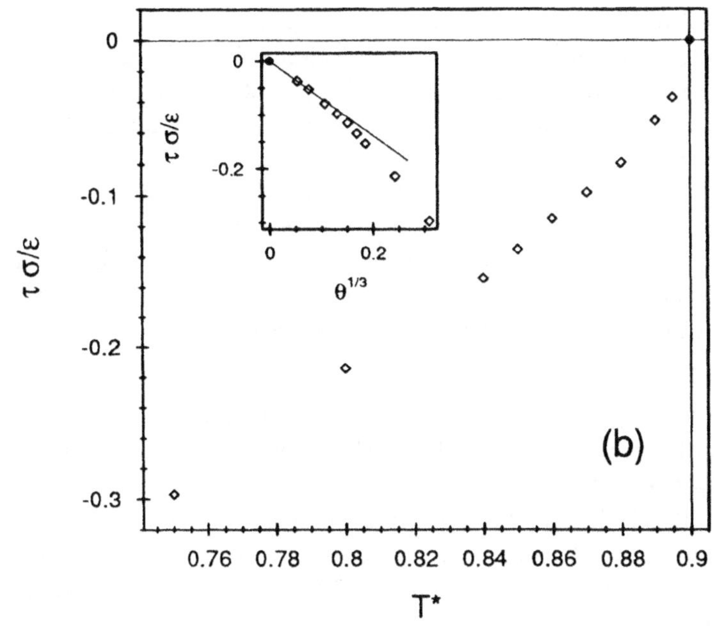

Figure 6. Line tension τ within the local theory for first-order wetting (a) and critical wetting (b). The data correspond to Figs. 4 and 5. In (a) the inset probes the logarithmic divergence $\tau \sim -\ln\theta$ for $T \to T_w$. For critical wetting τ vanishes $\sim \theta^{1/3}$ upon approaching T_w (see the inset). θ is measured in rad. In both cases $T_w^* = 0.9$. For first-order wetting τ is positive near T_w, whereas it is negative for critical wetting. Away from T_w τ is of the order of ϵ/σ.

the actual nonlocal description instead of the local approximation thereof (see Eqs. (13), (14), (20) and (21)). However, it turns out that this improvement is only relevant as long as the profile $l(x)$ exhibits significant curvature contributions which are omitted in Eq. (14). This only occurs at low temperatures in the case of first-order wetting [33].

3. Curved surfaces

As mentioned in the Introduction curvature can play an important role as a structural element of solid surfaces. As paradigms in the following spheres and cylinders are considered which are exposed to the vapor phase of a simple one-component fluid close to gas-liquid coexistence such that their surfaces are covered by thin liquidlike wetting films.

If the vapor and liquid phases of a one-component fluid are replaced by the A-rich and B-rich liquid phases of a binary liquid mixture the analysis of the aforementioned wetting phenomena is relevant for flocculation in colloidal suspensions. If the solvent of such a suspension consists of a binary liquid mixture one of these two components is preferentially adsorbed on the spherical surfaces of the suspended colloidal particles. Near a first-order demixing transition of the bulk solvent this adsorption may lead to the coating of the colloids by wetting films which snap into a bridgelike structure if two colloids come close to each other. This coagulation can become so pronounced that it results in flocculation (see, e. g., Refs. [34] and [35]). These experimental observations have triggered numerous theoretical investigations (see, e. g., Refs. [36] and [37]) which have shed light on various features of these complex phenomena; a complete overview of this experimental and theoretical body of literature is given in Ref. [38].

The basic step in understanding this process consists of the wetting of a single sphere or cylinder. Although several phenomenological models for describing the wetting behavior on curved surfaces have been studied in the past [39, 40, 41, 42, 43, 44, 45] a fully microscopic treatment on the basis of density functional theory with all forces properly taken to be long-ranged has been carried out only recently [38] and will be presented below. (A description in terms of the *D*zyaloshinskii-*L*ifshitz-*P*itaevskii theory [46] is reliable and helpful for specific experimental systems [47] but it lacks the transparency of describing phase transitions even in the bulk. Moreover, DLP theory only describes the *leading* asymptotic behavior of the effective interface potential in the limit of thick films but not the subdominant ones which are relevant for wetting phase transitions.)

3.1. EFFECTIVE INTERFACE POTENTIAL

Following the same line of arguments as in Sec. II one can apply the sharp-kink approximation

$$\rho(\mathbf{r}) = [\rho_l \Theta(h-r) + \rho_g \Theta(r-h)]\,\Theta(r-r_1). \tag{22}$$

for the radial density distribution and insert it into Eq. (1) where h and r_0 denote the radius of the emerging gas-liquid interface and of the substrate, respectively, so that the thickness l of the wetting film is given by $l = h - r_0 \geq d_w$ (see Fig. 7). The repulsive part of the substrate potential leads to an excluded volume close to the substrate. This is taken into account by the radius $r_1 = r_0 + d_w$. With the superscripts s, c, and p corresponding to a spherical, cylindrical, and a planar substrate, respectively, Eq. (22) leads to the following effective interface potential $\Omega_s(l)$ as the analogue of Eq. (5) [38]:

$$\Omega_s(l) = \frac{V_{liq}}{A(r_0)}\Delta\Omega_b + \frac{A(r_0+l)}{A(r_0)}\sigma_{lg}(r_0+l) + \sigma_{wl}(r_0) + \omega(l, r_0). \tag{23}$$

$\Omega_s(l)$ is the extra contribution to the bulk free energy due to the presence of the sphere or the cylinder per area $A(r_0)$ of the substrate surface. The volume V_{liq} of the liquid film is $V_{liq}^{(s)} = \frac{4\pi}{3}(h^3 - r_0^3)$ and $V_{liq}^{(c)} = \pi M (h^2 - r_0^2)$ for a sphere and a (long) cylinder of height M, respectively. In Eq. (23) the geometrical prefactors $V_{liq}/A(r_0)$ and $A(r_0+l)/A(r_0)$ have the following explicit form:

$$\frac{V_{liq}}{A(r_0)} = \frac{r_0}{\tau+1}\left(\left(1+\frac{l}{r_0}\right)^{\tau+1} - 1\right) \tag{24}$$

and

$$\frac{A(r_0+l)}{A(r_0)} = \left(1+\frac{l}{r_0}\right)^{\tau} \tag{25}$$

with $\tau = 2$ for a sphere and $\tau = 1$ for a cylinder. For a planar substrate these geometrical prefactors are equal to l in Eq. (24) and 1 in Eq. (25) (see Eq. (5)). These considerations already demonstrate that a negative substrate curvature prohibits the formation of macroscopically thick wetting films even at liquid-gas coexistence, i. e., $\Delta\Omega_b = 0$ because – unlike a planar substrate – the increase of the wetting film thickness is linked to an increase of the area of the emerging liquid-gas interface which is penalized by the corresponding surface tension. For positive curvatures, such as the inner surfaces of spherical or cylindrical shells, the opposite is true. In

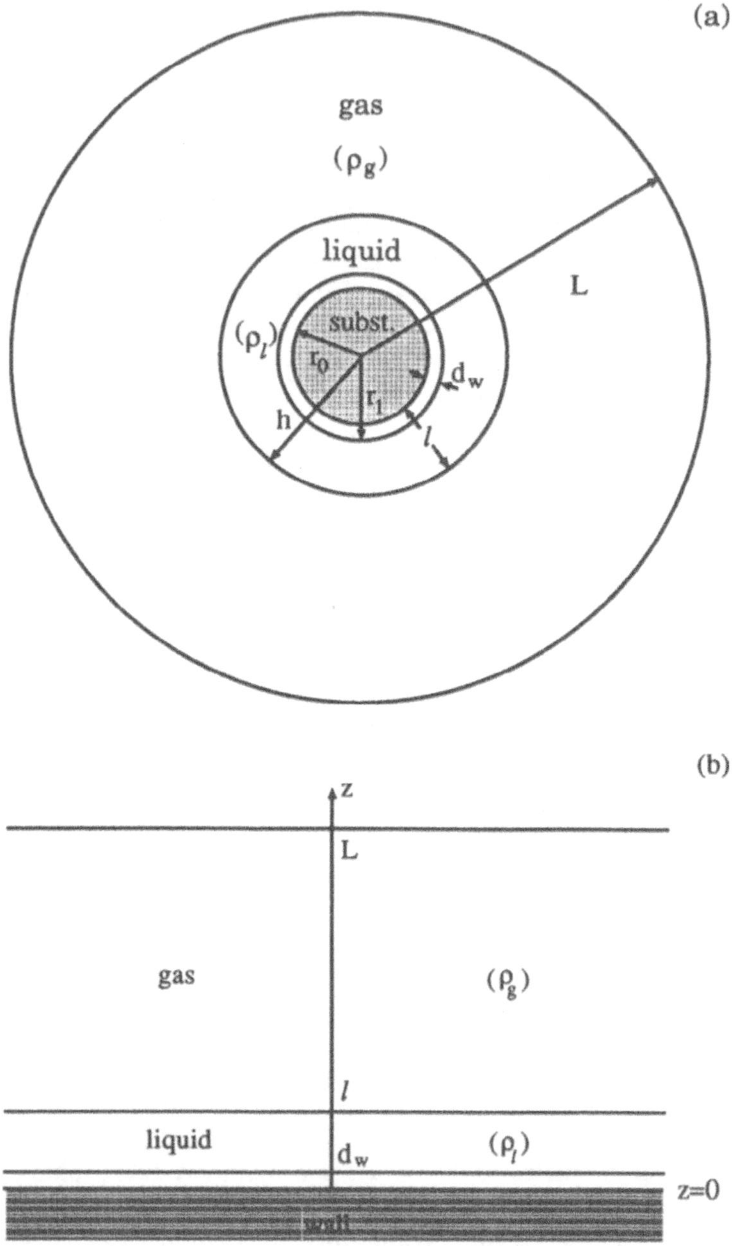

Figure 7. (a) Cross section of the density configuration around a sphere or a cylinder of radius r_0 with a number density ρ_w. The gas phase with number density ρ_g is truncated for $r > L$. The liquidlike film with number density ρ_l has a radius of curvature given by h and a thickness $l = h - r_0$. The density distribution vanishes within an excluded volume $r_0 < r < r_1 = r_0 + d_w$. (b) Corresponding density configuration close to a planar wall.

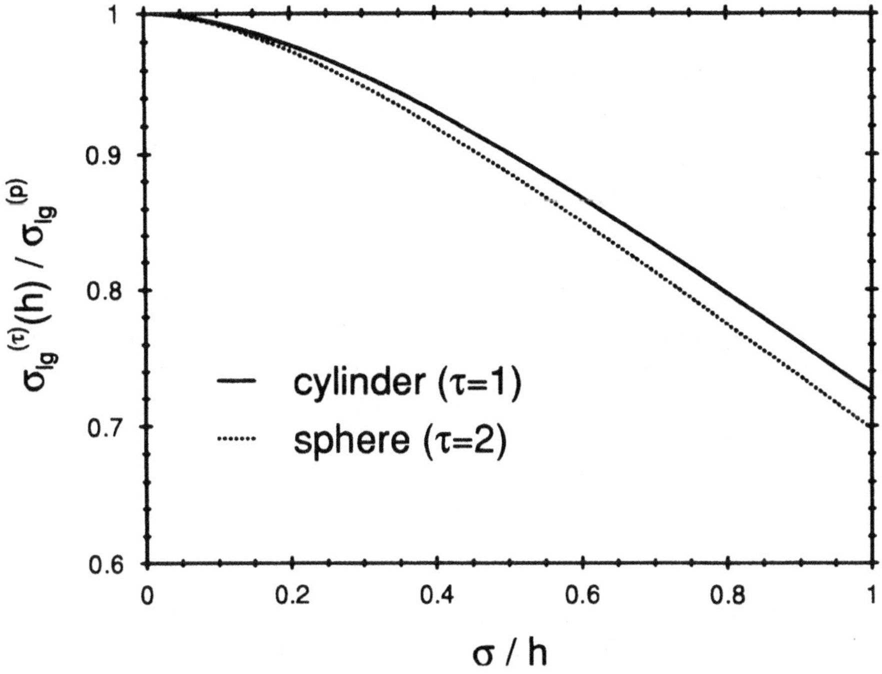

Figure 8. Liquid-vapor surface tension of a liquid drop (s) and thread (c) as compared with the planar surface tension (p) as function of their radius h in units of the diameter σ of the Lennard-Jones fluid particles. Within the sharp-kink approximation the ratios $\sigma_{lg}^{(\tau)}(h)/\sigma_{lg}^{(p)}$ are constant as function of the temperature. For $\sigma/h < 1.89 \cdot 10^{-7}$, one has $\sigma_{lg}^{(c)}(h) < \sigma_{lg}^{(s)}(h)$; otherwise $\sigma_{lg}^{(s)}(h)$ is smaller than $\sigma_{lg}^{(c)}(h)$. Beyond the sharp-kink approximation the ratio $\sigma_{lg}^{(s)}(h)/\sigma_{lg}^{(p)}(h)$ for a sphere is expected to approach 1 linearly as function of $\sigma/h \to 0$ with a slope $-2\delta/\sigma$ given by the Tolman length δ (see Ref. [38]).

that case curvature promotes the formation of wetting films, but wetting transitions are undercut by capillary condensation within the finite volume.

Before turning to the quantitative discussion of the thickness of the wetting films it is worth noting that negative curvature reduces the liquid-gas surface tension, which according to Eq. (23) depends on the radius $h = r_0 + l$ of curvature (see Fig. 8). For large h one finds for a Lennard-Jones fluid

$$\sigma_{lg}^{(c)}(h) = \sigma_{lg}^{(p)} \left(1 - \frac{1}{6}\frac{\ln(h/\sigma)}{(h/\sigma)^2} - \frac{1}{(h/\sigma)^2}\left(\frac{5}{3}\ln 2 + \frac{1}{144}\right) + \mathcal{O}\left((h/\sigma)^{-4}\right)\right), \quad (26)$$

and

$$\sigma_{lg}^{(s)}(h) = \sigma_{lg}^{(p)}\left(1 - \frac{2}{9}\frac{\ln(h/\sigma)}{(h/\sigma)^2} - \right.$$
$$\left. - \frac{1}{(h/\sigma)^2}\left(\frac{2}{9}\ln 2 + \frac{4}{27}\right) - \mathcal{O}\left((h/\sigma)^{-8}\right)\right) \tag{27}$$

with the planar surface tension (i. e., $h \to \infty$) $\sigma_{lg}^{(p)}$ given by Eq. (10). Since the density profile in Eq. (22) is antisymmetric around $r = h$, within our approach the Tolman length δ is zero which gives rise to a term $-2\delta/h$ in Eq. (27). Equations (26) and (27) demonstrate that the long-ranged forces between the fluid particles give rise to logarithmic corrections which dominate the usual subdominant term $\sim h^{-2}$. The wall-liquid surface tension $\sigma_{wl}(r_0)$ entering Eq. (23) only depends on r_0 and thus it is constant with respect to the minimization of $\Omega_s(l)$. Therefore it does not have to be considered further.

The last term in Eq. (23) describes the effective interaction between the wall-liquid and the liquid-vapor interface if they are a finite distance l apart. In the planar case it has the structure given by Eq. (6). Critical, first-order, and complete wetting transitions are captured properly by considering systems for which the coefficients a_2, a_3, and a_8 in $\omega^{(p)}(l)$ are nonzero; the latter arises from the repulsive parts of the interaction potentials. For such systems, if they are curved, the third term in Eq. (23) turns out to exhibit the following scaling form:

$$\omega^{(\tau)}(l, r_0) = \frac{a_2}{l^2}\mathcal{S}^{(\tau)}\left(\frac{r_0}{l}\right) + \frac{a_3}{l^3}\mathcal{R}^{(\tau)}\left(\frac{r_0}{l}\right) + \frac{a_8}{l^8}\mathcal{T}^{(\tau)}\left(\frac{r_0}{l}\right). \tag{28}$$

These scaling functions show the following asymptotic behaviors:

$$\mathcal{S}_\tau(x \to \infty) = 1 + \frac{\tau}{2x} + \mathcal{O}\left(x^{-2}\right), \tag{29}$$

$$\mathcal{S}_\tau(x \to 0) = \alpha_\tau x + \beta_\tau x^2 + \mathcal{O}\left(x^3\right), \tag{30}$$

with $\alpha_c = \frac{3\pi}{2}$, $\beta_c = -\frac{9\pi}{2}$, $\alpha_s = \frac{16}{3}$, and $\beta_s = -16$

$$\mathcal{R}_\tau(x \to \infty) = 1 + \frac{3\tau}{4x} + \mathcal{O}\left(x^{-2}\right) \tag{31}$$

and

$$\mathcal{R}_\tau(x \to 0) = \gamma_\tau + \delta_\tau x + \mathcal{O}\left(x^{-2}\right) \tag{32}$$

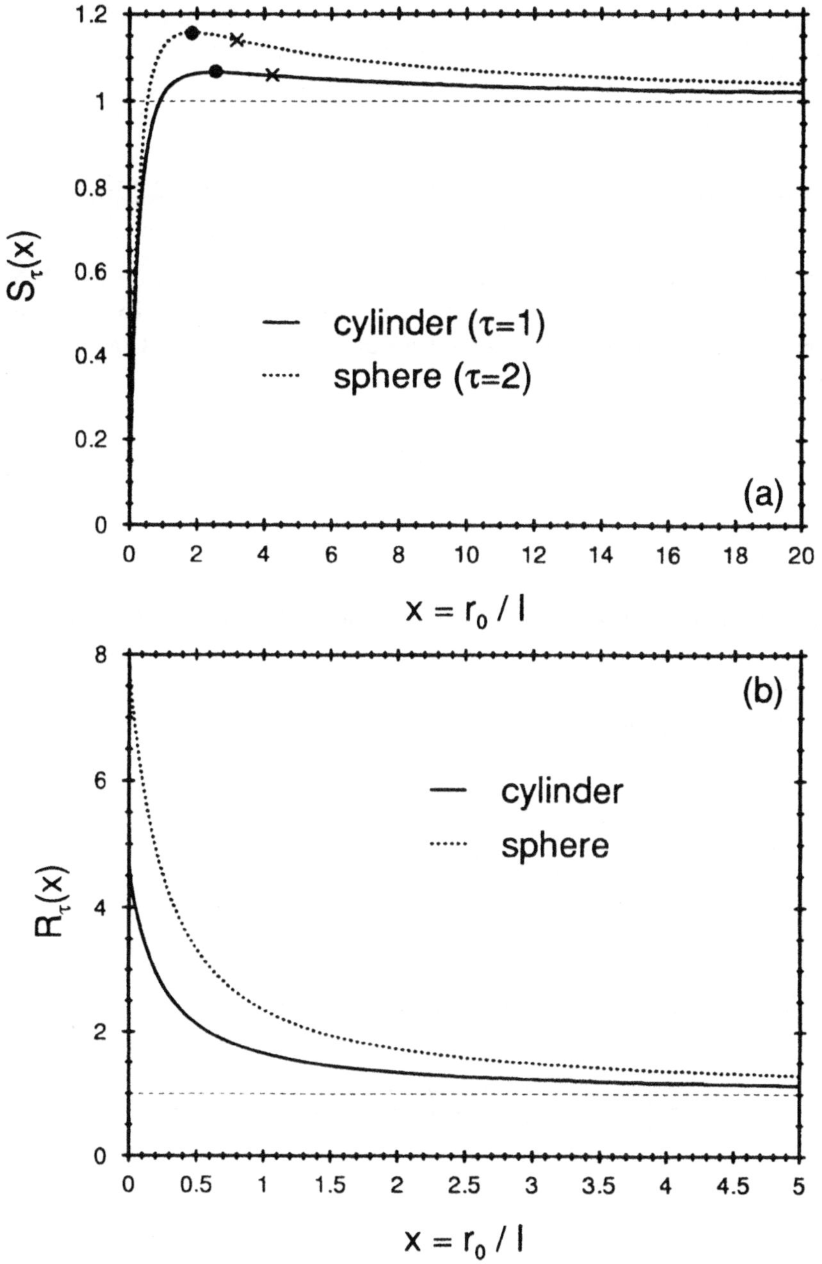

Figure 9. Scaling functions $\mathcal{S}_\tau(x)$ (a) and $\mathcal{R}_\tau(x)$ (b) entering into the effective interface potential $\omega^{(\tau)}(l, r_0)$ in Eq. (28). $\mathcal{S}_\tau(x)$ vanishes linearly for $x \to 0$ (Eq. (30)) and approaches 1 for $x \to \infty$ proportional to x^{-1} (Eq. (29)). Their maxima (•) and the inflection points (×) are located at $x_{max}^{(c)} \simeq 2.56$, $x_{max}^{(s)} \simeq 1.85$, and $x_t^{(c)} \simeq 4.24$, $x_t^{(s)} \simeq 3.18$, respectively. $\mathcal{R}_\tau(x)$ rises linearly to a finite value for $x \to 0$ ($\mathcal{R}_c(x = 0) = \frac{3}{2}\pi$ and $\mathcal{R}_s(x = 0) = 8$, see Eq. (32)) and approaches 1 for $x \to \infty$ proportional to x^{-1} (Eq. (31)).

with $\gamma_c = \frac{3\pi}{2}$, $\delta_c = -\frac{9\pi}{2}$, $\gamma_s = 8$, and $\delta_s = -24$. The full scaling curves are shown in Fig. 9. For a first-order wetting transition the behavior of $\omega(l)$ for l close to d_w may become important; this contribution cannot be cast in a simple scaling form and can be studied only numerically [38].

3.2. FILM THICKNESS AT COEXISTENCE AND ABOVE THE WETTING TRANSITION TEMPERATURE

With the knowledge of the ingredients in Eqs. (24) - (32) one can determine the minimum l_0 of $\Omega_s(l)$ in Eq. (23). This leads to a surprisingly rich and intricate behavior which depends sensitively on the dimensionless parameter

$$\kappa = \left(\frac{r_0^2 \sigma_{lg}^{(p)}}{a_2} \right)^{1/3} . \tag{33}$$

One of the main consequences of the negative curvature of the substrate is that, in contrast to a planar geometry, the film thickness attains a maximum but *finite* value l_{max} at coexistence and above the wetting transition temperature. This is shown in Fig. 10. For a given system, i. e., $\sigma_{lg}^{(p)}$ and a_2 fixed and for large r_0 the maximum thickness of the wetting film is given by

$$l_{max}^{(\tau)}(r_0 \to \infty) = \left(\frac{2a_2 r_0}{\tau \sigma_{lg}^{(p)}} \right)^{1/3} = \left(\frac{2}{\tau} \right)^{1/3} \kappa^{-1} r_0, \quad \kappa \gg 1. \tag{34}$$

For a small liquid-vapor surface tension $\sigma_{lg}^{(p)}$ (e. g., close to T_c) and a large Hamaker constant a_2 the maximum thicknesses are

$$l_{max}^{(c)} \simeq \left(\frac{9\pi a_2 r_0^2}{2\sigma_{lg}^{(p)}} \right)^{1/4} = \left(\frac{9\pi}{2} \right)^{1/4} \kappa^{-3/4} r_0, \quad \kappa \ll 1, \tag{35}$$

and

$$l_{max}^{(s)} \simeq \left(\frac{8a_2 r_0^3}{\sigma_{lg}^{(p)}} \right)^{1/5} = 8^{1/5} \kappa^{-3/5} r_0, \quad \kappa \ll 1. \tag{36}$$

In principle, κ can have any value. Typically, at low temperatures, $\sigma_{lg}^{(p)}$ ranges between 10^{-3} and 10^{-2}J/m^2 and a_2 between 10^{-20} and 10^{-19}J [48]

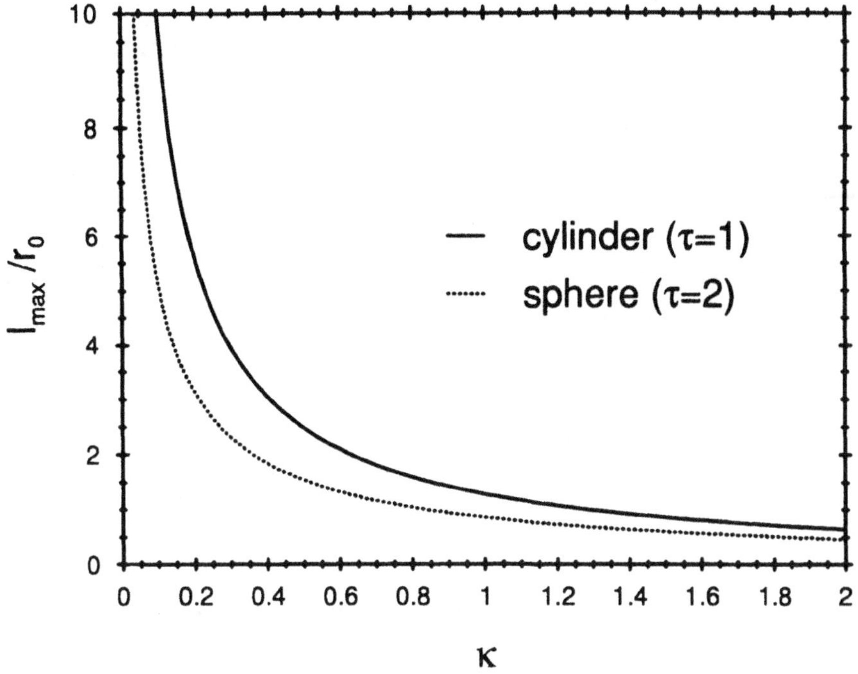

Figure 10. Maximum film thickness l_{max} above T_w and at coexistence in units of r_0 as function of $\kappa = (r_0^2 \sigma_{lg}^{(p)}/a_2)^{1/3}$. Both curves vanish for $\kappa \to \infty$ proportional to κ^{-1} whereas for $\kappa \to 0$ they diverge proportional to $\kappa^{-3/4}$ and $\kappa^{-3/5}$ for the cylinder and the sphere, respectively.

so that $\kappa = \kappa_0 (r_0/\text{Å})^{2/3}$ with $\kappa_0 = 0.05 - 0.20$. With $\kappa_0 = 0.05$ according to Fig. 10 this implies that for $r_0 = 100\text{Å}$, $l_{max}^{(c)} \simeq 130\text{Å}$ and $l_{max}^{(s)} \simeq 90\text{Å}$, whereas for $r_0 = 1\mu\text{m}$, $l_{max}^{(c)} \simeq 550\text{Å}$ and $l_{max}^{(s)} \simeq 430\text{Å}$. Due to finite resolutions for pressure and temperature equilibrium film thicknesses on a planar substrate can be followed up to at most a few hundred Å, say 500Å. In this case and for the above quoted values of $\sigma_{lg}^{(p)}$ and a_2, r_0 must be smaller than $50\mu\text{m}$ for a cylinder and $100\mu\text{m}$ for a sphere in order to cause a deviation from the corresponding planar film thickness of more than 5%.

3.3. COMPLETE WETTING OFF COEXISTENCE AND ABOVE THE WETTING TRANSITION TEMPERATURE

For complete wetting, i. e., off coexistence and above T_w the equilibrium film thickness has the form

$$l_0 = \mathcal{Y}(\kappa, \lambda) r_0 \tag{37}$$

where the dimensionless scaling function $\mathcal{Y}$ depends on κ (Eq. (33)) and on the dimensionless parameter λ measuring the distance from liquid-vapor coexistence:

$$\lambda = \frac{\Delta\rho r_0}{\sigma_{lg}^{(p)}} \Delta\mu. \tag{38}$$

For general values of κ and λ the scaling function $\mathcal{Y}(\kappa, \lambda)$ exhibits a surprisingly rich and complex behavior which is described in detail in Ref. [38]. In the special case $\Delta\mu \to 0$ and large κ, e. g., large spheres and cylinders exposed to small undersaturations, the behavior of l_0 can be expressed in terms of that on a planar substrate with an effective undersaturation $\Delta\mu_{eff}$:

$$l_0(r_0, \Delta\mu \to 0) = l_o^{(p)}(\Delta\mu_{eff}) \tag{39}$$

where $l_0^{(p)}(\Delta\mu \to 0) = (2a_2/(\Delta\rho\Delta\mu))^{1/3}$ and

$$\Delta\mu_{eff}(\Delta\mu; r_0, \tau) = \Delta\mu + \delta\mu(\kappa, \lambda) = \Delta\mu + \frac{\tau\sigma_{lg}^{(p)}}{\Delta\rho r_0}, \quad \kappa \gg 1, \quad \lambda \ll 1. \tag{40}$$

Thus, in the limit $\kappa \gg 1$, the curved substrates are covered by liquid films whose thicknesses resemble the same film thicknesses as the corresponding planar substrates if the liquid-vapor coexistence curve $\mu_0(T)$ is shifted upwards by the amount $\delta\mu_\infty^{(\tau)} = \tau\sigma_{lg}^{(p)}/(r_0\Delta\rho)$ (see Fig. 11). This is in accordance with the previous result at coexistence in Eq. (34).

3.4. FIRST-ORDER WETTING TRANSITIONS

Figure 12(a) shows the phase diagram for a first-order wetting transition of a suitably chosen model system. In accordance with Refs. [49] and [50] the prewetting line $\mu_{pw}^{(p)}(T)$ of the planar substrate joins the gas-liquid coexistence curve $\mu_0(T)$ tangentially,

$$\mu_{pw}^{(p)}\left(T \to T_W^{(p)}\right) - \mu_0(T) = k_\mu \left|T - T_w^{(p)}\right|^{\gamma(\zeta)}, \quad \gamma(\zeta) = \frac{\zeta}{\zeta - 1}, \tag{41}$$

if for large distances the pair potentials in the system decay like $r^{-(3+\zeta)}$ [50]; $\zeta = 3$ for Lennard-Jones fluids. Upon crossing the prewetting line the wetting film thickness undergoes a discontinuous thin-thick transition. The corresponding jump $\Delta l(T)$ diverges at the first-order transition temperature $T_w^{(p)}$ as

$$\Delta l(T \to T_w^{(p)}) = k_l \left|T - T_w^{(p)}\right|^{-\gamma(\zeta)/\zeta}. \tag{42}$$

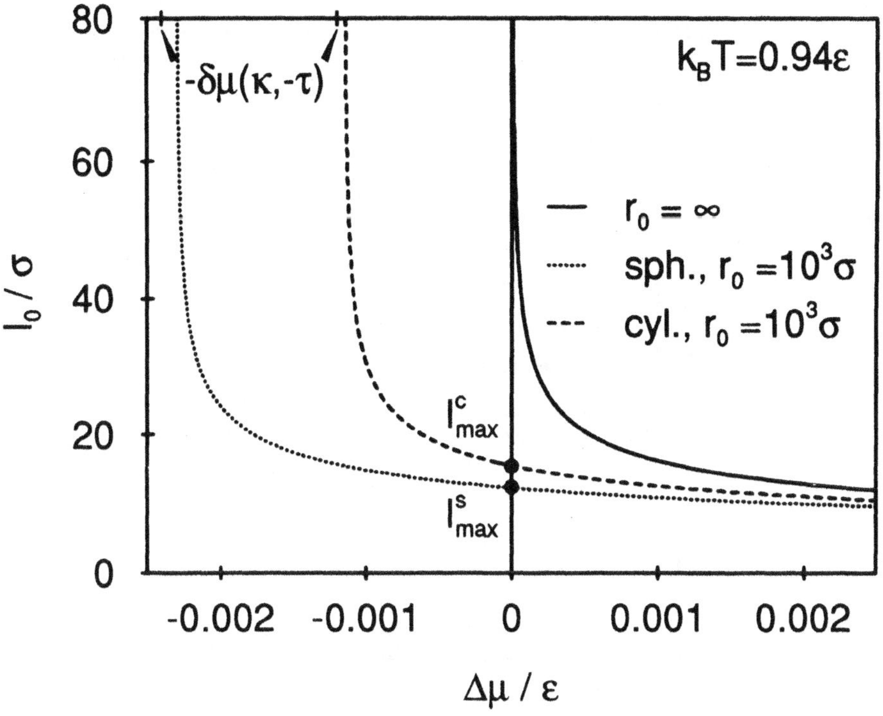

Figure 11. Complete wetting film thickness along an isotherm $k_BT/\epsilon = 0.94$ for a sphere and a cylinder, respectively, of radius $10^3\sigma$, compared with that of a planar wall ($r_0 = \infty$). For this system $\kappa \simeq 75.5$ so that Eq. (40) applies. Therefore the complete wetting film thickness on a curved substrate resembles that on the corresponding planar substrate as if liquid-vapor coexistence would occur at $\Delta\mu(\kappa,\lambda) = -\delta\mu(\kappa,\lambda = -\tau) < 0$ instead of at $\Delta\mu = 0$; $\Delta\mu_{eff} = 0$ implies $\lambda = -\tau$.

In accordance with the expectation based on Eqs. (39) and (40) the prewetting lines on curved substrates resemble the planar one shifted upward. However, since the prewetting lines also depend on the behavior of the effective interface potential for small l, this is not simply a rigid shift (see Fig. 12(a)). This implies that at coexistence the film thickness on curved substrates increases slightly up to a first-order transition temperature $T_w^{(\tau)}(\kappa)$ where the film thickness undergoes a finite jump $\Delta l^{(\tau)}(T_w^{(\tau)})$ and increases smoothly for higher temperatures (see Fig. 12(b)). For large κ, this finite jump at coexistence scales as [38]

$$\Delta l^{(\tau)}\left(T_w^{(\tau)}(\kappa\to\infty)\right) = k_l\left(\frac{1}{k_\mu}\delta\mu_\infty^{(\tau)}\right)^{-1/\zeta}. \tag{43}$$

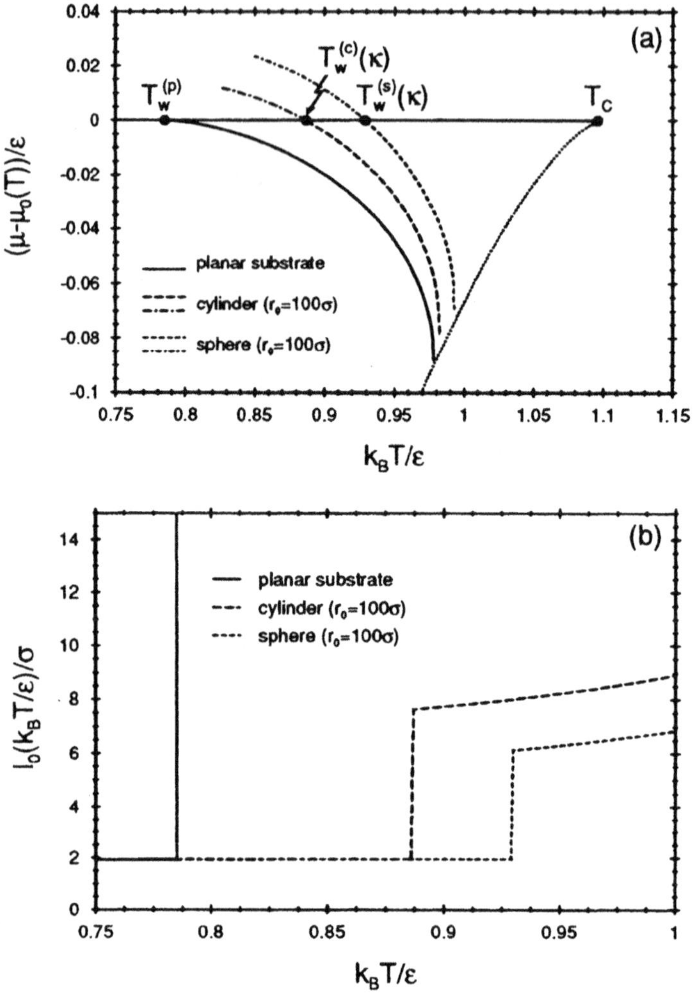

Figure 12. (a) Phase diagram of a first-order wetting transition on a cylindrical and a spherical substrate of radii $r_0/\sigma = 100$ obtained by a numerical minimization of Eq. (23) for a suitable set of interaction parameters [38]. The full and dashed curves correspond to the prewetting lines. For curved surfaces they intersect the gas-liquid coexistence curve μ_0 at a finite angle leading to a finite discontinuity of the film thickness at $T_w^{(\tau)}(\kappa)$. The thin-thick transition can be traced into the oversaturated vapor region (dashed-dotted lines) as long as the thermodynamically instable vapor phase can be maintained for these values of μ. At the upper end points of these curves the minimum in $\Omega_s(l)$ corresponding to the thicker film ceases to exist. Each prewetting line ends at its lower end in a critical point. However, in the present model and approximation the prewetting lines can only be followed up to the dotted curve where in the bulk the metastable liquid phase ceases to exist. Below the dotted curve the wetting films are so thin that the present sharp-kink approximation is no longer applicable. (b) Temperature dependence of the equilibrium film thickness l_0 at coexistence for the systems described by the phase diagram in (a). It illustrates the shift of the wetting transition temperature and the finite discontinuity induced by the curvature of the substrate.

Thus, $\zeta = 3$ implies $T_w^{(\tau)}(\kappa \to \infty) - T_w^{(p)} = (\tau/k_\mu)^{2/3}(\sigma_{lg}^{(p)}/(r_0\Delta\rho))^{2/3}$ and $\Delta l^{(\tau)}(T_w^{(\tau)}(\kappa \to \infty)) = k_l(k_\mu/\tau)^{1/3}(r_0\Delta\rho/\sigma_{lg}^{(p)})^{1/3}$.

3.5. CRITICAL WETTING TRANSITIONS

If the wetting transition is second order the prewetting line is absent and the film thickness on a planar substrate diverges at $T_w^{(p)}$ as $(T_w^{(p)} - T)^{-1}$ (see Fig. 13). On curved substrates and for $\kappa \gg 1$ the temperature dependence of the film thickness along the gas-liquid coexistence curve in the bulk resembles that on a planar substrate with a thermodynamic path at $\delta\mu_\infty^{(\tau)}$ off coexistence which passes the critical wetting transition. Consequently, planar substrates with a critical wetting transition lose it in the presence of curvature (see Fig. 13). Instead one observes a steep but smooth increase of the wetting film on a curved substrate upon passing $T_w^{(p)}$ at $\mu_0(T)$ and a levelling off at a finite value for $T > T_w^{(p)}$. Figure 13 demonstrates that near $T_w^{(p)}$ already very small curvatures ($r_0 = 10^8\sigma$ corresponds to a radius of about 4cm) considerably reduce and limit the formation of a thick wetting film. In the present example at the reduced temperature $(T_w^{(p)} - T)/T_w^{(p)} = 10^{-2}$ a radius of curvature of 4 cm already causes a halfening of the corresponding film thickness on a planar substrate. In this sense it is no surprise that so far the only critical wetting transition which has been found experimentally occurs on a liquid substrate which is flat [51].

On a planar substrate a critical wetting transition is brought about by the competition between the two leading contributions a_2/l^2 and a_3/l^3 in the effective interface potential $\omega(l)$ (see Eq. (6)). Therefore in this case the film thickness on curved substrates is governed by a scaling function which depends on three dimensionless scaling variables: $\kappa = (r_0^2\sigma_{lg}^{(p)}/a_2)^{1/3}$ (Eq. (33)), $\lambda = \Delta\rho r_0\Delta\mu/\sigma_{lg}^{(p)}$ (Eq. (38)), and $\gamma = 3a_3/(2a_2r_0)$. Note that both κ and γ can be negative. Up to a sign the parameter $\gamma = -l_0^{(p)}(\Delta\mu = 0, T \to T_w^{(p)})/r_0$ is the ratio of the planar film thickness in units of r_0. The interplay between these three dimensionless parameters, or the underlying length scales, gives rise to a remarkably rich behavior of the film thickness in the various limits that these parameters are small or large. For a complete discussion of this the reader is referred to Fig. 11 in Ref [38]. Figure 13 corresponds to the limiting case $\kappa \gg 1$ and $\gamma \gg 1$ which allows for a transparent interpretation in terms of an effective thermodynamic path corresponding to a fixed undersaturation $\delta\mu_\infty^{(\tau)}$ in the planar case.

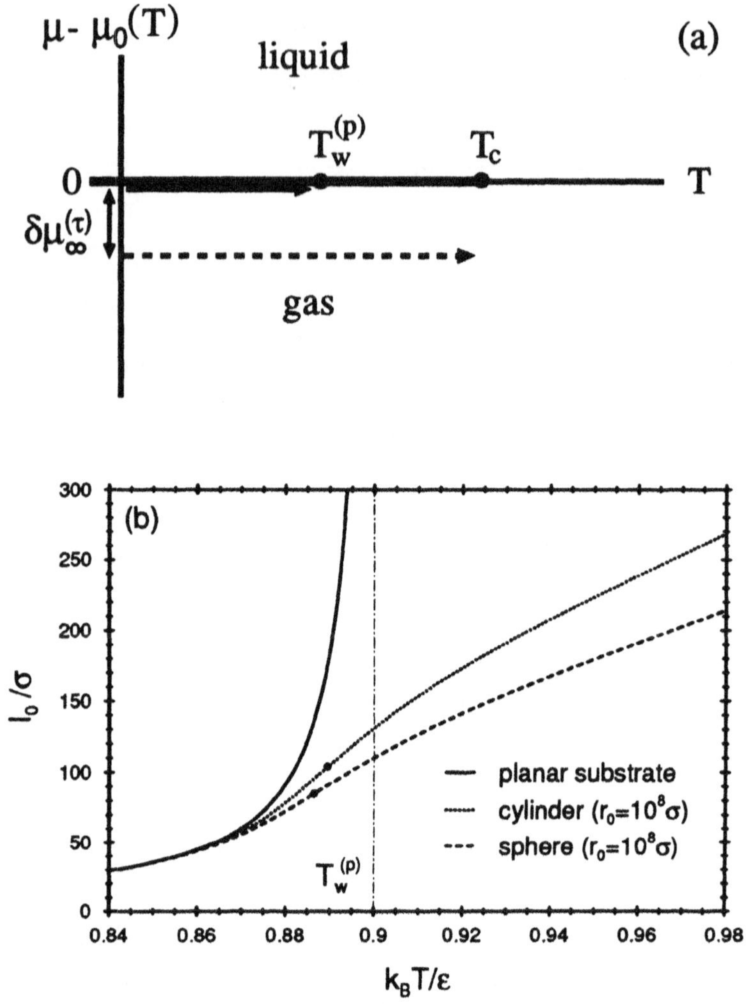

Figure 13. Phase diagram (a) and film thickness (b) in the case of critical wetting. On a planar substrate along the thermodynamic path denoted by the thin full line in (a) the film thickness diverges as $(T_w^{(p)} - T)^{-1}$ upon approaching the critical wetting transition temperature $T_w^{(p)}$ (full line in (b)). For large but fixed values of r_0 and $T \to T_w^{(p)}$, i. e., for $|\kappa| = (r_0^2 \sigma_{lg}^{(p)}/|a_2|)^{1/3} \gg 1$ and $|\gamma| = 3a_3/(2|a_2|r_0) \gg 1$ the film thickness on the curved substrates equals that on the corresponding planar one with an undersaturation $\delta\mu_\infty^{(\tau)}$. Thus the temperature dependence of the film thickness on curved surfaces resembles that on a planar substrate if one passes $T_w^{(p)}$ off coexistence (dashed path in (a)). Note that close to $T_w^{(p)}$ even very small curvatures ($r_0 = 10^8\sigma$) cause a significant reduction of the film thickness as compared with the planar case. In (b) the full dots indicate inflection points which occur *below* the wetting temperature $T_w^{(p)}$ of the planar substrate. For the choice of interaction potentials see Ref. [38].

3.6. FLUCTUATION EFFECTS

The above analysis has shown that the negative curvature of a substrate prevents the formation of infinitely thick wetting films. At coexistence a critical wetting transition is smeared out, whereas a first-order wetting transition is reduced to a discontinuous transition between a thin and a thick, but finite, film. However, the latter nonanalytic behavior as well as the discontinuities associated with the corresponding prewetting line are artefacts of the mean-field theory which has been used. Since the surface area of spheres is quasi-zero-dimensional and that of cylinders quasi-one-dimensional, fluctuations erase any phase transition and lead to a rounding of these discontinuities mentioned above. With increasing radii of the spheres and cylinders this rounding sharpens and a first-order transition emerges gradually. In this sense the present mean-field results become increasingly more accurate as the size of the spherical and cylindrical substrates becomes larger. The type of fluctuations which erase a discontinuous thin-thick transition consists of the breakup of a film with homogeneous film thickness into domains of coexisting patches of thin and thick films. Based on the results of Privman and Fisher [52] for the influence of finite-size effects on first-order transitions, the present density functional approach allows one to develop corresponding considerations in Refs. [41] and [42] into quantitative estimates for the width of the smearing of first-order wetting transitions on curved surfaces. According to the estimates derived in Ref. [38] the rounding of the discontinuous transitions shown in Fig. 12 are confined to a very narrow temperature interval δT around the prewetting line. For a typical spherical substrate with a radius $r_0 = 100\sigma$ one finds $\delta T/T_w^{(p)} \simeq 10^{-4}$ which is at the lower end of what can be resolved experimentally. For a cylindrical substrate with the same radius $\delta T/T_w^{(p)} \simeq 10^{-140}$ so that the rounding cannot be detected.

4. Chemical step

4.1. BOUNDARY CONDITIONS AND PHASE DIAGRAM

The simplest building block of a chemical heterogeneity is a planar substrate that consists of two quarter spaces w_- and w_+ which meet at the plane $x = 0$ and are composed of two different materials. This gives rise to a substrate potential $V(x,z) = V_{rep}(x,z) + V_{att}(x,z)$. For the repulsive part of the potential one can adopt the simple form $V_{rep}(x,z) = (\Theta(-x)u_9^- + \Theta(x)u_9^+)z^{-9}$. For a sharp transition region between the two halves of the substrate this is an acceptable approximation because the repulsive part of the pair potentials between the substrate and the fluid particles decays rapidly. The form of the long-ranged attractive part $V_{att}(x,z)$ can be obtained from summing pair potentials between the fluid particles and the two different substrate particles located on lattice sites in w_- and w_+. By using the Euler-MacLaurin summation formula one obtains an explicit expression for $V_{att}(x,z)$ which is rather lengthy and given in Eq. (2.19) in Ref. [53]. For large distances z from the substrate and x fixed, $V_{att}(x,z)$ vanishes proportional to z^{-3}. For $x \to \pm\infty$, $V(x,z)$ reduces to the substrate potential $V_\pm(z)$ of the corresponding semi-infinite, homogeneous, and flat substrate with $V_\pm(z \to \infty) = u_3^\pm z^{-3}$. For $|x| \gg z$ these limiting forms are attained proportional to x^{-3}, so that $V(x,z)$ exhibits the long-ranged van der Waals character both normal and parallel to the substrate surface.

If this substrate is exposed to the vapor phase of a simple fluid described, e. g., by a Lennard-Jones interaction potential, a laterally inhomogeneous number density distribution $\rho(x,z)$ arises which is translationally invariant in the y direction. This system differs from that considered in Sec. II in that here the boundary condition $\rho(x, z \to \infty) = \rho_g$ is homogeneous but the substrate inhomogeneous. Close to liquid-vapor coexistence in the bulk a liquidlike film will be formed on the substrate whose local thickness $l(x)$, defined as $\rho(x, z = l(x)) = (\rho_l + \rho_g)/2$, varies smoothly between the values $l(x \to \mp\infty) = l_\mp$ valid for the pure substrates $w_\mp$ (see Fig. 14). The thicknesses $l_\mp$ characterize the density profiles $\rho_\mp(z) = \rho(x \to \mp\infty, z)$. In Sec. II the competing boundary conditions $\rho(x = -\infty, z \to \infty) = \rho_g$ and $\rho(x = \infty, z \to \infty) = \rho_l$ only allow consideration of thermodynamic states at gas-liquid coexistence $\mu = \mu_0(T)$. In contrast the present system can be studied for any value of μ and T. The issue is to monitor the morphology $l(x)$ of the wetting film for the various thermodynamic paths shown in Fig. 15.

4.2. MORPHOLOGY OF THE WETTING FILM

As in Subsec. II.D within the sharp-kink approximation

$$\hat{\rho}(x,z) = \{\Theta(-x)\Theta(z-d_w^-) + \Theta(x)\Theta(z-d_w^+)\}\cdot \cdot\{\rho_l\Theta(l(x)-z) + \rho_g\Theta(z-l(x))\} \tag{44}$$

the minimization of Eq. (1) with respect to $l(x)$ leads to the analogue of Eqs. (13) and (14):

$$\Delta\Omega - \Delta\rho[\rho_l t(l(x)) - V(x,l(x))] = -(\Delta\rho)^2 \int_{-\infty}^{\infty} dx' \int_0^{l(x')-l(x)} dz' \int_{-\infty}^{\infty} dy'\, w(\sqrt{(x-x')^2 + y'^2 + z'^2})$$

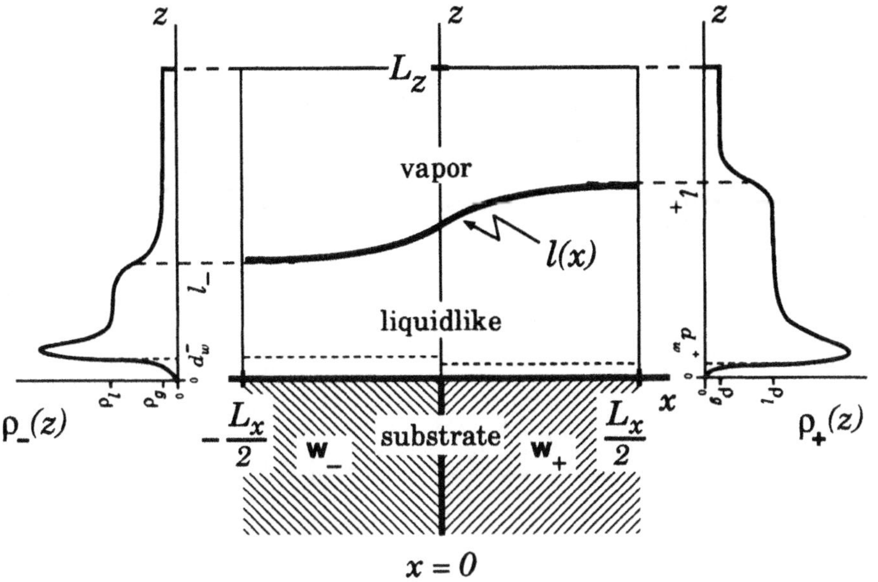

Figure 14. Schematic density distribution of the fluid exposed to the chemically structured substrate. At large lateral distances $|x|$ from the position $x = 0$ of the inhomogeneity the density profile $\rho(x, z)$ reduces to the density profiles $\rho_\pm(z)$ of the corresponding homogeneous cases. The profile $l(x)$ of the interface position is defined by $\rho(x, z = l(x)) = (\rho_l + \rho_g)/2$. The fluid is enclosed in the box $\Lambda_f = L_x \times L_y \times L_z$ surrounded by vacuum. The repulsive part of the substrate potential leads to the vanishing of $\rho(x, z)$ for $z \to 0$, which can be approximately described by excluded volumes d_w^- and d_w^+.

$$= \sigma_{lg} \frac{d^2 l(x)}{dx^2} \left[1 + \left(\frac{dl(x)}{dx} \right)^2 \right]^{-3/2} + \mathcal{O}((l'')^2, l''') \tag{45}$$

where the last equation is the local approximation of the former nonlocal expression. Equation (45) differs from Eqs. (13) and (14) in two respects: the boundary conditions for $l(x \to \pm\infty)$ as well as the substrate potentials V are different. In particular, in Eq. (45) V depends not only on z but also on x. Figures 16 and 17 illustrate the morphology of the wetting film for a complete wetting path as shown in Fig. 15 for $T_w^+ < T_0 < T_w^-$ where $T_w^+/T_c = 0.648$, $T_0/T_c = 0.712$, and $T_w^-/T_c = 0.842$ and obtained from the local approximation in Eq. (45) [53]. (For the comparison between local and nonlocal theory see Ref. [33].) Along the isotherm T_0 the substrate w_+ undergoes complete wetting whereas there is incomplete wetting of the substrate w_- in the limit $\Delta\mu \to 0$. The transition region around $x = 0$ can be characterized by the position of the turning point x_{tp} and the locations

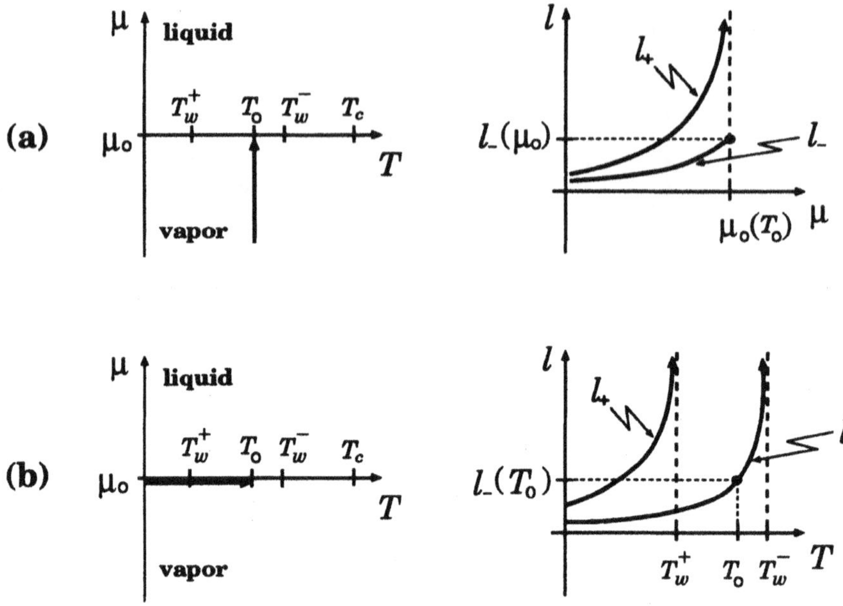

Figure 15. Various thermodynamic paths in the (μ, T) phase diagram leading to incomplete wetting of the substrate w_- but complete wetting of the substrate w_+ illustrated by the corresponding schematic behavior of the equilibrium film thicknesses $l_\pm$. The potential parameters are chosen such that $T_w^+ < T_w^-$. For simplicity the liquid-vapor coexistence curve $\mu_0(T)$ is straightened out. (a) If one approaches $\mu_0(T)$ along an isotherm with $T_w^+ < T_0 < T_w^-$, the substrate w_+ is completely wetted, i. e., $l_+ \to \infty$, whereas the film thickness l_- on the substrate w_- remains finite. (b) Upon an increase of the temperature along the liquid-vapor coexistence line the thickness l_+ of the liquidlike layer on the substrate w_+ diverges at T_w^+ and remains macroscopic for $T_w^+ < T < T_c$; because the temperature T_0 is smaller than T_w^-, the substrate w_- is only incompletely wetted. Both thermodynamic paths lead to the same final state of the system at $(\mu_0(T_0), T_0)$ and thus to the same profile $l(x; \mu_0(T_0), T_0)$. This also holds if the wetting transition at T_w^+ is first order, so that in (b) l_+ would jump to infinity at $T = T_w^+$.

x_0^+ and x_0^- at which $l(x)$ deviates by 20% from the asymptotic values l_- and l_+ on the left-hand and right-hand side, respectively. Whereas x_{tp} and x_0^- remain finite in the limit $\Delta\mu \to 0$, x_0^+ diverges proportional to $(\Delta\mu)^{-2/3}$. Thus as expected x_0^+ is proportional to the diverging lateral correlation length $\xi_\parallel \sim (\Delta\mu)^{-\nu_\parallel}$ for complete wetting where $\nu_\parallel = (1+\zeta)/(2\zeta) = 2/3$ for nonretarded dispersion forces $\sim r^{-(3+\zeta)}$ with $\zeta = 3$ (see Eq. (4.65) in Ref. [14]). In terms of l_+ one has $\xi_\parallel \sim l_+^{(1+\zeta)/2}$. This broadening of the lateral structure is clearly visible in Figs. 16 and 17. At coexistence $l(x \to \infty)$ must

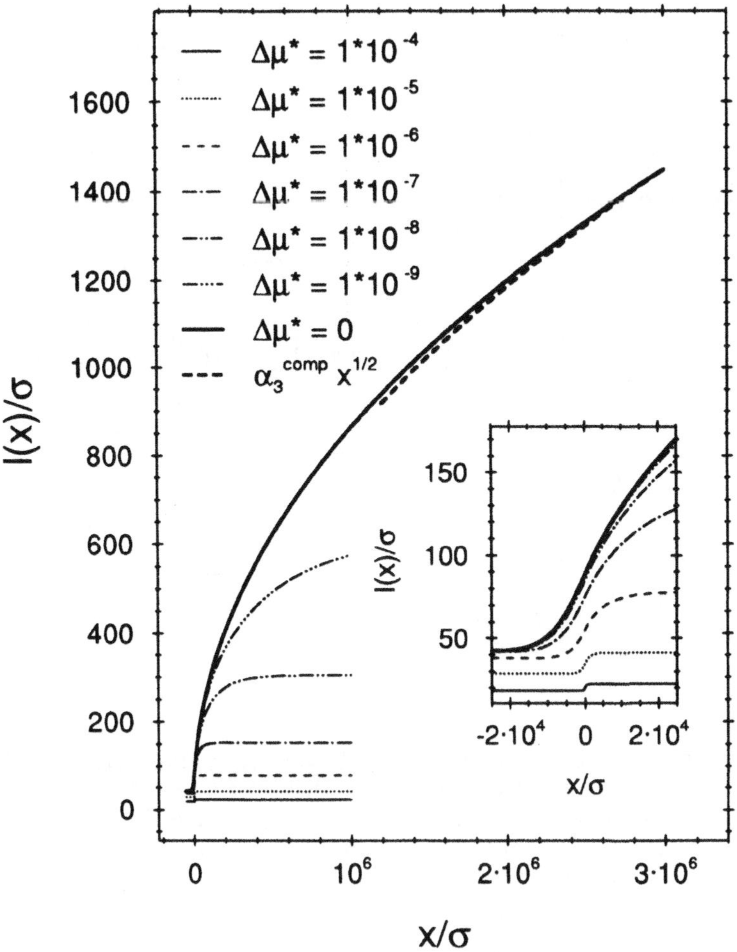

Figure 16. Profile $l(x)$ of the interface position as predicted by the local version of Eq. (45) for an isotherm at $T_0/T_c = 0.712$ as shown in Fig. 15(a) and complete wetting $\Delta\mu^* = \Delta\mu/\epsilon \to 0^+$. The substrate w_+ at $x > 0$ is completely wetted whereas the substrate w_- remains only partially wet. For the values of the interaction parameters see Ref. [53]. At coexistence, $\Delta\mu = 0^+$, the profile diverges as $\alpha_3^{comp} x^{1/2}$, as predicted analytically (thick dashed curve). Note that the scale of the ordinate is 2700 times larger than the scale of the abscissa.

diverge. One finds [53]

$$l(x \to \infty) = \alpha_\zeta^{comp} x^{2/(1+\zeta)}. \tag{46}$$

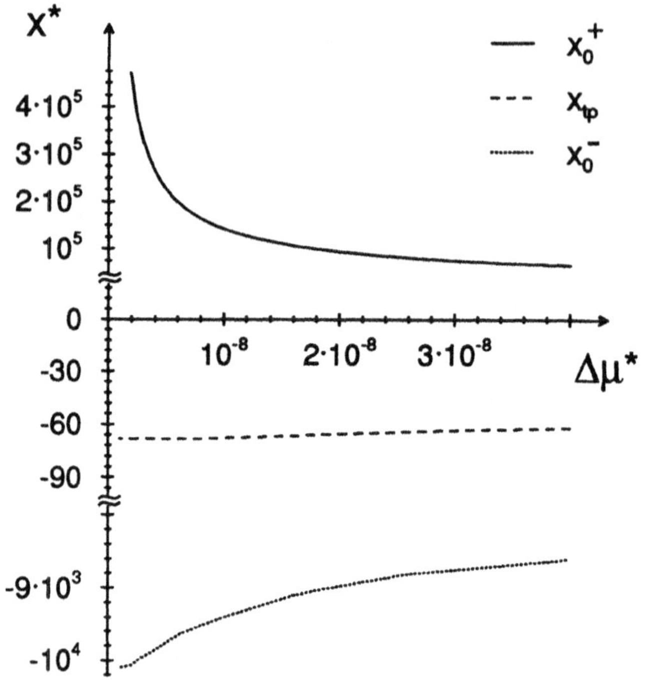

Figure 17. Positions $x^* = x/\sigma$ of the turning point x_{tp} and the locations x_0^+ and x_0^- at which $l(x)$ deviates by 20% from the asymptotic values l_- and l_+ on the left hand and right hand side, respectively, for $\Delta\mu \to 0$. x_0^+ diverges as $\Delta\mu^{-2/3}$ and leads to a clearly visible broadening of the profile around $x = 0$ for $\Delta\mu \to 0$ (see Fig. 16). x_{tp} and x_0^- attain finite values for $\Delta\mu \to 0$.

For the dispersion forces with $\zeta = 3$ as considered here the profile diverges like $x^{1/2}$ [54, 55]. The amplitude α_3^{comp} in Eq. (46) is given by

$$\alpha_3^{comp} = \left(8\frac{a_2^+}{\sigma_{lg}}\right)^{1/4} . \tag{47}$$

If T_0 coincides with T_w^+ and if this happens to be a critical wetting transition, a_2^+ is zero. In this case $l(x \to \infty)$ behaves as

$$l(x \to \infty) = \alpha_\lambda^{crit} x^{2/(1+\lambda)} \tag{48}$$

where λ characterizes the decay of the subdominant term in the effective interface potential: $\omega_\pm(l) = a_2^\pm l^{-\zeta+1} + a_3^\pm l^{-\lambda+1} + \ldots$. For nonretarded forces one typically has $\zeta = 3$ and $\lambda = 4$. In this case the profile diverges as

$x^{0.4}$, i. e., weaker as in the case of complete wetting. The corresponding amplitude is given by

$$\alpha_4^{crit} = \left(\frac{25}{2}\frac{a_3^+}{\sigma_{ly}}\right)^{1/5} . \tag{49}$$

For $x > 0$ and x sufficiently large compared with σ the broading of the lateral structure and the power law in Eq. (46) allow one to cast $l(x, \Delta\mu)$ into the scaling form

$$l(x, \Delta\mu) = l_+(\Delta\mu) f(x/\xi_{||}(\Delta\mu)) \tag{50}$$

where the scaling function $f(y)$ has the properties $f(\infty) = 1$ and $f(y \to 0) \sim y^{2/(1+\zeta)}$. The finite values $l_\pm$ are attained via van der Waals power law tails:

$$l(x \to \pm\infty) = l_\pm + \frac{\lambda_3^\pm}{x^3} + \mathcal{O}(x^{-4}). \tag{51}$$

For complete wetting one has

$$\lambda_{3,comp}^\pm = \frac{l_\pm^4}{6}\frac{a_2^- - a_2^+}{a_2^\pm} \tag{52}$$

whereas for approaching a critical wetting transition along coexistence one finds (see Eq. (7))

$$\lambda_{3,crit}^\pm = \frac{l_\pm^5}{2}\frac{a_2^- - a_2^+}{(u_4^\pm - t_4\rho_l)\Delta\rho}. \tag{53}$$

4.3. LINE TENSION

The density functional approach allows one to determine the line tension which is associated with the lateral inhomogeneity of the wetting films. Within the local theory the leading singularities of the line tension can be determined analytically. For complete wetting the line tension diverges logarithmically for $\Delta\mu \to 0$:

$$\tau_{comp}(\Delta\mu \to 0) = -\frac{\sqrt{2}}{4}\sqrt{a_2^\pm \sigma_{ly}} \ln\left(\frac{\Delta\mu}{2a_2^+/(\sigma^3\Delta\rho)}\right) + \mathcal{O}(\Delta\mu \ln \Delta\mu). \tag{54}$$

A numerical analysis shows [53] that this leading singular behavior dominates only for undersaturations $\Delta\mu^* \lesssim 10^{-10}$. In the case of critical wetting $\delta = (T_w^+ - T)/T_w^+ \to 0$ the line tension attains a maximum finite value via a cusplike singularity:

$$\tau_{crit}(\delta \to 0) = \text{const} - \frac{\alpha}{2}\frac{|a_2^-(T_w^+)|}{a_3^+(T_w^+)}\,\delta \ln\left(\frac{1}{\delta}\right) \tag{55}$$

where $a_2^+(T \to T_w^+) = -\alpha\delta + \mathcal{O}(\delta^2)$ with $\alpha > 0$ and $a_3^+(T_w^+) > 0$.

5. Fluids in a wedge

Unless much experimental care is applied in growing as well as cutting and polishing crystals, generally on an atomic scale their surfaces exhibit an irregular topography. The effect of this geometrical disorder on wetting phenomena has been studied experimentally (see, e. g., Refs. [56, 57, 58, 59]) and theoretically (see, e. g., Refs. [60, 61, 62, 63, 64, 65]) by coarse-grained approaches which average laterally over the local height variation of the substrate. (For adsorption on fractal surfaces see Refs. [66, 67, 68, 69].) These studies do not resolve the *local* microscopic structure of fluids filling the grooves and covering the tips of such disordered substrates.

As emphasized in the Introduction, in the context of microfluidics there is an increasing demand for understanding the structure and the spreading of fluids in individual and well-defined geometrical structures such as wedges [70, 71, 72, 73] and the imbibition and flow of wetting fluids in capillaries with an angular [74, 75, 76] and noncircular [77] cross-section. The systematic experimental study of the morphology of wetting films embedded in such structures is still in its infancy. A promising approach is to investigate a liquid film covering a parallel array of grooves with trapezoidal cross-section by X-ray reflectivity measurements [78, 79]. On the theoretical side there are also only few studies. They deal with the filling of square wells [80, 81] or with the formation of a liquid meniscus in corners [82] or wedges [39, 83, 84, 85, 86, 87, 88]. These studies have been focused on various aspects of these phenomena which turn out to be surprisingly rich with a comprehensive picture still having to emerge. In the following the fluid structures in a single wedge will be discussed as a paradigmatic case.

5.1. THEORY OF CAPILLARITY

So far only volatile liquids have been considered which can reach full thermodynamic equilibrium with their vapor phases within short time scales. For such liquids the number of particles in a liquid film is not fixed but can freely be exchanged with a reservoir. However, for many liquids such as oils and polymers this thermodynamic equilibrium cannot be reached via evaporation within experimentally relevant time scales so that for these nonvolatile liquids the liquid volume is a conserved quantity. It will turn out to be advantageous to consider the nonvolatile systems first before turning to the problem of wetting in wedges by volatile fluids in the next subsection.

There is a sizeable body of literature devoted to the determination of the shape of the liquid-vapor interface of a nonvolatile liquid of volume V exposed to a wedge, with and without gravity [89, 90, 91, 92, 93]. In a macroscopic picture the equilibrium shape of a nonvolatile liquid drop is determined by minimizing its surface free energy under the constraint of a prescribed volume V. This volume is taken to be sufficiently large so that the shape resulting from this generalized Wulff construction does not depend on the absolute value of V. The surface free energy for a given configuration is the sum of the liquid-vapor surface tension σ_{lg} and of the wall-liquid surface tension σ_{wl} multiplied with the respective areas of contact for this configuration. (In general, the two walls of the wedge can be characterized by different surface tensions $\sigma_{w_1 l}$ and $\sigma_{w_2 l}$).

Surprisingly, only recently the final and explicit solution of this minimization problem has been given (see Fig. 6 in Ref. [93]), seemingly without being aware of previous efforts [90, 91, 92] and vice versa. Due to the isotropy of the liquid-vapor surface tension the optimum shape is part of a sphere which intersects the walls of the wedge with the contact angle θ valid for the corresponding planar substrate:

$$\cos\theta = \frac{\sigma_{wg} - \sigma_{wl}}{\sigma_{lg}}. \tag{56}$$

θ is measured between the substrate surface and the liquid-vapor interface through the liquid phase. For $\theta > \frac{\pi}{2}$ there are two possible liquid configurations depending on the opening angle 2φ of the wedge. For $\varphi \geq \theta - \frac{\pi}{2} > 0$ an edge blob forms; this is part of a sphere such that the liquid is in contact with both walls and the edge of the wedge. The shape of the liquid-vapor interface is convex. If the opening angle of the wedge is reduced such that $\varphi < \theta - \frac{\pi}{2} > 0$, the liquid loses contact with the edge of the wedge and a spherical bridge connecting the two walls is formed. The transition between these two distinct configurations occurs at $\varphi = \theta - \frac{\pi}{2} > 0$. For $\theta < \frac{\pi}{2}$ a tubular bridge between the walls is not possible. For a wide wedge with $\varphi > \frac{\pi}{2} - \theta > 0$ one finds an edge blob in contact with both walls and with the edge of the wedge. Upon decreasing the opening angle of the wedge, at

$$\varphi = \frac{\pi}{2} - \theta \tag{57}$$

a transition to edge spreading occurs which persists for $\varphi \leq \frac{\pi}{2} - \theta > 0$. Edge spreading is not possible for $\theta > \frac{\pi}{2}$.

These findings can be reformulated and interpreted as follows. If the walls of the wedge exhibit a tendency for wetting, i.e., at a fixed temperature such that $\theta < \frac{\pi}{2}$ with $\theta(T)$ decreasing as function of temperature, there is always a sufficiently narrow wedge with opening angle $2\varphi_0$ such that for all $\varphi < \varphi_0$ edge spreading of the liquid occurs. Equivalently, for any wedge with a fixed opening angle 2φ there is always a temperature $T_\varphi < T_w$ such that for $T > T_\varphi$ edge spreading of the liquid occurs. T_w is given by $\theta(T = T_w) = 0$ whereas T_φ is defined implicitly by

$$\theta(T = T_\varphi) = \frac{\pi}{2} - \varphi. \tag{58}$$

In the planar limit, i.e., $\varphi \to \frac{\pi}{2}$, the temperature T_φ approaches T_w from below.

On the other hand, if the walls of the wedge exhibit a tendency for drying, i.e., at a fixed temperature such that $\theta > \frac{\pi}{2}$ with $\theta(T)$ increasing towards π as function of temperature, there is always a sufficiently narrow wedge with opening angle $2\varphi_0$ such that for all $\varphi < \varphi_0$ the liquid looses contact with the edge which means edge spreading by the vapor phase. Equivalently, for any wedge with a fixed opening angle 2φ there is always a temperature $\tilde{T}_\varphi < T_d$ such that for $T > \tilde{T}_\varphi$ edge spreading of the vapor occurs. T_d is given by $\theta(T = T_d) = \pi$ whereas $\tilde{T}_\varphi$ is defined implicitly by

$$\theta(T = \tilde{T}_\varphi) = \frac{\pi}{2} + \varphi. \tag{59}$$

In the planar limit, i.e., $\varphi \to \frac{\pi}{2}$, the temperature $\tilde{T}_\varphi$ approaches T_d from below.

The above considerations apply for nonvolatile liquids with a conserved volume of liquid. Following a similar line of arguments based on Laplace's and Young's equation for a liquid meniscus in a wedge, Hauge [87] suggested that at a temperature $T_\varphi < T_w$ defined by Eq. (58) a wedge exposed to the vapor phase of a volatile liquid at liquid-vapor coexistence in the bulk undergoes a filling transition, i.e., the meniscus height in the center of the wedge attains a macroscopic value. This is in accordance with the observation of edge spreading described above for nonvolatile liquids. Thus one expects that in a grand canonical ensemble at liquid-vapor coexistence a wedge is completely filled with liquid at temperatures $T_\varphi < T < T_w$ for which the corresponding planar substrate is still nonwet. First simulation data for a microscopic model support this conclusion [94].

5.2. DENSITY FUNCTIONAL THEORY

Wetting of a wedge which is exposed to the vapor reservoir of a volatile liquid can be analyzed within density functional theory along the same

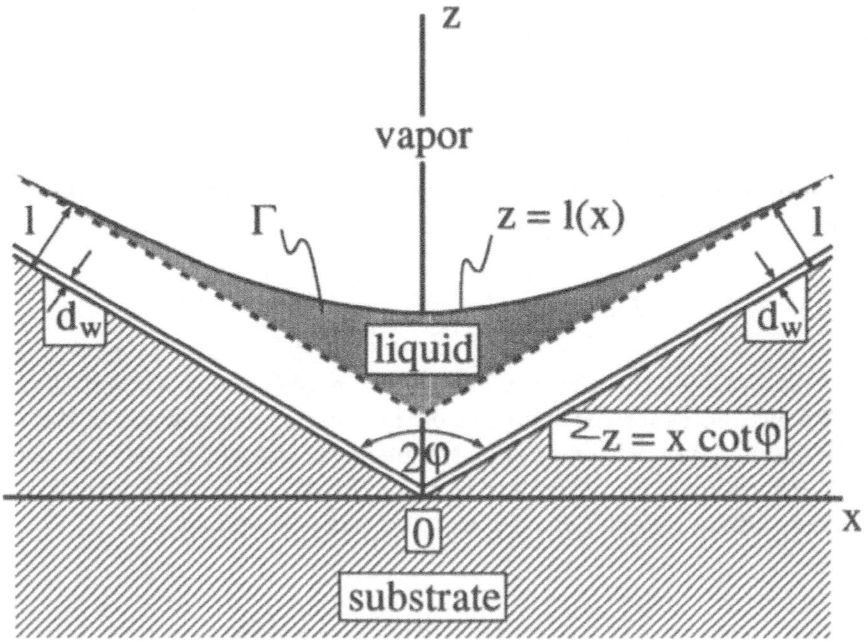

Figure 18. Vapor phase of a volatile liquid in contact with a homogeneous substrate forming a wedge with opening angle 2φ. The substrate surface is located at $z = |x| \cot \varphi$. A liquid meniscus of shape $z = l(x) = l(-x)$ is formed which is symmetric for the homogeneous substrate considered here. For $|x| \to \infty$ the substrate surface is covered by a wetting film of thickness l as for a single planar substrate. Therefore as boundary condition one has $l(x \to \pm\infty) = \pm x \cot \varphi + l/\sin \varphi$. Near the substrate the repulsive part of the substrate potential causes a depletion layer of thickness d_w. $\Gamma = 2\int_0^\infty dx\,[l(x) - x\cot\varphi - l\sin\varphi]$ denotes the excess coverage caused by the wedge as compared with a planar substrate; $\Gamma(\varphi \to \frac{\pi}{2}) = 0$.

lines as in Secs. II and IV by determining the equilibrium meniscus shape $l(x)$ (Fig. 18) from Eq. (45). In the present case $V(x, z)$ is the substrate potential provided by the material forming the wedge. Explicit expressions for $V(x, z)$ are given in Refs. [86] and [95]. (If the influence of the distant wall on the shape of the meniscus near one wall is neglected the local version of Eq. (57) leads to the so-called augmented Young equation [96].) Far from the edge of the wedge the substrate surfaces, which are located at $z = |x| \cot \varphi$, are covered by a wetting film of thickness l with the same value as for the corresponding single planar substrate. Therefore as boundary condition one has

$$l(x \to \pm\infty) = l_{asymp}(x) = \pm x \cot \varphi + l/\sin \varphi. \qquad (60)$$

For a complete wetting path ($\Delta\mu \to 0, T > T_w$) this asymptotic behavior is attained via van der Waals tails [86]:

$$l(x \to \infty) - l_{asymp}(x) = \frac{f_0}{x^3}. \tag{61}$$

In the limit $l/\sigma \to \infty$, $x/\sigma \to \infty$ and $l/\sigma \ll (x/\sigma)^{3/5}$ the amplitude f_0 can be determined analytically:

$$f_0^{nloc} = \frac{1}{12} l^4 \frac{\rho_g t_3 - \rho_w u_3}{\rho_l t_3 - \rho_w u_3} \geq \frac{1}{12} l^4 > 0. \tag{62}$$

f_0^{nloc} follows from the nonlocal version of Eq. (57) and it is larger than $f_0^{loc} = \frac{1}{12} l^4$ which holds for the solutions of the local approximation in Eq. (45). In both cases f_0 is positive so that the meniscus attains its asymptotic linear form from above. For complete wetting the wedge is filled with liquid such that the filling height $l(x = 0)$ and the excess coverage

$$\Gamma = \int_{-\infty}^{+\infty} dx \ [l(x) - l_{asymp}(x)] \tag{63}$$

diverge in the limit $\Delta\mu \to 0$. One finds

$$l(0) \sim l^{\zeta} \tag{64}$$

and

$$\Gamma \sim l^{2\zeta} \tag{65}$$

for interaction potentials decaying as $r^{-(3+\zeta)}$. Thus the excess coverage diverges like the square of the filling height which demonstrates that for large l the morphology of the wetting film in a wedge is characterized by a single length scale as for a flat substrate. Since for complete wetting $l \sim (\Delta\mu)^{-1/3}$ the above results imply $l(0) \sim (\Delta\mu)^{-1}$ and $\Gamma \sim (\Delta\mu)^{-2}$ [85, 86, 87], independent of the decay exponent ζ. Since the coverage Γ_{planar} of a single planar substrate scales with l, Eq. (65) demonstrates that $\Gamma_{wedge} \sim (\Gamma_{planar})^{-6}$ for dispersion forces with $\zeta = 3$. This result underscores how strongly the wedge promotes the covering by liquid as compared with the behavior on a planar surface.

5.3. PACKING EFFECTS IN A WEDGE

The density functional ansatz in Eq. (1) and the description of the fluid structure in the wedge in terms of an emerging liquid-vapor interface profile $l(x)$ are only valid for distances sufficiently far away from the solid surface because in the close vicinity of the solid substrate pronounced oscillations of the local number density occur which are not captured by the aforementioned approaches. Near a structureless planar wall these oscillations appear normal to the substrate and decay within a few molecular diameters (see, e. g., Refs. [97] and [98] for theory and experiments, respectively). These oscillations are caused by the repulsive parts of the substrate potential and of the interparticle pair potential of the fluid particles. Therefore the essence of these features is captured by considering a hard sphere fluid exposed to a structureless hard wall. These packing effects are not only relevant on an atomic scale for molecular liquids but also on a mesoscopic scale for colloidal particles confined by walls. By suitably tuning the permittivities of the walls, solute, and solvent the model of hard spheres near hard walls can closely resemble a colloidal suspension in which the colloidal particles effectively behave as hard spheres. In this sense the predictions of the hard sphere model, which is governed only by entropy, are directly experimentally accessible through optical methods or by force microscopy [48].

In a wedge the entropy driven ordered structures normal to each wall interfere and give rise to additional ordering phenomena in the lateral direction. These local structural properties have been investigated by means of grand canonical ensemble Monte Carlo simulations [99], advanced density functional theory [100], and exact sum rules [101]. The Monte Carlo simulations [99] have been carried out for a periodic array of parallel wedges resulting in a sawtooth-shaped, corrugated substrate with an opening angle $2\varphi = \gamma$. Due to the periodicity each wedge unit has a finite width S_x from tip to tip and a finite depth $D = (S_x/2)\cot(\gamma/2)$ from the tip to the bottom edge. The surface area of each unit is $S' \times S_y$ with $S' = S_x/\sin(\gamma/2)$ where S_y is the length of the edge of each wedge, repeated by periodic boundary conditions. For computational reasons the simulation cell in the z-direction is confined by a mirror corrugated substrate. The opposing tips of these substrates are separated from each other by a distance S_z. The origin is located half way between the bottom edges of the two opposing substrates. Figure 19 demonstrates the onset of the lateral packing effects upon decreasing the dihedral angle γ by showing at each point (x, z) the difference $\Delta\rho^* = \Delta\rho\,\sigma^3$ between the actual density $\rho(x, y)$ in the wedge and the corresponding density at that point if the distant wall would be absent and the close wall would have an infinite lateral extension. These simulations are carried out for a bulk density $\rho_b\sigma^3 = \rho_b^* = 0.7016$, i.e., sufficiently far below bulk freezing of the hard spheres, and for $S_z^* = 12$ and $S'^* = S_y^* = 10$ in units of σ. For this choice of parameters the vertical finite size of the simulation cell $S_z + 2D$ is so large that the two opposing substrates do not interfere with each other and the system reaches its bulk properties at the

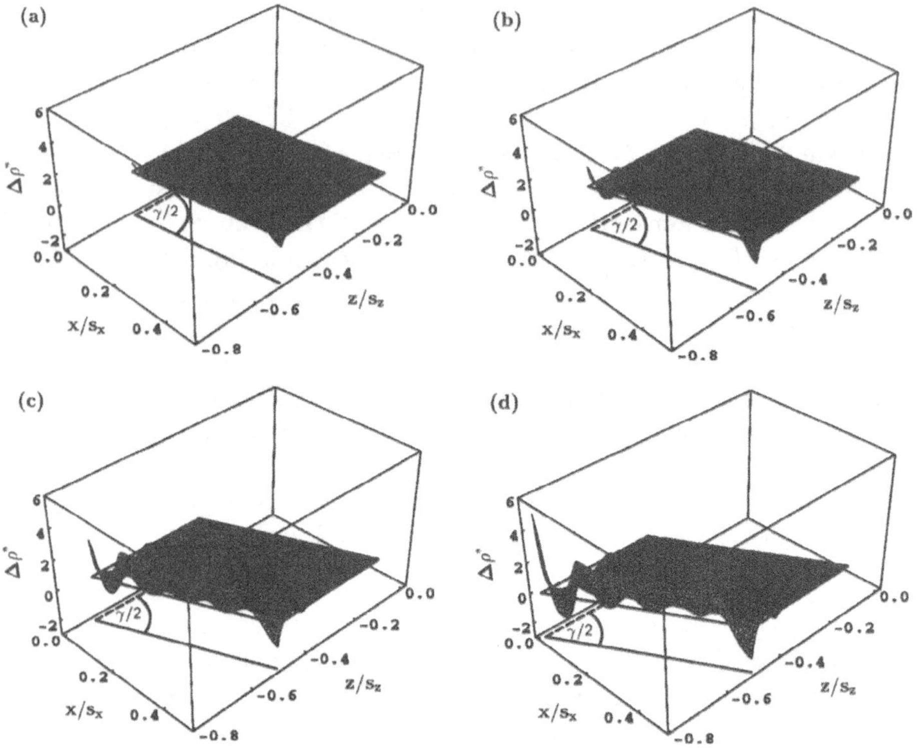

Figure 19. **The deviation $\Delta\rho^*$ of the local density in a wedge of dihedral angle $\gamma = 8\pi/9$ (a), $\gamma = 7\pi/9$ (b), $\gamma = 2\pi/3$ (c), and $\gamma = \pi/2$ (d) from the corresponding local density near a single planar hard wall. The solid line on the bottom of the box indicates the line of fluid-wall contact and the dashed line represents the bisector of the dihedral angle. Because of the symmetry of the system data are shown only for one quadrant for which the reference plane $\Delta\rho = 0$ is also indicated.**

center $z = 0$.

For $\gamma = 8\pi/9$ the substrate is almost planar, so that $\Delta\rho$ vanishes nearly everywhere except in the close vicinity of the corner ($|x|/S_x \ll 1$) and the tips ($|x|/S_x \simeq 0.5$) of the wedge (Fig. 19(a)). $\Delta\rho$ is positive in the corner while $\Delta\rho$ is negative at the tip. The positive deviation indicates that the two walls forming the wedge squeeze the hard spheres into the corner; this effect is more pronounced for small values of γ. On the other hand, around a tip the hard spheres are spatially less constrained and the fluid is more disordered as in the corresponding reference planar case, so that $\Delta\rho < 0$. Fig. 19(b) demonstrates the onset of a modulation of $\Delta\rho(x, z)$ along the line of liquid-wall contact which slightly extends into fluid regions farther away from the wall. These lateral oscillations decay upon moving away from the corner of the wedge towards its tip where the amplitude of the oscillations increases again. Thus at $\gamma = 7\pi/9$ both structural elements (i.e., corners and tips) give rise to lateral ordering without strong interference. Therefore this case closely resembles the behavior of a fluid in a single wedge or at a single tip. For smaller dihedral angles $\gamma = 2\pi/3$ in Fig. 19(c) and $\gamma = \pi/2$ in Fig. 19(d), packing effects induced by corner and tip merge, so that the oscillations in $\Delta\rho(x, z)$ at the fluid-wall contact persist from corner to tip and can no longer be attributed unambiguously to one or the other as in Fig. 19(b). For smaller dihedral angles the lateral extension of packing effects increases together with their amplitude. Both effects lead to more pronounced positive deviations $\Delta\rho$ (i.e., "squeezing") in the corner region. Increasingly negative deviations $\Delta\rho$ in the tip regime, on the other hand, reflect the more disordered character of the fluid as the tip sharpens with decreasing γ; but the effect of "squeezing" in the corner is stronger than the impact of the tip.

The density values at contact with hard walls fulfill interesting sum rules. At a single hard wall the so-called "wall theorem" [21] relates the contact value $\rho(z \to 0^+)$ to the bulk pressure p ($\beta = 1/(k_B T)$):

$$\beta p = \lim_{z \to 0^+} \rho(z). \tag{66}$$

In a single rectangular corner the integral over the deviation of the density at contact with the walls of the wedge from the corresponding contact value $\rho(Z = 0)$ at a flat surface can be expressed in terms of the wall-fluid interfacial tension σ_{wf} of a hard-sphere fluid at a flat hard wall [101]:

$$\int_0^\infty \left[\rho(x, z = x; \gamma = \frac{\pi}{2}) - \rho(Z = 0)\right] dx = -\beta \sigma_{wf} \sigma^2. \tag{67}$$

Thus *at contact*, local *surface* properties (i.e., $\rho(Z = 0)$) are related to *bulk* properties (i.e., βp) whereas *corner* properties are related to *surface* properties (i.e., σ_{wf}). This represents a hierarchy in which certain *local* properties are tied to *global* properties of the next higher thermodynamic level. The simulation data in Fig. 19(d) favorably compare with the sum rule in Eq. (67), but the interference between the corner and tip structures prohibits a precise quantitative test.

An even more detailed picture of the structure of a hard-sphere fluid in a hard wedge emerges in terms of the contour plots for $\rho(x, z)$ as shown in Fig. 20. This clearly demonstrates the packing effects both in lateral direction as well as normal to the surfaces. The entropically induced order in the corner becomes increasingly more pronounced upon decreasing the dihedral angel and significantly extends upwards in the center $x = 0$.

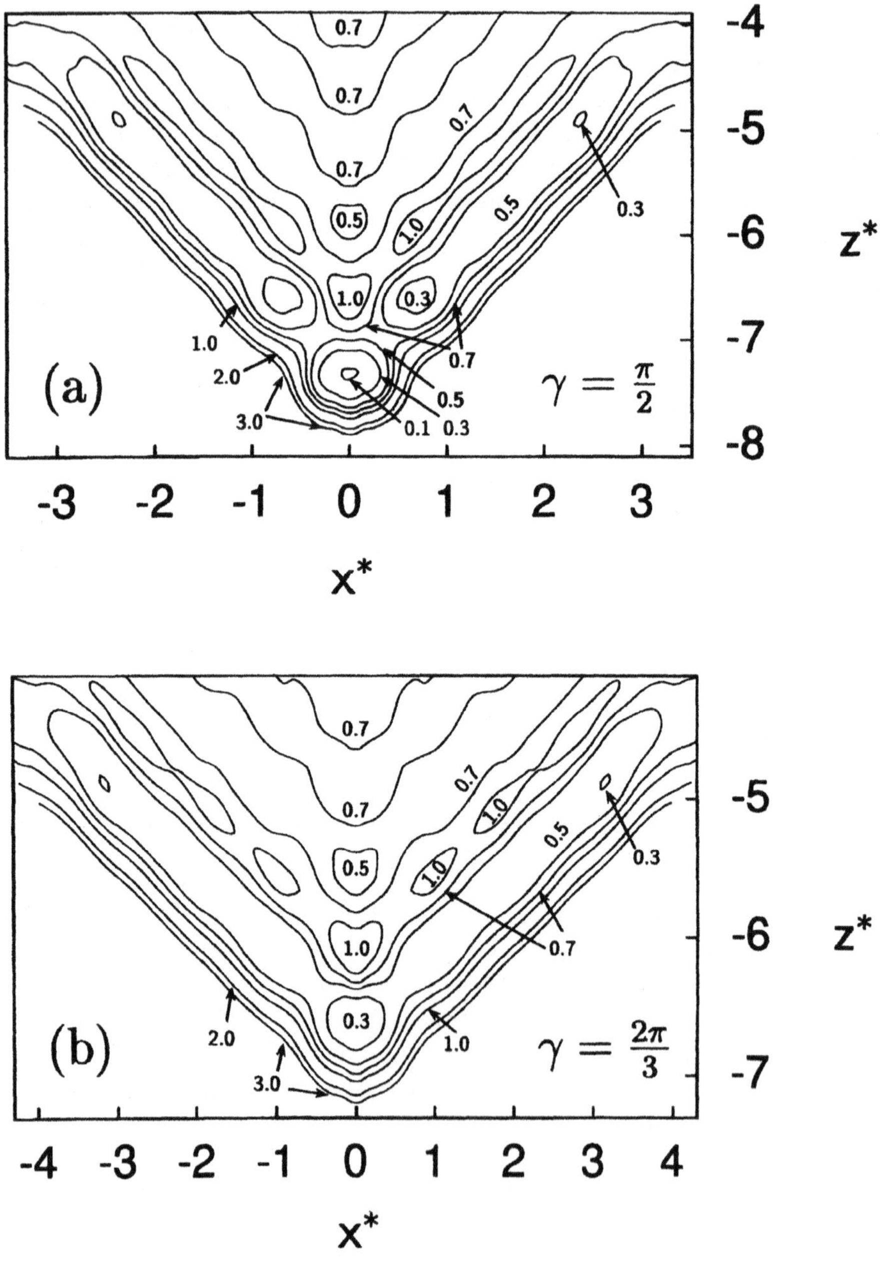

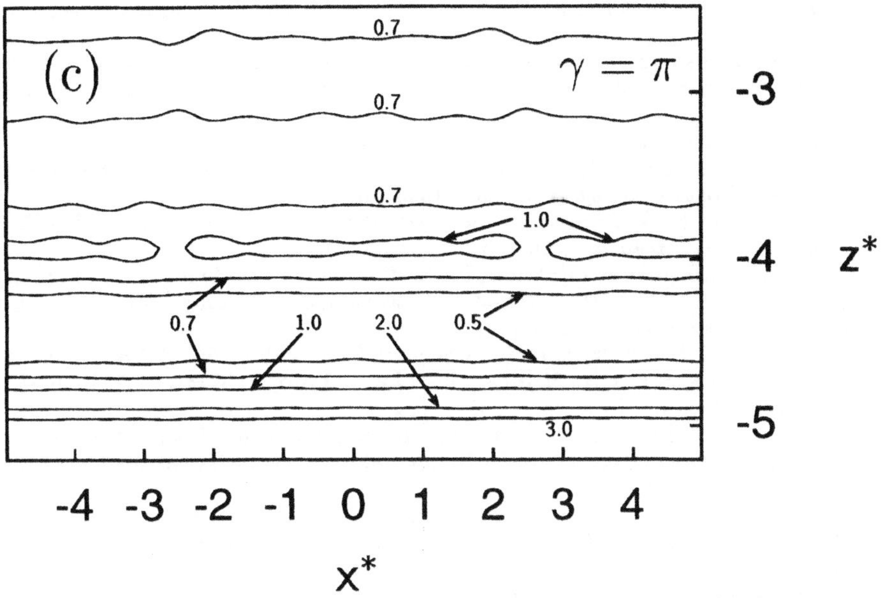

Figure 20. Contour plots of $\rho(x, z; \gamma)$ for three selected dihedral angles $\gamma = \pi/2$ (a), $\gamma = 2\pi/3$ (b), and $\gamma = \pi$ (c). Numbers and arrows are inserted to identify the various lines of constant $\rho^*(x, z; \gamma)$. For reasons of clarity the lines of fluid-wall contact are not shown. They are close to the isodensity curve corresponding to $\rho^* = 3.0$; $x^* = x/\sigma$; $z^* = z/\sigma$. In (c) the deviations from perfectly straight horizontal lines reflect the fluctuations induced by the Monte Carlo sampling.

References

1. W. Hansen, J. P. Kotthaus, and U. Merkt, in *Semiconductors and Semimetals*, edited by M. Reed (Academic, London, 1992), Vol. 35, p. 279; see also P. Bönsch, D. Wüllner, T. Schrimpf, A. Schlachetzki, and R. Lacmann, J. Electrochem. Soc. **145**, 1273 (1998), and J. Wang, D. A. Thompson, and J. G. Simmons, J. Electrochem. Soc. **145**, 2931 (1998).
2. D. W. L. Tolfree, Rep. Prog. Phys. **61**, 313 (1998).
3. Y. Xia and G.M. Whitesides, Annu. Rev. Mater. Sci. **28**, 153 (1998)
4. F. Burmeister, C. Schäfle, B. Keilhofer, C. Bechinger, J. Boneberg, and P. Leiderer, Adv. Mater. **10**, 495 (1998).
5. J. L. Wilbur, A. Kumar, E. Kim, and G. M. Whitesides, Adv. Mater. **6**, 600 (1994).
6. R. J. Jackman, J.L. Wilbur, and G. M. Whitesides, Science **269**, 664 (1995).
7. C. Casagrande, P. Fabre, E. Raphaël, and M. Veyssié, Europhys. Lett. **9**, 251 (1989); P. G. de Gennes, Rev. Mod. Phys. **64**, 645 (1992).
8. J. B. Knight, A. Vishwanath, J. P. Brody, and R. H. Austin, Phys. Rev. Lett. **80**, 3863 (1998).

9. E. Delamarche, A. Bernard, H. Schmid, B. Michel, and H. Biebuyck, Science **276**, 779 (1997).
10. E. M. Blokhuis and B. Widom, Curr. Opin. Colloid Interface Sci. **1**, 424 (1996).
11. G. H. Findenegg and S. Herminghaus, Curr. Opin. Colloid Interface Sci. **2**, 301 (1997).
12. P. G. de Gennes, Rev. Mod. Phys. **57**, 827 (1985).
13. D. E. Sullivan and M. M. Telo da Gama, in *Fluid Interfacial Phenomena*, edited by C. A. Croxton (Wiley, New York, 1986), p. 45.
14. S. Dietrich, in *Phase Transitions and Critical Phenomena*, edited by C. Domb and J. L. Lebowitz (Academic, London, 1988), Vol. 12, p. 1.
15. M. Schick, in *Liquids at Interfaces*, Proceedings of the Les Houches Summer School Lectures, Session XLVIII, edited by J. Chavrolin, J. F. Joanny, and J. Zinn-Justin (Elsevier, Amsterdam, 1990), p. 415.
16. S. Dietrich, in *Phase Transitions in Surface Films 2*, Vol. 267 of *NATO Advanced Study Institute, Series B: Physics*, edited by H. Taub, G. Torzo, H. J. Lauter, and S. C. Fain (Plenum, New York, 1991), p. 391.
17. G. Forgacs, R. Lipowsky, and T. M. Nieuwenhuizen, in *Phase Transitions and Critical Phenomena*, edited by C. Domb and J. L. Lebowitz (Academic, London, 1991), Vol. 14, p. 135.
18. T. Getta and S. Dietrich, Phys. Rev. E **57**, 655 (1988); and references therein.
19. R. Evans, in *Liquids at Interfaces*, Proceedings of the Les Houches Summer School Lectures, Session XLVIII, edited by J. Chavrolin, J. F. Joanny, and J. Zinn-Justin (Elsevier, Amsterdam, 1990), p. 1.
20. J. A. Barker and D. Henderson, J. Chem. Phys. **47**, 4714 (1967).
21. J. P. Hansen and I. R. McDonald, *Theory of Simple Liquids* (Academic, London, 1986).
22. S. Dietrich and M. Napiórkowski, Phys. Rev. A **43**, 1861 (1991).
23. S. Dietrich and M. Napiórkowski, Physica A **177**, 437 (1991).
24. M. Napiórkowski and S. Dietrich, Z. Phys. B **89**, 263 (1992).
25. M. Napiórkowski and S. Dietrich, Phys. Rev. E **47**, 1836 (1993).
26. M. Napiórkowski and S. Dietrich, Z. Phys. B **97**, 511 (1995).
27. G. Wiegand, T. Jaworek, G. Wegner, and E. Sackmann, J. Colloid Interface Sci. **196**, 299 (1997).
28. J. O. Indekeu, Physica A **183**, 439 (1992).
29. J. Drelich, Colloids Surf. A **116**, 43 (1996).
30. F. Bresme and N. Quirke, Phys. Rev. Lett. **80**, 3791 (1998).
31. A. S. Clarke, J. Chem. Phys. **96**, 9073 (1992); *ibid* **96**, 9078 (1992).
32. J. O. Indekeu and A. Robledo, Phys. Rev. E **47**, 4607 (1993).
33. C. Bauer and S. Dietrich (unpublished). The large deviations of the nonlocal theory from the local one reported in Ref. [18] are significantly reduced by using a suitable minimization algorithm for the nonlocal theory instead of solving the Euler-Lagrange equation $\delta\Omega_l[l(x)]/\delta l(x) = 0$ on a lattice as in Ref. [18]. Whereas the latter approach requires an enormous density of lattice sites, which could not be reached in Ref. [18], the minimization algorithm turns out to be far more efficient in obtaining the minimum of Ω_l.
34. D. Beysens and D. Estève, Phys. Rev. Lett. **54**, 2123 (1985).
35. M. L. Kurnaz and J. V. Maher, Phys. Rev. E **51**, 5916 (1995).
36. H. T. Dobbs, G. A. Darbellay, and J. M. Yeomans, Europhys. Lett. **18**, 439 (1992).
37. H. Löwen, Phys. Rev. Lett. **74**, 1028 (1995).
38. T. Bieker and S. Dietrich, Physica A **252**, 85 (1998).
39. J. R. Philip, J. Chem. Phys. **66**, 5069 (1977); *ibid* **67**, 1732 (1977).
40. R. Holyst and A. Poniewierski, Phys. Rev. B **36**, 5628 (1987); Physica A **149**, 622 (1988).
41. M. P. Gelfand and R. Lipowsky, Phys. Rev. B **36**, 8725 (1987).

42. J. O. Indekeu, P. J. Upton, and J. M. Yeomans, Phys. Rev. Lett. **61**, 2221 (1988); P. J. Upton, J. O. Indekeu, and J. M. Yeomans, Phys. Rev. B **40**, 666 (1989).
43. M. Swift and J. M. Yeomans, J. Phys. A **23**, 5655 (1990).
44. T. Gil and L. V. Mikheev, Phys. Rev. E **52**, 772 (1995).
45. I. Hadjiagapiou, J. Chem. Phys. **105**, 2927 (1996).
46. I. E. Dzyaloshinskii, E. M. Lifshitz, and L. P. Pitaevskii, Adv. Phys. **10**, 165 (1961).
47. V. A. Parsegian and G. H. Weiss, J. Chem. Phys. **60**, 5080 (1974); M. W. Cole and M. Schmeits, Surf. Sci. **75**, 529 (1978); N. S. Witte, J. Chem. Phys. **99**, 8168 (1993).
48. J. Israelachvili, *Intermolecular and Surface Forces*, Chs. 11.5 and 15.1 (Academic, London, 1985).
49. E. H. Hauge and M. Schick, Phys. Rev. B **27**, 4288 (1983).
50. M. P. Nightingale, W. F. Saam, and M. Schick, Phys. Rev. B **30**, 3830 (1984).
51. K. Ragil, J. Meunier, D. Brosseta, J. O. Indekeu, and D. Bonn, Phys. Rev. Lett. **77**, 1532 (1996).
52. V. Privman and M. E. Fisher, J. Stat. Phys. **33**, 385 (1983); J. Appl. Phys. **57**, 3327 (1985).
53. W. Koch, S. Dietrich, and M. Napiórkowski, Phys. Rev. E **51**, 3300 (1995).
54. J. F. Joanny and P. G. de Gennes, J. Chem. Phys. **81**, 552 (1984).
55. Y. Pomeau and J. Vannimenus, J. Colloid Interface Sci. **104**, 477 (1985).
56. D. Beaglehole, J. Chem. Phys. **93**, 893 (1989).
57. S. Garoff, E. B. Sirota, S. K. Sinha, and H. B. Stanley, J. Chem. Phys. **90**, 7505 (1989).
58. I. M. Tidswell, T. A. Rabedeau, P. S. Pershan, and S. D. Kosowsky, Phys. Rev. Lett. **66**, 2108 (1991).
59. P. S. Pershan, Ber. Bunsenges. Phys. Chem. **98**, 372 (1994).
60. D. Andelman, J.-F. Joanny, and M. O. Robbins, Europhys. Lett. **7**, 731 (1988).
61. M. O. Robbins, D. Andelman, and J.-F. Joanny, Phys. Rev. A **43**, 4344 (1991).
62. J. L. Harden and D. Andelman, Langmuir **8**, 2547 (1992).
63. A. O. Parry, P. S. Swain, and J. A. Fox, J. Phys.: Condens. Matter **8**, L659 (1996).
64. R. Netz and D. Andelman, Phys. Rev. E **55**, 687 (1997).
65. P. S. Swain and A. O. Parry, J. Phys. A: Math. Gen. **30**, 4597 (1997).
66. E. Cheng, M. W. Cole, and A. L. Stella, Europhys. Lett. **8**, 527 (1989).
67. P. Pfeifer, Y. J. Wu, M. W. Cole, and J. Krim, Phys. Rev. Lett. **62**, 1997 (1989).
68. M. Kardar and J. O. Indekeu, Europhys. Lett. **12**, 161 (1990).
69. G. Guigliarelli and A. L. Stella, Phys. Scr. T **35**, 34 (1991).
70. J. A. Mann, Jr., L. Romero, R. R. Rye, and F. G. Yost, Phys. Rev. E **52**, 3967 (1995).
71. R. R. Rye, F. G. Yost, and J. A. Mann, Langmuir **12**, 555 (1996); *ibid* 4625 (1996).
72. S. Gerdes and G. Ström, Colloids Surf. A **116**, 135 (1996).
73. S. Gerdes, A.-M. Cazabat, and G. Ström, Langmuir **13**, 7258 (1997).
74. M. Dong, F. A. L. Dullien, and I. Chatzis, J. Colloid Interface Sci. **172**, 21 (1995).
75. M. Dong and I. Chatzis, J. Colloid Interfaces Sci. **172**, 278 (1995).
76. D. Zhou, M. Blunt, and F. M. Orr, Jr., J. Colloid Interface Sci. **187**, 11 (1997).
77. E. Kim and G. M. Whitesides, J. Phys. Chem. B **101**, 855 (1997).
78. P. Müller-Buschbaum, M. Tolan, W. Press, F. Brinkop, and J. P. Kotthaus, Ber. Bunsenges. Phys. Chem. **98**, 413 (1994).
79. Z. Li, M. Tolan, T. Höhr, D. Kharas, S. Qu, J. Sokolov, M. H. Rafailovich, H. Lorenz, J.P. Kotthaus, J. Wang, S. K. Sinha, and A. Gibaud, Macromolecules **31**, 1915 (1998).
80. G. A. Darbellay and J. M. Yeomans, J. Phys. A: Math. Gen. **25**, 4275 (1992).
81. E. Cheng, J. R. Banavar, M. W. Cole, and F. Toigo, Surf. Sci. **261**, 389 (1992).
82. P. M. Duxbury and A. C. Orrick, Phys. Rev. B **39**, 2944 (1989).
83. B. V. Derjaguin and N. V. Churaev, J. Colloid Interface Sci. **54**, 157 (1976).
84. Y. Pomeau, J. Colloid Interface Sci. **113**, 5 (1985).

85. E. Cheng and M. W. Cole, Phys. Rev. B **41**, 9650 (1990).
86. M. Napiórkowski, W. Koch, and S. Dietrich, Phys. Rev. A **45**, 5760 (1992).
87. E. H. Hauge, Phys. Rev. A **46**, 4994 (1992).
88. Y. Song, Phys. Lett. A **180**, 361 (1993).
89. J. K. Lee and H. I. Aaronson, Surf. Sci. **47**, 692 (1975).
90. R. K. P. Zia, J. E. Avron, and J. E. Taylor, J. Stat. Phys. **50**, 727 (1988).
91. J. De Coninck, J. Fruttero, and A. Ziermann, Physica A **196**, 320 (1993); *ibid* **199**, 243 (1993).
92. L.-H. Tang and Y. Tang, J. Phys. II (France) **4**, 881 (1994); D. Langbein, J. Fluid Mech. **213**, 251 (1990).
93. P. Concus and R. Finn, Phys. Fluids **10**, 39 (1998); and references therein.
94. A. Lipowski, Phys. Rev. E **58**, R1 (1998).
95. P. Lambin, A. A. Lukas, I. Derycke, J.-P. Vigneron, and E. G. Derouane, J. Chem. Phys. **90**, 3814 (1989).
96. M. Kagan, W. V. Pinczewski, and P. E. Oren, J. Colloid Interface Sci. **170**, 426 (1995).
97. M. Schoen, *Computer Simulations of Condensed Phases in Complex Geometries* (Springer, Heidelberg, 1993); and references therein.
98. W. J. Huisman, J. F. Peters, M. J. Zwanenburg, S. A. de Vries, T. E. Derry, D. Abernathy, and J. F. van der Veen, Nature **390**, 379 (1997).
99. M. Schoen and S. Dietrich, Phys. Rev. E **56**, 499 (1997).
100. D. Henderson, S. Sokołowski, and D. Wasan, Phys. Rev. E **57**, 5539 (1998).
101. A. J. P. Nijmeijer and J. N. J. van Leeuwen, J. Phys. A **23**, 4211 (1990).

LIQUID STATE METHODS FOR DISORDERED SYSTEMS

M. L. ROSINBERG
Laboratoire de Physique Théorique des Liquides
Université Pierre et Marie Curie
4 Place Jussieu, 75252 Paris Cedex 05, France

1. Introduction

Theoretical work during the last decade has shown that one can extend the statistical-mechanical methods of equilibrium liquid-state theory to investigate the behavior of fluids adsorbed in disordered porous materials. The purpose of these two lectures is to provide an introduction to the subject. This is a vast domain which covers fundamental research as well as practical applications but given the short time and space available, some severe selection of material will be needed. Emphasis will be put on theoretical issues of a general nature rather than on the application to specific systems. We shall be mainly concerned with the problem of understanding fluid phase equilibria in these systems. There are some important questions for which we seek answers: are there true phase transitions in heterogeneous materials ? If so, how are they related to phase transitions in the bulk system ? Are there phase transitions that differ from those in the bulk ? What is the shape of the phase diagram ? Thanks to molecular simulations, lattice-gas models and density functional theory, satisfactory answers to these questions can be now given in the case of idealized pore geometries [1], e.g., fluids confined to narrowly spaced plane walls, cylindrical or spherical pores, and specific zeolite structures. In particular, the mechanism of capillary condensation i.e., the shift of the bulk gas-liquid transition due to confinement, and the surface transitions arising from interactions with the solid (layering and prewetting) are now well understood. Unfortunately, real porous solids are most often *disordered*, containing an interconnected network of pores of various sizes and shapes which creates a complex distribution of adsorption energy [2]. This is the case for instance of silica gels or activated carbons which are important adsorbents from a practical point of view. In these media, the very existence of macroscopic

C. Caccamo et al. (eds.), New Approaches to Problems in Liquid State Theory, 245–278.

phase equilibria is controversial and our fundamental understanding of the experimental facts is still far from complete. On the other hand, the effect of quenched disorder (such as spatially fixed impurities or fields) on phase transitions and critical phenomena in magnetic systems has been studied by the physics community for many years [3]. Therefore, one should not be surprised that we shall borrow the concepts and methods that have been introduced in this field. What may come as a surprise is that, in return, liquid-state methods can be also useful in the study of random magnets. We hope to show in these lectures that the possibility exits of a fruitful cross-fertilization between the two fields. It must be stressed that we shall deal only with static properties i.e., thermodynamic quantities and microscopic structure as described by equilibrium spatial correlation functions. All dynamical effects will be left aside although they are always present in experiments and play sometimes a predominant role (for instance in forbidding the setting up of a true equilibrium situation). Attention will be also restricted to atomic single-component classical fluids (or spin 1/2 lattice models). Therefore, we shall not discuss the interesting questions that arise in the behavior of ionic fluids, polymers, liquid crystals and quantum fluids in porous media. Generalization of the formal theory to these systems seems possible (and has been done in some cases), but of course technical complications arise. The problem of freezing is excluded too. Clearly, we make no pretense of covering every aspects of the topic outlined by the title of these lectures. We thus apologize in advance for not referring to all studies which have been completed in recent years. The material covered by these lectures has been developed over the past years by E. Kierlik, G. Tarjus and myself, and this work has also benefited from our collaboration with P. Monson and G. Stell. More details can be found in the publications that are listed in the References.

2. Some experimental results and their interpretation

Because of the importance of porous materials in many applications such as in catalysis, gas separation and purification, a large amount of experimental data on gas adsorption on porous substrates has been accumulated over the years through academic and industrial research [2]. Although measured adsorption isotherms display a great diversity of behavior, there is a wide class of adsorbents, referred to as mesoporous solids (where the average pore size is in the range from 2 to 50 nm), for which one observes at low temperature a rapid increase in the mass adsorbed at a pressure that is less than the bulk saturated vapor pressure, this phenomenon being accompanied by a pronounced hysteresis loop which usually shrinks with increasing temperature. This is illustrated in Figure 1 which shows the adsorption-

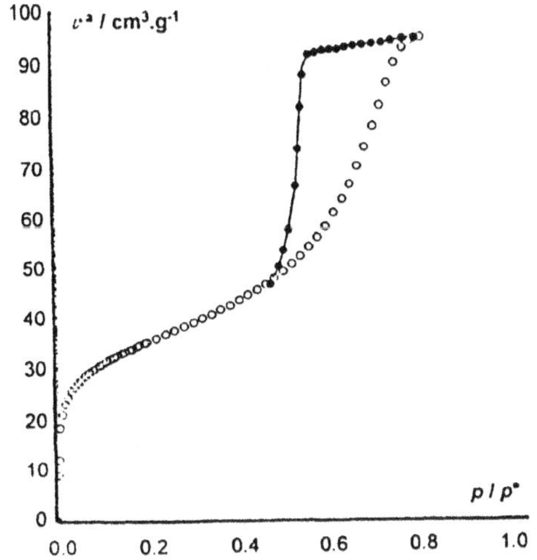

Figure 1. Adsorption-desorption isotherm of nitrogen on Vycor glass at 77.4 K (from Rouquerol *et al.* [4]).

desorption isotherm of N_2 on Vycor, a porous glass with pore diameters of $\sim$ 6nm and a porosity of $\sim$ 30%.

In the adsorption community, such behavior is conventionally attributed to capillary condensation of the gas in the pores, with the tacit assumption that the processes taking place in a given pore occur independently of those taking place in other pores. The total isotherm is then the sum of the adsorption in the individual pores, weighted according to some pore size distribution. This implies that first-order or critical transitions that may occur in each pore - and which are already rounded by finite-size effects - are smeared out. Therefore, despite the fact that one can map out a pseudo phase diagram of the confined fluid, as illustrated in Figure 2, (see also Figs. 1 and 2 in Ref. [6] and Figs. 3 and 7 in Ref. [7]), the possibility of collective phenomena between the fluid molecules in the entire three-dimensional porous structure is usually neglected in the theoretical analysis. On the other hand, the tortuous and interconnected nature of the pore network is often invoked as a cause of hysteresis, the evaporation of the condensate in a given pore depending on the presence or the absence of liquid in the neighboring pores (pore blocking effects) [8, 9].

A quite different point of view has motivated the interest of the physics community in studying fluids in porous media. The focus here has been on investigating the influence of quenched disorder on the critical behavior and on the process of phase separation. Indeed, as was suggested by de Gennes [10], near-critical binary mixtures (and by extension single component fluids) adsorbed in porous media may be thought as experimental realizations of the random field Ising model (RFIM), a spin system that has

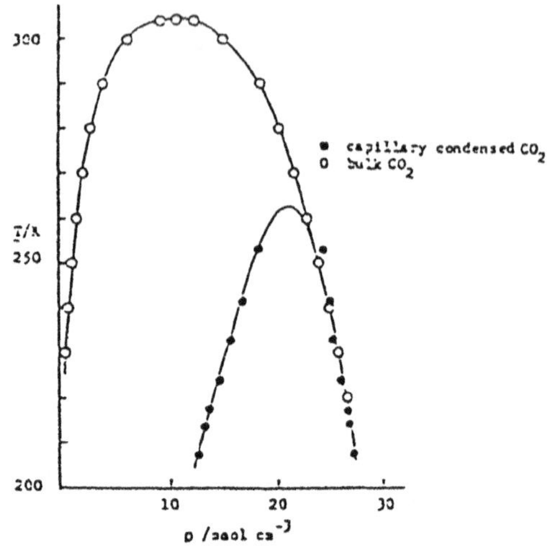

Figure 2. Comparison of the hysteresis phase diagram for CO_2 in Vycor with the phase diagram of bulk CO_2 (from Burgess *et al.* [5]).

been extensively studied for the past twenty years [11]. The argument is very simple: the solid surface attracts the fluid molecules or preferentially attracts one species of a mixture, resulting in a shift in the concentration or the composition of the free fluid in the middle of the pores; because of the disordered character of the pore structure, this shift corresponds to a spatially random perturbation of the chemical potential, which can be further modeled as a local random magnetic field in an Ising spin representation. Confinement plays no role in this picture. Of course, the correlation length in the fluid has to be much larger than the mean pore size in order that concentration or composition fluctuations sample the randomness. This is why it is important to consider near-critical fluids. Although the RFIM does not incorporate some important features that might be relevant for a realistic description of a fluid inside a porous solid, this interpretation has prompted a considerable number of experiments on binary fluid mixtures in porous glasses and single component or fluid mixtures in silica gels (for a recent review and references, see [12]).

It is in aerogels that clear evidence of a true thermodynamic phase separation has been seen. These materials are formed by very dilute disordered networks of silica strands that occupy only a very small portion of the total volume (the porosity can be as high as 99.9 %). The structure has also a mass-fractal character with no dominant length scale between 2 and typically 200 nm. Mesoscopic equilibrium phase separation of ^4He in aerogel has been found from heat capacity measurements and vapor-pressure isotherm studies [13]. As shown in Figure 3, there is a well-defined liquid-vapor coexistence curve which is much narrower than that in the bulk fluid and the

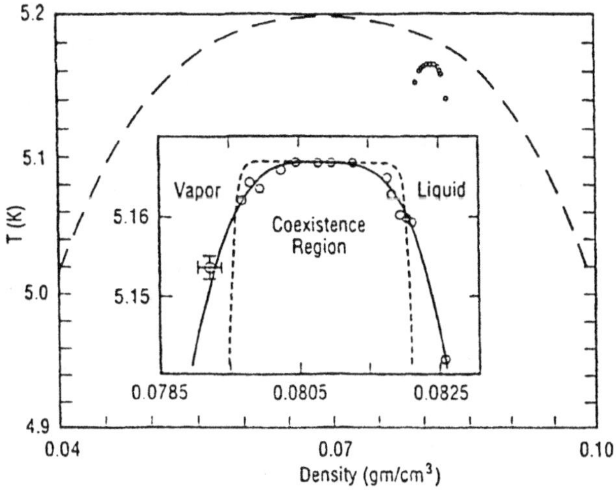

Figure 3. Coexistence curve of ^{4}He in an aerogel of 95 % porosity (circles and inset). The dashed line is the coexistence curve of bulk ^{4}He (from Wong and Chan [13]).

critical point is shifted to a lower temperature and a higher density.

Similar results have been observed with N_2 using light-scattering [14]. As noted in Ref. [14], the low density "vapor" phase consists of the vapor *and* a liquid film that coats the silica strands. What is remarkable is that such a small amount of impurity can alter drastically the phase diagram. In these sytems, confinement plays obviously a marginal role compared to that of disorder and wetting, and the RFIM picture should be more applicable than in the case of low porosity solids. However, the measured values of the exponent β characterizing the shape of the coexistence curve ($\beta \sim 0.28$ for ^{4}He, $\beta \sim 0.35$ for N_2) are more consistent with that of the bulk fluid than with that of the RFIM ($\beta \sim 0$ [15]). More recently, the coexistence curve of the isobutyric-water mixture in aerogel has been also determined [16]. Again, it is narrow, below that of the pure system and shifted to the side of the phase rich in the species preferentially attracted by the silica (here water). Note that this result has been obtained by placing the fluid/gel system in equilibrium with a reservoir of the fluid. Previous studies on liquid mixtures were done at fixed overall composition, and although very slow mesoscopic phase separation was observed with large domains of each phase, no well-defined phase transition could be detected.

In the case of mixtures adsorbed in Vycor, only microscopic (rather that macroscopic) separation has been observed, even far inside the coexistence region of the bulk system. The existence of a frozen domain state and the appearance in the shape of the structure factor of a Lorentzian-squared term that may arise from the field created by the pore surface, are consistent with features predicted by the RFIM. However, it is questionable to apply the

RFIM to systems where the effects of confinement may be more important than the effects of disorder. Accordingly, an alternative explanation has been proposed in terms of wetting phenomena in a single pore under the constraints of fixed composition [17]. In this picture, which is supported by the results of numerical simulations, the confinement and the competition between the preferential adsorption of one component and the interfacial tension between the two phases are at the origin of the very slow kinetics.

As it may appear from this brief survey (for another recent review, see Ref. [18]), the experimental observations are quite rich and the theoretical situation still rather unsettled. Although there is a striking resemblance between the genuine phase diagrams obtained in materials which have an open pore structure like aerogels and the "pseudo" phase diagrams obtained in materials which have a narrow distribution of pore sizes like Vycor, it is not obvious that these systems can be described within a single theoretical framework. This is however what can be tried by developing a liquid-state formalism that encompasses the effects of confinement, preferential adsorption and randomness. This task first requires some reasonable modeling of the solid/fluid system.

3. Modeling the solid/fluid system

Roughly speaking, a disordered porous material is just a mixture of voids and solid regions which can be viewed as two interpenetrating and percolating phases [19]. The detailed shapes of the pores depend on the shape of the solid particles, their size distribution and their mode of packing. But the overall structure can be considered as statistically homogeneous and isotropic at length scales larger than the dimensions of the constituent particles. This means that the material can be characterized by statistical quantities such as an average density, a specific void volume (or a porosity), a specific interfacial area between void and solid, and also a static structure factor which may be obtained from scattering experiments or image analysis. In practice, the microstructure is determined by the chemical composition and the procedure used to manufacture the material. For instance, the pore space in Vycor is formed by a process of spinodal decomposition of a borosilicate glass melt, the borate-rich phase being chemically etched out. This results in a sponge-like structure where the pores have a short-range correlation in position characteristic of a material formed through spinodal decomposition [20]. Silica gels, on the other hand, are made by aggregation of colloidal silica or by polymerization of a molecular precursor in a solvent which is eventually eliminated [21]. When hypercritically dried, alcogels produce aerogels which retain the delicate fractal structure of the original alcogels [22]. Modeling such amorphous materials is a challenging task. The essence of the approach that has been used in recent years both in computer simulations and in statistical mechanical theories is to treat the solid as a collection of particles in some predefined microstructure which is assumed to be rigid, i.e., it is not affected by the introduction of the solute. Therefore, the solid is simply viewed as a disordered array of quenched (or frozen in place) "obstacles" within which the fluid establishes itself in a state of thermal equilibrium [23, 24]. The simplest example of such a "quenched-annealed" (Q-A) mixture, where only excluded-volume interactions are taken into account, is a hard sphere fluid inside a matrix of hard spheres arranged in an equilibrium structure [25, 26, 27]. More realistic models have been considered. For instance, the gel particles in microporous silica xerogel have been modelled as a collection of oxygen centers smeared out over non-overlapping microspheres of approximately 1.5 nm in radius [28]. The influence of the microstructure on the behavior of an adsorbed fluid has then been investigated by varying the arrangement of the silica "grains" between that of an equilibrium hard sphere system and that of a more realistic random connected structure, or by placing them on the sites of a crystal lattice. Diffusion and partition coefficients [28], Henry's law constants and adsorption isotherms [29], heats of adsorption [30], adsorption equilibria for binary mixtures [31], and phase transitions [32] have been studied in these systems via computer simulations. Similarly, a molecular model for activated carbons has been proposed, which consists of an isotropic collection of three-dimensional platelets, each representing a stack of carbon basal plane [33]. This model is obviously more realistic than the widely used single slit-pore model which does not incorporate the connectedness of the pore phase [34]. More recently, realistic computer models of Vycor have been introduced, using quench molecular dynamics to mimic the actual procedure used to manufacture glasses [35] or reconstructing the mesoporous network from a non-porous silica solid [36], in such a way that the pore morphology is consistent with that obtained from scattering data [20]. Several studies have also used lattice models. The advantage of this type of models is that they allow a detailed investigation, especially by computer simulations. It is generally believed that the phase diagrams and the critical behavior observed in lattice models are similar to that of continuum systems. The Ising model [37, 38, 39], the q-state Potts model [40] and the Blume-Emery-Griffith model [41] have been simulated inside random lattice structures that mimic Vycor glass or aerogels. In the following, we shall describe in more details a simple lattice gas model of a Q-A mixture that contains some key features of real solid/fluid systems but allows to perform analytical calculations [42, 43, 44].

4. Thermodynamics and statistical mechanics of a quenched-annealed mixture

In a system with quenched variables, it may seem impossible and even useless to define the observable quantities for a particular sample because they depend apparently on a macroscopic number of random variables. In the case of a fluid in interaction with an ensemble of fixed obstacles, one has a continuous set of such random variables which are the values of the external field created by the local structure of the obstacles. As a consequence, the standard methods for treating fluids in external potentials, like density functional theory [45], can no longer be used. It is clear however that the observable quantities, such as the free energy of the adsorbed fluid, its density, its structural properties measured by scattering experiments, etc..., should only depend on some averaged characteristics of the material. More precisely, these quantities are expected to be *self-averaging*, which means that they have a unique value in the thermodynamic limit $V \to \infty$ which can be obtained by averaging over a large number of macroscopic subsamples or "realizations", each having a fixed but different arrangement of the host particles according to some statistical distribution. One thus has to perform two successive averages, first a thermal average over the fluid configurations in a given realization of the solid matrix, and then a second average over all possible realizations. A major consequence of this average over disorder is that translational invariance is recovered in the statistical mechanical treatment. As first suggested by Madden and Glandt [23], this allows to extend to Q-A systems the conventional methods of liquid-state theory. On the other hand, this procedure introduces some subtilities in the thermodynamics, and there has been some confusion in the litterature, especially about the concept of pressure [24, 25, 46].

The key point is that the averaged grand potential density $\omega_f = \Omega_f/V$, which is the thermodynamic potential conjugated to the total volume V of the ensemble fluid + matrix, i.e., $\omega_f = \partial\Omega_f/\partial V|_{T,\mu_f,\rho_m}$ (note that the derivative of Ω w.r.t. V is performed while keeping the matrix density ρ_m constant, as it must be [47]), does not coincide with the "mechanical" pressure, defined by averaging the microscopic stress tensor over an isotropic distribution of matrix particles [48]. Moreover, only this latter quantity has a simple virial expression (assuming that the fluid particles interact via a central pair potential); which can be also obtained by considering the mechanical equilibrium with an external reservoir and integrating the averaged force balance equation [49], i.e., the first equation of the Yvon-Born-Green hierarchy [50]. The virial expression of ω_f which can be obtained for instance from a careful application of the Green-Bogoliubov affine transformation, is complicated and can be derived only when the distribution of the matrix particles is that of an equilibrium system [47, 48]. Although this may appear as a surprising feature of Q-A systems - in a bulk fluid, $-\Omega/V$ identifies with the mechanical pressure -, it should not be a great concern since only ω_f is necessary for a fully self-consistent thermodynamic treatment. For instance, the Gibbs-Duhem equation for the adsorbed fluid writes [47],

$$S_f dT + N_f d\mu_f + V d\omega_f = 0 \tag{1}$$

where S_f is the averaged fluid entropy and N_f is the averaged number of fluid particles inside the matrix (there is an extra contribution to this equation if one allows for variations of the matrix density ρ_m. It is partly this contribution which complicates the virial expression of ω_f [47]). It is ω_f which determines a possible equilibrium between two phases of the adsorbed fluid from the conditions

$$\omega_f' = \omega_f'' \; ; \quad \mu_f' = \mu_f'' \tag{2}$$

at a given temperature. The Gibbs-Duhem equation also tells us that ω_f can be calculated by numerical integration of an adsorption isotherm,

$$\omega_f = -\int \rho_f(T, \mu_f) d\mu_f \tag{3}$$

where $\rho_f = N_f/V$ is the averaged fluid density inside the solid. One can derive a whole set of thermodynamic relations (isosteric, differential, and equilibrium heats of adsorption, Clapeyron expressions, etc...) which are relevant to experimental adsorption measurements and can be calculated by computer simulation using fluctuation theory [30]. Noteworthy is the compressibility sum-rule [47, 51]

$$\rho_f k_B T \chi_f = 1 + \rho_f \int h_{con}(r) d\mathbf{r} \tag{4}$$

where $\chi_f = 1/(\rho_f^2 k_B T)\partial\rho_f/\partial(\beta\mu_f)$ is the isothermal compressibility of the fluid and $h_{con}(r)$ is the so-called "connected" correlation function defined by [47]

$$\rho_f^2 h_{con}(r_{12}) = \overline{\rho_{ff}^{(2)}(\mathbf{r}_1, \mathbf{r}_2) - \rho_f(\mathbf{r}_1)\rho_f(\mathbf{r}_2)} \tag{5}$$

where $\rho_f(\mathbf{r}_1)$ and $\rho_{ff}^{(2)}(\mathbf{r}_1, \mathbf{r}_2)$ are, respectively, the one-body and two-body densities of the fluid for a given realization of the matrix, and the overbar denotes an average over all different realizations. Similarly, one defines a "disconnected" (or "blocking") function by

$$\rho_f^2 h_{dis}(r_{12}) = \overline{\rho_f(\mathbf{r}_1)\rho_f(\mathbf{r}_2)} - \rho_f^2 \; . \tag{6}$$

The total correlation function $h_{ff}(r_{12})$ defined by

$$\rho_f^2 h_{ff}(r_{12}) = \overline{\rho_{ff}^{(2)}(\mathbf{r}_1, \mathbf{r}_2)} - \rho_f^2 \ , \tag{7}$$

is thus the sum of h_{con} and h_{dis} (the Fourier transform of $\rho_f h_{ff}(r)$ is directly related to the structure factor $\hat{S}(\mathbf{k})$ of the adsorbed fluid). This decomposition of $h_{ff}(r)$ is quite similar to that of the spin-spin correlation function in the RFIM [11]. It is thus expected that $h_{con}(r)$ and $h_{dis}(r)$ behave differently near a possible critical point, with $h_{dis}(r)$ being longer-ranged. This leads us to the problem of calculating the structural properties of the adsorbed fluid using statistical mechanical techniques.

In modern equilibrium liquid-state theory [52], pair correlation functions are obtained by solving integral equations which correspond to approximate closures to the Ornstein-Zernike (OZ) equations. The whole machinery can be easily extended to Q-A systems by using the replica method, as suggested by Given and Stell [53, 54]. This standard method in the theory of random magnetic systems, such as spin glasses [55], allows to replace the study of the original Q-A mixture by that of a special equilibrium mixture of $n+1$ components. Formally, the replicas are introduced as follows. In order to obtain the physical (self-averaging) grand potential Ω_f, one has to compute

$$\Omega_f = -\frac{1}{\beta} \int d\mathbf{q}^{N_m} P_m(\mathbf{q}^{N_m}) \ln \Xi_f(\mathbf{q}^{N_m}) \tag{8}$$

where $\mathbf{q}^{N_m} = \{\mathbf{q}_1, \mathbf{q}_2, ...\mathbf{q}_{N_m}\}$ denotes a configuration of N_m matrix particles, $P_m(\mathbf{q}^{N_m})$ is the probability of this configuration, and $\Xi_f(\mathbf{q}^{N_m})$ is the matrix-dependent grand partition function defined by

$$\Xi_f(\mathbf{q}^{N_m}) = \sum_N \frac{z_f^N}{N!} \int d\mathbf{r}^N \exp[-\beta[H_{mf}(\mathbf{r}^N; \mathbf{q}^{N_m}) + H_{ff}(\mathbf{r}^N)]] \tag{9}$$

where $\mathbf{r}^N = \{\mathbf{r}_1, \mathbf{r}_2, ...\mathbf{r}_N\}$ denotes the positions of the N fluid particles, z_f is the fluid activity and $H_{mf} + H_{ff}$ is the total potential energy of the fluid in interaction with the matrix. For convenience, $P_m(\mathbf{q}^{N_m})$ may be assumed to have the form of a canonical Boltzmann-Gibbs distribution

$$P_m(\mathbf{q}^{N_m}) = \frac{1}{Z_m} \exp[-\beta H_{mm}(\mathbf{q}^{N_m})] \tag{10}$$

with $Z_m = \int d\mathbf{q}^{N_m} \exp[-\beta H_{mm}(\mathbf{q}^{N_m})]$. The H_{mm} so defined is an effective potential energy that may include irreducible many-body interactions and it need not be the true potential energy of the N_m particles, as assumed originally in Ref. [23]. H_{mm} is a pairwise additive Hamiltonian only when the solid has been produced by a quench (without structural relaxation) from an equilibrium distribution at some higher temperature. This simplifying assumption (as well as the independence of H_{mm} with temperature) is not necessary for the derivation of the OZ equations and in practice H_{mm} need not be known [24].

To perform the averaging of the logarithm in Eq. (8), one uses the following trick:

$$\begin{aligned}\Omega_f &= \lim_{n\to 0} -\frac{1}{\beta n}\int d\mathbf{q}^{N_m} P_m(\mathbf{q}^{N_m})[\exp(n\ln\Xi_f)-1] \\ &= \lim_{n\to 0} -\frac{1}{\beta n}(\Omega_f^{rep}-1) \end{aligned} \tag{11}$$

where

$$\begin{aligned}\Omega_f^{rep} &= \int d\mathbf{q}^{N_m} P_m(\mathbf{q}^{N_m})\Xi_f^n(\mathbf{q}^{N_m}) \\ &= \frac{1}{Z_m}\sum_{N_1,N_2,\ldots N_n}\frac{z_f^{N_1+N_2\ldots+N_n}}{N_1!N_2!\ldots N_n!}\int d\mathbf{q}^{N_n}(\prod_{a=1}^{n} d\mathbf{r}^{N_a})\exp[-\beta[H_{mm}(\mathbf{q}^{N_m}) \\ &+ \sum_{a=1}^{n}(H_{mf}(\mathbf{r}^{N_a};\mathbf{q}^{N_m})+H_{ff}(\mathbf{r}^{N_a}))]] \,. \end{aligned} \tag{12}$$

Ω_f^{rep} has the form of an *equilibrium* grand partition function of n *noninteracting identical* replicas of the original fluid ($a = 1\ldots n$), each interacting with the matrix particles. One can then apply to this $(n+1)$-components system the standard methods of liquid-state theory and in particular derive the properly generalized form of the OZ equations and of the corresponding approximate closures (of course, some care must be taken when H_{mm} is temperature-dependent and/or contains many-body interactions). There are three steps in the calculations which have to be made successively. First, the equations must be written for an integer n. Then, the analytical continuation must be made for arbitrary noninteger values of n. Finally, the limit $n \to 0$ has to be taken. Assuming that the permutation symmetry between the n replicas is not broken during the last step (as it occurs for instance in the Sherrington-Kirkpatrick spin glass model [55]), one then obtains the Replica Symmetric Ornstein-Zernike equations (RSOZ) [54], which relate the total pair correlation functions of the Q-A mixture to the corresponding direct correlation functions,

$$h_{mm} = c_{mm} + \rho_m c_{mm} \otimes h_{mm} \tag{13}$$

$$h_{fm} = c_{fm} + \rho_m c_{fm} \otimes h_{mm} + \rho_f c_{con} \otimes h_{fm} \tag{14}$$

$$h_{ff} = c_{ff} + \rho_m c_{fm} \otimes h_{mf} + \rho_f c_{con} \otimes h_{ff} + \rho_f c_{dis} \otimes h_{con} \tag{15}$$

$$h_{con} = c_{con} + \rho_f c_{con} \otimes h_{con} \tag{16}$$

where the symbol $\otimes$ denotes a convolution in **r**-space, $h_{mf} = h_{fm}$ and $c_{mf} = c_{fm}$ by symmetry. The disconnected and connected functions $h_{dis}(r)$ and $h_{con}(r)$ defined by Eqs. (5) and (6) are related to the replica correlation functions by [47]

$$h_{dis}(r) = \lim_{n\to 0} h_{ab}(r) \tag{17}$$

whith $a \neq b$, and

$$h_{con}(r) = \lim_{n\to 0} h_{aa}(r) - h_{dis}(r) \ . \tag{18}$$

The direct correlation functions $c_{dis}(r)$ and $c_{con}(r)$ are defined in the same fashion, with $c_{ff}(r) = c_{con}(r) + c_{dis}(r)$. Note that the correlation functions of the replicated mixture have standard graph series expansions only when the matrix configurations are those of an equilibrium quenched system. In the general case, one may just consider the RSOZ equations as the definition of the c's in terms of the h's. One may then introduce the actual structure factor of the porous solid as an input to Eqs. (14-16). The usual approximate closures to the OZ equations can still be used, but their validity must be checked at least for some simple systems.

5. Analytical studies of model systems: an overview

5.1. HARD SPHERE SYSTEMS

The hard sphere (HS) fluid plays a central role in our present understanding of equilibrium properties of dense fluids [52]. This is because packing effects are dominant at liquid-like densities so that the structure of the HS fluid can be used as a zeroth-order approximation to that of real liquids. This observation is the starting point of the successful perturbation theories, which are the modern version of van der Waals theory, and which represent the major achievement of liquid state research in the 1970s. The fact that the OZ equation in the Percus-Yevick (PY) approximation can be solved exactly for HS has been also very important. It is therefore not surprising that the first analytical studies of Q-A mixtures have dealt with HS systems. It is clear moreover that exclusion (confinement) effects are quite important in disordered solids with low porosity. In these studies [25, 26, 27], a HS fluid is placed in a quenched HS matrix or is contained in the interstices of a bed of randomly placed spherical obstacles (a quenched "ideal gas").

Closure relations to the RSOZ equations are based on hypernetted-chain (HNC) or PY-type approximations. In PY, one takes $c_{ij}(r) = 0$ for $r > \sigma_{ij}$, where σ_{ij} is the hard-sphere diameter. Since there is no direct interaction between different replicas (i.e., $v_{ab}(r) = 0$ for $a \neq b$), one has $c_{ab}^{PY}(r) = 0$ and thus $c_{dis}^{PY}(r) = 0$ in the limit $n \to 0$. That this is only an approximate result can be checked in the exactly solvable case of a random matrix in which the fluid particles do not interact with each other [54]. It is worth knowing that no analytical solution to the RSOZ equations within the PY approximation has been found so far for the Q-A HS mixture. In fact, it seems unlikely that a closed-form solution even exists, as can be anticipated from the density expansion of Baxter's $q_{ij}(r)$ functions[1].

Comparison with Monte Carlo simulations shows that the PY predictions for $h_{ff}(r)$ and $h_{mf}(r)$ are at their best at low to moderate fluid and matrix densities in the case of a HS matrix whereas the HNC appears to be best in an overall way in the random-matrix case [26]. Somewhat better results can be obtained by using Reference PY (PY) [26] or Reference HNC (RHNC) [27] closure relations. In particular, by adjusting the bridge functions so to impose self-consistency between the virial and compressibility routes to the pressure in the $n = 1$ (annealed) mixture, one can reach a degree of accuracy for all distribution functions (including $h_{dis}(r)$) which is almost similar to that found in ordinary liquid systems [27]. It may be noticed that $h_{ff}(r)$ and $h_{mf}(r)$ in the porous medium are almost indistiguishable from the corresponding distribution functions in the fully annealed mixture when the matrix particles are much larger than the fluid molecules (this is also true for thermodynamical properties: at a given fixed density and compostion, the chemical potential of the fluid in the annealed mixture and in the quenched matrix are quite similar). This result has been partly explained by the cancellation of some cluster diagrams in the density expansion of the distribution functions [46].

5.2. CONTINUUM MODELS WITH ATTRACTIVE INTERACTIONS: PHASE DIAGRAMS

As discussed previously, one of the most interesting and controversial problem concerning single-component fluids and fluid mixtures in disordered porous material is to understand phase equilibria. This issue can be investigated systematically using the liquid-state formalism which we have just presented.

[1] At first order in ρ_f, one finds that $q_{mf}(r)$ is a polynomial of order 8 in r, whereas $q_{dis}(r)$ is of order 17. At second order in ρ_f, $q_{mf}(r)$ is a polynomial of order 26 ! [56]. In the annealed mixture, $q_{ij}(r)$ is a quadratic polynomial at all densities [57].

To be more specific, we now consider a Q-A mixture in which all interactions, matrix-fluid and fluid-fluid, contain a hard-sphere part and an attractive tail $w_{ij}(r)$. The simplest approximation which can predict the occurence of a vapor-liquid equilibrium in the system consists in treating both w_{ff} and w_{mf} as perturbative potentials, writing

$$c_{ij}(r) = c_{ij}^{ref}(r) - \beta w_{ij}(r) \tag{19}$$

where $c_{ij}^{ref}(r)$ is a direct correlation function in the (reference) Q-A HS mixture, whose properties are assumed to be known. When using the compressibility route to the free energy (i.e. by integrating Eq. (4) or its equivalent form $\rho_f k_B T \chi_f = [1 - \rho_f \int c_{con}(r) d\mathbf{r}]^{-1}$), this random-phase approximation (RPA) is equivalent to a mean-field or generalized van der Waals theory in which the expression of the excess Helmholtz free energy density of the fluid is

$$-\frac{\beta A_{ex}^{MF}}{V} = -\frac{\beta A_{ex}^{ref}}{V} + \frac{1}{2}\hat{\phi}_{ff}(0)\rho_f^2 + \hat{\phi}_{mf}(0)\rho_m\rho_f \tag{20}$$

where A_{ex}^{ref} corresponds to the reference system and $\hat{\phi}_{ij}(0) = -\beta \int w_{ij}(r) d\mathbf{r}$. Application of this theory to model systems [58, 59] has shown that both the critical temperature and the critical density decrease with increasing matrix concentration, whereas the width of the coexistence curve is only slightly modified. Only the first feature is in agreement with the experimental phase diagrams in aerogels, and this phenomenon is quite similar to the decrease of the capillary condensation critical temperature with increasing confinement observed in studies of single, geometrically ideal pores [1]. In fact, the main weakness of this mean-field approximation is that the shape of the coexistence envelope does not change when varying $w_{mf}(r)$. In principle, one can remedy to this shortcoming by including $w_{mf}(r)$ in the reference system and treating only $w_{ff}(r)$ as a perturbation. This will be done below in the case of a lattice-gas model.

Another flaw of the RPA/mean-field theory is that the pair distribution functions $g_{ff}(r) = 1 + h_{ff}(r)$ and $g_{mf}(r) = 1 + h_{mf}(r)$ are non-zero inside the hard cores. This means that exclusion effects are not treated correctly. Integral equations like PY or HNC satisfy the "core-conditions" exactly, but they are notoriously inadequate for describing phase equilibria because of their singular mathematical behavior in and around the two phase region. For instance, the solution of the RSOZ equations within the PY approximation for a model of methane adsorbed in a silica xerogel [46] has shown that this approximation yields very good predictions for the pair distribution functions at high temperature (typically, above the bulk critical temperature), but the predictions deteriorate at lower temperatures, and it becomes increasingly difficult to obtain a solution to the equations at low temperature. This type of problem does not occur with the optimized random phase approximation (ORPA) and the closely related mean-spherical approximation (MSA) when the free energy is obtained by integration of the Gibbs-Helmoltz equation $U_{ex} = \partial\beta A_{ex}/\partial\beta$ (U_{ex} is the configurational internal energy of the fluid that can be calculated from the pair correlation functions). More generally, except very close to the critical point, accurate predictions for the coexistence curve of bulk simple fluids are obtained within the framework of the optimized cluster theory (OCT) introduced by Andersen and Chandler [60]. Another advantage of this class of approximations is that closed-form expressions of the free energy are available when using the energy route. The extension of the OCT to Q-A mixtures using the replica formalism does not present any major difficulty and we refer the reader to Ref. [61] for technical details. For instance, the ORPA free energy is given by

$$\begin{aligned} -\frac{\beta A_{ex}^{ORPA}}{V} &= -\frac{\beta A_{ex}^{ref}}{V} + \frac{1}{2}\rho_f^2 \hat{c}_{con}(0) + \rho_f \rho_m \hat{c}_{fm}(0) - \frac{\rho_f}{2} c_{ff}(0) \\ &- \frac{1}{2(2\pi)^3} \int d\mathbf{k} \{\ln[1 - \rho_f \hat{c}_{con}(k)] - \rho_f h_{dis}(k)[1 - \rho_f \hat{c}_{con}(k)]\} \\ &- \{\text{same terms in the reference system}\} \end{aligned} \quad (21)$$

where

$$\begin{aligned} c_{mf}(r) &= c_{mf}^{ref}(r) - \beta w_{mf}(r) \quad \text{for } r > \sigma_{mf} \\ c_{con}(r) &= c_{con}^{ref}(r) - \beta w_{ff}(r) \quad \text{for } r > \sigma_{ff} \\ c_{dis}(r) &= c_{dis}^{ref}(r) \end{aligned} \quad (22)$$

and the hat denotes the Fourier transform of a function. The MSA is obtained from the ORPA by replacing the exact reference correlation functions by their Percus-Yevick counterparts. One derives the EXP (exponential) approximation for the correlation functions and the corresponding "ORPA +B_2" approximation for the free energy by summing additional diagrams in the OCT [2]. In Ref. [61], these approximations have beeen applied to a Q-A system in which the matrix-fluid and fluid-fluid interactions decompose into a hard-sphere part and a Lennard-Jones 12-6 tail *à la* Weeks-Chandler-Andersen i.e., $w_{ij}(r) = -\epsilon_{ij}$ for $\sigma_{ij} < r < 2^{1/6}\sigma_{ij}$ and

[2]Note that in Ref. [61], $h_b(r)$ (i.e., $h_{dis}(r)$) should be replaced by $g_b(r) = h_b(r) + 1$ in the expressions of B_2 and B_3. This change, however, does not significantly alter the numerical results and the qualitative conclusions. The figures presented here correspond to the original calculations of Ref. [61].

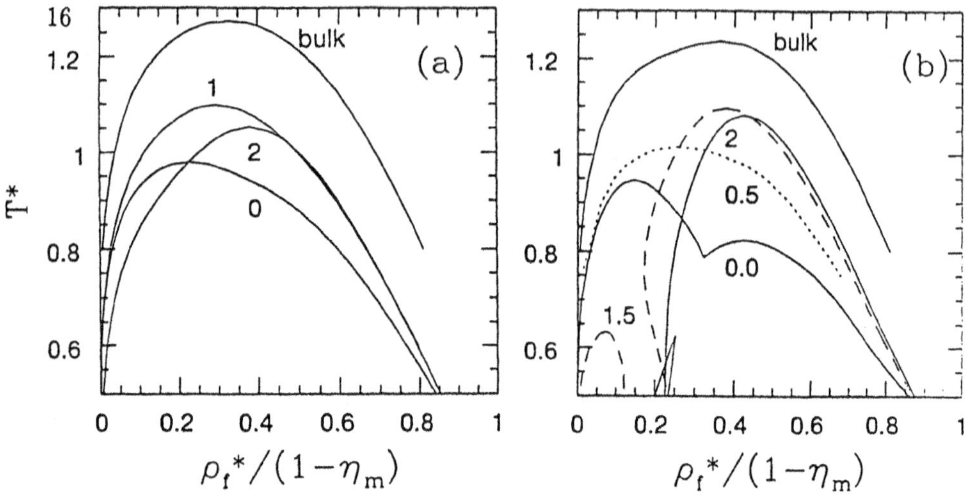

Figure 4. Coexistence envelopes in a matrix with reduced density $\rho_m\sigma^3 = 0.15$ ($T^* = k_BT/\epsilon_{ff}$, $\rho_f^* = \rho_f\sigma^3$). Each curve is labelled with the corresponding value of the interaction ratio y. (a) Predictions from the MSA. (b) Predictions from the ORPA+B_2 theory (from Kierlik *et al.* [61]).

$w_{ij}(r) = 4\epsilon_{ij}[(\sigma_{ij}/r)^{12} - (\sigma_{ij}/r)^6]$ for $r > 2^{1/6}\sigma_{ij}$ (the LJ tails are cut at $r = 2.5\sigma_{ij}$).

For simplicity, we only discuss here the predictions for the equal-size case, $\sigma_{mf} = \sigma_{ff} = \sigma$. The model is then characterized by only two parameters, the matrix density ρ_m which fixes the porosity ($\phi = 1 - \eta_m$ for an equilibrium HS matrix and $\phi = \exp(-\eta_m)$ for a random matrix, with $\eta_m = (\pi/6)\rho_m\sigma^3$), and the interaction ratio $y = \epsilon_{mf}/\epsilon_{ff}$ which fixes the relative strength of the solid-fluid and fluid-fluid interactions. Properties of the reference system have been calculated in the PY approximation. Figure 4 shows exemples of coexistence envelopes using the MSA or the "ORPA +B_2" approximations.

Like mean-field theory, MSA predicts that the critical temperature decreases as one decreases the matrix porosity for a fixed value of y. But the shape of the coexistence envelope now changes with y. Figure 4(a) shows that the variations of T_c with y may be non-monotonic: for $\rho_m^* = \rho_m\sigma^3 = 0.15$, T_c first increases and then decreases as one increases y. On the other hand, the critical density (renormalized by the matrix porosity) increases monotonically and becomes eventually larger than the bulk value. It is indeed expected, and observed experimentally [13, 14] that $\rho_{f,c}$ is displaced towards the liquid phase when the interaction with the solid is sufficiently attractive.

Surprisingly, the phase diagrams obtained with the "ORPA +B_2" theory look rather different from those predicted by MSA. The most salient feature is the appearance of a second coexistence region (or only a shoulder) for a certain range of values of ρ_m and y. For the purely repulsive matrix ($y = 0$), this phenomenon occurs on the high-density side of the main coexistence region, whereas it appears on the vapor side for a sufficiently attractive solid-fluid interaction. It is then accompanied by a significant narrowing of the main coexistence envelope which has some resemblance to the experimental observations discussed previously (cf. Figures 2 and 3). It is of course disturbing, from a theoretical perspective, that the results are so sensitive to the approximations in the theory (the global shape of the phase boundaries is again modified when one includes in the OCT free energy the contribution of the higher order term B_3 [61]). In fact, huge modifications in the phase diagram can be driven by changes in the free energy of the order of a few % in the intermediate-density region, and the intermediate phases may or may not exist depending on slight changes in the temperature or the matrix density. This convergence problem in the OCT can be attributed to the large values of the "optimized" renormalized potentials in the intermediate-density region where the mechanism of "repulsive force screening" [62] is less efficient than at high density. Such a problem does not exist in bulk fluids where these intermediate densities correspond to the two-phase region and only one van der Waals loop is expected.

In view of these difficulties, we cannot conclude definitely as to the existence of two critical points in this model system. However, although not observed so far in experiments, such "pre-evaporation" or "pre-condensation" transitions have been also obtained in the Monte Carlo simulation studies of a Lennard-Jones fluid adsorbed in a model silica xerogel [32]. In the intermediate phases, snapshots of the fluid particles configurations show that the fluid has evaporated or condensed in the dense regions of the matrix, while looking liquid-like (resp. gas-like) in the remaining void space. All phases are inhomogeneous, disordered and span the entire void space of the porous solid[3]. Interestingly, the adsorption isotherms obtained in the simulation exhibit an hysteresis loop which resemble closely that seen in experiments [63]. The same sort of phase diagrams is also predicted for the Q-A lattice gas model which we present in the next section. On the other hand, a theoretical calculation for a (non-disordered) periodic array of cylindrical strands, has attributed the appearance of a second critical point to fluctuation-induced effects [64].

[3]Note that the determination of phase diagrams via Monte Carlo simulations for systems in the presence of quenched disorder is not as well established as in the case of pure systems. Indeed, finite-size effects, that should be taken into account in the study of phase transitions, and averaging over disorder, that must a priori include a large number of matrix realizations, are not easily handled.

5.3. A Q-A LATTICE GAS MODEL

As we hope that there is some universality in the phase behavior of fluids in disordered porous media, we now propose to consider a simple lattice gas version of a Q-A mixture [42, 43, 44] that incorporates the influence of randomness, matrix-fluid interaction, and confinement (here described in terms of dilution of the lattice sites available for the fluid "particles").

We thus consider a d-dimensional lattice where the sites can be occupied either by matrix or fluid particles and the pair potentials all contain a point "hard core" which excludes multiple occupancy of a site. The Hamiltonian is

$$\mathcal{H} = -w_{ff} \sum_{<ij>} \tau_i \tau_j \eta_i \eta_j - w_{mf} \sum_{<ij>} [\tau_i \eta_i (1-\eta_j) + \tau_j \eta_j (1-\eta_i)] \tag{23}$$

where τ_i is the fluid occupancy variable at site i ($\tau_i = 0$ or 1) and η_i is the (quenched) site disorder variable taking the values 0, 1 at random such that on average $\bar{\eta}_i = 1 - \rho_m$; $w_{ff} > 0$ and w_{mf} are the fluid-fluid and fluid-matrix nearest-neighbor (n.n.) interactions and the sums are over n.n. pairs of sites. Like in the previous continuum model, the matrix is only specified by its density ρ_m and by the interaction ratio $y = w_{mf}/w_{ff}$. By using the usual equivalence between a lattice gas and a spin 1/2 model ($\tau_i = (1+\sigma_i)/2$, with $\sigma_i = \pm 1$), the above system can be transformed into an Ising model with Hamiltonian

$$\mathcal{H} = -J \sum_{<ij>} \sigma_i \sigma_j \eta_i \eta_j - K \sum_{<ij>} [\sigma_i \eta_i (1-\eta_j) + \sigma_j \eta_j (1-\eta_i)] - h \sum_i \sigma_i \eta_i \tag{24}$$

with $J = w_{ff}/4$, $K = J(2y-1)$, and $h = 1/2(\mu_f + q w_{ff}/2)$ where q is the coordination number of the lattice. Except for the case $K = 0$, i.e., $y = 1/2$, where one recovers the usual site-diluted Ising model [65], this Hamiltonian does not have up-down symmetry. What we have is a site-diluted Ising model with (locally) correlated random fields $h_i = h + 2Kn_i$, where n_i is the number of n.n. sites of site i that are occupied by the matrix. In contrast, in the standard RFIM, described by the Hamiltonian

$$\mathcal{H} = -J \sum_{<ij>} \sigma_i \sigma_j - \sum_i h_i \sigma_i \ , \tag{25}$$

the local random fields have a symmetric probability distribution and are delta-correlated in space [11]. An asymmetric RFIM where a fraction p of sites have a magnetic field h_0 and the rest of the sites have the field h_1, has been suggested to be more appropriate for the study of phase transitions in

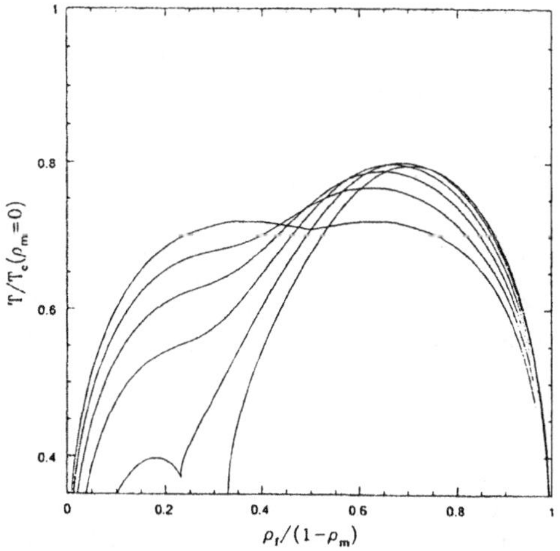

Figure 5. MSA predictions for the coexistence envelopes in a BCC lattice with $\rho_m = 0.25$ and various values of the interaction ratio. From top to bottom, y= 0.5, 0.625, 0.75, 0.875, 1 and 1.25 (from Kierlik *et al.* [44])

porous media [66]. We think that the model described by Eq. (23) is more faithful to the local inhomogeneities and correlations in the field induced by a disorder matrix. There have been two limited Monte Carlo studies of this model in two [38] and three dimensions [37] (in the latter work, only the case $y = 1$ was considered). The global phase diagram of the model has been calculated in the MSA in Refs. [43, 44] using the energy route to the thermodynamics. The MSA closure to the RSOZ equations is defined in Fourier space by

$$\begin{aligned}\hat{c}_{mf}(\mathbf{k}) &= c^0_{mf} + (\beta w_{mf} q)\ \hat{\lambda}(\mathbf{k}) \\ \hat{c}_{con}(\mathbf{k}) &= c^0_{con} + (\beta w_{ff} q)\ \hat{\lambda}(\mathbf{k}) \\ \hat{c}_{dis}(\mathbf{k}) &= 0 \end{aligned} \qquad (26)$$

where $\hat{\lambda}(\mathbf{k})$ is the characteristic function of the lattice and c^0_{mf} and c^0_{con} are state-dependent functions that can be completely determined by the core conditions $h_{mf}(\mathbf{r} = 0) = h_{ff}(\mathbf{r} = 0) = -1$ (since the matrix is totally random, one has $h_{mm}(\mathbf{r}) = -\delta_{\mathbf{r},0}$ and thus, from Eq. (13), $c_{mm}(\mathbf{r}) = -\delta_{\mathbf{r},0}/(1-\rho_m)$). The remarkable feature of the MSA for n.n. lattice gas or spin models is that the OZ equations can be solved analytically in terms of the lattice Green's function $P(\mathbf{r}, z)$, where $z = \rho_f(\beta w_{ff} q)/(1 - \rho_f c^0_{con})$. $P(\mathbf{r}, z)$ is a well-known mathematical object defined in Fourier space by $P(\mathbf{k}, z) = [1 - z\hat{\lambda}(\mathbf{k})]^{-1}$ [67] . We refer to Refs. [43, 44] for technical details.

Figure 5 shows the coexistence envelopes for a BCC lattice with matrix density $\rho_m = 0.25$ and various values of the interaction ratio y. As in the continuum model discussed previously, the most striking feature is the appearance of a shoulder on the gas side that eventually becomes a true second transition with a shrinking of the main coexistence region [4]. This transition, however, only appears in a narrow range of the parameters and depends on the lattice type. A better understanding of the origin of this phenomenon can be reached by using an ORPA approach (cf. Eqs. (22)) where the reference system now includes the matrix-fluid n.n. interaction w_{mf} [68]. This reference system can be treated exactly and displays correlations that extend to second nearest neighbors (in particular $c_{dis}^{ref} \neq 0$). At zero temperature, the adsorption isotherm exhibits sharp steps which corresponds to the successive filling of the sites available to the fluid particles and having $q, q-1, q-2$, etc... n.n. matrix sites. What is predicted by MSA and ORPA (and also by a mean-field treatment that includes w_{mf} in the reference system), is that some of these transitions persist at finite temperatures when the fluid-fluid attractive interaction is turned on. Therefore, the underlying physical picture is the following: depending on the values of ρ_m and y, one observes a direct vapor $\rightarrow$ liquid transition (i.e., the "capillary condensation" occurs before any "wetting" phenomena are observed), or two (or may be more) transitions from the vapor phase to an intermediate phase where the liquid has condensed in some regions of the matrix, and from the intermediate phase(s) to the liquid phase. The experimentally observed phase transitions of ^{4}He and N_2 in aerogels could correspond to this latter transition. A similar scenario has been proposed from a mean-field treatment of the asymmetric RFIM [66]. Preliminary comparison between the ORPA and MSA predictions and Monte-Carlo simulations shows good agreement for the adsorption isotherms. Determination of the phase diagram by simulation is in progress. Application of the Self-Consistent Ornstein-Zernike Approximation (SCOZA) to this system has been also considered, as we shall now discuss.

6. Application to magnetic systems with quenched disorder

On several occasions we have referred to the random field Ising model or to the site-diluted Ising model. These are spin models of random magnets which have been extensively studied over the last two decades, together with spin glasses. However, despite considerable effort, the understanding of the influence of quenched disorder on phase transitions in these systems is still incomplete. Attention has been focused on the critical behavior.

[4]Note that the coexistence envelope for the site-diluted Ising model ($y = 1/2$) displays a double hump around $\rho_f = (1 - \rho_m)/2$. This spurious behavior of the MSA when the Hamiltonian has up-down symmetry does not exist in higher dimensions. On the other hand, the coexistence envelopes for $y \neq 1/2$ still have an asymmetric shape and the second transition still exists [43, 44].

Series expansions, field-theoretic Renormalization Group (RG) techniques and numerical simulations have been applied to this problem which remains a controversial issue. It would certainly be presumptuous to hope that liquid state methods may yield answers to all issues in the realm of random magnets. In particular, these methods overlook the asymptotic critical regime (with the exception of the Hierarchical Reference Theory (HRT) to be discussed below in section 7, that includes the treatment of long-range fluctuations in an RG fashion). On the other hand, liquid state methods represent an easy way to include *some* fluctuations into an analytical treatment and therefore to go beyond the mean-field approximation which is often the only theory that is used to describe phase diagrams. As we shall see, liquid state methods are especially valuable to calculate non-universal quantities such as the critical temperature. In the perspective of designing a good theory for fluids in disordered materials, it is also interesting to study random spin models because they can be used as benchmarks. We may gain more confidence into our theoretical methods if they are able to reproduce faithfully what is already known for the RFIM, the site-diluted Ising model, etc... In the following we shall focus on a promising approach, the Self-Consistent Ornstein-Zernike Approximation (SCOZA) which was originally developed by Hoye and Stell [69] for pure fluids and lattice gas systems. Recently, it has been shown that the SCOZA provides a very good description of the properties of the three-dimensional Ising model, even in the close vicinity of the critical point [70] We think that this approximation can be useful in the case of disordered systems as well.

6.1. THE SITE-DILUTED ISING MODEL

The simplest model of a random magnet, which corresponds to alloys with formula $A_pB_{1-p}C$ where some non-magnetic atoms of type B replace at random positions magnetic atoms of type A in the pure magnet AC, is the site-diluted Ising model [65] whose Hamiltonian is

$$\mathcal{H} = -J \sum_{<ij>} \sigma_i \sigma_j \eta_i \eta_j \tag{27}$$

where $\sigma_i = \pm 1$ and $\eta_i = 0, 1$ are site (uncorrelated) disorder variables such that $\overline{\eta}_i = p$, the concentration of magnetic atoms. We have seen in the previous section that this model may be considered as a Q-A binary mixture. In lattice gas language it is described by the Hamiltonian (23) with $y = w_{mf}/w_{ff} = 1/2$. We can thus apply to this model the whole machinery of the RSOZ equations (with $\rho_m = 1 - p$) and of the corresponding closure approximations.

As noted before, the MSA for the n.n. Q-A lattice gas, described by the closure relations, Eqs. (26), can be solved analytically. The SCOZA strategy for this system is the following [71]. We keep an Ornstein-Zernike ansatz, assuming that the direct correlation functions have the range of the corresponding interaction potentials, but we now consider that the values at $\mathbf{r} = \mathbf{0}$ and/or $\mathbf{r} = \mathbf{e}$ (where $\mathbf{e}$ denotes a vector from the origin to one of its nearest neighbors) are state- and dilution-dependent. Eqs. (26) are thus replaced by

$$\begin{aligned} \hat{c}_{mf}(\mathbf{k}) &= c^0_{mf}(\rho_f, \rho_m, T) + c^1_{mf}(\rho_f, \rho_m, T)\ \hat{\lambda}(\mathbf{k}) \\ \hat{c}_{con}(\mathbf{k}) &= c^0_{con}(\rho_f, \rho_m, T) + c^1_{con}(\rho_f, \rho_m, T)\ \hat{\lambda}(\mathbf{k}) \\ \hat{c}_{dis}(\mathbf{k}) &= 0\,. \end{aligned} \tag{28}$$

It is easy to see that that the solution to the RSOZ equations is the same as that of the MSA except that z is now equal to $\rho_f q c^1_{con}/(1 - \rho_f c^0_{con})$ and w_{mf}/w_{ff} is replaced by the state-dependent interaction ratio $y = c^1_{mf}/c^1_{con}$. Again, the unknown quantities c^0_{mf} and c^0_{con} can be determined by the core conditions for $g_{mf}(\mathbf{r})$ and $g_{ff}(\mathbf{r})$. The remaining unknown functions $z(\rho_f, \rho_m, T)$ and $y(\rho_f, \rho_m, T)$ are then obtained by imposing thermodynamic self-consistency. Indeed, in the MSA, like in most approximate closures to the OZ equations, the free energy is not the same when calculated from the energy or the compressibility route (in lattice systems, the virial route does not exit). In the Q-A binary mixture, the self-consistency requirement is embodied in two "Maxwell relations". We have, on one hand, the Gibbs-Helmholtz relation $u = \partial\beta a/\partial\beta|_{\rho_f,\rho_m}$ where the internal energy per site is given by

$$u = -q[\frac{1}{2}\rho_f^2 g_{ff}(\mathbf{r}=\mathbf{e}) + \rho_m\rho_f g_{mf}(\mathbf{r}=\mathbf{e})] \tag{29}$$

and, on the other hand, the *two* (compressibility) sum-rules

$$\begin{aligned} \hat{c}_{con}(0) &= -\left.\frac{\partial^2(\beta a_{ex})}{\partial\rho_f^2}\right|_\beta \\ \hat{c}_{mf}(0) &= -\left.\frac{\partial^2(\beta a_{ex})}{\partial\rho_f\partial\rho_m}\right|_\beta . \end{aligned} \tag{30}$$

This yields the two relations

$$\left.\frac{\partial^2 u}{\partial\rho_f^2}\right|_\beta = -\left.\frac{\partial\hat{c}_{con}(0)}{\partial\beta}\right|_{\rho_f,\rho_m}$$

$$\left.\frac{\partial^2 u}{\partial \rho_f \partial \rho_m}\right|_\beta = -\left.\frac{\partial \hat{c}_{mf}(0)}{\partial \beta}\right|_{\rho_f,\rho_m} . \tag{31}$$

After replacing the correlation functions by their analytical expressions, we finally obtain a pair of coupled partial differential equations for $z(\rho_f, \rho_m, T)$ and $y(\rho_f, \rho_m, T)$ which generalize the SCOZA PDE of Ref. [70] to the Q-A mixture (see Ref. [71] for more details).

Unfortunately, these PDE's are technically difficult to solve and, up to now, the only results available have been obtained from a high-temperature series expansion of the solution. The numerical study has been limited to the case of the site-diluted Ising model for which one knows that there is only one critical point located at $\rho_f = (1 - \rho_m)/2$ (moreover, it is only for this model that the up-down symmetry of the Hamiltonian allows for a critical point in three dimensions despite the use of an OZ ansatz for solving the RSOZ equations, cf. the discussion in section 7). The predictions for the critical temperature and the critical exponent γ are given in Table 1. These results have been obtained from a [6,5] Dlog Padé analysis of the high-T series expansion of the compressibility on the critical isochore, i.e., the zero-field susceptibility for the spin system.

TABLE 1. Critical temperature T_c and effective exponent γ for the site-diluted Ising model on a simple cubic lattice (from Kierlik *et al.* [71]).

	kT_c/J		γ	
p	SCOZA	MC	SCOZA	MC
1	4.526	4.512	1.240	1.24
0.95	4.279	4.262	1.253	1.28
0.90	4.027	4.011	1.273	1.31
0.80	3.509	3.499	1.342	1.35
0.70	2.967	-	1.471	-
0.60	2.401	2.422	1.684	1.51
0.50	1.841	1.849	1.851	1.49
0.40	1.300	1.209	2.163	-

Very good agreement with MC results is reached for T_c. To our knowledge, this is the only theory that can predict accurately the variations of the critical temperature in a large range of dilutions. γ must be interpreted here as an *effective* exponent since we expect it to reach asymptotically the value for the spherical model. γ increases as p decreases, as observed in the simulations, but we must wait until a numerical solution of the PDE's is available to get more reliable estimates. Finally, we wish to point out that the theory predicts that T_c goes to zero for some value p_c of the dilution which should correspond to the geometric percolation threshold of the lattice [65]. The value $p_c = 0.3405$ obtained for the SC lattice is in reasonable agreement with the exact value, 0.310.

6.2. THE RANDOM FIELD ISING MODEL

Random field systems, as described by the Hamiltonian (25), have been under intensive investigation both experimentally and theoretically for more than twenty years [11]. They have a number of interesting realizations in nature, with diluted antiferromagnets in a homogeneous external field being the most studied experimental system. They are still some important theoretical questions which are under debate, mostly related to the critical behavior. The main problem is that there is so far no satisfying RG analysis, the standard perturbative theory being inadequate because of the large number of local minima in the energy landscape. This may explain the breakdown of dimensional reduction[5] and might lead to the existence of a "glassy" phase signaled by a replica symmetry breaking (RSB) mechanism [72]. The influence of the random field distribution on the order of the paramagnetic to ferromagnetic transition is also one of the important issues. Whereas mean-field theory which should be valid in high dimensions, predicts that some distributions give rise to a tricritical point, neither numerical simulations nor high-temperature series expansions have been able to yield a clear-cut answer yet in finite-dimensional systems. Liquid state methods for Q-A systems are well suited to the study of this problem and the SCOZA has been also recently applied to the RFIM [73]. To transform the RFIM into a Q-A mixture, we consider probability distributions $\mathcal{P}(h)$ for the (uncorrelated) random fields that depend on a single positive parameter h_0 which measures the strength of disorder. This includes the much studied Gaussian distribution $\mathcal{P}(h) = (2\pi h_0^2)^{-1/2}\exp[-h^2/(2h_0^2)]$ and the bimodal distribution $\mathcal{P}(h) = 1/2[\delta(h-h_0)+\delta(h+h_0)]$. One then introduces the reduced quenched variables $\tau_i = h_i/h_0$. One has $\mathcal{P}(h)dh = \mathcal{P}(\tau)d\tau$ and $\mathcal{P}(\tau)$ is now independent of h_0. In the Gaussian case, τ_i is a continuous (soft) spin variable whereas it only takes the values ± 1 (Ising-like) in the bimodal case. We thus have both annealed and quenched variables, τ_i and σ_i, and the Hamiltonian (25) writes

$$\mathcal{H} = -J\sum_{<ij>}\sigma_i\sigma_j - h_0\sum_i \sigma_i\tau_i - H\sum_i \sigma_i \tag{32}$$

[5] Dimensional reduction means that the critical exponents of the RFIM in dimension d should be equal to those of the pure system in dimension $d-2$

where a uniform magnetic field H has been also added. Therefore, we can apply again to this system the machinery of the RSOZ equations. Keeping spin model notations, they can be written in Fourier space as

$$\begin{aligned}
\hat{G}_{00}(\mathbf{k}) &= \frac{1}{\hat{C}_{00}(\mathbf{k})} \\
\hat{G}_{01}(\mathbf{k}) &= -\frac{\hat{C}_{01}(\mathbf{k})}{\hat{C}_{00}(\mathbf{k})\hat{C}_{con}(\mathbf{k})} \\
\hat{G}_{con}(\mathbf{k}) &= \frac{1}{\hat{C}_{con}(\mathbf{k})} \\
\hat{G}_{dis}(\mathbf{k}) &= [\frac{\hat{C}_{01}^2(\mathbf{k})}{\hat{C}_{00}(\mathbf{k})} - \hat{C}_{dis}(\mathbf{k})]\frac{1}{\hat{C}_{con}^2(\mathbf{k})} \qquad (33)
\end{aligned}$$

with $G_{00}(\mathbf{r}) = \overline{\tau_{\mathbf{0}}\tau_{\mathbf{r}}} - \overline{\tau}^2$, $G_{01}(\mathbf{r}) = \overline{\tau_{\mathbf{0}} < \sigma_{\mathbf{r}} >_T} - \overline{\tau}m$, where $m = \overline{< \sigma_i >_T}$ is the averaged magnetization, $G_{con}(\mathbf{r}) = \overline{< \sigma_{\mathbf{0}}\sigma_{\mathbf{r}} >_T - < \sigma_{\mathbf{0}} >_T < \sigma_{\mathbf{r}} >_T}$, and $G_{dis}(\mathbf{r}) = \overline{< \sigma_{\mathbf{0}} >_T < \sigma_{\mathbf{r}} >_T} - m^2$ ($< ... >_T$ denotes the thermal average). One has $G_{00}(\mathbf{r}) = (\overline{\tau^2} - \overline{\tau}^2)\ \delta_{\mathbf{r},\mathbf{0}}$ for uncorrelated random fields, but, in principle, the case of correlated fields can be considered too. As for Eqs. (13-16), these RSOZ equations are obtained as the limit when $n \to 0$ of a set of OZ equations in replica space. The direct correlation functions are defined as the second functional derivatives of the Gibbs potential $\mathcal{G}_{rep}$ w.r.t. the local magnetizations m_i, where $\mathcal{G}_{rep}$ is obtained from the free energy $\mathcal{F}_{rep}$ by a Legendre transform that takes the local fields into the local magnetizations (see Ref. [73] for details). In the case of the Gaussian distribution, these equations take a simpler form [73] because, as is usually done, one can integrate out the disorder from the beginning to introduce an effective Hamiltonian that now contains an interaction among replicas. This cannot be done in general, however, and the advantage of the present formulation is that it allows for the study of a large class of field distributions.

The application of the ORPA to the RFIM represents a first improvement over mean field theory. The ORPA closure writes

$$\begin{aligned}
\hat{C}_{01}(\mathbf{k}) &= \hat{C}_{01}^{ref}(\mathbf{k}) \\
\hat{C}_{con}(\mathbf{k}) &= c_c(m, \tilde{J}, \tilde{h}_0) - \tilde{J}\ \hat{\lambda}(\mathbf{k}) \\
\hat{C}_{dis}(\mathbf{k}) &= \hat{C}_{dis}^{ref}(\mathbf{k}) \qquad (34)
\end{aligned}$$

where $\tilde{J} = \beta qJ$, $\tilde{h}_0 = \beta h_0$ and the reference system corresponds to $J = 0$ (all the properties of this system can be calculated exactly since this is a noninteracting spins problem). $c_c(m, \tilde{J}, \tilde{h}_0)$ is calculated by imposing

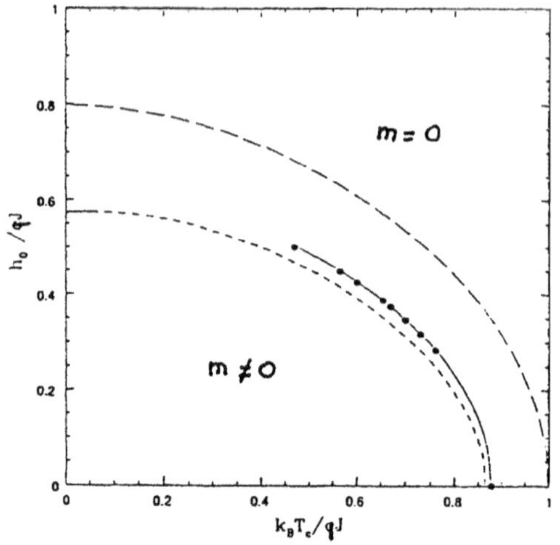

Figure 6. Phase diagram of the RFIM with a Gaussian distribution in five dimensions. Long-dashed line: mean-field theory; short-dashed line: ORPA; solid line: SCOZA; dots: "exact" phase boundary [74]. In principle, the SCOZA line should continue down to $T_c = 0$ but no reliable Padé approximant has been found at low temperatures (from Kierlik *et al.* [73]).

that the hard-spin condition is satisfied, i.e., $G_{11}(\mathbf{r}=\mathbf{0}) = 1 - m^2$ where $G_{11}(\mathbf{r}) = G_{con}(\mathbf{r}) + G_{dis}(\mathbf{r})$ (this condition for the spin 1/2 model is equivalent to the core condition in a lattice gas).

On the other hand, the SCOZA closure is

$$\begin{aligned}\hat{C}_{01}(\mathbf{k}) &= c_{01}(m,\tilde{J},\tilde{h}_0)\\ \hat{C}_{con}(\mathbf{k}) &= c_c(m,\tilde{J},\tilde{h}_0)[1 - z(m,\tilde{J},\tilde{h}_0)\hat{\lambda}(\mathbf{k})]\\ \hat{C}_{dis}(\mathbf{k}) &= c_d(m,\tilde{J},\tilde{h}_0) \qquad (35)\end{aligned}$$

and the additional unknown functions z, c_d and c_{01} are determined by imposing self-consistency in the calculation of the Gibbs potential. This yields a set of Maxwell equations relating $\hat{C}_{con}(\mathbf{k}=\mathbf{0})$ and $\hat{C}_{01}(\mathbf{k}=\mathbf{0})$ to the values of $G_{11}(\mathbf{r})$ and $G_{01}(\mathbf{r})$ at $\mathbf{r}=\mathbf{e}$ or $\mathbf{r}=\mathbf{0}$. One finally obtains a set of coupled PDE's for the unknown variables, as in the case previously discussed of the site-diluted Ising model.

The predictions for the phase diagram of the Gaussian RFIM in five dimensions are shown in Figure 6 for illustration[6]. As can be seen, the

[6]Because of the OZ ansatz, the SCOZA again cannot predict a true critical point in $d \leq 4$. However, in three dimensions, the critical temperatures extrapolated from the high-T series expansions of the solution are still in good agreement with the "exact" results, at least for the small fields that are relevant for experimental realizations of the RFIM in magnetic systems.

ORPA is already much better than the mean-field theory and excellent agreement with the "exact" results [74] is obtained with SCOZA (like before, the SCOZA predictions have been obtained from a high-temperature series expansion of the solution to the PDE's). Both theories predict that a tricritical point exists in finite dimensions for the bimodal distribution and that it moves to weaker fields as the dimension is reduced. For the Gaussian distribution, $1/d$ series expansion of the SCOZA solution suggests that the transition may also become first order when d is around 4 and below. This must be confirmed, however, by the numerical solution of the PDE's.

An important remaining question is the stability of the replica symmetric solution with respect to RSB. This problem is currently under investigation. The basic idea is to perform a second Legendre transform that introduces the standard spin-glass parameter $q_{SG} = \overline{< \sigma_i >_T^2}$ as a control variable instead of h_0. One then derives a new set of RSOZ equations involving direct correlation functions that are generated by the new Legendre transformed Gibbs potential. Instability w.r.t. RSB is then indicated by the vanishing at $\mathbf{k} = \mathbf{0}$ of the combination of direct correlation functions that corresponds to the so-called "replicon" sector of the correlation functions matrix. The advantage of writing the RSOZ equations in the second Legendre transform formalism is that it is also a natural theoretical framework to study spin glass models.

7. Critical behavior of Q-A systems

To conclude, we would like to present some preliminary calculations for the critical behavior of Q-A systems. This is a difficult problem. We have already pointed out that the only experimental results available for fluids in disordered materials [13, 14] give an exponent β which is bulk-like and inconsistent with the value predicted for the standard RFIM by Monte-Carlo simulations [15]. This may be due to an abrupt crossover [75] or to the fact that the aerogel imposes a correlated disorder on the fluid (along the strands), a feature that is not taken into account in the standard RFIM. This may be also related to the fact that the actual phase diagram is more complex, as suggested in Ref. [66] and by the calculations presented in section 5. But the critical behavior of the RFIM itself is not well understood. One cannot exclude that a RSB scenario similar to that proposed for the RFIM also occurs for liquids in disordered porous media.

In this respect, standard liquid-state integral equations are not of great help because they fail badly whenever long-range fluctuations play an essential role, as is the case near criticality. Moreover, since they are thermodynamically inconsistent, different critical exponents are obtained depending on the chosen route. For instance, the MSA and the ORPA for the Q-A lattice gas yield the mean-field (classical) critical exponents via the energy route and the critical exponents of the spherical model in a random field for $4 < d \leq 6$ via the compressibility route [42]. The unfortunate fact that $d = 4$ is the lower critical dimension via the compressibility route comes from the use of an Ornstein-Zernike ansatz that assumes that the direct correlation functions have the same range as the interaction potentials. As a consequence, the exponents η and $\overline{\eta}$ defined by $h_{con}(r) \sim r^{-d+2-\eta}$ and $h_{dis}(r) \sim r^{-d+4-\overline{\eta}}$ when $r \to \infty$ at the critical point are zero. This implies that no finite critical temperature exists for $d \leq 4$ since this would lead to the unacceptable result that $h_{dis}(r)$ and thus $h_{ff}(r)$ do not decay to zero when $r \to \infty$ at the critical point[7].

There are, however, two liquid-state approaches that allow for a better treatment of both short-range and long-range fluctuations. These are the SCOZA that we have discussed previously and the hierarchical reference theory (HRT) of Parola and Reatto [76]. In what follows, we would like to show briefly how the HRT may be applied to Q-A systems. Like in RG, the basic idea in the HRT is to turn on gradually fluctuations over longer and longer length scales (or, alternatively, in Fourier space, over shorter and shorter wave vectors). Technically, this is done by introducing a sequence of "Q-systems" defined by a modified attractive interaction

$$v_{ff}^{Q}(r) = v_{ff}^{ref}(r) + w_{ff}^{Q}(r) \tag{36}$$

where Q is a cutoff wave vector and the Fourier transform of $w_{ff}^{Q}(r)$ is defined by

$$\hat{w}_{ff}^{Q}(k) = \begin{cases} \hat{w}_{ff}(k) & \text{for } k > Q \\ 0 & \text{for } k < Q\ . \end{cases} \tag{37}$$

Q plays the role of an infrared cutoff whose effect consists in depressing fluctuations on length scales larger than $1/Q$. The difference in free energy between two Q-systems infinitesimally close to each other can be determined exactly by perturbation theory. This enables one to derive an exact hierarchy of integro-differential equations describing the evolution of the free energy and of the many-particle direct correlation functions. The extension of the HRT to a Q-A binary mixture is detailed in Ref. [77]. Following Ref. [76] and using the one-loop expression of the free energy (which is similar the ORPA formula in Eq. (21)), we find that the first equation of the hierarchy, which gives the evolution of the excess Helmholtz free energy of the Q-system, reads

[7] More generally, the RSOZ equations imply that $\overline{\eta} = 2\eta$, assuming that $\hat{c}_{dis}(0)$ and $\hat{c}_{mf}(0)$ remain finite at the critical point, a condition generally expected for direct correlation functions. This relation has been proposed to hold exactly in the RFIM [11] and it is confirmed by high-temperature series expansions [74].

$$-\frac{\partial \mathcal{A}^Q}{\partial Q} = \frac{\Omega_d}{2(2\pi)^d} Q^{d-1}\{\ln[1+\frac{\rho_f\hat{\phi}_{ff}(Q)}{\hat{\mathcal{C}}^Q_{con}(Q)}] + \rho_f[\hat{c}^Q_{dis}(Q) + \rho_m \frac{\hat{c}^Q_{mf}(Q)^2}{1-\rho_m\hat{c}_{mm}(Q)}] \\ [\frac{1}{\hat{\mathcal{C}}^Q_{con}(Q)} - \frac{1}{\hat{\mathcal{C}}^Q_{con}(Q)+\rho_f\hat{\phi}_{ff}(Q)}]\} \tag{38}$$

where Ω_d is the solid angle in d dimensions ($\Omega_d = 4\pi$ in three dimensions) and $\hat{\phi}_{ff}(k) = -\beta\hat{w}_{ff}(k)$. As in Ref. [76], we have introduced a modified free energy $\mathcal{A}^Q = -\beta A^Q_{ex}/V - \rho_f/2[\phi_{ff}(0) - \phi^Q_{ff}(0)] + \rho_f^2/2[\hat{\phi}_{ff}(0) - \hat{\phi}^Q_{ff}(0)]$ and a modified direct correlation function $\mathcal{C}^Q_{con}(k) = 1 - \rho_f[\hat{c}^Q_{con}(k) + \hat{\phi}_{ff}(k) - \hat{\phi}^Q_{ff}(k)]$ in order to eliminate some discontinuous contributions in the actual physical quantities due the sharp cutoff adopted in the definition of $\hat{w}^Q_{ff}(k)$ (see Refs. [76, 77] for technical details). Analogous to the case of bulk fluids, the flow equation simplifies in the critical region and at small cutoff wave vectors Q where $\hat{\mathcal{C}}^Q_{con}(Q) \to 0$. Asymptotically, when $T \to T_c$ and $Q \to 0$, we have

$$\frac{\partial \mathcal{A}^Q}{\partial Q} \sim \frac{\Omega_d}{2(2\pi)^d} Q^{d-1}\{\ln[\hat{\mathcal{C}}^Q_{con}(Q)] - \frac{\rho_f}{\hat{\mathcal{C}}^Q_{con}(Q)}[\hat{c}^Q_{dis}(Q) + \rho_m \frac{\hat{c}^Q_{mf}(Q)^2}{1-\rho_m\hat{c}_{mm}(Q)}]\} \tag{39}$$

where the explicit dependence on the pair potential $w_{ff}(r)$ has disappeared.

To go further and get an approximate theory that gives non-trivial critical exponents, we need a closure relation for $\hat{\mathcal{C}}^Q_{con}(k)$. Following again Ref. [76], we use an Ornstein-Zernike ansatz

$$\hat{\mathcal{C}}^Q_{con}(k) \sim \hat{\mathcal{C}}^Q_{con}(0) + b_Q k^2 + \ldots , \tag{40}$$

where $\hat{\mathcal{C}}^Q_{con}(0)$ satisfies the exact compressibility sum-rule, Eq. (4), which has the modified form $-\rho_f\partial^2\mathcal{A}^Q/\partial\rho_f^2 = \hat{\mathcal{C}}^Q_{con}(0)$, and b_Q is a regular function of Q, ρ_f and T with a finite limit, $b > 0$, for $Q \to 0$ and $T \to T_c$. As discussed before, this ansatz implies unfortunately that $\eta = \overline{\eta} = 0$ and that there is no critical point for $d \leq 4$. We assume moreover that $\rho_f[\hat{c}^Q_{dis}(Q) + \rho_m\hat{c}^Q_{mf}(Q)^2]/[1-\rho_m\hat{c}_{mm}(Q)]$ has a finite limit $g > 0$ when $Q \to 0$ and $T \to T_c$. We then obtain

$$\frac{\partial \mathcal{A}^Q}{\partial Q} = \frac{\Omega_d}{2(2\pi)^d} Q^{d-1}[\ln(-\rho_f\frac{\partial^2\mathcal{A}^Q}{\partial\rho_f^2} + bQ^2) - \frac{g}{-\rho_f\partial^2\mathcal{A}^Q/\partial\rho_f^2 + bQ^2}] . \tag{41}$$

In a bulk fluid, only the first term in the r.h.s. of Eq. (41) exists and we recover the result of Ref. [76] which is also equivalent to the RG equation obtained in field theory using the local potential approximation (LPA) [78]. On the other hand, when $g \neq 0$, the r.h.s of Eq. (41) is dominated by the second term when $Q \to 0$ in the critical region[8]. To remove the model-dependent constants b and g and eliminate the explicit dependence on Q, we then define the rescaled variables

$$\begin{aligned} H^Q &= \frac{2b}{gK_d} Q^{-(d-2)} [\mathcal{A}^Q(\rho_{f,c}) - \mathcal{A}^Q(\rho_f)] \\ x &= b(\frac{2}{gK_d})^{1/2} Q^{-(d-4)/2} (\rho_f - \rho_{f,c}) \end{aligned} \tag{42}$$

where $K_d = \Omega_d/(2\pi)^d$ and $\rho_{f,c}$ is the critical value of the fluid density. This yields a universal RG-like equation

$$\begin{aligned} \frac{\partial H^Q}{\partial(\ln Q)} &= \frac{d-4}{2} x \frac{\partial H^Q}{\partial x} - (d-2) H^Q \\ &+ [\frac{1}{\partial^2 H^Q/\partial x^2 + 1} - \frac{1}{\partial^2 H^Q/\partial x^2|_{x=0} + 1}] \end{aligned} \tag{43}$$

Note that the exact same equation is obtained if one repeats this approximate HRT treatment for the RFIM [77]. To get the stationary (i.e. Q-independent) solution $H^*(x)$ of Eq. (43) which corresponds to a fixed point, it is convenient to introduce the variable $\mathcal{M}(x) = \partial H^*(x)/\partial x$ and to derive the fixed point equation w.r.t. x. We finally obtain

$$\frac{d-4}{2} x \mathcal{M}'(x) - \frac{d}{2} \mathcal{M}(x) = \frac{\mathcal{M}''(x)}{[1 + \mathcal{M}'(x)]^2} \tag{44}$$

We have calculated the solution of this equation in five dimensions using a shooting method as in Ref. [79] and found only one non-trivial fixed point. From the linearization of Eq. (43) around the fixed-point solution we have found that the first two largest eigenvalues are $\lambda_1 = 1.5395$ and $\lambda_2 = -0.6557$. This gives for the critical exponents $\nu = 1/\lambda_1 = 0.6496$ and $\gamma = 2\nu = 1.299$. This can be compared to the value of γ for the 5-d RFIM obtained from high-temperature series (1.13 ± 0.02 [74]) and to the value obtained by the same approximate theory for the 3-d *bulk* fluid (1.378 [76]). We can thus see that in this non-perturbative approach the critical exponents of the random system in dimension d are not equal to those of

[8]The case of the site-diluted Ising model is more complicated. Although this model can still be described by the Q-A Hamiltonian, Eq. (23), with $w_{mf} = w_{ff}/2$, the up-down symmetry implies that $g \to 0$ on the critical isochore $\rho_f = (1 - \rho_m)/2$ [71]. Therefore, to apply the HRT to this system one must also make an approximation for $\hat{c}_{dis}(k)$ and $\hat{c}_{mf}(k)$ when $k \to 0$.

the pure system in dimension $d-2$. On the other hand, at first order in $\epsilon = 6-d$, one checks that dimensional reduction ($d \to d-2$) holds.

Of course, it would be very interesting to study Eq. (38) outside the asymptotic region to keep track of the correlations on all length scales. This is in fact the main *raison d'être* of the HRT. It is certainly possible to generalize the approximate method proposed for the pure fluid in Ref. [76], but obtaining a numerical solution to the resulting equations may be a formidable problem. An important question which is perhaps easier to investigate is the crossover determined by the competition between the pure and random fixed points (a percolation fixed point should be also expected). One should also try to improve the method in two directions: first, to bypass the OZ approximation (40) by considering closures at the level of the three- and four body direct correlation functions [76] or by using an analog of the derivative expansion in field theory [79]; second, to look for a possible replica symmetry breaking mechanism.

Acknowledgments

I wish to acknowledge my co-workers E. Kierlik and G. Tarjus with whom this work has been developed. Our innumerable lively and friendly discussions over the past years have led to the views which are presented here. I have also taken great pleasure and considerable profit from our collaboration with P. Monson and G. Stell.The Laboratoire de Physique Théorique des Liquides is UMR No. 7600 au CNRS.

References

1. Evans, R. (1990) Fluids adsorbed in narrow pores: phase equilibria and structure, *J. Phys.: Condens. Matter* **2** 8989.
2. Gregg, S. J. and Sing, K.S.W. (1982) *Adsorption, Surface Area and Porosity*, Academic press, London.
3. See e.g., Fisher, D.S., Grinstein, G.M. and Khurana, A. (1988) Theory of random magnets, *Physics Today* (December) 56.
4. Rouquerol, F., Chaillot, D., Denoyel, R., Llewellyn, P. and Rouquerol, J. (1998) An alternative to the BJH treatment of adsorption-desorption data, Proceedings of the *Sixth International Conference on Fundamentals of Adsorption* (to appear).
5. Burgess, C.G.V., Everett, D.H. and Nuttall, S. (1989) Adsorption hysteresis in porous materials, *Pure and Applied Chem.* **61** 1845.
6. Wilkinson, N.J., Alam, M.A., Clayton, J. M., Evans R. , Fretwell, H. M., and Umar, S. G. (1992) Positron annihilation study of capillary condensation of nitrogen gas in a mesoporous solid", *Phys. Rev. Lett.* **69**, 3535.
7. Thommes, M. and Findenegg, G.H. (1994) Pore condensation and critical-point shift of a fluid in controlled-pore glass, *Langmuir* **10**, 4270.
8. Mason, G. (1982) The effect of pore space connectivity on the hysteresis of capillary condensation in adsorption-desorption isotherms, *J. Colloid Interface Sci.* **88**, 36.
9. Ball., P.C and Evans, R. (1989) Temperature dependence of gas adsorption on a mesoporous solid:capillary criticality and hysteresis, *Langmuir* **5**, 714.

10. Brochard, F. and de Gennes, P. G. (1983) Phase transitions of binary mixtures in random media, *J. Phys. Lett. (Paris)* **44** L785; de Gennes, P. G. (1984) Liquid-liquid demixing inside a rigid network. Qualitative features, *J. Phys. Chem.* **88**, 6469.
11. For recent reviews, see Belanger, D.P. (1998) Experiments on the random field Ising model, in A.P. Young (ed.), *Spin glasses and random fields*, World Scientific, Singapore, p. 251; Nattermann, T. (1998) Theory of the random field Ising model, *ibid*, p. 277.
12. Frisken, B.J., Liu, A.J. and Cannell, D.S. (1994) Critical fluids in porous media, *MRS Bull.* (May) 19.
13. Wong, A.P.Y. and Chan, M.H.W. (1990) Liquid-vapor critical point of ^{4}He in aerogel, *Phys. Rev. Lett.* **65**, 2567
14. Wong, A.P.Y., Kim, S.B., Goldburg, W.I. and Chan, M.H.W. (1993) Phase separation, density fluctuation, and critical dynamics of N_2 in aerogel, *Phys. Rev. Lett.* **70**, 954.
15. Reiger H. (1995) Critical behavior of the 3d random field Ising model: two-exponents scaling or first order phase transition ?, *Phys. Rev. B* **52**, 5659.
16. Zhuang, Z., Casielles; A.G. and Cannell, D.S. (1996) Phase diagram of Isobutyric acid and water in dilute silica gels, *Phys. Rev. Lett.* **77**, 2969.
17. Monette, L., Liu, A.J. and Grest, G.S. (1992) Wetting and domain-growth kinetics in confined geometries, *Phys. Rev. A* **46**, 7664.
18. Thommes M. and Gammon R. W., in preparation.
19. Reiss, H. (1992) Statistical geometry in the study of fluids and porous media, *J. Phys. Chem.* **96**, 4736.
20. Levitz, P., Ehret, G., Sinha, S.K. and Drake, J. M. (1991) Porous Vycor glass: the microstructure as probed by electron microscopy, direct energy transfer, small angle scattering, and molecular adsorption, *J. Chem. Phys.* **95**, 6151.
21. Ferri, F., Frisken, B.J. and Cannell, D.S. (1991) Structure of silica gels, *Phys. Rev. Lett.* **67**, 3626.
22. Vacher, R., Woignier, T., Pelous, J. and Courtens, E. (1988) Structure and self-similarity of silica aerogels, *Phys. Rev. B* **37**, 6500.
23. Madden, W.G. and Glandt, E.D. (1988) Distribution functions for fluids in random media, *J. Stat. Phys.* **51**, 537.
24. Madden, W.G. (1992) Fluid distributions in random media: arbitrary matrices, *J. Chem. Phys.* **96**, 5422.
25. Fanti, L.A, Glandt, E.D and Madden, W.G. (1990) Fluids in equilibrium with disordered porous materials. Integral equation theory, *J. Chem. Phys.* **93**, 5945.
26. Lomba, E., Given, J.A., Stell, G., Weis, J.J. and Levesque, D. (1993) Ornstein-Zernike equations and simulation results for hard-sphere fluids adsorbed in porous media, *Phys. Rev. E* **48**, 233.
27. Meroni, A., Levesque, D. and Weiss, J.J. (1996) Correlation functions of hard sphere fluids in porous media, *J. Chem. Phys.* **105**, 1101.
28. MacElroy, J.M.D. and Raghavan, K. (1990) Adsorption and diffusion of a Lennard-Jones vapor in microporous silica, *J. Chem. Phys.* **93**, 2068.
29. Kaminsky, R.D and Monson, P.A. (1991) The influence of adsorbent microstructure upon adsorption equilibria: investigations of a model system, *J. Chem. Phys.* **95**, 2936.
30. Vuong, T. and Monson, P.A. (1996) Monte Carlo simulation studies of heats of adsorption in heterogeneous solids *Langmuir* **12**, 5425.
31. Kaminsky, R.D and Monson, P.A. (1994) Modeling the influence of theterogeneous adsorbent microstructure upon adsorption equilibria for binary mixtures, *Langmuir* **10**, 530.
32. Page, K.S. and Monson, P.A. (1996) Monte Carlo calculations of phase diagrams for a fluid in a disordered porous media, *Phys. Rev. E* **54**, 6557.
33. Segarra, E.I and Glandt, E.D. (1994) Model microporous carbons: microstructure, surface polarity and gas adsorption, *Chem. Eng. Sci.* **49**, 2953.
34. See e.g., Lastokie, C., Gubbins, K.E. and Quirke, N. (1993) Pore size heterogeneity

and the carbon slit pore: a density functional theory model, *Langmuir* **9**, 2693.
35. Gelb, L.D. and Gubbins (1998) Characterization of porous glasses: simulation models, adsorption isotherms, and the Brunauer-Emett-Teller analysis method, *Langmuir* **14**, 2097.
36. Pellenq, R.J.-M., Rodts, S., Pasquier, V., Delville A. and Levitz, P. (1998) A grand canonical Monte-Carlo simulation study of xenon adsorption in a vycor-like matrix, Proceedings of the *Sixth International Conference on Fundamentals of Adsorption* (to appear).
37. Stauffer, D. and Pandey, R.B. (1992) Simulation of demixing transition for binary fluid within a gel, *J. Phys. A: Math. Gen.* **25**, L1079.
38. Lee, J.C. (1993) Ordering processes in porous media, *Phys. Rev. lett.* **70**, 3599.
39. MacFarland, T. Barkema, G.T. and Marko, J.F. (1996) Equilibrium phase transitions in a porous medium, *Phys. Rev. B* **53**, 148.
40. Uzelac, K., Hasmy, A. and Jullien, R. (1995) Numerical study of phase transitions in the pores of an aerogel, *Phys. Rev. Lett.* **74**, 422.
41. Falicov, A. and Berker, A.N. (1995) Correlated random-chemical-potential model for the phase transitions of helium mixtures in porous media, *Phys. Rev. Lett.* **74**, 426.
42. Pitard, E., Rosinberg, M.L., Stell, G. and Tarjus, G. (1995) Critical behavior of a fluid in a disordered porous matrix. An Ornstein-Zernike approach, *Phys. Rev. Lett.* **74**, 4361.
43. Pitard, E., Rosinberg, M.L. and Tarjus, G. (1996) Thermodynamics of fluids in disordered porous materials, *Molecular Simulation* **17**, 399.
44. Kierlik, E., Rosinberg, M.L., Tarjus, G. and Pitard, E. (1998) Mean spherical approximation for a lattice model of a fluid in a disordered matrix, *Mol. Phys.* (in press).
45. Evans, E. (1992) Density functionals in the theory of non-uniform fluids, in D. Henderson (ed.), *Fundamentals of inhomogeneous fluids*, Dekker, New York, p. 85.
46. Vega, C., Kaminsky, R.D. and Monson, P. A. (1993) Adsorption of fluids in disordered porous media from integral equation theory, *J. Chem. Phys.* **99**, 3003.
47. Rosinberg, M.L., Tarjus, G. and Stell, G. (1994) Thermodynamics of fluids in quenched disordered matrices, *J. Chem. Phys.* **100**, 5172.
48. Kierlik, E., Rosinberg, M.L., Tarjus, G. and Monson, P. A., (1995) The pressure of a fluid confined in a disordered porous material, *J. Chem. Phys.* **103**, 4256.
49. Dong, W. (1995) Mechanical route to the pressure of a fluid adsorbed in a random porous medium, *J. Chem. Phys.* **102**, 6570.
50. Dong, W., Kierlik, E. and Rosinberg, M.L. (1994) Integral equations for a fluid near a random substrate, *Phys. Rev. E* **50**, 4752.
51. Ford, D.M. and Glandt, E. (1994) Compressibility equation for fluids in random microporous matrices, *J. Chem. Phys.* **100**, 2391.
52. Hansen, J.P. and Mc Donald, J.R. (1986) *Theory of simple liquids*, Academic, London.
53. Given, J.A. (1992) Liquid-state methods for random media: I. random sequential adsorption, Phys. Rev. A **45**, 816.
54. Given, J.A. and Stell, G. (1992) Comment on: Fluid distributions in random media: arbitrary matrices, J. Chem. Phys. **97**, 4573.
55. See e.g., Mezard, M., Parisi, G. and Virasoro, M. (1987) *Spin Glass Theory and Beyond*, World Scientific, Singapore.
56. Rosinberg, M.L. (unpublished).
57. Baxter, R.J. (1970) Ornstein Zernike relation and Percus-Yevick approximation for fluid mixtures, *J. Chem. Phys.* **52**, 4559.
58. Ford, D.M. and Glandt, E. (1994b) Vapor-liquid phase equilibrium in random microporous matrices, *Phys. Rev. E* **50**, 1280.
59. Kaminsky, R. D. and Monson, P.A. (1994) A simple mean field theory of adsorption in disordered porous materials, *Chem. Eng. Sci.* **49**, 2967.
60. Andersen, H.C. and Chandler, D. (1972) Optimized cluster expansions for classical

fluids: I. General theory and variational formulation of the mean-spherical model and hard-sphere Percus-Yevick equations, *J. Chem. Phys.* **57**, 1918.
61. Kierlik, E., Rosinberg, M.L., Tarjus, G. and Monson, P.A. (1997) Phase diagrams of single-component fluids in disordered porous materials: predictions form integral equation theory, *J. Chem. Phys.* **106**, 264.
62. Sung, S.H. and Chandler, D. (1974) Optimized cluster theory, the Lennard-Jones fluid, and the liquid-gas phase transition, *Phys. Rev. A* **9**, 1688.
63. Sarkisov, L., Page, S.K. and Monson, P.A. (1998) Molecular modeling of fluid phase equilibrium in disordered porous materials, Proceedings of the *Sixth International Conference on Fundamentals of Adsorption* (to appear).
64. Donley, J.P. and Liu, A.J. (1996) Phase behavior of near-critical fluids confined in periodic gels, *Phys. Rev. E* **55**, 539.
65. Stinchcombe, R.B. (1983) Dilute magnetism, in C. Domb and M. S. Green (eds.), *Phase transitions and critical phenomena*, Academic, London, Vol. 7, p. 151.
66. Maritan, A., Swift, M.R., Cieplak, M., Chan, M.H.W, Cole, M.W. and Banavar, J.R. (1991) Ordering and phase transitions in random-field systems, *Phys. Rev. Lett.* **67**, 1821.
67. Joyce, G.S. (1972) Critical properties of the spherical model, in C. Domb and M. S. Green (eds.), *Phase transitions and critical phenomena*, Academic, London, Vol. 2, p. 375.
68. Kierlik, E., Rosinberg, M.L. and Tarjus, G. (in preparation).
69. Hoye, J.S. and Stell, G. (1984) Ornstein-Zernike equation for a two-Yukawa $c(r)$ with core condition. III. A self-consistent approximation for a pair potential with hard core and a Yukawa tail, *Mol. Phys.* **52**, 1071.
70. Pini, D., Stell, G. and Dickman, R. (1998) A thermodynamically self-consistent theory of structure for three-dimensional lattice gases, *Phys. Rev. E* (**57**, 2862.
71. Kierlik, E., Rosinberg, M.L. and Tarjus, G. (1997) A self-consistent Ornstein-Zernike approximation for the site-diluted Ising model, *J. Stat. Phys.* **89**, 215.
72. Mezard, M. and Young, A.P. (1992) Replica symmetry breaking in the random field Ising model, *Europhys. Lett.* **18**, 653; Mezard, M. and Monasson, R. (1994) Glassy transition in the three-dimensional random field Ising model, *Phys. Rev. B* **50**, 7199; De Dominicis, C., Orland, H. and Temesvari, T. (1996) Random field Ising model: dimensional reduction or spin-glass phase ?, *J. de Physique I* **5**, 987.
73. Kierlik, E., Rosinberg, M.L. and Tarjus, G. (1998) A self-consistent Ornstein-Zernike approximation for the random field Ising model, preprint cond-mat/9801060.
74. Gofman, M., Adler, J., Aharony, A., Harris, A.B. and Schwartz, M. (1993) Evidence for two exponent scaling in the random field Ising model, *Phys. Rev. Lett.* 71, 1569; (1996) Critical behavior of the random-field Ising model, *Phys. Rev. B* **53**, 6362.
75. Cao M.S. and Machta J. (1993) Migdal-Kadanoff study of the random field Ising model, *Phys. Rev. B* **48**, 3177.
76. For a review, see Parola, A. and Reatto, L. (1995) Liquid state theories and critical phenomena, *Adv. in Phys.* **44**, 211.
77. Kierlik, E., Rosinberg, M.L. and Tarjus, G. (in preparation)
78. Nicoll J.F., Chang T.S. and Stanley H.E. (1974) Approximate Renormalization Group based on the Wegner-Houghton differential generators, *Phys. Rev. Lett.* **33**, 540.
79. Morris T. R. (1994) On truncations of the exact renormalization group, *Phys. Lett. B* **334**, 355; Morris T. R. (1994), Derivative expansion of the exact renormalization group, *Phys. Lett. B* **329**, 241.

INHOMOGENEOUS FLUIDS IN AN EXTERNAL FIELD

F. LADO
Department of Physics
North Carolina State University
Raleigh, NC, 27695-8202, USA

AND

E. LOMBA
Instituto de Química Física Rocasolano, CSIC
Serrano 119
28006 Madrid, Spain

1. Introduction

The Gibbsian N-body density function of a Hamiltonian H_N that is rotationally and translationally invariant must itself be rotationally and translationally invariant, as must then also be all reduced n-body density functions of this Hamiltonian. In particular, the one-body density is a constant and the two-body density depends only on relative coordinates. An external field destroys this homogeneity, producing anisotropy or nonuniformity [1, 2], and so makes necessary the joint calculation of the coupled one-body and two-body density functions. A striking if familiar example of the response of a bulk system to an external field is ferromagnetism. We shall use this particular case here to present a general procedure to compute the coupled one-body and two-body density functions of an inhomogeneous classical fluid in an external field. Remarkably, the procedure is no more difficult to carry through than similar calculations for ordinary homogeneous systems.

Perhaps the simplest model of a disordered continuum system exhibiting ferromagnetic behavior is a fluid of hard spheres with embedded Heisenberg spins described using classical statistical mechanics [3, 4, 5, 6, 7]. On one hand, the simplifications of this model compared to magnetic dipole-dipole interactions are significant for simulation. The dipole-dipole model presents substantial conceptual difficulties in connection with the nonexistence of the thermodynamic limit for orientationally ordered phases. Moreover, its

C. Caccamo et al. (eds.), New Approaches to Problems in Liquid State Theory, 279–291.

simulation results are strongly dependent on the boundary conditions. Simulations for the Heisenberg spin model are free of these problems. (The integral equation formalism developed below serves equally well for either potential.) On the other hand, disordered Heisenberg spin systems are of interest in their own right. It has been known for some time that Co/P alloys [8] and Co/Au melts [9] have a tendency to form amorphous ferromagnets, although their "true liquid" nature has been questioned because of the technical difficulties posed by undercooling below the Curie temperature. While these systems might be more representative of quenched spin fluids, very recently Albrecht and coworkers [10] have finally managed to undercool a $Co_{80}Pd_{20}$ melt below its Curie temperature at zero field, thus obtaining the first evidence of ferromagnetic behavior in a liquid metal under conditions where the Heisenberg exchange interaction absolutely dominates the magnetic dipole-dipole contribution.

The Heisenberg spin fluid in an external magnetic field $\mathbf{B}_0$ is defined by the canonical partition function $Z = Z^{id}Z^{ex}$, where

$$\begin{aligned} Z^{id} &= \frac{1}{N!\Lambda^{3N}} \int \prod_{j=1}^{N} [d\mathbf{r}_j d\omega_j] \exp\left(\beta \sum_j \boldsymbol{\mu}_j \cdot \mathbf{B}_0\right) \\ &= \frac{1}{N!}\left(\frac{V}{\Lambda^3}\right)^N \left[4\pi \frac{\sinh \beta\mu B_0}{\beta\mu B_0}\right]^N, \end{aligned} \tag{1}$$

$$\begin{aligned} Z^{ex} &= \frac{1}{(4\pi V)^N} \int \prod_{j=1}^{N} [d\mathbf{r}_j d\omega_j f_0(\omega_j)] \\ &\times \left(-\beta \sum_{i<j} u_0(r_{ij}) - \beta \sum_{i<j} u_{ss}(r_{ij}, \omega_i, \omega_j)\right). \end{aligned} \tag{2}$$

Here $\beta = 1/k_B T$ is the inverse Kelvin temperature, with k_B Boltzmann's constant, Λ the de Broglie thermal wavelength, and $\omega = (\theta, \phi)$ the orientation of the magnetic dipole moment $\boldsymbol{\mu}$, referred to the uniform field $\mathbf{B}_0$ as the z direction. The interaction energies are written as pairwise sums of the hard sphere potential $u_0(r)$ for spheres of diameter σ and the Heisenberg spin-spin potential $u_{ss}(r, \omega_1, \omega_2)$ for $r > \sigma$,

$$u_{ss}(r, \omega_1, \omega_2) = -J(r)\,\hat{\mathbf{s}}_1 \cdot \hat{\mathbf{s}}_2, \qquad \beta J(r) = K \frac{e^{-\kappa(r/\sigma - 1)}}{r/\sigma}, \tag{3}$$

where $\hat{\mathbf{s}}$ is the unit spin vector in the direction of $\boldsymbol{\mu}$. In this last expression, the dimensionless coupling strength K may be read as the inverse reduced temperature, $K = 1/T^*$, while κ is a dimensionless range parameter. Finally, in Eq. (2), $f_0(\omega)$ is the normalized one-body orientational distribution of a noninteracting spin system,

$$f_0(\omega) = \frac{e^{\beta\mu B_0 \cos\theta}}{\sinh(\beta\mu B_0)/\beta\mu B_0}. \tag{4}$$

2. Integral equation formulation

The key quantities needed for a complete magnetic and thermodynamic description of this system are the one-body and two-body density functions,

$$\rho^{(1)}(\mathbf{r},\omega) = \left\langle \sum_{j=1}^{N} \delta(\mathbf{r}-\mathbf{r}_j)\delta(\omega-\omega_j) \right\rangle = \frac{\rho}{4\pi} f(\omega), \tag{5}$$

$$\begin{aligned} \rho^{(2)}(\mathbf{r},\omega,\mathbf{r}',\omega') &= \left\langle \sum_{i\neq j} \delta(\mathbf{r}-\mathbf{r}_i)\delta(\omega-\omega_i)\delta(\mathbf{r}'-\mathbf{r}_j)\delta(\omega'-\omega_j) \right\rangle \\ &= \frac{\rho^2}{(4\pi)^2} f(\omega) f(\omega') g(|\mathbf{r}-\mathbf{r}'|,\omega,\omega'), \end{aligned} \tag{6}$$

where $\rho = N/V$ is the density and $f(\omega)$ the one-body orientational distribution in the interacting fluid. Equation (6) defines the generalized pair distribution function $g(r,\omega,\omega')$ of the inhomogeneous spin system in an external magnetic field. The angular brackets in these definitions denote a canonical ensemble average with the Hamiltonian of Z above.

The basic equations that determine the distribution functions $f(\omega)$ and $g(r,\omega,\omega')$ are well known [11, 12]. The one-body density can be differentiated with respect to $x = \cos\theta$ to give

$$\frac{d}{dx} \ln\left[\frac{f(\omega)}{f_0(\omega)}\right] = -\frac{\rho}{4\pi} \int d\mathbf{r}\, d\omega'\, f(\omega') g(r,\omega,\omega') \frac{d\beta u_{ss}(r,\omega,\omega')}{dx}, \tag{7}$$

the first member of a Kirkwood-Born-Green-Yvon hierarchy. Calculation of $f(\omega)$ from this equation requires knowing g. In classical liquid state theory, the pair distribution function is obtained from the Ornstein-Zernike (OZ) equation and a closure relation [11, 12]. The first of these, generalized for anisotropy, reads

$$\begin{aligned} \gamma(r_{12},\omega_1,\omega_2) &= \frac{\rho}{4\pi} \int d\mathbf{r}_3 d\omega_3\, f(\omega_3) \\ &\quad \times [\gamma(r_{13},\omega_1,\omega_3) + c(r_{13},\omega_1,\omega_3)] c(r_{32},\omega_3,\omega_2) \end{aligned} \tag{8}$$

for the indirect correlation function $\gamma = g - 1 - c$, where c is the direct correlation function. The second, or closure, relation expresses c back in terms of γ and the model's pair interactions,

$$\begin{aligned} c(r,\omega_1,\omega_2) &= \exp[-\beta u_0(r) - \beta u_{ss}(r,\omega_1,\omega_2) + \gamma(r,\omega_1,\omega_2) \\ &\quad + b(r,\omega_1,\omega_2)] - 1 - \gamma(r,\omega_1,\omega_2). \end{aligned} \tag{9}$$

This relation must be supplemented with an approximation for b, the so-called bridge function.

We have recently described [13, 14] a practical approach to solving Eqs. (7)-(9) for any classical fluid in an external field using the same numerical procedures already employed for *homogeneous* systems . The essential ingredient turns out to be orthogonal polynomials tailored to the specific one-body distribution of the fluid [15]. The usual procedure for orientation-dependent functions such as $\gamma(r,\omega_1,\omega_2)$ is to expand in spherical harmonics $Y_{lm}(\omega)$ [16],

$$\gamma(r,\omega_1,\omega_2) = 4\pi \sum_{l_1,l_2,m} \gamma_{l_1 l_2 m}(r) Y_{l_1 m}(\omega_1) Y_{l_2 \bar{m}}(\omega_2), \tag{10}$$

$$Y_{lm}(\omega) = \frac{1}{\sqrt{4\pi}}(-1)^m e^{im\phi} P_{lm}(\cos\theta), \tag{11}$$

where $P_{lm}(\cos\theta)$ is the familiar associated Legendre function, but here renormalized to two, and $\bar{m} = -m$. This expansion in fact makes the solution of the inhomogeneous liquid equations very difficult. We will instead expand in *modified* spherical harmonics,

$$\gamma(r,\omega_1,\omega_2) = 4\pi \sum_{l_1,l_2,m} \gamma_{l_1 l_2 m}(r) \mathcal{Y}_{l_1 m}(\omega_1) \mathcal{Y}_{l_2 \bar{m}}(\omega_2), \tag{12}$$

$$\mathcal{Y}_{lm}(\omega) = \frac{1}{\sqrt{4\pi}}(-1)^m e^{im\phi} \mathcal{P}_{lm}(\cos\theta). \tag{13}$$

The modified Legendre functions $\mathcal{P}_{lm}(\cos\theta)$ are explicitly constructed, beginning with $\mathcal{P}_{00}(\cos\theta) = 1$, using the Gram-Schmidt method [17, 18] with the orthonormality condition

$$\frac{1}{2}\int_{-1}^{1} dx\, f(x)\mathcal{P}_{lm}(x)\mathcal{P}_{l'm}(x) = \delta_{ll'}, \tag{14}$$

where $f(\cos\theta)$ is the one-body distribution of the fluid. For an isotropic fluid with $f(x) = 1$, this construction produces the standard Legendre functions $P_{lm}(x)$ (normalized to two). For an *anisotropic* fluid with $f(x) \neq 1$, the difference between $P_{lm}(x)$ and $\mathcal{P}_{lm}(x)$ is crucial.

The numerical evaluation of the OZ equation is simpler in Fourier transform representation, which deconvolutes the $\mathbf{r}$ integral. Thus, in transform space, Eq. (8) becomes

$$\tilde{\gamma}(k,\omega_1,\omega_2) = \frac{\rho}{4\pi}\int d\omega_3\, f(\omega_3)[\tilde{\gamma}(k,\omega_1,\omega_3) + \tilde{c}(k,\omega_1,\omega_3)]\tilde{c}(k,\omega_3,\omega_2), \tag{15}$$

with one remaining integration. This final integral can also be eliminated and the evaluation of the OZ equation reduced to algebra by now expanding the pair functions as in (12); Eq. (15) then becomes

$$\tilde{\gamma}_{l_1 l_2 m}(k) = (-1)^m \rho \sum_{l_3} [\tilde{\gamma}_{l_1 l_3 m}(k) + \tilde{c}_{l_1 l_3 m}(k)] \tilde{c}_{l_3 l_2 m}(k). \tag{16}$$

The significant feature here is that this OZ equation for an inhomogeneous fluid in an external field is *identical to that of an ordinary homogeneous fluid* [16] and so, along with some closure equation familiar from simple liquids, can be solved for $g(r, \omega_1, \omega_2)$ with the same algorithms already used for homogeneous systems [19, 20].

Equations (7)-(9) being nonlinear, their solution is iterative. Given a set of approximate coefficients $\gamma_{l_1 l_2 m}(r)$, an improved set is produced by the following steps:

1. *Closure:* $\gamma_{l_1 l_2 m}(r) \rightarrow c_{l_1 l_2 m}(r)$.

$$\gamma(r, x_1, x_2, y) = \sum_{l_1, l_2, m} \gamma_{l_1 l_2 m}(r) \mathcal{P}_{l_1 m}(x_1) \mathcal{P}_{l_2 m}(x_2) (-1)^m T_m(y), \tag{17}$$

$$\begin{aligned} c(r, x_1, x_2, y) &= \exp[-\beta u_0(r) - \beta u_{ss}(r, x_1, x_2, y) + \gamma(r, x_1, x_2, y) \\ &\quad + b(r, x_1, x_2, y)] - 1 - \gamma(r, x_1, x_2, y), \end{aligned} \tag{18}$$

$$\begin{aligned} c_{l_1 l_2 m}(r) &= \frac{1}{4\pi} \int_{-1}^{1} \frac{dy}{\sqrt{1 - y^2}} \int_{-1}^{1} dx_2 \int_{-1}^{1} dx_1 f(x_1) f(x_2) \\ &\quad \times c(r, x_1, x_2, y) \mathcal{P}_{l_1 m}(x_1) \mathcal{P}_{l_2 m}(x_2) (-1)^m T_m(y) \end{aligned} \tag{19}$$

2. *Transform:* $c_{l_1 l_2 m}(r) \rightarrow \tilde{c}_{l_1 l_2 m}(k)$.

$$\tilde{c}_{l_1 l_2 m}(k) = \frac{4\pi}{k} \int_0^\infty dr \, r c_{l_1 l_2 m}(r) \sin kr \tag{20}$$

3. *OZ equation:* $\tilde{c}_{l_1 l_2 m}(k) \rightarrow \tilde{\gamma}_{l_1 l_2 m}(k)$.

$$\tilde{\Gamma}_m(k) = (-1)^m \rho \tilde{C}_m(k) \tilde{C}_m(k) [I - (-1)^m \rho \tilde{C}_m(k)]^{-1} \tag{21}$$

4. *Inverse transform:* $\tilde{\gamma}_{l_1 l_2 m}(k) \rightarrow \gamma_{l_1 l_2 m}(r)$.

$$\gamma_{l_1 l_2 m}(r) = \frac{1}{2\pi^2 r} \int_0^\infty dk \, k \tilde{\gamma}_{l_1 l_2 m}(k) \sin kr \tag{22}$$

In these equations, $x = \cos\theta$, $y = \cos\phi$, and $T_m(y)$ is a Chebyshev polynomial; further, in Eq. (21), I is the unit matrix while $\tilde{\Gamma}_m(k)$ and $\tilde{C}_m(k)$ are symmetric matrices with elements $\tilde{\gamma}_{l_1 l_2 m}(k)$ and $\tilde{c}_{l_1 l_2 m}(k)$, respectively, with $l_1, l_2 \geq m$. The inversion (19) is performed using n-point Gaussian quadratures with the zeroes of $\mathcal{P}_{n0}(x)$ and $T_n(y)$. Upon convergence of the $\gamma_{l_1 l_2 m}(r)$ coefficients, the generalized pair distribution function g is finally computed as

$$g(r, \omega_1, \omega_2) = \exp[-\beta u_0(r) - \beta u_{ss}(r, \omega_1, \omega_2) + \gamma(r, \omega_1, \omega_2) + b(r, \omega_1, \omega_2)].$$

This solution for $g(r, \omega_1, \omega_2)$ is obtained using the polynomials $\mathcal{P}_{lm}(x)$ generated with the current $f(x)$. One now returns to Eq. (7) and updates the one-body distribution $f(x)$. In expanded form, this equation is

$$\frac{d}{dx} \ln \left[\frac{f(x)}{f_0(x)} \right] = \sum_{l_1, l_2, m} \xi_{l_1 l_2 m} \mathcal{P}_{l_1 m}(x) \frac{d\mathcal{P}_{l_2 m}(x)}{dx}, \tag{23}$$

$$\xi_{l_1 l_2 m} = -\rho \int dr \sum_{l_3} g_{l_1 l_3 m}(r) u_{l_3 l_2 \bar{m}}(r), \tag{24}$$

where the $u_{l_1 l_2 m}(r)$ are the (known) coefficients of the spin-spin interaction $u_{ss}(r, \omega_1, \omega_2)$, so that finally

$$\ln f(x) = \ln f_0(x) + \sum_{l=0}^{\infty} a_l \mathcal{P}_{l0}(x). \tag{25}$$

Here a_l for $l > 0$ is determined by numerical (Gaussian) integration of Eq. (23) and a_0 by normalization. The n-point Gaussian quadrature is exact for polynomial integrands of degree less than $2n$, a condition that always holds for the computed a_l coefficients. For the calculations reported in Ref. [14], it turns out that the term linear in x dominates the series expansion in Eq. (25), so that an effective field B acting on a particle that includes the magnetic effects of the particle's surroundings through the coefficient a_1 can be calculated from this equation. In the absence of an interaction ($a_1 = 0$) this field is of course just B_0.

The iterations for $f(\omega)$ and $g(r, \omega_1, \omega_2)$ are continued until both functions are self-consistently determined. The only approximation in the calculation is the unavoidable one in every liquid state calculation to date, the representation of the bridge function $b(r, \omega_1, \omega_2)$ [11].

The angular projection coefficients of Eq. (12) for, say, $g(r, \omega_1, \omega_2)$ cannot be directly compared with the standard projections typically calculated in a simulation, since these correspond to the usual spherical harmonic basis. However, it is straightforward to recover one set of coefficients from the other given the moments of the one-particle distribution function $f(x)$. For example, the (unnormalized) projection of $g(r,\omega_1,\omega_2)$ onto the rotational invariant $\hat{\mathbf{s}}_1 \cdot \hat{\mathbf{s}}_2$ is found to be

$$\begin{aligned} h^{110}(r) &= 3\{\langle x\rangle^2 g_{000}(r) + \sigma_x \langle x\rangle [g_{100}(r) + g_{010}(r)] \\ &\quad + \sigma_x^2 g_{110}(r) - \left[1 - \left\langle x^2\right\rangle\right] g_{111}(r)\}. \end{aligned} \tag{26}$$

For an isotropic system, with $\langle x\rangle = 0$ and $\langle x^2\rangle = 1/3$, this reduces to the usual $h^{110}(r) = g_{110}(r) - 2g_{111}(r)$.

We can further compute the complete magnetic and thermodynamic properties of a Heisenberg spin fluid in a uniform magnetic field $\mathbf{B}_0$. The net magnetization in the direction $\alpha = x, y, z$ is

$$M_\alpha = \mu \left\langle \sum_j s_{j\alpha} \right\rangle = \frac{1}{\beta}\frac{\partial \ln Z}{\partial B_{0\alpha}}, \tag{27}$$

and so the magnetic susceptibility components are

$$\begin{aligned} \chi_{\alpha\beta} &\equiv \frac{1}{V}\frac{\partial M_\alpha}{\partial B_{0\beta}} = \rho\beta\mu^2\{\langle s_{1\alpha}s_{1\beta}\rangle - \langle s_{1\alpha}\rangle\langle s_{1\beta}\rangle \\ &\quad + (N-1)[\langle s_{1\alpha}s_{2\beta}\rangle - \langle s_{1\alpha}\rangle\langle s_{2\beta}\rangle]\}. \end{aligned} \tag{28}$$

Only the diagonal elements are nonvanishing. With $\mathbf{B}_0$ defining the z direction, we get the longitudinal and transverse magnetic susceptibilities as

$$\chi_{zz}/\rho\beta\mu^2 = \sigma_x^2\left[1 + \rho\tilde{h}_{110}(0)\right] + \sigma_x\langle x\rangle\left[\rho\tilde{h}_{100}(0) + \rho\tilde{h}_{010}(0)\right], \tag{29}$$

$$\chi_{yy}/\rho\beta\mu^2 = \frac{1}{2}\left(1 - \left\langle x^2\right\rangle\right)\left[1 - \rho\tilde{h}_{111}(0)\right]. \tag{30}$$

For an isotropic system, these susceptibility expressions reduce to familiar forms.

The internal energy U, pressure p, and isothermal compressibility K_T are calculated with the standard formulas of homogeneous fluids [11, 16], again thanks to the special orthogonality of the $\mathcal{Y}_{lm}(\omega)$.Finally, an expression for the Helmholtz free energy A is given in Ref. [14].

3. Computed results

We have obtained the distribution functions $f(\omega)$ and $g(r,\omega_1,\omega_2)$ for the Heisenberg spin fluid in a magnetic field $\mathbf{B}_0$ using a reference version of the Zerah-Hansen (RZH) closure [21]; for $r > \sigma$, this is

$$\begin{aligned} c^{RZH}(r,\omega_1,\omega_2) &= \{\exp\left(m(r)\left[-\beta u_{ss}(r,\omega_1,\omega_2) + \gamma(r,\omega_1,\omega_2)\right.\right. \\ &\quad \left.\left. + b_{\mathrm{HS}}(r;\sigma)\right]\right) - 1\}/m(r) - \gamma(r,\omega_1,\omega_2). \end{aligned} \tag{31}$$

The exact relation $c(r,\omega_1,\omega_2) = -1 - \gamma(r,\omega_1,\omega_2)$ holds for $r < \sigma$.

The RZH closure features a mixing function $m(r) = 1 - \exp(-\alpha r)$ with a parameter α that is fixed by requiring consistency between the virial and compressibility bulk moduli. (However, see below for the case $B_0 \to 0$.) The reference hard sphere bridge function $b_{\rm HS}(r;\sigma)$, added by us in Eq. (31) to the usual ZH closure [21], ensures the RZH closure yields good hard sphere results in the limit of small K and improves the thermodynamic results overall.

To be consistent with earlier work [4, 5], we have truncated the spin-spin interaction at $r = r_c \equiv 2.5\sigma$; this hardly affects the thermodynamics (aside from the pressure) but does considerably raise the critical K value. (Some results for the untruncated potential obtained using density functional theory and Monte Carlo simulation can be found in Refs. [6] and [7].)

We report here calculations for $\kappa = 1$, $\rho\sigma^3 = 0.7$, $K = 1/T^* = 0$ to 0.5, and two values of the external field, $\beta\mu B_0 = 1$ and 0, using the RZH closure. As noted earlier, we find for these cases that the one-body orientational distribution function $f(x)$ continues to be well described by the functional form of $f_0(x)$, but with an *effective field* B,

$$f(x) = \frac{e^{\beta\mu B x}}{\sinh(\beta\mu B)/\beta\mu B}. \tag{32}$$

In the limit $B_0 \to 0$, the effective field B is zero for $K < K_c$ and *finite* for $K > K_c$, where $K_c = 1/T_c^*$ is then the computed Curie point. This zero field limit has recently been the focus of several works [4, 5, 6, 7], but until now no theoretical approximation has been able to completely describe the ordered phase having spontaneous magnetization.

For $K > K_c$, there is a singularity when $B_0 \to 0$ that is different from the second order paramagnetic-ferromagnetic transition at $K = K_c$. In fact, the $B_0 = 0$ line corresponds in the B_0–M plane (where M is the magnetization per particle) to the spinodal line that indicates the equilibrium between phases with positive and negative magnetization. Consequently, the transverse susceptibility χ_{yy} (but not the longitudinal component χ_{zz}) will diverge [22] as $B_0 \to 0$, reflecting the negligible cost of rotating an ordered sample in the absence of an external field. It turns out that the optimization of α in (31) to achieve thermodynamic consistency leads to a χ_{yy} that diverges at small but non-zero fields. This deficiency is easily cured if one relaxes the thermodynamic consistency requirement and instead optimizes the α parameter by displacing the divergence of χ_{yy} to the $B_0 = 0$ line. We find that the thermodynamics is hardly affected by the new choice of α.

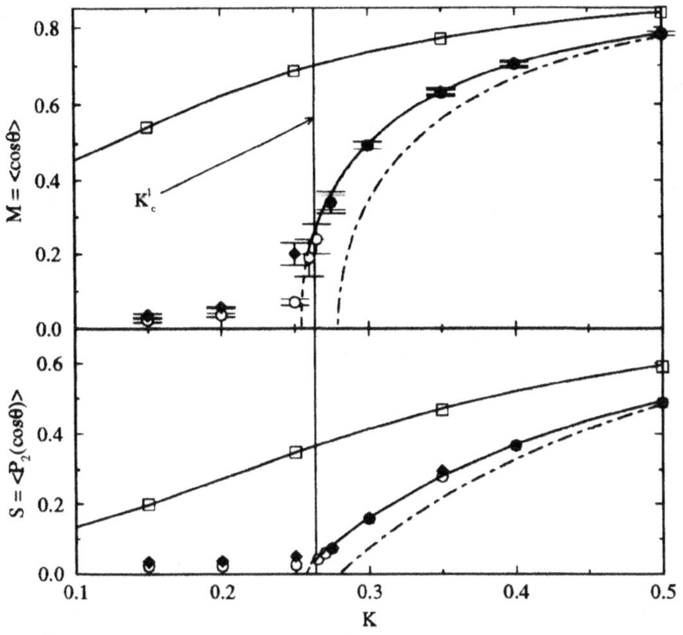

Figure 1. Magnetization per particle M and second order parameter S as functions of inverse temperature $K = 1/T^*$ obtained from MC simulation [squares for $\beta\mu B_0 = 1$ and white circles (2048 particle sample) and black diamonds (1372 particle sample) for $\beta\mu B_0 = 0$]. The RZH integral equation results appear as solid lines. The dashed lines correspond to a power law fit to the RZH data just above K_c^u. The dash-dot curves represent density functional theory estimates [7].

Using Monte Carlo simulation, Nijmeijer and Weis [4] found that the paramagnetic-ferromagnetic transition for $\rho\sigma^3 = 0.7$ with the truncated potential occurs at $K_c = 0.264 \pm 0.001$. Our calculation in the paramagnetic phase and zero field limit yields a divergence in χ_{zz} at $K_c{}^l = 0.2645$, in excellent agreement with the simulation value.

In Fig. 1 we show the values of the magnetization per particle $M \equiv M_z = \langle \cos\theta \rangle$ and the second order parameter $S = \langle P_2(\cos\theta) \rangle$ obtained from the theory and from a standard NVT Monte Carlo simulation using 864 (for $B_0 = 1$), 1372 and 2048 particles (for $B_0 = 0$), and averages over 40 000 configurations. Size effects are noticeable for $B_0 = 0$ close to and below (*i.e.*, $K < K_c$) the critical point. We include in Fig. 1 some zero-field results from density functional theory in the modified mean field approximation [7]. In the vicinity of the critical temperature, one encounters convergence difficulties in solving the integral equation as spin-spin correlations become long ranged. (Both χ_{zz} and χ_{yy} in Eqs. (29) and (30) diverge, with χ_{zz} exhibiting the characteristic λ-type divergence of second-

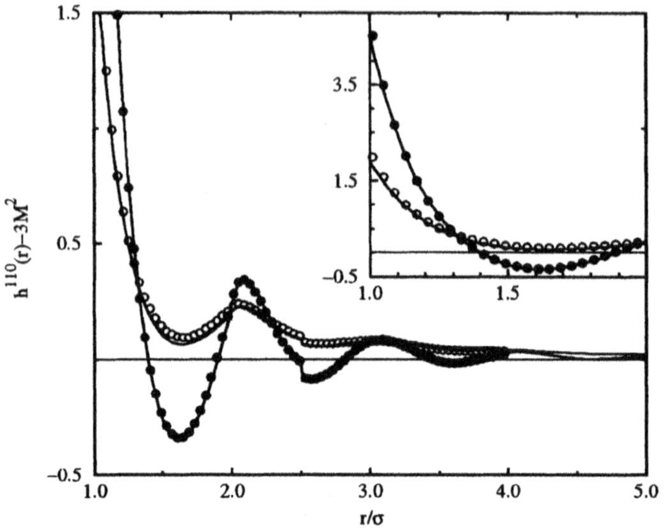

Figure 2. Behavior of the angular correlation function $h^{110}(r)$ as the Curie temperature is approached in the ferromagnetic phase at zero field obtained from MC simulation using 1372 particles (white circles correspond to $K = 0.275$ and black circles to $K = 0.5$) for a density $\rho\sigma^3 = 0.7$. The RZH results are shown as solid lines in both cases. The long-range limiting value $3M^2$ has been subtracted from both theory and simulation results to facilitate the comparison between the two temperatures. The discontinuity at $r = 2.5\sigma$ is due to the potential truncation.

order transitions.) In Fig. 2 we illustrate the increase in the range of the relevant angular correlation function, $h^{110}(r)$, as the Curie temperature is approached. The function $h^{110}(r)$, defined in Eq. (26), describes most of the orientational behavior of the spin fluid, since it is the projection of $g(r, \omega_1, \omega_2)$ onto the rotational invariant $\hat{\mathbf{s}}_1 \cdot \hat{\mathbf{s}}_2 = \cos\theta_{12}$. From Eq. (26), we have $\lim_{r\to\infty} h^{110}(r) = 3M^2$, a limiting behavior that is exactly fulfilled in our calculations by both simulated and theoretical results. In the figure, we have subtracted the corresponding limiting values of the correlation functions so that the plots present the same asymptotic limit of zero for the two temperatures, which facilitates their comparison. The increase in the range of the correlations with the lowering of the temperature is readily appreciated. The agreement between theory and simulation is remarkable in both cases.

To get an estimate of the critical inverse temperature from the ordered-phase results, we fit the RZH magnetization in the vicinity of the critical temperature ($K < 0.3$) to a power law, $M = a(K - K_c^u)^\beta$, which leads to $a = 1.694$, $\beta = 0.397$, and $K_c^u = 0.254$. The fitted value of β is close to the value $\beta = 0.387$ reported by Nijmeijer and Weis [4] and to the critical

TABLE 1. Magnetic and thermodynamic properties of the Heisenberg spin fluid for $\kappa = 1$ and $\rho\sigma^3 = 0.7$ calculated in the RZH approximation for $\beta\mu B_0 = 0$. The values of χ_{yy} for $K > K_c$ correspond to the last converged solution. The MC value in parentheses is a rough estimate since, due to the vicinity of the critical point, the results at this temperature are particularly sensitive to system size. $\chi^* \equiv \chi/\rho\beta\mu^2$.

K	$\beta\mu B$	$\beta U/N$ RZH	$\beta U/N$ MC	$\beta p/\rho$	$\rho k_B T K_T$	χ^*_{zz}	χ^*_{yy}	α
0.10	0	-0.013		5.685	0.0574	0.574	0.574	10.0
0.15	0	-0.034	-0.034	5.668	0.0576	0.877	0.877	10.0
0.20	0	-0.074	-0.073	5.635	0.0580	1.736	1.736	10.0
0.25	0	-0.156	(-0.175)	5.539	0.0601	8.854	8.854	4.00
0.30	1.778	-0.597	-0.597	4.927	0.0618	1.655	8.534	1.88
0.35	2.637	-1.009	-1.008	4.465	0.0657	0.666	7.733	1.92
0.40	3.358	-1.384	-1.384	4.039	0.0694	0.366	6.312	1.90
0.45	4.021	-1.739	-1.740	3.633	0.0734	0.245	5.424	1.86
0.50	4.647	-2.084	-2.082	3.237	0.0779	0.209	5.841	1.81

TABLE 2. Magnetic and thermodynamic properties of the Heisenberg spin fluid for $\kappa = 1$ and $\rho\sigma^3 = 0.7$ calculated in the RZH approximation for $\beta\mu B_0 = 1$. $\chi^* \equiv \chi/\rho\beta\mu^2$.

K	$\beta\mu B$	$\beta U/N$ RZH	$\beta U/N$ MC	$\beta p/\rho$	$\rho k_B T K_T$	χ^*_{zz}	χ^*_{yy}	α
0.10	1.575	-0.140		5.539	0.0584	0.310	0.454	7.0
0.15	2.018	-0.297	-0.298	5.365	0.0596	0.281	0.540	7.0
0.20	2.548	-0.517		5.121	0.0615	0.227	0.618	7.0
0.25	3.122	-0.780	-0.782	4.826	0.0638	0.174	0.679	7.0
0.30	3.712	-1.068		4.505	0.0666	0.134	0.725	7.0
0.35	4.304	-1.367	-1.365	4.170	0.0698	0.106	0.760	7.0
0.40	4.896	-1.673		3.828	0.0734	0.088	0.786	7.0
0.45	5.486	-1.983		3.482	0.0774	0.075	0.806	7.0
0.50	6.073	-2.295	-2.295	3.133	0.0819	0.066	0.821	7.0

exponent of the 3D lattice Heisenberg model, $\beta = 0.362 \pm 0.004$ [23] and $\beta = 0.3639 \pm 0.0035$ [24]. However, the available data from the integral equation magnetization are too far away from K_c to draw here any conclusion with confidence and this agreement might simply be fortuitous. Nevertheless, a similar fit carried out on the second-order parameter S produces the same estimate of K_c^u and so we might conclude that this is the theoretical

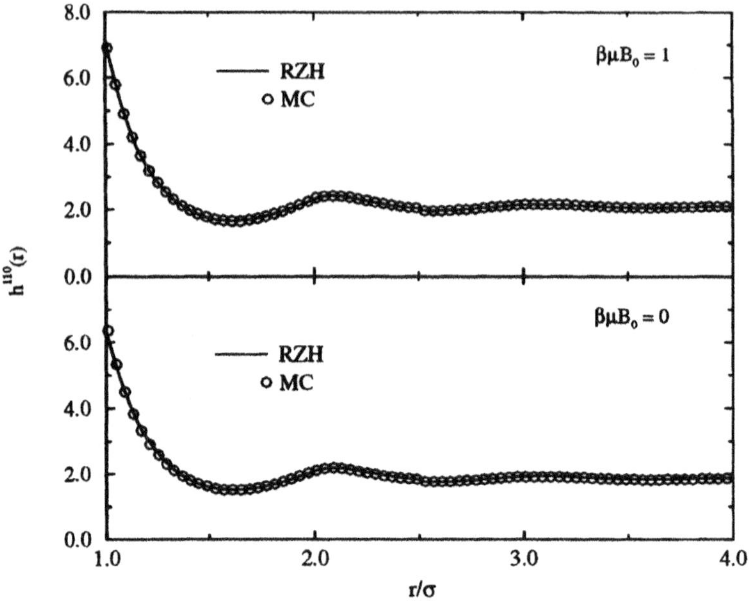

Figure 3. The $h^{110}(r)$ angular correlation function from MC simulation (white circles) and the RZH integral equation (solid lines) for $K = 0.5$ and $\rho\sigma^3 = 0.7$ with and without an external magnetic field. The discontinuity at $r = 2.5\sigma$ is due to the potential truncation.

estimate of the critical inverse temperature derived from the ordered-phase data. The agreement between the critical point estimates $K_c^l = 0.2645$ obtained in the disordered phase and $K_c^u = 0.254$ obtained in the ordered phase can be regarded as good. Clearly, there is room for improvement, perhaps by imposing further consistency conditions. In particular, one would want $\chi_{zz} = \rho\mu\partial\langle x\rangle/\partial B_0$ to hold numerically.

The magnetic and thermodynamic properties of this set of states, including the effective field B, are summarized in Tables 1 and 2 for external fields $\beta\mu B_0 = 0$ and 1, respectively. Direct comparison with Monte Carlo data is available only for the internal energy U and here we find excellent agreement with the results obtained using the RZH closure. That this excellent agreement between simulation and theory also extends to the correlation functions when the external field is turned on can be seen in Fig. 3. We note in Table 2 that for $\beta\mu B_0 = 1$ the consistency parameter α in the RZH closure is insensitive to changes in magnetic interaction strength; the determining factor is the hard sphere interaction.

References

1. N.D. Mermin, Phys. Rev. A **137**, 1441 (1965).
2. J.R. Henderson, in *Fundamentals of Inhomogeneous Fluids* (Dekker, New York, 1992), edited by D. Henderson, Chap. 2.
3. J. S. Høye and G. Stell, Phys. Rev. Lett. **36**, 1569 (1976).
4. M. J. P. Nijmeijer and J. J. Weis, Phys. Rev. Lett. **75**, 2887 (1995) ; Phys. Rev. E **53**, 591 (1996).
5. J. J. Weis, M. J. P. Nijmeijer, J. M. Tavares, and M. M. Telo da Gama, Phys. Rev. E **55**, 436 (1997).
6. E. Lomba, J. J. Weis, N. G. Almarza, F. Bresme, and G. Stell, Phys. Rev. E **49**, 5169 (1994).
7. J. M. Tavares, M. M. Telo da Gama, P. I. C. Teixeira, J. J. Weis, and M. J. P. Nijmeijer, Phys. Rev. E **52**, 1915 (1995).
8. See G. S. Cargill II and R. W. Cochrane, in *Amorphous Magnetism*, edited by H. O. Hooper and A. M. de Graaf (Plenum, New York, 1973), p. 313.
9. G. Bush and H. J. Guentherodt, Phys. Lett. **27A**, 110 (1968); B. Kraeft and H. Alexander, Phys. Konden. Mater. **16**, 281 (1973).
10. T. Albrecht, C. Bührer, M. Fähnle, K. Maier, D. Platzek, and J. Reske, Appl. Phys. A **65**, 215 (1997).
11. J. P. Hansen and I. R. McDonald, *Theory of Simple Liquids* (Academic, London, 1986).
12. D. Henderson, in *Fundamentals of Inhomogeneous Fluids* (Dekker, New York, 1992), edited by D. Henderson, Chap. 4.
13. F. Lado and E. Lomba, Phys. Rev. Lett. **80**, 3535 (1998).
14. F. Lado, E. Lomba, and J. J. Weis, Phys. Rev. E (to be published).
15. F. Lado, Phys. Rev. E **55**, 426 (1997).
16. C. G. Gray and K. E. Gubbins, *Theory of Molecular Fluids* (Clarendon, Oxford, 1984), Volume 1.
17. See, for example, G. Arfken, *Mathematical Methods for Physicists* (Academic, Orlando, 1985), Chap. 9.
18. N. I. Akhiezer, *The Classical Moment Problem* (Hafner, New York, 1965), Chap. 1.
19. F. Lado, Mol. Phys. **47**, 283 (1982).
20. Since the interatomic vector $\mathbf{r}_{12}$ is completely decoupled from the spin orientations $\hat{\mathbf{s}}_1$ and $\hat{\mathbf{s}}_2$ in the Heisenberg fluid, the solution of the (OZ + closure) equations here is actually simpler than that described in Ref. [19].
21. G. Zerah and J. P. Hansen, J. Chem. Phys. **84**, 2336 (1986).
22. L. D. Landau, E. M. Lifshitz, and L. P. Pitaevskii, *Electrodynamics of Continuous Media*, (Pergamon, Oxford, 1984), Second Edition, Section 46.
23. C. Holm and W. Janke, Phys. Rev. B **48**, 936 (1993).
24. K. Chen, A. M. Ferrenberg, and D. P. Landau, Phys. Rev. B **48**, 3249 (1993).

FREE ENERGY DENSITY FUNCTIONAL FROM 0D CAVITIES

P. TARAZONA
Departamento de Física Teórica de la Materia Condensada, Universidad Autónoma de Madrid, E-28049 Madrid, Spain

AND

Y. ROSENFELD
Nuclear Research Centre Negev, PO Box 9001, Beer-Sheva 84190, Israel

Abstract. Density functional approximations for the Helmholtz free energy in systems of hard spheres and hard disks are built, within the Fundamental Measure Theory, to reproduce the exact free energy of a cavity which cannot hold more than one particle. These systems, known as the zero-dimensional ($0D$) limit of hard particles, have an excess free energy from the impossibility of multiple occupation of the cavity, which depends only on the average number of particles in the cavity, at equilibrium with a reservoir. The structure of the FMT density functionals may be completely determined by this $0D$ limit. The bulk equation of state is a particular output of the approximation, rather than an input to build the density functional.

1. Introduction

The density functional formalism (DFF) provides a generic framework to study the equilibrium properties of systems with many particles [1, 2]. The formalism gives useful relations between the thermodynamics, the correlation structure, and the interactions in these systems; however it does not provide us with any explicit way to obtain the central object of the formalism: the Helmholtz free energy density functional $\mathcal{F}[\rho(\mathbf{r})]$. This functional is explicitly known for ideal classical systems, but it is in general unknown for systems of interacting particles, the one-dimensional hard rod system being the most relevant exception [11]. The practical use of the DFF is based

C. Caccamo et al. (eds.), New Approaches to Problems in Liquid State Theory, 293–302.

mainly on approximations for $\mathcal{F}[\rho(\mathbf{r})]$, in which one uses the knowledge, gained by other methods, about the thermodynamics and the correlation structure of homogeneous bulk fluids to, approximate the free energy of a generic inhomogeneous system.

Here we discuss a different route to construct approximations for $\mathcal{F}[\rho(\mathbf{r})]$ for hard-core classical particles [4]. The approach gives approximations for $\mathcal{F}[\rho(\mathbf{r})]$ based only on the study of a very narrow cavity, which cannot hold more than one hard-core particle. The equation of state and the correlation structure of the bulk fluid are obtained as a particular output, rather than being an input, of the approximation. This approach leads in a natural way to a class of free energy density functionals which had been developed in the last decade under the generic name of Fundamental Measure Theory (FMT) [5, 6, 7]. In the FMT it is assumed that the fundamental geometrical object to describe the packing restrictions of a hard-core fluid is not the Mayer function (whose integral gives the exclude volume) but the molecular volume itself. Thus, for hard spheres with diameter σ, the Mayer function is a sphere of radius σ and it would play a central role in any virial expansion or in non-local density approximations like the WDA [9, 10]; while in the FMT the fundamental geometry would be given by a sphere of radius $R = \sigma/2$. The original version of the FMT was inspired in the Scaled Particle (SP) approximation [3]. In our recent work [4] we presented a new point of view leading to new density functional approximations for hard disks and hard spheres.

A cavity which cannot hold more than one hard-core particle may be considered as the zero-dimensional (0D) limit of a hard-sphere fluid. From any functional, $\mathcal{F}[\rho(\mathbf{r})]$, for a 3D system we may get the free energy of a two-dimensional system (with the same interactions) considering the fluid inside a very narrow slit pore. The one-dimensional free energy may be obtained from a fluid in a very narrow cylindrical capillary, and the very narrow cavity should provide the ultimate dimensional crossover to the 0D free energy. Of course, any of the reduced-dimension systems are extremely inhomogeneous, from the 3D point of view, so that a good description of the dimensional crossover requires a very good density functional approximation, taking into account the non-local dependence of the free energy on the density distribution. As we show here, the requirement that $\mathcal{F}[\rho(\mathbf{r})]$ should give the exact 0D limit is strong enough to determine the functional dependence within the FMT class [4].

2. Exact 0D-free energy excess

It may appear at first that a 0D system of hard-core molecules should be purely ideal, there is not enough room for more than one particle so that there should be no role for the interactions and the system should behave like an ideal gas in the external potential described by the cavity. This is indeed the case in the canonical description of the system with one particle, N=1. The density distribution follows the Boltzmann factor of the external potential in the cavity, $\rho_1(\mathbf{r}) = \exp(-\beta V(\mathbf{r}))/Z_1$, with the canonical partition function $Z_1 = \int d\mathbf{r} \exp(-\beta V(\mathbf{r}))$, and the free energy is $F_1 = -\beta^{-1}\log(Z_1)$ both for ideal and for interacting particles.

However, the DFF is based on a grand canonical description and hence we should consider that our narrow cavity is connected (through a much narrower passage) to a reservoir of particles, with chemical potential μ. The average occupation of the cavity, $\bar{N}$, goes from 0 to 1, as μ goes from minus to plus infinity. The density distribution inside the cavity will be $\rho(\mathbf{r}) = \bar{N}\rho_1(\mathbf{r})$, in terms of the canonical ($N = 1$) distribution. The grand-potential energy for the hard-spheres in the 0D cavity is the superposition of the empty and the occupied cavity,

$$\Omega = F - \bar{N}\mu = -\beta^{-1}\log(1 + Z_1 \exp(\beta\mu)). \tag{1}$$

An the ideal gas in the same external potential may have the the same density distribution (with the appropriate chemical potential μ_{id}), but in that case the value of $\bar{N}$ is the average of all the possible occupations from $N = 0$ to infinity, and the grand potential energy is

$$\Omega_{id} = F_{id} - \bar{N}\mu_{id} = -\beta^{-1}\log\left(\sum_{N=0}^{\infty} \frac{Z_1^N e^{\beta\mu_{id}N}}{N!}\right) = -\beta^{-1} Z_1\, e^{\beta\mu_{id}}. \tag{2}$$

The chemical potentials are obtained directly from the derivatives of Ω:

$$\beta\mu = \log\left(\frac{\bar{N}}{Z_1\,(1-\bar{N})}\right) \quad \text{and} \quad \beta\mu_{id} = \log\left(\frac{\bar{N}}{Z_1}\right). \tag{3}$$

For a given value of $\bar{N}$ there is a 0D excess Helmholtz free energy given exactly by $F(\bar{N}) - F_{id}(\bar{N}) = \beta^{-1}\varphi_0(\bar{N})$ with

$$\varphi_0(\bar{N}) = (1-\bar{N})\log(1-\bar{N}) + \bar{N}. \tag{4}$$

This result should be recovered when the exact excess free energy density functional $\Phi^{(D)}[\rho(\mathbf{r})] \equiv \beta F[\rho(\mathbf{r})] - \beta F_{id}[\rho(\mathbf{r})]$, in dimension D, is evaluated for any density distribution compatible with a 0D cavity (embedded in that dimension), i.e. such that that $\rho(\mathbf{r})\rho(\mathbf{r}') = 0$ for any pair of points separated by more than a HS diameter ($|\mathbf{r} - \mathbf{r}'| \geq \sigma = 2R$). Notice that, within this restriction, the excess free energy depends only on the integral of the density distribution $\bar{N} \leq 1$, it is independent of the shape of the distribution $\rho(\mathbf{r})$ and of the embedding dimension D.

3. The 1D contribution to $\Phi^{(D)}[\rho]$

We now search for functional approximations to $\Phi^{(D)}[\rho(\mathbf{r})]$, compatible with the 0D limit, and we start from the exact density functional for hard rods (1D) of length $2R$, which is [11]

$$\Phi^{(1)}[\rho(x)] = -\int_{-\infty}^{\infty} dx \; \frac{\rho(x+R)+\rho(x-R)}{2} \; \log\left[1 - \int_{x-R}^{x+R} dx' \rho(x')\right] . \quad (5)$$

The integral within the log represents the 1D version of the local packing fraction, which for hyper-spheres in a generic dimension D is

$$\eta(\mathbf{r}) = \int d\mathbf{r}' \rho(\mathbf{r}') \Theta(|\mathbf{r}-\mathbf{r}'| - R), \quad (6)$$

with the usual Heaviside step function, $\Theta(x)$, so that the function $\eta(\mathbf{r})$ gives the probability that the point $\mathbf{r}$ is inside a molecule.

To recover the 0D limit we may write (5) as

$$\Phi^{(1)}[\rho(x)] = \int_{-\infty}^{\infty} dx \; \varphi_1(\eta(x)) \left[\frac{1}{2} \frac{\partial \eta(x)}{\partial R} \right], \quad (7)$$

where $\varphi_1(\eta) = d\varphi_o(\eta)/d\eta = -\log(1-\eta)$ is the first derivative of the function (4). In any 0D cavity (embedded in 1D) we have that, for any point x with $\rho(x) \neq 0$,

$$\eta(x) = \int_{x-R}^{x+R} dx' \rho(x') = \int_{-\infty}^{\infty} dx' \rho(x') = \bar{N} \quad (8)$$

is constant and equal to the total occupation of the cavity. With this result (7) is integrated by parts to recover (4) exactly.

This exact dimensional crossover, from 0D to 1D, is easy to generalize for higher dimensions in the *strict* 0D limit of a cavity with vanishing accessible volume. The density distribution is $\rho(\mathbf{r}) = \bar{N}\delta(\mathbf{r})$, where the origin is taken at the center of the cavity. The exact 0D limit for any mean occupation with $0 \leq \bar{N} \leq 1$ is obtained from the excess free energy functional

$$\Phi_1^{(D)}[\rho] = \int d\mathbf{r} \; \varphi_1(\eta(\mathbf{r})) \left[s_D^{-1} \frac{\partial \eta(\mathbf{r})}{\partial R} \right] =$$

$$= \int d\mathbf{r} \; \varphi_1(\eta(\mathbf{r})) \int d\mathbf{R}_1 \rho(\mathbf{r}+\mathbf{R}_1) w_D(R_1), \quad (9)$$

where s_D is the *molecular surface area* $s_1 = 2$, $s_2 = 2\pi R$, $s_3 = 4\pi R^2$, and the weight function $w_D(|\mathbf{R}_1|) = s_D^{-1}\delta(R - |\mathbf{R}_1|)$ is a normalized shell, restricting the vector $\mathbf{R}_1$ to have radius R.

In the FMT approach, the excess free energy density functional for hard molecules in dimension D is written with the generic form

$$\Phi^{(D)}[\rho] = \sum_{k=1}^{D} \Phi_k^{(D)}[\rho], \tag{10}$$

and the first contribution $\Phi_1^{(D)}[\rho]$ has precisely the form (9). For hard rods it is the only (and exact) term, while for hard disks and hard spheres the sum in (10) has one or two more contributions respectively. To keep the exact result (4), in the strict 0D limit, the contributions of $\Phi_k^{(D)}[\rho]$, for $k > 1$, have to vanish when the density distribution is a delta-function. The most serious drawback of the earlier FMT versions for hard spheres was due to a divergence of $\Phi_3^{(3)}[\rho]$. Hence the exact result for the strict 0D limit, given by $\Phi_1^{(3)}[\rho]$, was completely spoiled and the description of density distributions with narrow peaks (like in the hard sphere crystal) was completely wrong. Empirical changes of $\Phi_3^{(3)}[\rho]$ to avoid the divergence were enough to stabilize the crystal phase in more recent versions of the FTM functional for hard spheres. Here we try to set the form of these contributions, still based only on the 0D limit, but for cavities with more complex shapes.

4. The $k = 2$ contribution to $\Phi^{(D)}[\rho]$

Consider now a density distribution embedded in dimension D and made of two delta-peaks, $\rho(\mathbf{r}) = \bar{N}_1\delta(\mathbf{r}-\mathbf{r}_1)+\bar{N}_2\delta(\mathbf{r}-\mathbf{r}_2)$, with $r_{12} = |\mathbf{r}_1-\mathbf{r}_2| < 2R$, so that the two subcavities at $\mathbf{r} = \mathbf{r}_1$ and $\mathbf{r} = \mathbf{r}_2$ cannot be occupied at the same time and $\bar{N} = \bar{N}_1 + \bar{N}_2 \leq 1$. The excess free energy is still given by $\varphi_0(\bar{N})$ and it does not depend on r_{12} nor on the mean occupation of each subcavity. For 1D embedding dimension it is straightforward to check that the exact result is recovered by $\Phi^{(1)}[\rho] = \Phi_1^{(1)}[\rho]$ in (9); however for higher embedding dimension the functional $\Phi_1^{(D)}[\rho]$ gives

$$\Phi_1^{(D)}[\rho] = \varphi_o(N) - \xi_D(r_{12})\ [\varphi_o(\bar{N}) - \varphi_o(\bar{N}_1) - \varphi_o(\bar{N}_2)]. \tag{11}$$

The deviation from the exact 0D limit factorizes in a function of subcavity occupations and a function $\xi_D(r_{12})$ which depends on the geometry of the cavity and the embedding dimension. For 2D and 3D we have

$$\xi_2(\mathbf{r}) = \frac{2}{\pi}\sin^{-1}\left(\frac{r}{2R}\right), \quad \text{and} \quad \xi_3(\mathbf{r}) = \frac{r}{2R}. \tag{12}$$

We may now search for a second contribution to the density functional, $\Phi_2^{(D)}[\rho]$ that cancels the second term in the right hand side of (11). This functional has to vanish for density distributions made of a single delta function or when the two delta-functions in $\rho(\mathbf{r})$ are separated by more than a molecular diameter. Within the FMT scheme we expect the functional to include the non-local dependence through the function $\eta(\mathbf{r})$. The exact form required for $\Phi_2^{(D)}[\rho]$ is given by

$$\Phi_2^{(D)}[\rho] = \int d\mathbf{r}\ \varphi_2(\eta(\mathbf{r})) \prod_{i=1}^{2} \int d\mathbf{R}_i \rho(\mathbf{r}+\mathbf{R}_i) w_D(R_i)\ P_D(\mathbf{R}_1, \mathbf{R}_2), \quad (13)$$

with $\varphi_2(\eta) = \partial^2 \varphi_o(\eta)/\partial\eta^2 = 1/(1-\eta)$. The kernel $P_D(\mathbf{R}_1, \mathbf{R}_2)$, couples the integrals over $\mathbf{R}_1$ and $\mathbf{R}_2$, and it is determined by the geometrical factor $\xi_D(r_{12})$ in (12), with $r_{12} = |\mathbf{R}_1 - \mathbf{R}_2|$.

In terms of the reduced variable $x = r_{12}/(2R)$, we find

$$P_2(\mathbf{R}_1, \mathbf{R}_2) = 4\pi R^2 x (1-x^2)^{1/2} \sin^{-1}(x), \quad (14)$$

for hard disks and

$$P_3(\mathbf{R}_1, \mathbf{R}_2) = 8\pi R^3\ x^2 = 4\pi R\ (R^2 - \mathbf{R}_1 \cdot \mathbf{R}_2), \quad (15)$$

for hard spheres.

It is interesting to observe that for 3D hard spheres the functional (13), with the kernel (15), is precisely of the form $\Phi_2^{(3)}[\rho]$ used as the second contribution in the original FMT approach. In this term we may factorize the dependence on the density distribution by defining a vector weight function:

$$\mathbf{w}_D(\mathbf{R}_1) = s_D^{-1}\ \delta(r - |\mathbf{R}_1|)\ \frac{\mathbf{R}_1}{R}.$$

The scalar weight $w_D(R_1)$, used in (9), is the modulus of $\mathbf{w}_D(\mathbf{R}_1)$ and we may define scalar and vector weighted densities,

$$n_D(\mathbf{r}) = \int d\mathbf{r}' \rho(\mathbf{r}') w_D(\mathbf{r}-\mathbf{r}') \quad \text{and} \quad \mathbf{n}_D(\mathbf{r}) = \int d\mathbf{r}' \rho(\mathbf{r}') \mathbf{w}_D(\mathbf{r}-\mathbf{r}'),$$

to rewrite the 3D versions of (9) and (13) as

$$\Phi_1^{(3)}[\rho] = \int d\mathbf{r}\ \varphi_1(\eta(\mathbf{r}))\ n_3(\mathbf{r}), \quad (16)$$

and

$$\Phi_2^{(3)}[\rho] = 4\pi R^3 \int d\mathbf{r}\ \varphi_2(\eta(\mathbf{r}))\ [(n_3(\mathbf{r}))^2\ - \mathbf{n}_3(\mathbf{r}) \cdot \mathbf{n}_3(\mathbf{r})]; \quad (17)$$

which are the usual forms within the FMT [5].

5. Density functional for hard disks

The kernel (14) to be used in $\Phi_2^{(2)}[\rho]$ for hard disks does not factorizes as in the case of 3D hard spheres. Thus, the double convolution of the density distribution in (13) cannot be written in terms of scalar and vector weighted densities as was assumed in the FMT [6]. This impossibility is certainly a drawback for the practical use of the approximation but it has to be accepted as a necessary requirement for a density functional with the exact 0D limit in the case of two delta-peaks cavities. In fact, it may be shown that the functional $\Phi^{(2)}[\rho] = \Phi_1^{(2)}[\rho]+\Phi_2^{(2)}[\rho]$ gives the exact 0D limit in a much broader class of narrow cavities in two dimensions: all those in which there is at least one *pivot point* at which $\eta(\mathbf{r}) = \bar{N}$, meaning that whenever there is a hard disk in the cavity that *pivot point* is inside the disk.

The condition to recovering the exact $\varphi_0(\bar{N})$ is satisfied in many 0D cavities, like those with circular or elongated shapes, in which the geometrical center of the cavity is always a *pivot point*, as long as the cavity is 0D. However, there are some peculiar shapes which although they are 0D (i.e. without enough room enough for two hard disks) do not have such *pivot point*: e.g. an equilateral triangle with physical walls of length L in the range $3\sqrt{3} < L/R < 2(1+\sqrt{3})$. In these *lost cases* the exact $\varphi_0(\bar{N})$ cannot be recovered from any density functional of the type (13) because $\eta(\mathbf{r})$ never reaches the value $\bar{N}$ so that this value cannot appear in the thermodynamic functions $\varphi_i(\eta)$.

We may explore now the performance of the density functional $\Phi^{(2)}[\rho] = \Phi_1^{(2)}[\rho] + \Phi_2^{(2)}[\rho]$ to describe the free energy with any density distribution. First of all we may apply it to any one-dimensional distribution of the type $\rho(x,y) = \delta(y)\rho(x)$, so that the hard disks are aligned and behave as hard rods. In this case, $\Phi^{(2)}[\rho] = \Phi_1^{(2)}[\rho] + \Phi_2^{(2)}[\rho]$ recovers the exact density functional for hard rods with density distribution $\rho(x)$.

If we use $\Phi^{(D)}[\rho] = \Phi_1^{(D)}[\rho] + \Phi_2^{(D)}[\rho]$ to evaluate the free energy of a homogeneous 2D hard disk fluid, we get $\Phi/N = \varphi_1(\eta) + \eta\ \varphi_2(\eta)$ and the equation of state is $\beta p = \rho/(1-\eta)^2$, precisely the SP approximation for this system. This equation of state was imposed in the original derivation of the FMT for hard disks [6], in our case it comes out of the density functional built to satisfy the 0D limit of a narrow cavity. The difference between the SP and the exact thermodynamics may be related to the existence of the *lost cases* for which the FMT density functional are unable to give the exact 0D free energy.

It is instructive to see how the original FMT approximation for hard disks may be related to the functional $\Phi_2^{(D)}[\rho]$ in (13). The kernel $P_2(\mathbf{R}_1, \mathbf{R}_2)$

may be expanded as $P_2(\mathbf{R}_1, \mathbf{R}_2) = 4\pi R^2 x^2 + \mathcal{O}(x^4)$, for low values of $x = |\mathbf{R}_1 - \mathbf{R}_2|/(2R)$. The truncation at order x^2 allows us to express $\Phi_2^{(D)}[\rho]$ in terms of vector weighted densities, as in (17). The functional with this truncated kernel would give a good description of a 0D cavity which is not point like but still very narrow ($|\mathbf{r}_1 - \mathbf{r}_2| << R$). However, when that truncated functional is used to get the bulk free energy of a hard disk fluid, it gives an extra factor of 2 for the term $\varphi_2(\eta)$. The empirical compromise, in the original FMT version for hard disks [6], was to take $P_2^{appr}(\mathbf{R}_1, \mathbf{R}_2) = 2\pi R^2 x^2$, in order to recover the SP equation of state and allow for the factorization of the functional $\Phi_2^{(2)}[\rho]$, as in (17).

6. Density functional for hard spheres

We now consider a density distribution consisting of three delta-function peaks, $\rho(\mathbf{r}) = \sum \bar{N}_i \delta(\mathbf{r} - \mathbf{r}_i)$, with $i = 1, 2, 3$ and $|\mathbf{r}_j - \mathbf{r}_i| < 2R$, so that it still corresponds to a 0D cavity with $\bar{N} = \sum \bar{N}_i \leq 1$. Unless the points $\mathbf{r}_i$ are located to give one of the peculiar *lost cases* described above, the functional $\Phi^{(D)}[\rho] = \Phi_1^{(D)}[\rho] + \Phi_2^{(D)}[\rho]$, for 2D, would give the exact free energy $\varphi(\bar{N})$. However, when the embedding dimension is 3D there is still a difference, which may be compensated with a third functional term:

$$\Phi_3^{(3)}[\rho] = \int d\mathbf{r}\, \varphi_3(\eta(\mathbf{r})) \prod_{i=1,3} \int d\mathbf{R}_i\, \rho(\mathbf{r}+\mathbf{R}_i)\, w_3(R_i)\, Q(\mathbf{R}_1, \mathbf{R}_2, \mathbf{R}_3), \quad (18)$$

where $\varphi_3(\eta) = \partial^3 \varphi_o(\eta)/\partial \eta^3 = (1 - \eta)^{-2}$. Notice that this contribution cannot appear in 2D, since there would be no overlap between the three circular shells. This gives a clear geometrical account of why this type of density functional is expressed as a series of exactly D terms (10).

Previous versions of the $\Phi_3^{(3)}$ contribution to the FMT functional for hard spheres [5] have the structure (18), but the geometrical dependence in the kernel $Q(\mathbf{R}_1, \mathbf{R}_2, \mathbf{R}_3)$ was set (implicitly [5] or empirically [7]) to recover the equation of state and the correlations in the bulk fluid, as given by the Percus-Yevick approximation. However, in those bulk-based approaches to $\Phi^{(3)}[\rho]$ the 0D limit was not considered, and in fact the functional $\Phi^{(3)}[\rho]$ had a negative divergence in the strict 0D limit. A version with a *regularized* kernel (still written as a scalar products of the vectors $\mathbf{R}_i$), removed the divergence and gave the first FMT functional able to describe the hard sphere crystal [8]. In the present approach the kernel $Q(\mathbf{R}_1, \mathbf{R}_2, \mathbf{R}_3)$ is described in geometrical terms [4] and vanishes exactly for the strict 0D limit and also for any 1D density distribution, giving back the exact hard rods density functional (5) for any density distribution like $\rho(\mathbf{r}) = \rho(x)\delta(y)\delta(z)$.

However, the kernel $Q(\mathbf{R}_1, \mathbf{R}_2, \mathbf{R}_3)$ required to satisfy the 0D limit does not factorize, making the functional much more cumbersome to use than the empirical versions of the FMT. Moreover, the bulk equation of state for 3D hard spheres, obtained within this approach, is very poor. The failure may be tracked down to the failure in the description of some 0D cavities. The functional (18) does contribute only when there is a triple overlap between the spherical shells, and this is the condition which fails in the *lost cases*. It was verified [4] that adding the free energy correction from these *lost cases* would yields back the SP equation of state, which for 3D is also the Percus-Yevick compressibility equation [1].

Our main conclusion is that the FMT approach, based on the local packing fraction $\eta(\mathbf{r})$ and in the molecular surface weight functions $w_D(r)$, is a natural way to build free energy density functionals for hard-core particles in low dimensions. The 0D limit free energy, $\varphi_0(\bar{N})$ may be used as a *generating function* and the equation of state is obtained in terms of the successive derivatives of this function, as in the SP approximation. This view of $\varphi_0(\bar{N})$ as a *generating function* has been explored independently by other authors [12], but based only on the strict 0D limit. We have shown [4] that considering more general types of 0D cavities is enough to fully specify the density functional form within the FMT class. This generalization fixes the kernel in the k-shell overlap integrals with a form which does not factorize in terms of scalar and vector weighted densities. These kernels provide the *regularized* versions of the FMT, which have been sought empirically in previous versions, to avoid the divergences in the correlation structure which appear with functional forms like (13) and (18), with arbitrary kernels.

However, the exact forms of the 0D-based kernels not only create difficulties in the practical use of the density functional approximations but also give bulk equations of state which rapidly degradate with increasing dimension. We obtain the exact result for 1D, a fairly-good (SP) approximation for 2D, but a very poor approximation for 3D. The geometry associated to the *FMT-lost cases* does not appear in 1D, but it becomes more important with increasing dimension. In this sense, the FMT may be regarded as a low-dimension expansion to $\Phi[\rho]$, while the other type of non-local density functionals like the WDA (related to the virial expansion and based on the Mayer function rather than on the molecular shape) become more accurate in high-dimensions. The success of the 0D-based FMT free energy density functionals for hard-cubes [13] may also be understood as a results of the low *effective dimensionality* of these systems, in which the molecular overlaps factorize into three functions of x, y and z.

With these considerations, we have proposed a semi-empirical FMT density functional [4] for HS with an approximate kernel $Q^{appr}(\mathbf{R}_1, \mathbf{R}_2, \mathbf{R}_3)$ based on the Taylor expansion of the exact one for very narrow cavities and renormalized to give the PY equation of state. This kernel is

$$Q^{appr}(\mathbf{R}_1, \mathbf{R}_2, \mathbf{R}_3) = 12\pi^2 \, [\mathbf{R}_1 \cdot (\mathbf{R}_2 \times \mathbf{R}_3)]^2. \tag{19}$$

The approximation for $\Phi_3^{(3)}[\rho]$ with this choice will be free of divergences appearing in previous versions. It vanishes for any linear density distribution, yielding the exact density functional in for hard rods when the spheres are fully confined in two directions, it may be expressed in terms of tensor weighted densities, nearly as simple to handle as the vector weighted densities of the previous versions.

Acknowledgments

This work was supported by the DGICT (Spain)-grant PB94-005-C02-01.

References

1. J.P. Hansen and I.R. McDonald, "Theory of Simple Liquids", Second Edition, Academic Press, London, (1986).
2. See the recent collection of reviews, *Fundamentals of Inhomogeneous Fluids*, edited by D. Henderson (Dekker , 1992), and in particular the review by R. Evans.
3. H. Reiss, H. Frisch, and J.L. Lebowitz, J.Chem.Phys. **31**, 369 (1959); H. Reiss, J.Phys.Chem. **96**, 4736(1992); Y. Rosenfeld, J.Chem.Phys. **89**, 4272 (1988).
4. P. Tarazona and Y. Rosenfeld, Phys. Rev. E **55**, R4873 (1997).
5. Y. Rosenfeld, Phys.Rev.Lett. **63**, 980 (1989); J.Chem. Phys. **98**, 8126 (1993); Phys.Rev.E **50**, R3318 (1994); J.Phys.Cond.Matter **8**, 9287 (1996); *ibid*, L795 (1996).
6. Y. Rosenfeld, Phys.Rev.A **42**, 5978 (1990).
7. E. Kierlik and M.L. Rosinberg, Phys.Rev.A **42**, 3382 i (1990); **44**, 5025 (1991). S. Phan, E. Kierlik, M.L. Rosinberg, B. Bildstein, and G. Kahl , Phys.Rev.E **48**, 618 (1993).
8. Y. Rosenfeld, M. Schmidt, H. Löwen and P. Tarazona, J. Phys.: Cond. Matter **8**, L577 (1996); Phys.Rev.E **55**, 4245 (1997); M.Schmidt, Ph.D. Thesis, 1996.
9. P. Tarazona, Molec.Phys. **52**, 81 (1984); P.Tarazona, Phys.Rev.A **31**, 2672 (1985).
10. W.A. Curtin and N.W. Ashcroft, Phys.Rev.A 32, 2909 (1985).
11. J.K. Percus, J.Stat.Phys. **15**, 505 (1976).
12. A. Gonzalez, J.A. White and R.Evans, J.Phys.: Cond. Matter **9**, 2375 (1997).
13. J.A. Cuesta and Yuri Martinez-Raton, Phys.Rev.Letters **78**, 3681 (1997).

GEOMETRICALLY BASED FREE ENERGY FUNCTIONAL FOR CONFINED HARD SPHERES AND THE DENSITY FUNCTIONAL THEORY FOR SIMPLE CLASSICAL SYSTEMS

Y. ROSENFELD
Nuclear Research Center Negev, P.O.Box 9001,
Beer-Sheva 84190, Israel

1. Introduction

In the last two decades there has been a continuous progress in the theory of nonuniform classical fluids [1], as described by Bob Evans in his first lecture. Important approximations and models were developed for the Helmholtz free energy functional, $F[\rho(\mathbf{r})]$, of a inhomogeneous density distribution, $\rho(\mathbf{r})$. The geometrical character of the hard-sphere interactions, which has been a major reason for their long standing central role in the microscopic theory of classical fluids [2], also simplifies the construction of model functionals [3]. Pedro Tarazona, in his first lecture, described the geometrically-based so called fundamental-measure functionals (FMFs) [4]-[8], emphasizing several very recent analyses [9]-[13] that revealed the important role played by the dimensional cross-over properties of the functional. This supplementary lecture is divided into two main parts. In the first part (*Section II*) I list some of the basic physical properties expected from the exact (but unknown!) free-energy functional for the hard-sphere system (especially when applied to densely packed hard-spheres), and then explain how these are achieved by the FMFs. These properties are also important in applications for general interactions. The second part of this lecture concerns the possibility to apply the hard-sphere FMFs to other types of pair interactions, and is divided into three sections: In the first (*Section III*) I indicate how the functional for additive hard-sphere mixtures can be directly generalized for other types of systems, specifically: *(i)* parallel hard ellipsoids, *(ii)* isotropic mixtures of non-spherical hard-particle fluids, *(iii)* non-additive hard-spheres, and *(iv)* adhesive hard-spheres. In the second (*Section IV*) I consider the applications of the "*bridge functional*" in relation to: *(i)* How can a fundamental-measure functional be self-consistently

C. Caccamo et al. (eds.), New Approaches to Problems in Liquid State Theory, 303–320.

improved? and *(ii)* How can it be extended for general pair interactions? In the third (*Section V*) I consider another general method, namely variational thermodynamic perturbation theory, and show how the fundamental-measure hard-sphere *reference-functional* predicts correctly the asymptotic high density expansion of the free energy of bulk solids and fluids. Systems which cannot be treated satisfactorily by these methods which are based on the hard-sphere functional are briefly discussed in the concluding part (*Section VI*). An updated and hopefully complete list of references to applications of the fundamental-measure functionals is provided.

2. Densely packed hard spheres and the fundamental-measure functionals

2.1. EXPECTED PROPERTIES OF THE EXACT FUNCTIONAL

The exact free energy functional is composed of two terms, $F[\rho(\mathbf{r})] = F_{id}[\rho(\mathbf{r})] + F_{ex}[\rho(\mathbf{r})]$. The ideal-gas free energy functional is given by the exact relation, $F_{id}[\rho(\mathbf{r})] = k_BT \int d\mathbf{r}\rho(\mathbf{r})\{\ln(\rho(\mathbf{r})\lambda^D) - 1\}$, where $\lambda = \left(\frac{h^2}{2\pi m k_B T}\right)^{1/2}$ is the de Broglie wave-length, and T is the temperature. The important generally unknown part is the (excess) contribution of the interactions between the particles, F_{ex}. Configurations of densely packed hard spheres, in different effective dimensionalities D, provide the ultimate test for model free energy functionals. The exact functional is expected to have the following main properties: **(1)** Correct overall *dimensional crossover*, namely the description of bulk systems of reduced dimension, as strongly inhomogeneous density distributions in a larger dimension, e.g. the functional for $(D-1)$, should come out from the functional for D when the density profile is a Dirac delta function along one of the coordinates. **(2)** Solid-solid transitions for confined hard-spheres are dictated by considerations of the close-packed density for configurations of different symmetries [14]. The functional must include a mechanism for locating these configurations of *close packing*. **(3)** The equation of state should *diverge* near these close-packed configurations. **(4)** For correct description of densely packed hard-spheres in a small cavity (the solid in particular, when each particle can be viewed as confined in a fluctuating cage of its nearest neighbours), the free energy functional should provide reliable results for the "$0D$ *limit*" corresponding to a cavity that cannot hold more then one particle [9]. Correct $0D$ limit is also required in order to predict the *vacancy concentration* in the solid [9]. **(5)** The functional must exhibit *symmetry-breaking* which separates between solutions of the density profile equations with different spatial structures. Since the free energy diverges near any close-packing, then, near this situation, it is higher than that for another non-close-packed structure with the same average density. It is thus only a local minimum in the space of density profiles, unless the density-profile equations, obtained from the minimization of the grand potential, feature a "symmetry-breaking" which separates the solutions with different symmetries. **(6)** Finally, the *cell-theory* free-volume picture is the physically accepted behavior of the exact functional near situations of close-packing [15]-[17]. The free-volume cell-theory result for the pressure [15] diverges as $P \propto (\overline{\eta}_c - \overline{\eta})^{-1}$, where $\overline{\eta}$ is the average packing fraction and $\overline{\eta}_c$ is the average packing fraction at some close-packed configuration. It is the exact equation of state for $D = 1$, and numerical simulations indicate [15]-[17] that it is correct near *any* configuration of close-packing at $\overline{\eta}_c$ also in $D = 2, 3$.

The fundamental-measure functionals are at present unique in possessing these desirable physical properties, and provide a scenario for how these are achieved by the exact (but unknown!) functional. These results are obtained as general properties of the FMFs, independently of the detailed specific form of the functional.

2.2. FUNDAMENTAL-MEASURE FUNCTIONALS: GENERAL FORM, CELL-THEORY, AND DIMENSIONAL CROSS-OVER

The derivation of the first fundamental-measure functional [4] was achieved by two *interpolations*: (1) Scaled-particle theory [18] thermodynamics which interpolates analytically between the limits of small and large cavities (or insertion particles) in a hard-sphere fluid. (2) Direct correlation functions (like in the Percus-Yevick [19] approximation) which interpolate between the low density limit where they are dominated by the convolutions of pair exclusion volumes, and the asymptotic high density limit where the pair direct correlations are dominated by the convolutions of single particle geometries (e.g. overlap volume) [20]. It was based on the *convolution decomposition* of the excluded volume for a pair of spheres in terms of characteristic functions for the geometry of the individual spheres, which then play the role of the weight functions for calculating the weighted densities [4]. The resulting theory thus manifestly unifies scaled-particle and Percus-Yevick theories as a special case for bulk $3D$ fluids, it is exact in $1D$, it is accurate also in $2D$, and yields accurate 3-particle direct correlation functions analytically [21],[22]. The Fundamental-Measure Functionals (FMFs) [4]-[13],[9] have the following form for the excess free energy,

$$F_{ex}[\rho(\mathbf{r})] = k_B T \int d\mathbf{r} \varphi^{(D)}[\{n_i(\mathbf{r})\}] \quad , \tag{1}$$

where the approximation consists of the assumption that $\varphi^{(D)}$ is a function of a certain limited set of weighted densities,

$$n_i(\mathbf{r}) = \int \rho(\mathbf{r}')w^{(i)}(\mathbf{r}-\mathbf{r}')d\mathbf{r}' \ . \tag{2}$$

The weight functions, $w^{(i)}(\mathbf{r})$, are characteristic functions for the geometry of a sphere of radius R, like its volume, $w^{(Volume)}(r) = \theta(R-r)$, or its surface, $w^{(Surface)}(r) = \delta(R-r)$, where $\theta(r)$ and $\delta(r) = -d\theta(r)/dr$ are the Heaviside and Dirac functions, respectively. In particular, $n_{Volume}(\mathbf{r}) \equiv \eta(\mathbf{r}) = \int d\mathbf{r}'\rho(\mathbf{r}')\theta(R-|\mathbf{r}-\mathbf{r}'|)$, is a *local packing fraction* equal (e.g.) to $\overline{\eta} = \overline{\rho}\frac{4\pi}{3}R^3$ for the uniform (bulk) $3D$ fluid of N spheres in a volume V with average density $\overline{\rho} = \frac{N}{V}$.

These weighted densities have the property that given a density distribution of width Δ at the origin, $\rho_\Delta(r > \Delta) = 0$, then $n_\alpha(r > R+\Delta) = 0$. Thus, if the total density is composed of an arbitrary configuration of localized peaks with nearest neighbor distance d, then as long as $d > 2(R+\Delta)$ the FMF excess free energy contributions from each site are completely independent of the others, and the FMFs in the appropriate limit thus contain the corresponding cell-theory picture. Moreover, if every localized peak is normalized then the excess free-energy contribution of every such peak corresponds to the $0D$ limit, namely a cavity with at most one particle. Thus, the *cell picture* as contained by the FMFs becomes self-consistent if the functional satisfies the $0D$ limit for which the exact result is known: If $\overline{\eta} \leq 1$ is the average occupation (packing fraction) of the cavity, then the *exact* $0D$ excess (over ideal gas contribution) free energy is given by [9]

$$\frac{F_{ex}}{k_B T} = \varphi^{(D=0)}(\overline{\eta}) = \varphi_0(\overline{\eta}) = \overline{\eta} + (1-\overline{\eta})\ln(1-\overline{\eta}) \leq 1 \ , \tag{3}$$

independent of the detailed structure of the cavity.

Using also the "surface" weighted density, $\zeta(\mathbf{r}) = \int d\mathbf{r}'\rho(\mathbf{r}')\delta(R-|\mathbf{r}-\mathbf{r}'|)$, and the dimensionless vector $\vec{\xi}(\mathbf{r}) = \frac{\nabla\eta(\mathbf{r})}{\zeta(\mathbf{r})}$ with norm $\xi(\mathbf{r}) = \left|\vec{\xi}(\mathbf{r})\right|$, the following $3D$ excess free energy density was eventually derived [9]: $\Phi^{(D=3)}(\eta(\mathbf{r}),\zeta(\mathbf{r}),\xi(\mathbf{r})) = \Phi_1 + \Phi_2 + \Phi_3$, where $\Phi_1 = -\frac{\zeta}{4\pi R^2}\ln(1-\eta)$, $\Phi_2 = \frac{\zeta^2}{4\pi R}(1-\xi^2)\frac{1}{1-\eta}$, $\Phi_3 = \frac{\zeta^3}{24\pi}(1-\xi^2)^3\frac{1}{(1-\eta)^2}$. This specific form features the Percus-Yevick [19] - scaled-particle [18] description of the bulk fluid, $\rho(\vec{r}) = \rho_0$, and is exact (gives (3)) in the strict 0D limit, $\rho(\vec{r}) = \eta\delta(\vec{r})$ [9]. Various functions of the components of $\vec{\xi}$ can be invoked in order to obtain specific desireable properties of the functional [9],[13], but these do not affect the present discussion.

Recent studies revealed [12],[13] that the correct $0D$ crossover can be *systematically* imposed, with the exact $0D$ result playing the role of a generating functional, and the function $\varphi^{(D)}[\{n_i(\mathbf{r})\}]$ depends on $\eta(\mathbf{r})$ by a linear combination of the derivatives, $\varphi_k(\eta) = \frac{\partial \varphi_0(\eta)}{\partial \eta^k}, k \leq D$, which are *singular* at $\eta(\mathbf{r}) = 1$. In $1D$ the FMF is exact [23] and features the logarithmic form, $\ln(1 - \eta(\mathbf{r}))$. In $3D$ several FMFs where derived [4],[7],[9], containing also the $(1 - \eta(\mathbf{r}))^{-1}$ and $(1 - \eta(\mathbf{r}))^{-2}$ terms, which yield the Percus-Yevick [19] - scaled-particle [18] description of the bulk fluid, and with generally good dimensional cross-over properties which can be systematically improved. The most sophisticated forms [13] have correct crossover to *almost arbitrary* $0D$ distribution, and yield the*exact* $1D$ *functional* for the corresponding $3D$ density distribution, $\rho(\mathbf{r}) = \rho(x)\delta(y)\delta(z)$. With new forms regularized in the $0D$ limit, the FMFs give [9] accurate results for the $3D$ fluid-solid transition (predicting for the first time the correct vacancy concentration of the solid), and for the equation of state of both the *fcc* and *bcc* crystals all the way from freezing to near close packing. With correct $0D$ forms the FMF's agree well with simulations for hard spheres [24] and hard-disks [25] confined in spherical and circular (respectively) cavities. FMFs for parallel hard cubes were derived very recently [12] with completey correct dimensional crossover. The most important feature of the FMF's is their *singularity* at $\eta(\mathbf{r}) = 1$.

2.3. NEAR ARBITRARY HARD-SPHERE CLOSE-PACKING

Consider a FMF for configurations of highly localized and densely packed spheres characterized by density distributions of narrow width $\Delta \ll R$, around fixed sites $\{\mathbf{R}_l\}$: $\rho(\mathbf{r}) = \sum_l \rho_\Delta(|\mathbf{r} - \mathbf{R}_l|)$, $\rho_\Delta(r > \Delta) \ll 1$. Define $\delta = \frac{\Delta}{R} \ll 1$, and the "free distance" [15] $a = (d - 2R)/2 \ll R$. Using the narrowness of this $\rho(\mathbf{r})$, then (see, e.g. [3]) the "ideal gas" free energy takes the form $F_{id}/Nk_BT = -D\ln\delta + D\ln(\lambda/R) + const$. The weighted densities are given by $n_\alpha(\vec{x}) = \sum_l n_\Delta^{(\alpha)}(|\vec{x} - \vec{R}_l|)$, where the functions $n_\Delta^{(\alpha)}(t)$ are: $\eta_\Delta(t < R - \Delta) = 1$, $\eta_\Delta(t > R + \Delta) = 0$, $\partial\eta_\Delta(t)/\partial t \leq 0$, $\zeta_\Delta(t < R - \Delta) = 0$, $\zeta_\Delta(t > R + \Delta) = 0$, $\zeta_\Delta(R - \Delta < t < R + \Delta) = O(\delta^{-1})$, $\xi_\Delta(t) = (1 - O(\delta^2))$. The singular region where $\eta = 1$ is excluded from the integration in (1) by the definition of ζ. When $\Delta \leq a$ then the excess free energy contributions from each site are identical, completely independent of the others, and given by an integral over the spherical shell $R - \Delta < x < R + \Delta$ of width 2Δ, centered at a site chosen as origin, $\vec{x} = 0$. They correspond to a *cell model*, where each cell contribution is equal to the excess free energy in the $0D$ limit of a singly occupied cavity [9], which should be equal to $F_{ex}/Nk_BT = \varphi^{(D=0)}(\eta = 1) = 1$ for the exact functional. The specific form $\Phi^{(D=3)}$ given above introduces a small error [9] and gives $F_{ex}/Nk_BT = 1 - O(\delta^2)$. The total free energy with $\Delta \leq a$ is minimized when Δ is maximal, namely $\Delta = a$, and the total free-energy takes the familiar [15] free-volume form, $F/Nk_BT = -D\ln(a/R) + D\ln(\lambda/R) + const$. What is usually called the configurational free energy [15] is dominated by the *ideal-gas* part of the free energy functional that gives rise to the free-volume pressure, $PV/Nk_BT = d/(d-2R)$, which is close to the hard-sphere fcc crystal pressure down to the melting density [15]. The excess (non-ideal) FMF pressure is zero near close packing, and is relatively very small even near melting. If we use the step form, $\rho_\Delta(x) = \rho_\Delta^{step}(x) \equiv \Theta(\Delta - x)/\left(4\pi\Delta^3/3\right)$, and $\Delta = a$, the FMF functional gives exactly the self-consistent free volume result of Kirkwood and Wood [15]. Mathematically, the cell picture can be obtained only when the weight functions have a range $\leq R$.

This picture, with $\Delta \approx \frac{1}{\sqrt{\alpha}}$, also holds [9] for the commonly used and accurate [26] "Gaussian" model of the $3D$ solid, $\rho_\Delta(r) = (\frac{\alpha}{\pi})^{3/2}e^{-\alpha r^2}$. Defining the small parameter $\epsilon = \overline{\eta}_c - \overline{\eta}$, the FMFs give the same free-volume type form for the free energy near *any* configuration of close packing characterized by $\overline{\eta}_c$, namely $F = Nk_BT\left(-D\ln\epsilon + D\ln(\lambda/R) + const\right)$. The corresponding equation of state, with the pressure diverging as $P \propto (\overline{\eta}_c - \overline{\eta})^{-1}$, is in agreement with all the available simulation results [15]-[17] for $D = 2, 3$, while the FMF is exact for $D = 1$.

2.4. DETECTION OF TOUCHING SPHERES

How this physically correct result $\Delta \approx a$ arises as a *self consistent* solution for minimizing the free energy? How does the FMF "know" that two spheres are almost touching one another? In order to answer these related questions allow $\Delta - a > 0$, and consider the lense-shape overlap region of two neighbouring sherical shells described above. To facilitate the discussion I use $\rho_\Delta^{step}(x)$, for which $\eta_\Delta(t)$ has a simple geometric meaning: it is proportional to the overlap volume of two spheres, of radii R and Δ, at separation t. From simple geometrical considerations the shell contributions of two nearest neighbours will give rise to a singularity $\eta(\vec{x}) = 1$ only when $a \to 0$ and $\Delta \to 0$ simultaneously: when the big sphere of radius R is at distance R from the small sphere of radius Δ, then as long as R is finite, i.e. $\frac{\Delta}{R} \neq 0$, the overlap volume is smaller then half the small-sphere volume. This is the geometry that enables the $(1 - \eta(\vec{x}))$ singularities of the FMF to "detect" touching spheres in close packed arrangements. Tedious but straightforward calculation shows that setting $a = 0$ for $\delta = \frac{\Delta}{R} \ll 1$ leads to a non-integrable singularity in the terms Φ_2 and Φ_3, while Φ_1 is integrable (i.e. one needs stronger than logarithmic singularity for $D = 2, 3$). It is found that the Δ dependence of the exces free energy is only through the ratio $\frac{\Delta}{a}$. All the singularities are integrable if one sets $\Delta \propto a$, before letting $a \to 0$. Each of the three contributions to the FMF has a different $\frac{\Delta}{a}$ dependence, and the total integral behaves like $\left(\frac{\Delta}{a}\right) s$ with exponent s of the order of unity. Minimizing the total free energy one gets the optimal value $\frac{\Delta_{MIN}}{a} \approx O(1)$, while F_{ex}/Nk_BT is nearly constant of order unity. The divergence of the equation of state at closest packing, and the free volume cell picture, are thus obtained for the FMF by minimizing the total free energy with respect to Δ.

2.5. SYMMETRY-BREAKING MECHANISM

While the analysis of the behavior near closest packing involves only a pair of nearest neighbours (the two closest to the lense-shape shell-overlap region described above), the "*symmetry breaking*" mechanism arises [10] from contributions of a whole environment of neighboring density peaks which characterise the density distribution $\rho(\vec{x})$. We expect [10] that a random distribution of the centers of the Gaussians gives rise to many situations, for which the maximal value of $\eta(\vec{x})$ crosses unity from below to yield $\eta(\vec{x}) > 1$, which represent physically forbidden situations from the point of view of the FMF, separating solutions for the density profile equation, of two types of spatial symmetry. For values of the width δ which are relevant for describing the solid by Gaussians, only configurations of relatively high symmetry can exclude forbidden situations with $\eta(\vec{x}) > 1$. Thus, the FMF result of the stable fcc solid using the Gaussian representation for the density profile (i.e. obtained as the free-energy minimum as function of the Gaussian width), represents a "solid" branch of the solution of the density profile equations, separated from the "fluid" branch by a region of "forbidden" configurations of the centers of the Gaussians. This feature is not limited to fcc symmetry or to Gaussians, and represents a general feature of the FMF functional that will emerge as soon as various types of density profiles will be studied.

3. Direct generalizations of the hard-sphere functional

3.1. PARALLEL HARD-ELLIPSOIDS

The system of parallel hard ellipsoids (PHE's) is related to the hard-spheres by an anisotropic mapping [27]. By applying that mapping on the hard-sphere weight functions, to obtain the PHE-weight-functions, we find that the convolution decomposition holds exactly also for PHE's. Correspondingly, the resulting fundamental-measure functional [6] yields the exact analytic solution of the Percus-Yevick equations for the direct correlation functions for the PHE fluid, and thus *predicts correctly* the absence of a smectic phase. The PHE's provide a very useful reference system for aligned particles.

3.2. ISOTROPIC MIXTURES OF NON-SPHERICAL HARD PARTICLES: APPLICATION OF THE GAUSS-BONNET THEOREM

The convolution decomposition for spheres, which was the starting point for the derivation of the fundamental-measure functional, turns out [6] to be just a special case of the Gauss-Bonnet theorem for convex bodies. By relating that convolution decomposition for spheres with the *Gauss-Bonnet theorem* for general convex bodies, the fundamental-measure functional is made applicable [6] to fluids of asymmetric molecules. The resulting functional yields the *exact* second virial coefficient for the isotropic hard convex body bulk fluid. It predicts a third virial coefficient for the one-component isotropic hard spherocylinder bulk fluid which is only 10% smaller than the exact coefficient, even for a length over width ratio as large as about 6. As can be gleaned from related work [20],[28], it also yields an accurate equation of state for isotropic bulk fluids. Using the scaled field particle diagramatic representation it was possible to obtain analytic geometric approximations for the direct correlation functions and cavity distribution functions of the general isotropic bulk hard particle fluid, *conformal* to those for the hard sphere fluid mixture [20]. Using the second functional derivatives of the FMF these direct correlations are approximately re-expressed in terms of convolutions of the weight functions. This geometric representation is in accord with the solution of the Percus-Yevick equation and simulations [29]. When truncated at second order, the FMF for isotropic hard-particle fluids reduces to a version of a recent functional which was employed successfuly [30] for the inhomogeneous fluid of hard linear molecules (spherocylinders, ellipsoids). On the basis of these results it is expected that as long as the aspect ratio of the molecules (e.g. ratio between the longest and shortest distances across the molecule) is not too large (e.g. smaller than about five), then the FMF for isotropic hard-particle fluids will be of an accuracy comparable to the one achieved by this same functional for the spheres.

3.3. NON-ADDITIVE HARD-SPHERES

As suggested by Kierlik and Rosinberg [7], consider for example [9] a binary mixture of hard-spheres with radius R in the $2D$-limit with densities $\rho_1(z) = \rho^{(2D)}\delta(z+\omega)$, $\rho_2(z) = \rho^{(2D)}\delta(z-\omega)$ where $2\rho^{(2D)} = \frac{N}{A}$ is the $2D$ density, and $\pi(2\rho^{(2D)})R^2 = \eta$ is the $2D$ packing fraction. This system is equivalent to an equimolar mixture of *non-additive* hard-disks with negative non-additivity, $R_{12} = [R^2 - \omega^2]^{1/2} \leq R$. Similarly, employing the $4D$ FMF we can obtain the functional for $3D$ hard-spheres with negative non-additivity.

3.4. ADHESIVE HARD-SPHERES AND THE DEPLETION INTERACTION

The depletion interaction between the large spheres in a hard-sphere mixture with large size ratio is deeply connected [31] to the adhesive hard-sphere model of Baxter [32]. Taking the appropriate limit for the hard-sphere functional for the mixture, it gives rise to a functional for a one component system of adhesive hard-spheres. In a recent work [33] this general idea was formulated and then implemented to derive the fundamental-measure functional for *adhesive* hard-parallel-cubes.

3.5. FUNCTIONALS BASED ON THE ANALYTIC SOLUTIONS OF THE MEAN-SPHERICAL-APPROXIMATION

It will be interesting to obtain the corresponding results for adhesive hard-spheres, and compare with another functional which was derived recently [34] to correspond to the solution of the mean-spherical approximation (MSA). As indicated in the past [4],[5],[35], solutions of the MSA for Coulomb and Yukawa interactions can also be used to generate fundamental-measure free-energy functionals but, except mean-field type forms incorporating the charge contributions up to second-order (see below), based on the strong-coupling "Onsager" limit [5] of the solution, none has been explicitly derived as yet.

4. The Bridge functional

4.1. APPROXIMATE "UNIVERSALITY OF THE BRIDGE-FUNCTIONAL"

The starting point for the application of the density functional method for both uniform and non-uniform fluids are the density-profile equations [1]. The density profile $\rho(\vec{r})$ for the fluid subject to external potential $u(\vec{r})$ is obtained by solving the Euler-Lagrange equation $\delta\Omega[\rho(\vec{r})] \;/\; \delta\rho(\vec{r}) = 0$ which correspond to the minimization of the grand potential $\Omega[\rho(\vec{r})] = F_{id}[\rho(\vec{r})] + F_{ex}[\rho(\vec{r})] + \int d\vec{r}\,\rho(\vec{r})[u(\vec{r}) - \mu]$, where μ is the chemical potential. The density profile equations can be written in the modified hypernetted-chain form [5],[36],[37], involving the *bridge functional* which is related to the sum of all terms *beyond second order* in the functional Taylor expansion of the *excess free energy* around some *reference density.* The fundamental measure free-energy functional provides explicit simple expressions for the bridge functional, involving only integrations of known functions. This enables to apply the hard-sphere functional for fluids with arbitrary interactions, going beyond the (standard) van der Waals approximation for the attractive interactions. This new free energy density functional [5],[36],[37] for general inhomogeneous simple ("atomic") fluid mixtures is based on that for the hard spheres and it is operationally equivalent to an ansatz of *universality* of the bridge functional: The bridge functional, namely the sum of all terms beyond second order in the functional Taylor expansion, is approximated by that for the hard-spheres. The *second-order* part of the functional can be *optimized* by imposing the *test-particle self-consistency*, namely by considering the density profile equations (i.e. the Euler-Lagrange equations for minimizing the Grand Potential), with the same universal functional, but in the special case when the external potential is generated by a test-particle at the origin of coordinates. These equations are solved coupled with the Ornstein-Zernike relations, to generate the bulk pair correlation functions, which are the required input for the second-order part of the functional. The choice of the hard-sphere diameters (for fluids with arbitrarily soft potentials) is such that an error term in the free energy is minimized. There is no attempt to impose any specific structural-thermodynamic consistency relations, *everything is predetermined* by the quality of the approximation for the initial free energy functional for the inhomogeneous hard sphere fluid mixture, from which the "universal" bridge functional is obtained. The application of this new general method for non-uniform fluids to the special case of the bulk fluid, corresponds to the well established, successful thermodynamically-consistent modified-hypernetted-chain theory [38], but with the bridge functions now generated by an explicit, "universal", hard-sphere bridge functional.

4.2. RESULTS AND COMPARISON WITH SIMULATIONS

Using this new method accurate results were obtained for the bulk pair correlation functions for a large variety of potentials, for both one component systems and mixtures [5],[37],[39]-[43]. Of particular importance is the possibility to solve accurately the inverse scattering problem (i.e. to obtain the pair potential from a known structure factor) [40],[41]. The systems considered include: non-additive hard sphere mixtures, penetrable spheres, Lennard-Jones mixtures, liquid metals, and strongly coupled binary plasma mixtures.

The test-particle self-consistency is related to the consistency between the direct correlation functions and the "cavity" distribution functions (i.e. potential of mean force) where the correlation functions are put on the free energy scale [22],[44]. Unless the initial hard-sphere functional is already the exact one, the optimized (via test-particle self consistency) hard-sphere functional is more accurate than the initial one [5]. In particular, the fundamental-measure initial hard-sphere functional predicts the Percus-Yevick closure result that hard-sphere fluid binary mixtures never phase separate, but when optimized it does predict phase separation for large size ratios [36]. The Percus-Yevick closure which is obtained from the second functional derivatives of the fundamental-measure functional, is much improved by the application of the density profile equations even without optimization [5],[36] since they employ only the first functional derivative of the functional. For approximate functionals we can expect that the deviations from the exact result are larger for higher functional derivatives. Another examples for this special rolc of the density profile equations for a given free energy functional is provided by their application [45] via the fundamental-measure functionals for a direct calculation of the depletion interaction between two large spheres in an asymmetric hard-sphere mixture, and by their subtle influence on the accuracy of the functional for adhesive hard-particles [33].

In addition to bulk fluids, the hard-sphere "universal" bridge-functional, and the corresponding optimized free energy functional have been tested (directly and also implicitly) successfully for a variety of hard and soft pair interactions and external potentials, by comparison with computer simulations of density profiles for a large variety of situations where size or packing effects play an important role [5],[37],[39],[46]-[50]. The cases considered include among others: (1) wetting transitions and capillary condensation of argon atoms in carbon slit-like pores, (2) adsorption of binary methane-argon mixtures on a graphite surface, (3) argon-krypton mixtures in carbon slit-like pores, (4) selective adsorption from binary mixtures, (5) plasmas near interfaces, and (6) strong electrolytes near a charged electrode. More recently the hard-sphere bridge functional was used to construct a generic density functional for electric double layers in a molecular solvent [52].

The separation into a "second order" part and "bridge" part does not introduce explicit dependence on the reference densities for the exact functional, but it does for the (approximate) functional of the test-particle consistent theory. As a result it does not satisfy [39] the Gibbs adsorption equation (relating the excess adsorption and the derivative of the excess grand free energy with respect to the chemical potential) while the simple mean-field theory does. Recall that the test-particle consistent theory does not impose any particular thermodynamic consistency relation, and a slight inconsistency (which is inevitable for an approximation) is apparently augmented when it comes to locating the phase boundaries for inhomogeneous fluids. Several remedies can be provided [39],[51] in order to improve consistency with respect to this Gibbs sum rule.

5. Variational thermodynamic perturbation theory with a fundamental-measure reference functional

5.1. VARIATIONAL PERTURBATION THEORY

The exact thermodynamic perturbation theory divides the interaction potential, $\varphi(r) = \varphi_{ref}(r) + \varphi_{pert}(r)$, and free energy, $F[\rho(\mathbf{r})] = F_{ref}[\rho(\mathbf{r})] + F_{pert}[\rho(\mathbf{r})]$, into reference and perturbation parts, where $F_{ref}[\rho(\mathbf{r})]$ is the functional for the reference system [53],[54]. To first order in the perturbation [54], $F^{(1)}_{pert}[\rho(\mathbf{r}')] = \frac{1}{2}N\overline{\rho}\int\limits_0^\infty \overline{g}_{ref}(r;\rho(\mathbf{r}'))\varphi_{pert}(r)d\mathbf{r}$, where $\overline{g}_{ref}(r;\rho(\mathbf{r}))$ is obtained by averaging the pair distribution function of the reference system, $\rho^{(2)}_{ref}(\mathbf{r}_1,\mathbf{r}_2;\rho(\mathbf{r}))$. In the bulk fluid, $\rho(\mathbf{r}) = \overline{\rho}$, $\rho^{(2)}(\mathbf{r}_1,\mathbf{r}_2) = \overline{\rho}^2 g(r_{12})$, and the structural information is produced by the pair correlation function $g(r)$. For the bulk solid, when the density is dominated by localized peaks at the lattice sites, previous calculations [53],[54] *assumed* $\rho^{(2)}(\mathbf{r}_1,\mathbf{r}_2) = \rho(\mathbf{r}_1)\rho(\mathbf{r}_2)g(\mathbf{r}_1,\mathbf{r}_2)$, where the product $\rho(\mathbf{r}_1)\rho(\mathbf{r}_2)$ almost exhausts the structural information; in the case of hard spheres (HS) the main purpose of $g(\mathbf{r}_1,\mathbf{r}_2) = g_{HS}(\mathbf{r}_1,\mathbf{r}_2)$ is to produce the correlation hole of pair exclusion between two spheres of radius R: $g_{HS}(\mathbf{r}_1,\mathbf{r}_2) \cong \theta(r_{12} - 2R)$. This result is obtained *automatically* in the uncorrelated cell model picture, which is the exact limit of the FMFs near any close packing.

The reference system parameters are still at our disposal, and for the hard-sphere reference system the optimal ("effective") hard-sphere radius $R(\rho(\mathbf{r}),T)$ is obtained from the *variation equation*, $\frac{\partial F[\rho(\mathbf{r})]}{\partial R} = \frac{\partial}{\partial R}(F_{HS}[\rho(\mathbf{r})] + F_{pert}[\rho(\mathbf{r})]) = 0$. Contrary to the exact variational principle for the free energy functional with respect to the local density profile, this variation equation represents only one particular method for determining the "best" reference system parameters (denoted as the *variational perturbation theory*), and it does not follow from a variational *principle*. The variational method [55] proved successful for the fluid and the solid [56], and it can be treated analytically in certain important cases [57]. It was shown [57] that the asymptotic high density or strong coupling limit of the integral equation theory for bulk fluids with repulsive pair interactions, coincides with the corresponding limit for the variational perturbation theory with the Percus-Yevick hard-sphere reference correlation functions. This result singles out the *variational perturbation theory* with a hard-sphere reference functional as perhaps more fundamental than other choices for a reference system, which is expected to yield the correct asymptotic high density limit. This choice also has the appealing feature that the entire density profile calculation for the system with soft interactions is made variationally.

5.2. BULK SOLID: LATTICE MADELUNG ENERGY AND THE HARMONIC EXPANSION

Adopting the Gaussian-peaks picture for a densely packed configuration of hard spheres, and using standard expressions for the uncorrelated cell model [55] we find [58] that the perturbation term can be expanded as $F^{(1)}_{pert}/N = u_M(\overline{\rho}) + u_\alpha(\overline{\rho})\alpha^{-1} + ...$, with the Gaussian width $\alpha^{-1/2}$ as the small parameter. Here $u_M(\overline{\rho})$, the static lattice Madelung energy for the particular configuration, and $u_\alpha(\overline{\rho})$, the force ("spring") constant, are functions of the average density, with forms that depend on the interparticle pair potential $\varphi(r)$. This picture can be related to the properties of the hard-sphere system near a situation of close packing only if the hard-sphere functional has a true divergence, which is the case for the FMFs. Indeed, by adopting the same set of density profiles for the reference hard-sphere system, then each value of α corresponds to a value of the packing fraction, and for the FMFs (see above) we have $\alpha \propto \Delta^{-2} \propto \epsilon^{-2}$. We can thus use the Gaussian representation of the hard-spheres (which is assumed known from a separate calculation of the hard-sphere solid) in order to *define* the function $u_\epsilon(\overline{\rho})$ by using the following relation: $u_\alpha(\overline{\rho})\alpha^{-1} = u_\epsilon(\overline{\rho})\epsilon^2$. In order to calculate the function $u_\epsilon(\overline{\rho})$ we need the function $u_\alpha(\overline{\rho})$ from the standard treatment [55], and the relation $\epsilon(\alpha)$ as obtained from the density functional calculation for the hard-sphere solid. Finally, together with the "ideal" contribution to the free energy functional, the variational principle for the density profiles yields (recall that here $\epsilon = \overline{\eta}_c - \overline{\eta}$): $\frac{F}{Nk_BT} = (-D\ln\epsilon \ + D\ln(\lambda/R) \ + const) + \frac{u_M(\overline{\rho})}{k_BT} + \frac{u_\epsilon(\overline{\rho})}{k_BT}\epsilon^2 + \dots$. The variation equation $\frac{\partial F}{\partial R} = 0$ (i.e. $\frac{\partial F}{\partial \epsilon} = 0$) for the optimal ("best") reference parameters gives $\epsilon = \left(\frac{D}{2}k_BT/u_\epsilon(\overline{\rho})\right)^{1/2} + ...$, so that [58] the variational perturbation functional for soft potentials, which is based on a FMF for the reference hard-sphere system, features the harmonic potential energy,

$$\frac{U}{N}(\overline{\rho}, T) = u_M(\overline{\rho}) + \frac{D}{2}k_BT \ + \dots \ , \tag{4}$$

as the leading terms in the asymptotic high density expansion. As obtained above this result is *general* for all D (for $D = 1$ it is exact !) and for all pair potentials, e.g. it applies equally well to the Lennard-Jones potential and to the one component Coulomb plasma. This perturbation theory is expected to be accurate because of the predominantly harmonic behavior of simple classical solids even near melting.

5.3. BULK FLUID: "FLUID MADELUNG ENERGY" AND POWER-LAW HEAT CAPACITY

The FMF for the $D = 3$ hard-sphere fluid features the Percus-Yevick [19] pair correlation function and the scaled-particle [18] equation of state, so that the present variational perturbation theory belongs to a well studied generic analytic form [56],[57]. Following that analysis we found [58] that the first order perturbation theory functional for the bulk fluid can be expanded from the $\eta = 1$ singularity (here $\epsilon = 1-\eta$), $\frac{F}{Nk_BT} = (\frac{D}{2}\epsilon^{-(D-1)} + ...+ D\ln(\lambda/R) + const) + \frac{u_{M,\eta=1}(\overline{\rho})}{k_BT} + \frac{u_{\epsilon,\eta=1}(\overline{\rho})}{k_BT}\epsilon^{\gamma} + ...$, where $u_{M,\eta=1}(\overline{\rho})$ and $u_{\epsilon,\eta=1}(\overline{\rho})$ are analogous to the corresponding terms for the solid, but evaluated in the bulk $\eta = 1$ limit of the functional (which in $D = 3$ is the $\eta = 1$ limit of the Percus-Yevick equation for hard-spheres [57]). The temperature independent leading part of the potential energy, $u_{M,\eta=1}(\overline{\rho})$, can be termed the *fluid Madelung energy* . It is an exact lower bound to the potential energy for repulsive potentials [57]. In $D = 3$ we have $\gamma = 3$, the variation equation $\frac{\partial F}{\partial \epsilon} = 0$ yields $\epsilon = (k_BT/u_{\epsilon,\eta=1}(\overline{\rho}))^{1/5} + ...$, so that [58]

$$\frac{U}{N}(\overline{\rho},T) = u_{M,\eta=1}(\overline{\rho}) + (u_{\epsilon,\eta=1}(\overline{\rho}))^{2/5}(k_BT)^{3/5} + ... \; , \tag{5}$$

and the leading term for the excess heat capacity behaves like $c_V \sim T^{-2/5}$.

It should be noted, however, that contrary to the solid for which the harmonic expansion (4) is a good approximation for general potentials, the situation is different for the fluid. The leading term, after the Madelung term, of the asymptotic expansion for the *fluid*, is expected to provide a good estimate of the thermal energy (over the Madelung, static-lattice, value) near freezing densities only for predominantly repulsive interactions. Moreover, within the class of (e.g.) inverse-power potentials (see below), r^{-n}, the leading thermal energy term is more dominant near freezing the longer the range of the potential (i.e. the smaller is the power n).

5.4. FREEZING AND MELTING "LAWS" AND COMPARISON WITH SIMULATIONS

The variational hard-sphere perturbation theory defines a mapping of the system with soft interaction on the hard-spheres via the effective radius $R(\overline{\rho},T)$. It is well known [1],[2] that by such mappings the hard-sphere fluid-solid transition correlates with the transition for soft interactions through Lindemann-type melting and freezing indicators. For our asymptotic expansions these indicators correspond to the *approximations* that $\epsilon_{melt} = \overline{\eta}_c - \overline{\eta}_{melt} = \left(\frac{3}{2}k_BT/u_\epsilon(\overline{\rho}_{melt})\right)^{1/2}$ and $\epsilon_{freez} = 1 - \eta_{freez} = \left(k_BT/u_{\epsilon,\eta=1}(\overline{\rho}_{freez})\right)^{1/5}$ are "universal" constants, the same for all *repulsive* potentials, along the melting line and along the freezing line. Furthermore, to within an accuracy equal to the relative density change upon freezing (which is very small ($\lesssim 4\%$) for soft repulsive potentials), we can estimate the melting densities simply by equating the solid and fluid free energies. In view of the approximate melting and freezing indicators above, we obtain [58] the following new universal (the same for all *repulsive* potentials) *approximate* melting "law" which involves only (*i*) the Madelung energy for the solid with the particular lattice structure, and (*ii*) the *fluid Madelung energy* (which is also an exact lower bound for the potential energy):

$$\frac{u_M(\overline{\rho}_{melt}) - u_{M,\eta=1}(\overline{\rho}_{melt})}{k_B T} = K_{melt} = 0.7 \tag{6}$$

The value of the "universal" constant K_{melt} is calibrated [58] by its value for the Coulomb one component plasma (OCP) which has exactly zero density change at melting.

The present model free energy functional, with the fundamental-measure hard-sphere reference functional, when applied to bulk fluids gives rise to one particular variant of the variational perturbations theory which was considered in the past [56]. The available simulation data for simple fluids is generally in good agreement [56] with the prediction of the variational perturbation theory for bulk fluids, and for our purpose here we emphasize several specific predictions of this theory, *as generated by the fundamental measure functionals*, which agree very well with the simulations. Consider [58] in particular the repulsive inverse power potentials, $\frac{\varphi(r)}{k_B T} = \frac{\Gamma}{(r/a_{WS})^n}$, with $n \leq 12$, where $a_{WS} = \left(\frac{3}{4\pi\overline{\rho}}\right)^{1/3}$ is the Wigner-Seitz radius, and Γ is the coupling parameter. The classical Coulomb one component plasma (OCP) corresponds to $n = 1$, and consider also the screened Coulomb (Yukawa) potential, $\frac{\varphi(r)}{k_B T} = \frac{\Gamma}{(r/a_{WS})} e^{-\alpha(r/a_{WS})}$, which for $\alpha = 0$ corresponds to $n = 1$. The available simulation data for simple fluids is in good agreement [56] with the prediction of the variational perturbation theory, one variant of which is constituted by the present model. In particular, for the repulsive potentials, when $u_{M,\eta=1}(\overline{\rho})$ is subtracted from the simulation energies they follow very well [58] the predicted $T^{3/5}$ behavior for temperatures up to many times the melting temperature. The melting "law" eq.(6) is well obeyed [58] by all available simulation results for repulsive soft interactions.

5.5. OVERVIEW

With a fundamental-measure hard-sphere reference functional we obtain a *unified analytic description* of classical bulk solids and fluids, predicting correctly major features of their equations of state and freezing parameters as obtained by simulations. Moreover, the fundamentally different fluid and solid asymptotic high density expansions for the potential energy, featuring a static-lattice Madelung term and the harmonic $\frac{3}{2}k_B T$ correction, on one hand, and a fluid Madelung energy with a $\sim T^{3/5}$ thermal energy correction, on the other, *both originate from the same singularity* in the *hard-sphere* free energy functional. In the past it was proposed [57] that the $\eta = 1$ singularity of the Percus-Yevick equation for hard-spheres, in view of its remarkable properties in describing the asymptotic high density limit of the hypernetted-chain equation for soft repulsive potentials, and in providing exact lower bound to their potential energy, plays the role of an "ideal liquid" starting point for describing the dense fluid (liquid) state (in analogy to the "ideal gas" for low density fluids). Indeed, "ideal liquid" properties provided important motivations [4] for deriving the fundamental-measure functionals. Yet, $\eta = 1$ for bulk fluids corresponds to an *unphysical state*, with the spheres filling the volume. The fundamental-measure functionals, which feature a symmetry breaking by which this unphysical state is not actually reached, now provide a realization for a *physically acceptable* free energy model (with many of the desirable properties for the exact functional) with the $\eta = 1$ singularity as an "ideal" starting point for obtaining the high density properties of *both* solids and fluids.

6. Concluding remarks

It appears from all these many investigations described above that, by capturing the correct geometrical features, the fundamental-measure hard-sphere functional and the corresponding bridge functional lead to accurate free energy density functionals for simple classical systems, with accurate description of the structure for a large class of inhomogeneous simple fluids. However, we should not forget the geometrical nature of the hard-sphere interaction, which means that (e.g.) systems with large non-additivity in their repulsions, and/or tendency to form pairs and higher order clusters due to their attractions (or peculiar repulsion), are not expected to be well treated by the methods discussed above. In particular, certain situations for penetrable spheres [59],[43] or the pairing in electrolytes [60], which are beyond the scope of the hypernetted-chain approximation, also cannot be addressed by invoking the hard-sphere bridge functional.

Acknowledgments

The insightful collaboration of Thierry Biben, Bob Evans, Jean-Pierre Hansen, Gerhard Kahl, Hartmut Löwen, Matthias Schmidt and Pedro Tarazona and the interesting correspondence with José Cuesta have been much appreciated.

References

1. R. Evans in, *Fundamentals of Inhomogeneous Fluids*, edited by D. Henderson (Dekker , 1992); H. Löwen, Phys.Rep. **237**, 249(1994).

2. J.A. Barker and D. Henderson, Rev.Mod.Phys. **48**, 587 (1976); J.P. Hansen and I.R. McDonald, "Theory of Simple Liquids", Second Edition, Academic Press, London, (1986).
3. P. Tarazona, Molec.Phys. **52**, 81 (1984); Phys.Rev.A **31**, 2672 (1985).
4. Y. Rosenfeld, Phys.Rev. Letters **63**, 980 (1989); a short review: Y. Rosenfeld, J.Phys.Cond.Matter **8**, 9289 (1996).
5. Y. Rosenfeld, J.Chem.Phys. **98**, 8126 (1993).
6. Y. Rosenfeld, Phys.Rev.E **50**, R3318 (1994); Molec.Phys. **86**, 637 (1995).
7. E. Kierlik and M.L. Rosinberg, Phys.Rev.A **42**, 3382(1990); **44**, 5025 (1991); S.Phan, E.Kierlik, M.L.Rosinberg, B.Bildstein, and G.Kahl , Phys.Rev.E **48**, 618 (1993).
8. J.A. Cuesta, Phys.Rev.Letters **76**, 3742 (1996).
9. Y. Rosenfeld, M.Schmidt, H. Löwen and P. Tarazona, J. Phys.: Cond. Matter **8**, L577 (1996); Phys.Rev.E. **55**, 4245 (1997).
10. Y. Rosenfeld, J.Phys.Cond.Matter **8**, L795 (1996).
11. A. Gonzalez, J.A. White and R.Evans, J.Phys.: Cond. Matter **9**, 2375 (1997).
12. J.A. Cuesta and Y. Martinez-Raton, Phys.Rev.Lett. **78**, 3681 (1997); J.Chem.Phys. **107**, 6379 (1997).
13. P. Tarazona and Y. Rosenfeld, Phys.Rev.E **55**, R4873 (1997).
14. B. Pansu, P. Pieranski, and P. Pieranski, J. Physique **45**, 331 (1984).
15. A. Münster, *Statistical Thermodynamics*, Volume 2, Springer-Verlag, Berlin, 1974, section 16.2, and references therein.
16. M. Schmidt and H. Löwen, Phys.Rev.Lett. **76**, 4522 (1996).
17. M.D. Rintoul and S. Torquato, Phys.Rev.Lett. **77**, 4198 (1996).
18. H. Reiss, J.Phys.Chem. **96**, 4736 (1992).
19. J.K. Percus and G.J. Yevick, Phys.Rev. **110**, 1 (1958); M.S. Wertheim, Phys.Rev.Lett. **10**, 321 (1963).
20. Y. Rosenfeld, J.Chem.Phys. **89**, 4272 (1988).
21. Y. Rosenfeld, D. Levesque, and J.J. Weis, J.Chem.Phys. **92**, 6818 (1990); Y. Rosenfeld, Phys.Rev. A **42**, 5978 (1990).
22. Y. Rosenfeld, J.Chem.Phys. **93**, 4305 (1990).
23. J.K. Percus, J.Stat.Phys. **15**, 505 (1976).
24. A. Gonzalez, J.A. White, F.L. Roman, S. Velasco, and R. Evans, Phys.Rev.Letters. **79**, 2466 (1997); A. Gonzalez, J.A. White, F.L. Roman, and R. Evans, J.Chem.Phys. **109**, 3637 (1998).
25. S-C. Kim and H. Löwen, preprint (1998).
26. R. Ohnesorge, H. Löwen, and H. Wagner, Europhys.Lett. **22**, 245 (1993).
27. J.L. Lebowitz and J.W. Perram, Molec.Phys. **50**, 1207 (1983); P. Harrowell and D. Oxtoby, Mol.Phys. **54**, 1325 (1985).
28. P.G. Bolhuis and H.N.W. Lekkerkerker, Physica A **196**, 375 (1993).
29. A. Chamoux and A. Perera, J.Chem.Phys. **104**, 1493 (1996).
30. P. Kalpaxis and G. Rickayzen, Molec.Phys. **80**, 391 (1993); M. Calleja and G. Rickayzen, Phys.Rev.Letters **74**, 4452 (1995); G. Rickayzen and M. Calleja, Molec.Phys. **90**, 869 (1997).
31. T. Biben and J.P. Hansen, Europhys.Lett. **12**, 347 (1990); Y. Heno and C. Regnout, J.Chem.Phys. **95**, 9204 (1991).
32. R.J. Baxter, J.Chem.Phys. **49**, 2770 (1968).
33. Y. Martinez-Raton and J.A. Cuesta, Phys.Rev.E **58**, R4080 (1998).
34. C. Tutschka and G. Kahl, preprint (1998).
35. L. Blum and Y. Rosenfeld, J.Stat.Phys. **63**, 1177 (1991).
36. Y. Rosenfeld, Phys.Rev.Letters **72**, 3831 (1994); J.Phys.Chem.**99**, 2857 (1995).
37. Y. Rosenfeld, Phys.Rev.E **54**, 2827 (1996).
38. Y. Rosenfeld and N.W. Ashcroft, Phys. Rev. A **20**, 1208 (1979).
39. M. Sweatman, M., July 1995, Ph.D. Thesis, Bristol University ; M. Sweatman and R. Evans, unpublished.

40. G. Kahl, B. Bildstein, and Y. Rosenfeld, Phys.Rev.E **54**, 5391 (1996).
41. Y. Rosenfeld and G. Kahl, J.Phys.Cond.Matter **9**, L89 (1997).
42. Y. Rosenfeld, Molec.Phys. **94**, 929 (1998).
43. M. Watzlawek, C.N. Likos, H. Löwen, and Y. Rosenfeld, to be published (1998).
44. Y. Rosenfeld, Phys.Rev.A **35**, 938 (1987).
45. B. Götzelman, Thesis, University of Wuppertal, Germany (1997); B. Götzelman, R.. Evans, and S. Dietrich, preprint (1998).
46. E. Kozak and S. Sokolowski, J.Chem.Soc.Faraday Trans. **87**, 3415 (1991).
47. E. Kierlik, M.L. Rosinberg, J.E. Finn, and P.A. Monson, Molec.Phys. **75**, 1435 (1992).
48. E. Kierlik, M.L. Rosinberg, Y. Fan, and P.A. Monson, J.Chem.Phys. **101**, 10947 (1994).
49. E. Kierlik, Y. Fan, P.A. Monson, and M.L. Rosinberg, J.Chem.Phys. **102**, 3712 (1995).
50. L.J.D. Frink and F. van Swol, J.Chem.Phys. **105**, 2884 (1996); **106**, 3782 (1997).
51. Y. Rosenfeld, in the ACS Symposium Series No. 629, *Chemical Applications of Density-Functional Theory*, Edited by B.B. Laird, R.B. Ross, and T. Ziegler, ACS publishing 1996, Chapter 14, pages 198-211.
52. T. Biben, J.P. Hansen, and Y. Rosenfeld, Phys.Rev.E **57**, R3727 (1998).
53. J.F. Lutsko and M. Baus, J.Phys.:Cond.Matter **3**, 6547 (1991).
54. C. Rascon, L. Mederos, and G. Nevascues, Phys.Rev.Lett. **77**, 2249 (1996); C. Rascon, G. Nevascues, and L. Mederos, Phys.Rev.E **53**, 5698 (1996).
55. G.A. Mansoori and F.B. Canfield, J.Chem.Phys. **51**, 4958, 4967 (1969); H.S. Kang, T. Ree, and F. Ree, J.Chem.Phys. **84**, 4547 (1986).
56. Y. Rosenfeld, Phys.Rev.A **29**, 2877 (1984); **28**, 3063 (1984); J.Chem.Phys. **73**, 5753, 5760 (1980).
57. Y. Rosenfeld, Phys.Rev.A **32**, 1834 (1985); *ibid* **33**, 2025 (1986).
58. Y. Rosenfeld and P. Tarazona, Molec.Phys. **95**, 141 (1998).
59. C.N. Likos, M. Watzlawek, and H. L"owen, Phys.Rev.E **58**, 3135 (1998).
60. D.M. Zuckerman, M.E. Fisher, and B.P. Lee, Phys.Rev.E **56**, 6569 (1997).

CHARGE-STABILIZED COLLOIDS CONFINED TO A SLIT: A CASE FOR ATTRACTION BETWEEN LIKE CHARGED PARTICLES

DAVID GOULDING AND JEAN-PIERRE HANSEN
Department of Chemistry
University of Cambridge, Lensfield Road, Cambridge, CB2 1EW (UK)

1. Introduction

Mineral or synthetic colloidal particles with diameters in the range $10^2 \leq d \leq 10^3 nm$, mutually attract through strong Hamaker-van der Waals interactions, which would lead to irreversible flocculation if the particles were neutral. However when dispersed in water they acquire surface charges due to the dissociation of ionizable radicals, amounting typically to a total effective valence $\mid Z \mid \simeq 10^4$. The released counterions as well as co and counterions from added salt, form electric double-layers near the colloid surfaces, with a width of the order of the Debye screening length:

$$\lambda = \left[\frac{4\pi(n_+z_+^2 + n_-z_-^2)e^2}{\epsilon k_B T}\right] \tag{1}$$

where z_+, z_- are the valences of microscopic cations and anions, n_+, n_- their number densities, and ϵ is the dielectric constant of the solvent (water), assumed to be a mere continuum. Except for at very high salt concentration the electrostatic repulsion between double-layers associated with neighbouring colloidal particles (polyions) will completely dominate the van der Waals attractive forces, thus stabilizing the dispersion. It is now generally accepted that in the bulk, i.e. for pairs of polyions far removed from any container surface, the effective electrostatic interaction between polyions is well descibed by the pair-wise additive, screened Coulomb potential proposed independently by Derjaguin and Landau [1], and by Verwey and Overbeek (DLVO)[2]:

$$v_{eff}(r) = (Z'e)^2 \frac{e^{-\kappa r}}{\epsilon r} \tag{2}$$

C. Caccamo et al. (eds.), *New Approaches to Problems in Liquid State Theory*, 321–336.

where r is the distance between polyions, $\kappa = 1/\lambda$ and Z' is an effective valence, renormalized to account for polyion-microion excluded volume, which within DLVO theory is:

$$Z'^2 = Z^2 \frac{e^{\kappa d}}{(1 + \kappa d/2)^2} \tag{3}$$

Despite the rather drastic approximations made to arrive at the DLVO potential (2), which involve neglect of microion correlations, and a Debye-Hückel - like linearization of the Poisson - Boltzmann (PB) equation, the general validity of the pair potential has been recently confirmed, both by direct experimental measurement [3, 4] and by first principles simulations [5]. The main uncertainty in the use of the DLVO potential lies in the choice of the effective charge Z, to account for the tightly bound counter-ions of the "Stern layer", which cannot be treated within a linearized PB theory [6]. For practical purposes, Z' should be regarded as an adjustable parameter. It is worth pointing out that the experimental observation of coexisting, colloidal microphases of very low and of high polyion concentration [7] does not necessarily imply attractive forces between polyions in the bulk of the dispersion, but can be explained on the basis of the purely repulsive DLVO potential, provided a state-dependant volume term in the free energy [8], which had been previously overlooked, is properly taken into account [9].

In view of the success of DLVO theory to describe bulk behaviour of charge-stabilised colloidal dispersions, it came as a surprise that recent experiments yielded clear evidence for a significant attractive force acting between two colloidal particles confined near a surface or in a slit [10, 11]. Additional evidence for an effective attraction between confined, like-charged colloidal particles is provided by numerical solutions of the non-linear PB equation for two polyions in a charged cylindrical pore [12].

The objective of this paper is a generalisation of DLVO theory to deal with the effective interaction between confined like charged-colloidal particles. The geometry which will be explicitly considered is that of a slit, while the limiting case of polyions near a single planar wall has already been considered elsewhere [13, 14].

2. Slit geometry

The geometry of the problem is summarized in Fig. 1. The colloidal dispersion is confined between two parallel planes situated at $z = \pm L$. The planes separate the suspension, of dielectric constant ϵ, from a dielectric medium (ϵ'), representing, e.g. glass plates, which carry a surface charge density σe, while the spherical polyions, carry a (positive) charge Ze, uniformly distributed over their surface. The local densities of the co and counterions

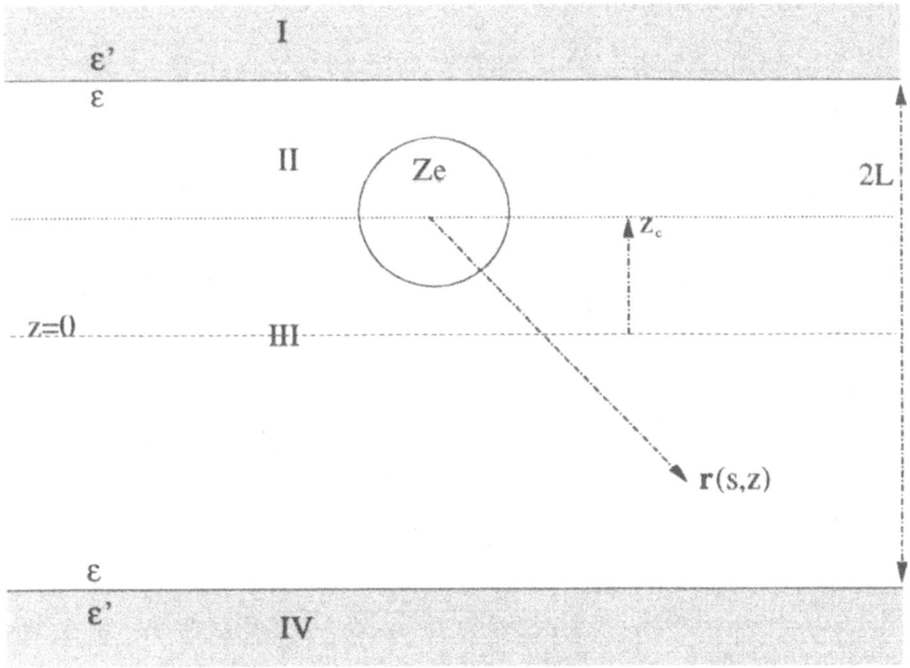

Figure 1. Geometry of slit problem showing a particle in a slit a height z_c from the midplane and a dielectric discontinuity inside and outside the slit (the walls are coloured grey). Also shown are the four regions the potential must be split up into to solve the LPB problem.

will be denoted by $\rho_+(\mathbf{r})$ and $\rho_-(\mathbf{r})$; they are assumed to be monovalent point ions which cannot penetrate the polyions, leading to the constraints:

$$\rho_\pm(\mathbf{r}) = 0; \quad |\, \mathbf{r} - \mathbf{R}_i \,| < R, \quad 1 \leq i \leq N \tag{4}$$

where $\mathbf{R}_i$ is the position of the centre of any one of the N polyions per unit area of the slit. In addition, confinement implies:

$$\rho_\pm(\mathbf{r}) = 0; \quad |\, z \,| > L \tag{5}$$

If correlations between microions are neglected, as in PB theory, the equilibrium density profiles for any given polyion configuration $\{\mathbf{R}_i\}$ are those which minimize the following free energy functional subject to the constraints (4) and (5);

$$F[\rho_+, \rho_-] = F_{id}[\rho_+, \rho_-] + F_{el}[\rho_+, \rho_-] \tag{6}$$

where the ideal contribution is:

$$F_{id}[\rho_+, \rho_-] = \sum_{\alpha=+,-} k_B T \int_{\wp} \rho_\alpha(\mathbf{r}) \Big\{ \log(\Lambda_\alpha^3 \rho_\alpha(\mathbf{r})) - 1 \Big\} d\mathbf{r} \tag{7}$$

while the electrostatic part reads:

$$F_{el} = \frac{1}{2} \int_{\wp} \rho_c(\mathbf{r}) \Psi(\mathbf{r}) d\mathbf{r} \tag{8}$$

In these equations $\wp$ is the domain to which the dispersion is confined, $\rho_c(\mathbf{r})$ is the total charge density:

$$\rho_c(\mathbf{r}) = \rho_{ext}(\mathbf{r}) + e[\rho_+(\mathbf{r}) - \rho_-(\mathbf{r})] \tag{9}$$

and $\Psi(\mathbf{r})$ is the local electrostatic potential, which is related to $\rho_c(\mathbf{r})$ by Poisson's equation:

$$\nabla^2 \Psi(\mathbf{r}) = -\frac{4\pi}{\epsilon} \rho_c(\mathbf{r}) \tag{10}$$

The latter must be solved subject to appropiate boundary conditions on the surface bounding $\wp$. The "external" charge density in eq.(9) is the sum of charge densities associated with each of the N polyions:

$$\rho_{ext}(\mathbf{r}) = \sum_{i=1}^{N} \rho_{ext}^{(i)}(\mathbf{r}) \tag{11}$$

In the spirit of DLVO theory, it is now assumed that the profiles $\rho_\alpha(\mathbf{r})$ are slowly varying, so that the ideal contribution F_{id} may be expanded in a functional Taylor expansion around the mean (macroscopic) densities n_α ($\alpha = +, -$):

$$\begin{aligned} F_{id}[\rho_+, \rho_-] = \quad & F_{id}(n_+, n_-) + k_B T \sum_\alpha \int_{\wp} \Delta\rho_\alpha(\mathbf{r}) \Big[\log(\Lambda_\alpha^3 n_\alpha) + \Delta\rho_\alpha(\mathbf{r}) \Big] d\mathbf{r} \\ & - k_B T \sum_\alpha \int_{\wp} \frac{\Delta\rho_\alpha^3(\mathbf{r})}{6 n_\alpha^2} d\mathbf{r} + O(\Delta\rho_\alpha^4) \end{aligned} \tag{12}$$

where $\Delta\rho_\alpha(\mathbf{r}) = \rho_\alpha(\mathbf{r}) - n_\alpha$, and $F_{id}(n_+, n_-)$ is the free energy of a homogeneous ideal mixture of non-interacting microions. If the expansion is truncated after second order, which amounts to linearisation of PB theory, the equilibrium density profiles (which minimize the free energy functional), and the resulting electrostatic potential reduce to sums of terms ("orbitals") centred on each of the N polyions [5]:

$$\Delta\rho_\alpha(\mathbf{r}) = \sum_{i=1}^{N} \rho_\alpha^{(i)}(\mathbf{r} - \mathbf{R}_i) \tag{13}$$

Upon substituting these linear superpositions into the quadratic functional, and using Poisson's equation, it is easily verified that, apart from a structure-independant volume term [8, 9], the total potential energy of the "dressed" polyions:

$$V_N(\{\mathbf{R}_i\}) = \sum_{i<j}\sum \frac{Z^2e^2}{\epsilon R_{ij}} + F[\rho_+, \rho_-] \tag{14}$$

reduces to a sum over all pairs of an effective pair potential:

$$v_{eff}(\mathbf{R}_i, \mathbf{R}_j) = e\int_{\wp} \rho_{ext}^{(i)}(\mathbf{r})\Psi^{(j)}(\mathbf{r})d\mathbf{r} \tag{15}$$

where $\Psi^{(j)}(\mathbf{r})$ is the electrostatic potential due to polyion j and its associated "orbital" $(\rho_+^{(j)} - \rho_-^{(j)})$.

In the bulk, i.e. the limit $L \to \infty$, this leads directly back to the DLVO potential (2) [5]. In the following sections the modifications brought about by confinement to a slit are analysed.

3. Uncharged Walls

First consider the case where the confining walls do not carry any electric charge, $(\sigma = 0)$. The main complication arises from the dielectric discontinuities at $z = \pm L$ which result in boundary conditions to be imposed on the electrostatic potential. The problem may be solved by taking the following steps:

(a) Since the solution of the linearised problem (corresponding to the quadratic functional introduced in the previous section) is a superposition of densities or potentials associated with individual polyions, it is sufficient to calculate the potential distribution around a single polyion satisfying the appropiate boundary conditions on the surfaces bounding the slit. This is achieved for a **point** polyion, by a generalisation of the elegant method devised by Stillinger for one polyion near a single planar surface [15].

(b) The resulting local densities $\rho_\alpha(\mathbf{r})$ are then renormalized, to account for the constraints (4) arising from the finite size of the polyion.

(c) The effective pair potential between two finite polyions is finally calculated by substituting the result in eq. (15).

To implement step (a), the domain $\wp$ of area A in the xy-plane, and of volume 2LA is periodically repeated in the x and y directions. The "external" charge of the point polyion situated at $(0, 0, z_c)$ is expanded in a Fourier series in the plane $z = z_c$. The potential $\Psi_{\mathbf{k}}(\mathbf{r})$ due to each of these Fourier components is calculated by solving the linearized PB equation in each of the four regions indicated in Fig.1 , and matching the solution on the boundary planes $z = -L$, $z = z_c$ and $z = L$. Finally the cell area A is taken to infinity, allowing the Fourier series to be replaced by Fourier integrals, which are evaluated in cylindrical coordinates. Details are given in Appendix A. The final form of the electrostatic potential around the polyion (in regions II and III) is :

$$\Psi(s, z; z_c) = 2\pi \int_0^\infty f(k \mid z, z_c) J_0(k_s s) k dk \tag{16}$$

where the function f is given explicitly by eq.(30), and J_0 is the zeroth order cylindrical Bessel function.

The potential Ψ, and the resulting local densities have cylindrical rather than spherical symmetry. For the sake of simplicity, the excluded volume constraints (4) are enforced by neglecting the anisotropy of the co and counterion distributions, which is a reasonable approximation as long as the polyion is not too close to one of the boundary planes. In other words the renormalized effective polyion valence is still given by eq.(3), which is, strictly speaking, valid only in the bulk (i.e. in the limit $L \to \infty$). Shifting the origin vertically to $-L$ (i.e. substituting $z_c' = z_c + L$) in the expression for the potential in Appendix A, and then lettting $L \to \infty$, one recovers Stillinger's result for a single plane [15].

An asymptotic analysis of the expression for the potential shows that for fixed z_c and z, it decreases like $1/s^3$ in the direction parallel to the planes bounding the slit. For $z = z_c = 0$, the explicit asymptotic form is:

$$\lim_{s \to \infty} \Psi(s, 0, 0) = \frac{4Z' e \epsilon_r e^{-2\kappa L}}{\epsilon \kappa^2 (1 - e^{-2\kappa L})^2 s^3} \tag{17}$$

The slow $1/s^3$ decay, i.e. the absence of exponential screening due to confinement is qualitatively similar to that predicted near a single plane [13]. The resulting effective pair potential between two polyions is finally obtained from eq.(15), and examples are shown in Fig. 2 and 3 for two different slit widths, and compared to the corresponding effective pair potential in the bulk ($L \to \infty$) or near a single wall. The confinement is seen to significantly enhance the repulsion between polyions compared to both bulk and single wall situations. For sufficient separation between the slit walls, the contributions of the two walls are essentially additive. At long range, the effective pair potential reflects the slow $1/s^3$ decay of the potential Ψ around a polyion (eq.(17)).

4. Charged Walls

Next consider the case where the two walls of the slit carry a uniform surface charge σ. We are interested in the effect of these surface charges, and their associated electric double layers, on the effective pair potential

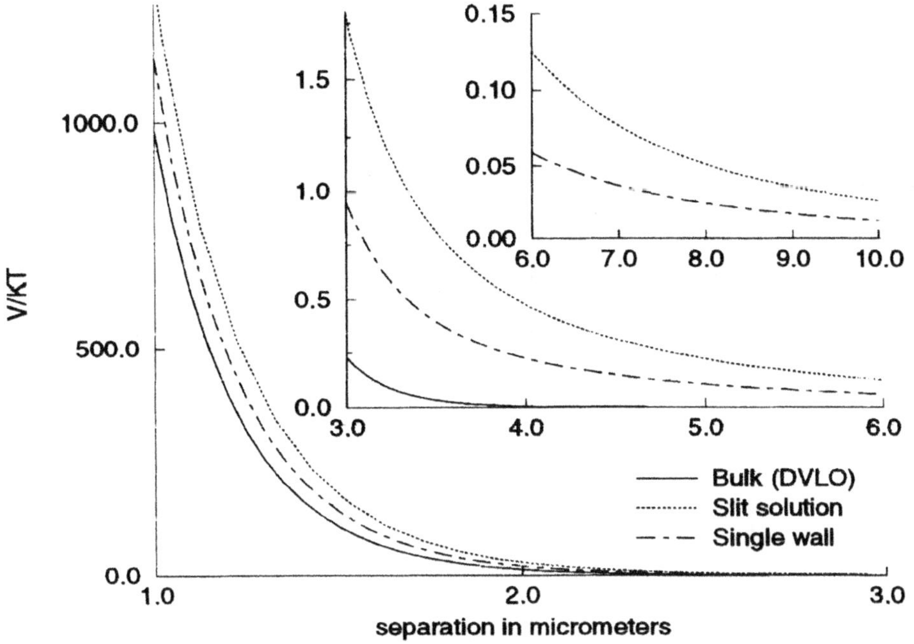

Figure 2. Dimensionless linear effective pair potential, $v_{eff}(z; R)/k_B T$ versus distance R between polyions situated in the middle of a silt of width 1 μm. The screening length is, $\lambda = 1/\kappa = 0.275\mu m$ and the polyions have bare valence of Z=5000. $T = 298K$; $\epsilon = 80$ and $\epsilon' = 6.5$. The full line is the effective pair potential in the bulk ($z_c = \infty$); the dotted curve is the result of the present calculation based on the slit solution solution (eq.30), while the dot-dashed curve is Stillinger's solution for a particle 0.5 μm from a single wall.

between polyions. This requires a further extension of DLVO theory. With the simple quadratic free energy functional, embodied in eqs.(6), (8) and (12) (limited to 2nd order $\Delta\rho_\alpha$), the optimal co and counterion densities $\rho_\pm(\mathbf{r})$ will then be of the form

$$\rho_\alpha(\mathbf{r}) = n_\alpha + \sum_{i=1}^{N} \rho_\alpha^{(i)}(\mathbf{r} - \mathbf{R}_i) + \rho_\alpha^{(w1)}(z) + \rho_\alpha^{(w2)}(z); \qquad \alpha = +, - \tag{18}$$

where the "orbitals" $\rho_\alpha^{(i)}$ around the polyions remain unchanged (except for an adjustment of κ to allow for the counterions released by the walls), and $\rho_\alpha^{(w1)}$, $\rho_\alpha^{(w1)}$ are the microion densities near the lower and upper walls, which depend only on the vertical coordinate z. Only the combinations $\rho_+^{(w)} - \rho_-^{(w)}$ will be needed below:

$$\rho^{(w1)} = [\rho_+^{(w1)}(z) - \rho_-^{(w1)}(z)] = \frac{\sigma\kappa}{1 - e^{-2\kappa L}} e^{-\kappa(L+z)} \tag{19-a}$$

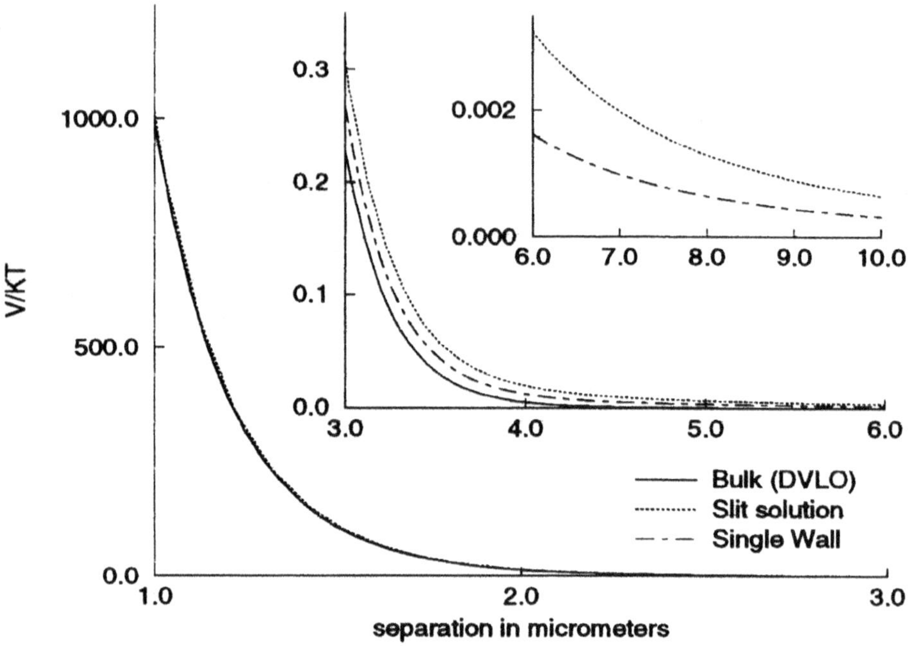

Figure 3. same as fig.2 but the slit width is $2\mu m$ and the polyions are $1\mu m$ from the wall.

$$\rho^{(w2)} = [\rho_+^{(w2)}(z) - \rho_-^{(w2)}(z)] = \frac{\sigma\kappa}{1 - e^{-2\kappa L}} e^{-\kappa(L-z)} \tag{19-b}$$

Within the quadratic functional, these additional densities will have no effect on the effective pair potential between the two polyions, a familiar shortcoming of linearised theories. To obtain the required interference effect to lowest approximation, the expansion of the ideal term of the free energy functional must be pushed one order further, i.e. the cubic term on the r.h.s. of eq.(12) must be included. This cubic term is treated as a perturbation, i.e. evaluated with the density profiles (18) from linearized theory (corresponding to the quadratic functional). Substituting (18) into the cubic term, the only term which contributes to the effective pair interaction between polyions is:

$$\Delta v_{eff}(\mathbf{R}_1, \mathbf{R}_2) = F_{pert}$$
$$= \frac{-k_B T(n_+ - n_-)}{(n_+ + n_-)^3} \int_{\wp} \rho^{(1)}(\mathbf{r} - \mathbf{R}_1)\rho^{(1)}(\mathbf{r} - \mathbf{R}_2)[\rho^{(w1)}(z) + \rho^{(w2)}(z)]d\mathbf{r} \tag{20}$$

where $\rho^{(i)}(\mathbf{r}-\mathbf{R}_i) = \rho_+^{(i)}(\mathbf{r}-\mathbf{R}_i) - \rho_-^{(i)}(\mathbf{r}-\mathbf{R}_i)$.

Using the results of section 3 for uncharged walls, as well as the wall double-layer profiles (19), Δv_{eff} may be cast in the form (with $\mathbf{R}_i = (\mathbf{s}_i, z_i)$):

$$\begin{aligned}\Delta v_{eff}(\mathbf{R}_1,\mathbf{R}_2) &= \Delta v_{eff}(\mathbf{s}_{12}, z_1, z_2) \\ &= -\frac{(n_+ - n_-)\kappa^5 Z'^2 \sigma e^{-\kappa L} k_B T}{8\pi(n_+ + n_-)^3(1-e^{-2\kappa L})} \int_0^\infty [\zeta_1(k \mid z_1, z_2) \\ &+ \zeta_2(k \mid z_1, z_2) + \zeta_3(k \mid z_1, z_2)] \frac{k J_0(k s_{12})}{k^2+\kappa^2} dk \qquad (21)\end{aligned}$$

where the functions $\zeta_n(k \mid z_1, z_2)$ are given in appendix B, and $s_{12} = \mid \mathbf{s}_1 - \mathbf{s}_2 \mid$ is the projection of the distance between the two polyions in the xy plane.

The contribution (21) to the effective interaction between polyions is attractive. Its assymptotic form is:

$$\begin{aligned}\lim_{s\to\infty} \Delta v_{eff}(s) &= -\frac{4(n_+ - n_-)e^{-2\kappa L}\sigma\epsilon_r k_B T \kappa Z'^2}{3\pi(1-e^{-2\kappa L})^2(n_+ + n_-)^3} \\ & \left[1 + \frac{e^{-\kappa L}}{(1-e^{-2\kappa L})}\right]\frac{1}{s^3} \qquad (22)\end{aligned}$$

The ratio of the asymptotic form of the perturbation over that of the unperturbed effective potential derived from eq.(17) is:

$$\begin{aligned}\lim_{s\to\infty} \frac{\Delta v_{eff}(s)}{v_{eff}(s)} &= -\frac{4\kappa\sigma(n_+ - n_-)}{3(n_+ + n_-)^2}\left(1 + \frac{e^{-\kappa L}}{1-e^{-2\kappa L}}\right) \\ &\simeq \frac{-4L\kappa(n_+ - n_-)^2}{3(n_- + n_+)^2} \qquad (23)\end{aligned}$$

where the charge neutrality constraint $(n_+ - n_-) = (\sigma/L)$, valid in the limit of infinite dilution of the polyions, compatible with the limit $s \to \infty$, was used. The total effective potential is hence attractive at long range (though this says nothing as to the depth of the attraction), provided:

$$4\kappa L(n_+ - n_-)^2 > 3(n_+ + n_-)^2 \qquad (24)$$

The concentrations of the co and counterions present can be extracted by considering the number of counterions required to ensure charge neutrality and then adding this to the concentration from any added salt. For an isolated pair of particles in a narrow slit (2L<3 μm) it is clear for even small wall charges this can lead to large concentrations of counterions

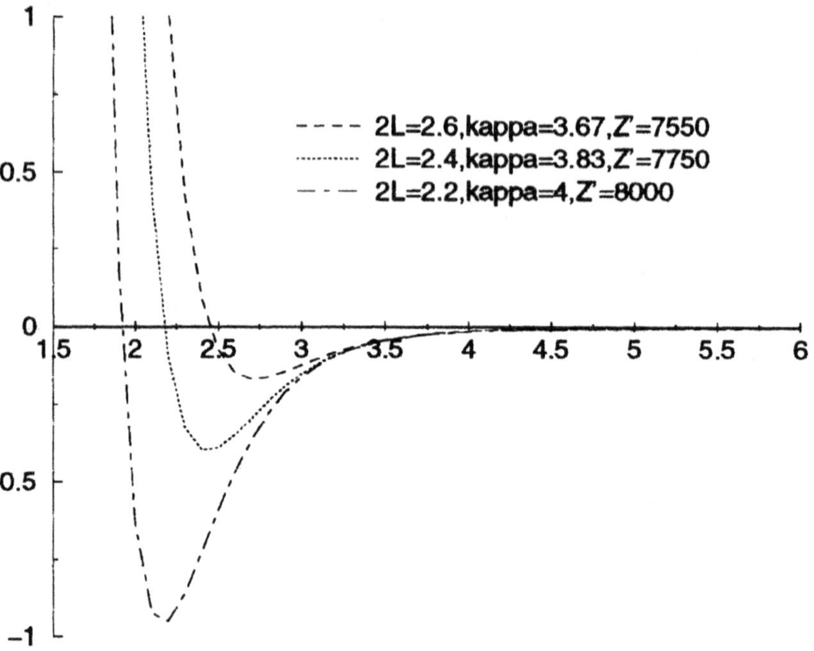

Figure 4. **Dimensionless total effective pair potential, $v_{eff}(z;R)/k_BT+\Delta v_{eff}(z;R)/k_BT$ versus distance R between polyions situated in the middle of a slit of width 2L (all distances are in μm) above the confining surface. The polyions have a bare valence of Z=5000 e, the surface carry a charge of 2000 e μm^{-2}. $T = 298K$; $\epsilon = 80$ and $\epsilon' = 6.5$.**

which, when chemical equilibrium of the species is taken into account, will likely lead to the lowering of the co-ion concentration to effectively zero. The counterion concentration will then just be the wall charge divided by half the slit width.

At intermediate distances v_{eff} and Δv_{eff} must be calculated numerically. Results for their sum are shown in Fig. 4 for a pair of polyions situated in mid-plane ($z_1 = z_2 = 0$). A significant attractive well, of depth $\sim k_BT$, is seen to develop for sufficiently narrow slits (typically for $2L \leq 2.6\mu m$ in the examples shown), at distances ($s \simeq 2.2 - 2.6\mu m$ are comparable to those measured in experiments carried out under similar physical conditions [11]. As expected, the attraction becomes negligible as the slit width $2L$ is increased. (beyond about 3 μm under the conditions explored in this work), again in semi-quantitative agreement with experiment.

The expressions given in the appendices simplify greatly in the limit where the upper wall is removed to infinity, and reduce to the formulae given in references [13] and [14].

5. Discussion and perspectives

The main conclusion from the present Density Functional Theory (DFT) analysis is that strong confinement of charged colloidal particles has a very significant influence on the effective interaction between the polyions. When a quadratic free energy functional is used, which corresponds to the DLVO level of approximation in the bulk, confinement by uncharged planar walls, which reduces the suspension to a quasi-2d system, leads to a significant enhancement of the repulsion, compared to the bulk. At large interparticle distances, the exponential screening is replaced by an inverse power law decay in agreement with general Statistical Mechanics considerations of Coulombic systems at interfaces [16].

The case where the confining surfaces are charged requires consideration of non-linear corrections beyond DLVO theory. Lowest order DFT peturbation theory [5] has been used here to include the interference effect of the electric double-layers near the charged walls. This is similar in spirit to recent attempts to account for three-body interactions between polyions in the bulk [17] (in the present case the charged walls provide the third body).

We are aware of the limitations of the perturbative approach. In particular the theory is unreliable for very short interparticle distances, because non-linear and correlation effects between microions are expected to play an increasingly important role as particles move closer together, particularly for smaller polyions, as illustrated by recent simulations [18]. The short-range attraction observed in that case, due to Coulomb depletion is, however , of a very different nature from the intermediate range attraction predicted for microion sized colloidal particles in the present work. The tunable long-range attraction, and the structure-independent volume terms in the free energy may lead to very interesting phase behaviour of these quasi-2d colloidal systems, which is being investigated.

We are also considering the case of a quasi-1d confinement of charge-stabilised colloidal particles to a cylindrical pore, carrying a surface charge. Numerical solutions of the full non-linear PB equation for this geometry are available [12], albeit with different boundary conditions; they point to the existence of an attractive force at intermediate distances. Our preliminary calculations, along lines similar to those sketched in this paper for a slit geometry, indicate that the attractive component is strongly reduced in the cylindrical geometry; stronger confinement , i.e. a cylinder radius comparable to the particle radius, is required to observe a sizeable effect.

References

1. B.V. Derjaguin and L.D. Landau, Acta Physicochim URSS **14**, 633 (1941).

2. E.J.W. Verwey and J.Th.G. Overbeek, "Theory of the Stability of Lyophobic Colloids",(Elsevier, Amsterdam, 1948).
3. J.C. Crocker and D.G. Grier, Phys. Rev. Letters **73**, 352 (1994).
4. K. Vondermasson, J. Bongers, A. Mueller and H.Versmold, Langmuir **10**, 1351 (1994).
5. H.Löwen, J.P. Hansen and P.A. Madden, J. Chem. Phys. **98**, 3275 (1993).
6. S. Alexander, P.M. Chaikin, P. Grant, S. J. Morales, P. Pincus and D. Hone, J. Chem. Phys. **80**, 5776 (1984)
7. K. Ito, H. Yoshida and N. Ise, Science **263**, 66(1994); B.V.R. Tata, E. Yamahara, P.V. Rajaniam and N. Ise, Phys. Rev. Letters **78**, 2660 (1997).
8. M.J. Grimson and M.Silbert, Mol. Phys. **74**, 397(1991).
9. R. van Roij and J.P. Hansen, Phys. Rev. Letters **79**, 3082(1997).
10. G.M. Kepler and S. Fraden, Phys. Rev. Letters **73**, 356(1994).
11. J.C. Crocker and D.G. Grier, Phys. Rev . Letters **77**, 1897 (1996; A.E. Larsen and D.G. Grier, Nature **385** 230 (1997).
12. W. R. Bowen and A. O. Sharif, Nature **393**, 663 (1998).
13. D. Goulding and J.P. Hansen, Mol. Phys., in the press (1998), see "publish.aps.org/eprint/gateway/eplist/aps1998may08_002" for an e-print.
14. D. Goulding and J.P. Hansen, to be published, see "publish.aps.org/eprint/gateway/eplist/aps1998may08_002" for an e-print.
15. F.H. Stillinger, J.Chem.Phys. **35**, 1584 (1961).
16. B. Jancovici, J. Stat. Phys. **28**, 43 (1982) and **29**, 263 (1982); M. Baus, Molec .Phys **48**, 347 (1983).
17. H. Löwen and E.Allayharov, J.Phys.Cond.Matter **10**, 4147 (1998).
18. E.Allayharov, I.D'Amico and H. Löwen, Phys.Rev.Letters **81**, 1334 (1998).

A. Electrostatic potential around a single point polyion in an infinite slit

The slit is divided into 4 horizontal regions indicated in Fig.1 . The relative dielectric constant of the two media is denoted by $\epsilon_r = \epsilon'/\epsilon$. In practical calculations the following values were adopted: $\epsilon' = 6.5$ (glass) and $\epsilon = 80$ (water). In regions I and IV the electrostatic potentials Ψ^I and Ψ^{IV} satisfy Laplace's equation;

$$\begin{aligned} \nabla^2\Psi^I(\mathbf{r}) &= 0 \\ \nabla^2\Psi^{IV}(\mathbf{r}) &= 0 \end{aligned} \tag{25}$$

while in regions II and III the electrostatic potentials satisfy the PB equation which will be approximated here by its linear form :

$$\begin{aligned} \nabla^2\Psi^{II}(\mathbf{r}) &= \kappa^2\Psi^{II} \\ \nabla^2\Psi^{III}(\mathbf{r}) &= \kappa^2\Psi^{III} \end{aligned} \tag{26}$$

Since the polyion point charge distribution is expanded in a Fourier series spanning a periodically repeated square area A in the plane $z = z_c$, the linearity of the above equations allows us to seek a solution in the form of a similar Fourier series. The solutions to eqs. (25) and (26) can thus be cast in the form:

$$\Psi^i(\mathbf{r}) = \sum_{\mathbf{k}} \Psi^i_{\mathbf{k}}(\mathbf{r}); \quad I \le i \le IV \tag{27}$$

where $\mathbf{k} = (k_x, k_y) = (n_x, n_y)2\pi/A^{1/2}$ is a reciprocal vector in the xy plane; with $\mathbf{r} = (\mathbf{s}, z)$, the components in the four regions read:

$$\begin{aligned}
\Psi^I_k(\mathbf{r}) &= B^I_k \exp\left(i\mathbf{k}.\mathbf{s} - \mid \mathbf{k} \mid z\right) \\
\Psi^{IV}_k(\mathbf{r}) &= B^{IV}_k \exp\left(i\mathbf{k}.\mathbf{s} + \mid \mathbf{k} \mid z\right) \\
\Psi^{II}_k(\mathbf{r}) &= B^{II(+)}_k \exp\left(i\mathbf{k}.\mathbf{s} + (\kappa^2 + k^2)^{\frac{1}{2}}(L - z)\right) \\
&+ B^{II(-)}_k \exp\left(i\mathbf{k}.\mathbf{s} + (\kappa^2 + k^2)^{\frac{1}{2}}(z - L)\right) \\
\Psi^{III}_k(\mathbf{r}) &= B^{III(+)}_k \exp\left(i\mathbf{k}.\mathbf{s} + (\kappa^2 + k^2)^{\frac{1}{2}}(L + z)\right) \\
&+ B^{III(-)}_k \exp\left(i\mathbf{k}.\mathbf{s} - (\kappa^2 + k^2)^{\frac{1}{2}}(z + L)\right)
\end{aligned} \tag{28}$$

The 6 coefficients B^i_k are determined by the following electrostatic boundary conditions: continuity of Ψ at every interface; continuity of the component of the electric displacement normal to the interface; $D_z = -\epsilon_i(\nabla\Psi^i).\hat{\mathbf{z}}$, at the I/II and III/IV interfaces; discontinuity of D_z across the II/III interface, by an amount $4\pi Ze\ e^{i\mathbf{k}.\mathbf{s}}$ in the positive z direction. These boundary conditions lead to 6 simultaneous equations in the B^i_k's:

$$\begin{aligned}
&B^I_k = B^{II(+)}_k + B^{II(-)}_k \\
&B^{IV}_k = B^{III(+)}_k + B^{III(-)}_k \\
&B^{II(+)}_k \exp((\kappa^2 + k^2)^{\frac{1}{2}}(L - z_c)) + B^{II(-)}_k \exp((\kappa^2 + k^2)^{\frac{1}{2}}(z_c - L)) = \\
&B^{III(+)}_k \exp((\kappa^2 + k^2)^{\frac{1}{2}}(L + z_c)) + B^{III(-)}_k \exp(-(\kappa^2 + k^2)^{\frac{1}{2}}(L + z_c))
\end{aligned}$$

$$\begin{aligned}
&-\epsilon' B^I_k \mid k \mid = \epsilon(\kappa^2 + k^2)^{\frac{1}{2}} \left\{B^{II(-)}_k - B^{II(+)}_k\right\} \\
&\epsilon' B^{IV}_k \mid k \mid = \epsilon(\kappa^2 + k^2)^{\frac{1}{2}} \left\{B^{III(+)}_k - B^{III(-)}_k\right\}
\end{aligned}$$

$$\begin{aligned}
&B^{II(+)}_k \exp((\kappa^2 + k^2)^{\frac{1}{2}}(L - z_c)) - B^{II(-)}_k \exp((\kappa^2 + k^2)^{\frac{1}{2}}(z_c - L)) = \\
&B^{III(-)}_k \exp(-(\kappa^2 + k^2)^{\frac{1}{2}}(z_c + L)) - B^{III(+)}_k exp((\kappa^2 + k^2)^{\frac{1}{2}}(L + z_c)) \\
&+ \frac{4\pi Ze}{\epsilon(\kappa^2 + k^2)^{\frac{1}{2}}}
\end{aligned} \tag{29}$$

The resulting expressions for the potential in the 4 regions are:

$$\Psi_k^I(s,z,z_c) = \frac{4\pi Ze\ e^{i\mathbf{k}.\mathbf{s}}}{\epsilon}$$

$$\frac{e^{-(\kappa^2+k^2)^{\frac{1}{2}}(L-z_c)-|k|z}}{\left(1-e^{-2(\kappa^2+k^2)^{\frac{1}{2}}}\right)(\kappa^2+k^2)^{\frac{1}{2}}+\epsilon_r|k|\left(1+e^{-2(\kappa^2+k^2)^{\frac{1}{2}}L}\right)}$$

$$\frac{\left(1+e^{2(\kappa^2+k^2)^{\frac{1}{2}}(L+z_c)}\right)(\kappa^2+k^2)^{\frac{1}{2}}+\epsilon_r|k|\left(1-e^{2(\kappa^2+k^2)^{\frac{1}{2}}(L+z_c)}\right)}{\left(1+e^{-2(\kappa^2+k^2)^{\frac{1}{2}}L}\right)(\kappa^2+k^2)^{\frac{1}{2}}+\epsilon_r|k|\left(1-e^{-2(\kappa^2+k^2)^{\frac{1}{2}}L}\right)}$$

$$\Psi_k^{II}(s,z,z_c) = \frac{2\pi Ze\ e^{i\mathbf{k}.\mathbf{s}}}{\epsilon(\kappa^2+k^2)^{\frac{1}{2}}}$$

$$\left\{\left[\frac{\left((\kappa^2+k^2)^{\frac{1}{2}}+\epsilon_r|k|\right)e^{-(\kappa^2+k^2)^{\frac{1}{2}}(z-z_c)}}{\left(1+e^{-2(\kappa^2+k^2)^{\frac{1}{2}}L}\right)(\kappa^2+k^2)^{\frac{1}{2}}+\epsilon_r|k|\left(1-e^{-2(\kappa^2+k^2)^{\frac{1}{2}}L}\right)}\right.\right.$$

$$\left.+\frac{\left((\kappa^2+k^2)^{\frac{1}{2}}-\epsilon_r|k|\right)e^{-(\kappa^2+k^2)^{\frac{1}{2}}(2L-(z+z_c))}}{\left(1+e^{-2(\kappa^2+k^2)^{\frac{1}{2}}L}\right)(\kappa^2+k^2)^{\frac{1}{2}}+\epsilon_r|k|\left(1-e^{-2(\kappa^2+k^2)^{\frac{1}{2}}L}\right)}\right].$$

$$\left.\left[\frac{\left(1+e^{-2(\kappa^2+k^2)^{\frac{1}{2}}(L+z_c)}\right)(\kappa^2+k^2)^{\frac{1}{2}}+\epsilon_r|k|\left(1-e^{-2(\kappa^2+k^2)^{\frac{1}{2}}(L+z_c)}\right)}{\left(1-e^{-2(\kappa^2+k^2)^{\frac{1}{2}}L}\right)(\kappa^2+k^2)^{\frac{1}{2}}+\epsilon_r|k|\left(1+e^{-2(\kappa^2+k^2)^{\frac{1}{2}}L}\right)}\right]\right\}$$

$$\begin{aligned}\Psi_k^{IV}(s,z,z_c) &= \Psi_k^I(s,-z,-z_c)\\ \Psi_k^{II}(s,z,z_c) &= \Psi_k^{III}(s,-z,-z_c)\end{aligned} \tag{30}$$

In the limit $A \to \infty$, the summation over wavevectors $\mathbf{k}$ goes over to an integral, which is most easily expressed in cylindrical coordinates:

$$\begin{aligned}\Psi^{(i)}(\mathbf{r}) = \Psi^{(i)}(s,z,z_c) &= \lim_{A\to\infty}\sum_{\mathbf{k}}\Psi_{\mathbf{k}}^{(i)}(\mathbf{r})\\ &= \int_{\wp} f(k \mid z,z_c)\exp\{i\mathbf{k}.\mathbf{s}\}\,d\mathbf{k}\\ &= \int_0^{\infty} f^{(i)}(k \mid z,z_c)k\,dk\int_0^{2\pi}\exp\{iks\cos\theta\}\,d\theta\\ &= \int_0^{\infty} f^{(i)}(k \mid z,z_c)J_0(ks)2\pi k\,dk\end{aligned} \tag{31}$$

where f^i is the coefficient of $e^{i\mathbf{k}.\mathbf{s}}$ in the expressions (30) for the potential components Ψ_k^i.

B. Calculation of the first non-linear correction to the effective pair potential

The integral over the domain $\wp$ appearing in eq.(20) may be reduced to the form given in eq.(21), with the following explicit expressions for the functions $\zeta_n(k \mid z_1, z_2)$:

$$\zeta_1(k \mid z_1, z_2) = d(k, z_1)d(k, z_2)$$

$$\left\{ \frac{((k^2+\kappa^2)^{\frac{1}{2}}+\epsilon_r k)^2}{2(\kappa^2+k^2)^{\frac{1}{2}}-\kappa} \left(e^{-(\kappa^2+k^2)^{\frac{1}{2}}(z_2-z_1)+\kappa z_2} - e^{-(\kappa^2+k^2)^{\frac{1}{2}}(2L-z_2-z_1)+\kappa L} \right) \right.$$

$$+ \frac{((k^2+\kappa^2)^{\frac{1}{2}}+\epsilon_r k)^2}{2(\kappa^2+k^2)^{\frac{1}{2}}+\kappa} \left(e^{-(\kappa^2+k^2)^{\frac{1}{2}}(z_2-z_1)-\kappa z_2} - e^{-(\kappa^2+k^2)^{\frac{1}{2}}(2L-z_2-z_1)-\kappa L} \right)$$

$$+ \frac{4((k^2+\kappa^2)^{\frac{1}{2}}+\epsilon_r k)((k^2+\kappa^2)^{\frac{1}{2}}-\epsilon_r k)}{\kappa} \left(\sinh(\kappa L) - \sinh(\kappa z_2)\right)$$

$$e^{-(\kappa^2+k^2)^{\frac{1}{2}}(2L-z_1-z_2)} + \frac{((k^2+\kappa^2)^{\frac{1}{2}}-\epsilon_r k)^2}{2(\kappa^2+k^2)^{\frac{1}{2}}+\kappa} \left(e^{-(\kappa^2+k^2)^{\frac{1}{2}}(2L-z_1-z_2)+\kappa L} \right.$$

$$\left. -e^{-(\kappa^2+k^2)^{\frac{1}{2}}(4L-z_1-3z_2)+\kappa z_2} \right) + \frac{((k^2+\kappa^2)^{\frac{1}{2}}-\epsilon_r k)^2}{2(\kappa^2+k^2)^{\frac{1}{2}}-\kappa}$$

$$\left. \left(e^{-(\kappa^2+k^2)^{\frac{1}{2}}(2L-z_1-z_2)-\kappa L} - e^{-(\kappa^2+k^2)^{\frac{1}{2}}(4L-z_1-3z_2)-\kappa z_2} \right) \right\}$$

$$\zeta_2(k \mid z_1, z_2) = d(k, z_1)d(k, -z_2)$$

$$\left\{ \frac{2((k^2+\kappa^2)^{\frac{1}{2}}+\epsilon_r k)^2}{\kappa} \left(\sinh(\kappa z_2) - \sinh(\kappa L)\right) e^{-(\kappa^2+k^2)^{\frac{1}{2}}(z_2-z_1)} \right.$$

$$+ \frac{2((k^2+\kappa^2)^{\frac{1}{2}}-\epsilon_r k)^2}{\kappa} \left(\sinh(\kappa z_2) - \sinh(\kappa L)\right) e^{-(\kappa^2+k^2)^{\frac{1}{2}}(4L+z_2-z_1)}$$

$$\frac{((k^2+\kappa^2)^{\frac{1}{2}}+\epsilon_r k)((k^2+\kappa^2)^{\frac{1}{2}}-\epsilon_r k)}{2(\kappa^2+k^2)^{\frac{1}{2}}+\kappa} \left(e^{-(\kappa^2+k^2)^{\frac{1}{2}}(2L-(z_2+z_1))+\kappa z_2} \right.$$

$$-e^{-(\kappa^2+k^2)^{\frac{1}{2}}(2L+z_2-3z_1)+\kappa z_1} - e^{-(\kappa^2+k^2)^{\frac{1}{2}}(2L+3z_2-z_1)-\kappa z_2}$$

$$\left. +e^{-(\kappa^2+k^2)^{\frac{1}{2}}(2L-(z_2+z_1))-\kappa z_1} \right)$$

$$\frac{((k^2+\kappa^2)^{\frac{1}{2}}+\epsilon_r k)((k^2+\kappa^2)^{\frac{1}{2}}-\epsilon_r k)}{2(\kappa^2+k^2)^{\frac{1}{2}}-\kappa} \left(e^{-(\kappa^2+k^2)^{\frac{1}{2}}(2L-(z_2+z_1))-\kappa z_2} \right.$$

$$-e^{-(\kappa^2+k^2)^{\frac{1}{2}}(2L+z_2-3z_1)-\kappa z_1} - e^{-(\kappa^2+k^2)^{\frac{1}{2}}(2L+3z_2-z_1)+\kappa z_2}$$

$$\left. \left. +e^{-(\kappa^2+k^2)^{\frac{1}{2}}(2L-(z_2+z_1))+\kappa z_1} \right) \right\}$$

$$\zeta_3(k \mid z_1, z_2) = d(k, -z_1)d(k, -z_2)$$

$$
\left\{ \frac{((k^2+\kappa^2)^{\frac{1}{2}}+\epsilon_r k)^2}{2(\kappa^2+k^2)^{\frac{1}{2}}-\kappa} \left(e^{-(\kappa^2+k^2)^{\frac{1}{2}}(z_2-z_1)-\kappa z_1} - e^{-(\kappa^2+k^2)^{\frac{1}{2}}(2L+z_2+z_1)+\kappa L} \right) \right.
$$

$$
+ \frac{((k^2+\kappa^2)^{\frac{1}{2}}+\epsilon_r k)^2}{2(\kappa^2+k^2)^{\frac{1}{2}}+\kappa} \left(e^{-(\kappa^2+k^2)^{\frac{1}{2}}(z_2-z_1)+\kappa z_1} - e^{-(\kappa^2+k^2)^{\frac{1}{2}}(2L+z_2+z_1)-\kappa L} \right)
$$

$$
+ \frac{2((k^2+\kappa^2)^{\frac{1}{2}}+\epsilon_r k)((k^2+\kappa^2)^{\frac{1}{2}}-\epsilon_r k)}{\kappa} (\sinh(\kappa L) - \sinh(\kappa z_2))
$$

$$
\left(e^{-(\kappa^2+k^2)^{\frac{1}{2}}(2L-(z_1-z_2))} + e^{-(\kappa^2+k^2)^{\frac{1}{2}}(2L+(z_1-z_2))} \right)
$$

$$
+ \frac{((k^2+\kappa^2)^{\frac{1}{2}}-\epsilon_r k)^2}{2(\kappa^2+k^2)^{\frac{1}{2}}+\kappa} \left(e^{-(\kappa^2+k^2)^{\frac{1}{2}}(2L+z_1+z_2)+\kappa L} \right.
$$

$$
\left. - e^{-(\kappa^2+k^2)^{\frac{1}{2}}(4L+3z_1+z_2)-\kappa z_1} \right) + \frac{((k^2+\kappa^2)^{\frac{1}{2}}-\epsilon_r k)^2}{2(\kappa^2+k^2)^{\frac{1}{2}}-\kappa}
$$

$$
\left. \left(e^{-(\kappa^2+k^2)^{\frac{1}{2}}(2L+z_1+z_2)-\kappa L} - e^{-(\kappa^2+k^2)^{\frac{1}{2}}(4L+3z_1+z_2)+\kappa z_1} \right) \right\} \tag{32}
$$

where the coefficients $d(k_y, z_i)$ are given by:

$$
d(k_y, z_i) = \frac{(1+e^{-2(\kappa^2+k^2)^{\frac{1}{2}}(L+z_i)})(\kappa^2+k^2)^{\frac{1}{2}}+\epsilon_r k(1-e^{-2(\kappa^2+k^2)^{\frac{1}{2}}(L+z_i)})}{(1-e^{-2(\kappa^2+k^2)^{\frac{1}{2}}L})(\kappa^2+k^2)^{\frac{1}{2}}+\epsilon_r k(1+e^{-2(\kappa^2+k^2)^{\frac{1}{2}}L})}
$$

$$
\frac{1}{(1+e^{-2(\kappa^2+k^2)^{\frac{1}{2}}L})(\kappa^2+k^2)^{\frac{1}{2}}+\epsilon_r k(1-e^{-2(\kappa^2+k^2)^{\frac{1}{2}}L})} \tag{33}
$$

WETTING PHASE TRANSITIONS AND PETROLEUM RESERVOIRS

J. O. INDEKEU
Laboratorium voor Vaste-Stoffysica en Magnetisme
Katholieke Universiteit Leuven, B-3001 Leuven, Belgium

K. RAGIL AND D. BROSETA
Institut Français du Pétrole, 1-3 Av. du bois Préau,
F-92500 Rueil-Malmaison, France

AND

D. BONN AND J. MEUNIER
Laboratoire de Physique Statistique, Ecole Normale Supérieure,
24, rue Lhomond, F-75231 Paris Cedex 05, France

Abstract. The wetting properties of oil on water are relevant to the industrial problem of oil recovery from porous reservoirs. The location of wetting phase transitions, a subject of fundamental interest for experimentalists and theorists, is also of practical importance for petroleum engineering because the hydrodynamics of oil flow is different on either side of the transition. This seminar treats the application of the phenomenological wetting theory of Cahn to fluids with van der Waals forces, and in particular to alkanes on water. Including the long-range substrate-adsorbate interaction in the theory allows one to reproduce the experimentally observed sequence of two wetting-like transitions. A critical wetting transition, which arises from the subtle action of the van der Waals forces, is preceded by a first-order transition between a thin and a thick film at a lower temperature. The three wetting regimes separated by these two transitions presumably lead to three different oil recovery efficiencies.

1. Introduction and motivation

The petroleum industry is faced with the problem of recovering oil (in the form of hydrocarbons) from porous rocks that contain water, oil and gas. The wetting properties of water on rock and oil on water influence the

C. Caccamo et al. (eds.), New Approaches to Problems in Liquid State Theory, 337–344.

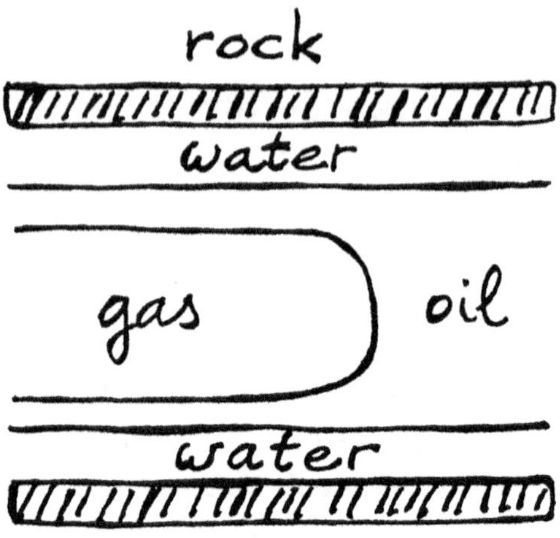

Figure 1. Complete wetting of oil on water.

hydrodynamic properties of oil drainage and these in turn determine the recovery efficiency. Typically, water wets the porous medium completely, whereas the oil wets the water either partially or completely. In recent experiments with sand reservoirs filled with salt water ("brine"), oil and air, it has been observed that *complete wetting of oil on water* significantly enhances the oil recovery rate [1]. Presumably this is due to the hydraulic continuity of the oil film (on water), induced by the perfect spreading (Fig.1). In contrast, *partial wetting* of oil on water leads to oil droplets (with nonzero contact angles) which are more difficult to displace and tend to stay behind, so that significant oil retention occurs (Fig.2).

Let us denote the surface tension by γ and use indices o, w and g for oil, water and gas. The spreading coefficient S, defined as

$$S = \gamma_{w,g} - (\gamma_{w,o} + \gamma_{o,g}) \tag{1}$$

is negative for partial wetting of oil on water and zero (or even positive in non-equilibrium spreading) for complete wetting. Consequently S plays a key role, so it appears, in oil recovery.

From the point of view of *wetting phase transitions*, which take place when S approaches zero from below [2], the challenge arises of locating possible wetting transitions of hydrocarbons on water, in order to be able to predict favorable conditions (in temperature, pressure, or chemical composition) for improved oil recovery. Thus a practical system of industrial

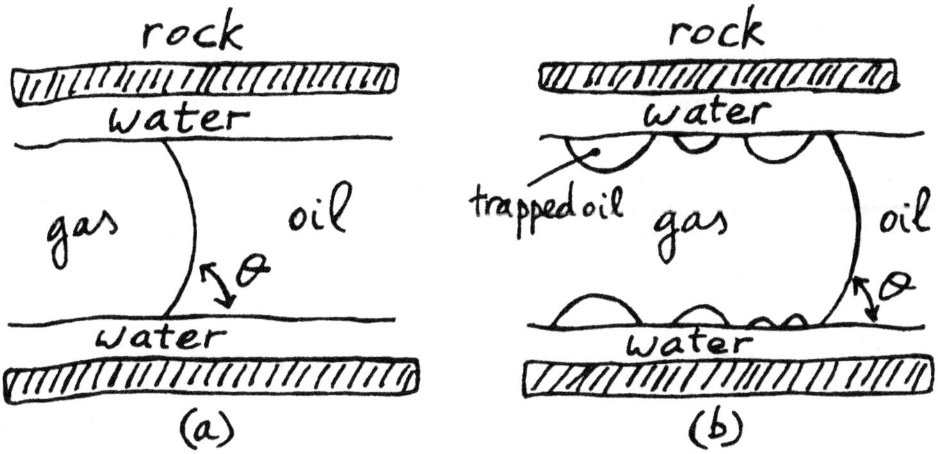

Figure 2. Partial wetting of oil on water, (a) before, and (b) after displacement.

interest presents a problem of fundamental interest to experimentalists and theorists in wetting phenomena.

For those unfamiliar with the subject, an introduction to wetting phenomena [3] can be summarized for our present purposes as follows. When $S < 0$ the oil forms droplets on water, with a contact angle given by Young's law

$$\gamma_{w,g} = \gamma_{w,o} + \gamma_{o,g} \cos\theta \tag{2}$$

Besides these droplets there is generally an oil film of *microscopic* thickness. At the transition from partial to complete wetting this film becomes of *macroscopic* thickness, in a singular manner. The type of singularity is connected with the *order* of the wetting phase transition. If the thickness makes a jump, the transition is of *first order*. In that case the spreading coefficient displays a discontinuous first derivative. On the other hand, if the film thickness diverges in a continuous manner, the wetting transition is said to be *critical*. The singularity of S is then weaker.

In the past few decades there was a consensus that in fluids with van der Waals forces wetting transitions should be expected to be of first order, as found experimentally [4], and this belief was strengthened by the fact that critical wetting remained elusive to experimental verification. However, precisely in the system of our present interest, alkanes on water, critical wetting has recently been observed [5]. This is certainly of fundamental interest, but the order of the wetting transition has no direct repercussions on oil recovery efficiency. There is another observation [6], made by the same experimental group, which has more important practical implications. The critical wetting transition appears to be preceded, at a lower temperature, by a first-order transition from a microscopic ($\approx 10 \mathring{A}$ thick) to a *mesoscopic* film ($\approx 100 \mathring{A}$ thick). Consequently, there are three instead of two regimes of wetting behaviour, and if this finding is robust to the presence of chemical or geometrical disorder, it may have interesting consequences for oil recovery, as sketched in the conclusions and prospects (section 3).

2. Summary of the theoretical predictions and experimental observations

In a first step, Ragil *et al.* adapted the phenomenological Cahn theory of wetting to the concrete system of linear alkanes adsorbed on water [7]. From this theory they predicted *first-order wetting transitions* and estimated the transition temperatures, T_{1w}, as a function of alkane chain length. For instance, for hexane on water $T_{1w} = 293K$, well in between the triple point ($178K$) and the critical point ($507K$) of hexane. For pentane this calculation predicts a wetting temperature below the freezing point of water, and consequently complete wetting at room temperature. These results are in reasonable accord with experiments, but can for instance not explain the "strange" behaviour of pentane on water, displaying drops with small but non-zero contact angles on top of a thick but not macroscopic adsorbed pentane film.

The major shortcoming of this simple theory so far is the complete neglect of forces with algebraic decay, such as van der Waals forces. The theory allows only "short-range forces", which in this context can be interpreted as the forces that remain after one "cuts off" the tails of the intermolecular potentials. This is easy to imagine, since precisely this operation is performed in most Molecular Dynamics simulations of fluids in order to keep the computations feasible. Typically, the cut-off radius is 2.5 molecular diameters, and the potential is then shifted to avoid a discontinuity. Although for many purposes this truncation may be irrelevant, for wetting phenomena it is not.

A wetting film, thick compared to the molecular size, will gain or lose free energy per unit area upon further thickening, depending on the net effect of all the tails of the intermolecular water-oil and oil-oil potentials. The effective Hamaker constant A is the leading amplitude of the net interaction between the water-oil and the oil liquid-vapour interfaces. Depending on the sign of A the oil wetting film will tend to thicken or to thin. It turns out that for pentane on water above T_{1w} the sign of A is such that the film is prevented from thickening to a macroscopic layer. Thus, although the

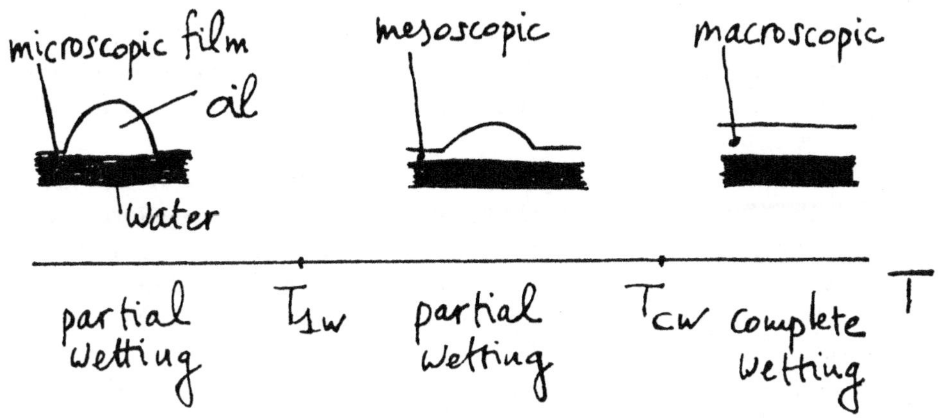

Figure 3. Three wetting regimes separated by two phase transitions.

short-range forces favour wetting, the long-range forces inhibit complete wetting. The spreading coefficient S remains negative. This state continues until at a higher temperature, $T_{cw} \approx 53°C$, the Hamaker constant changes sign, putting all forces in line to achieve complete wetting. The transition at T_{cw} is thus the genuine wetting transition. It is of critical character. The former transition at T_{1w} is still present, as a first-order transition from a thin to a thick but not macroscopic film.

An adaptation of the Cahn theory, allowing for the presence of a long-range substrate-adsorbate "field", can approximately take into account this crucial interaction between interfaces and reproduce the experimentally observed transitions and their order [8]. This development or "second step" can be viewed as a perturbation theory in which the effect of the van der Waals tails is added to the asymptotic part of the interaction between interfaces. It is very natural for this theory to predict two separate transitions. One is due solely to the subtle action of the long-range interactions and the other would be the only wetting transition if long-range forces were neglected.

The prediction from this theory is therefore a sequence of two transitions (Fig.3). Below T_{1w} the system is partially wet, the wetting film is microscopic, and droplets occur with finite contact angles. After a first-order jump, at T_{1w}, the wetting film is of mesoscopic thickness, and drops with small contact angles persist. Finally, the film thickness diverges continuously at T_{cw} and complete wetting occurs. As mentioned already, a similar sequence of transitions has recently been observed for hexane on brine, and the two transition temperatures were studied as a function of salt concentration [6].

There are various new aspects associated with the theoretical methods used in the Cahn theory adapted to systems with long-range forces. In particular, the use of phase portraits and equal-areas rules is modified in an interesting way relative to the case of short-range forces. Also the calculation of the interaction between interfaces, the *interface potential*, has been revised.

It should be stressed that the theory we have employed is a phenomenological one. This means that various parameters, which can in principle be calculated from a microscopic theory, are given approximate values based on simple estimations or experimental data. The development of a more microscopic density functional theory in which, for systems consisting of spherical intermolecular interaction potentials, the precise asymptotic form of the interface potential has been calculated, can be found in the work of Dietrich and Napiórkowski [9].

3. Conclusions and prospects

We have seen how the inclusion of long-range forces in the Cahn theory permits to take into account the subtle role played by these interactions in controlling the thickness of a mesoscopic wetting film. In "normal" situations the van der Waals forces promote wetting, since they favour a sequence of phases in which a phase with *intermediate* dielectric polarizability (per unit volume) wets the interface between two phases with *low* and *high* polarizability, respectively. This is usually the case for a sequence substrate/liquid/gas. Water, as substrate, is exceptional in that the dielectric functions are so complex that they lead to a change of sign of the effective Hamaker constant at some temperature T_{cw}. Below T_{cw} the van der Waals forces act antagonistically in that they prevent the (lighter) oil phase from wetting completely the water/gas interface. Above T_{cw} the van der Waals forces "behave" normally again.

Consequently, even though the short-range forces favour wetting above the first-order transition temperature T_{1w}, the film is prevented from achieving full macroscopic thickness. The system displays so to speak *frustrated complete wetting.* This state, in which drops coexist with a mesoscopic film (Fig.4), is different from the cases shown in Figs.1-2 which were envisaged in the oil recovery experiments.

The new state shows significant hydraulic continuity through the film, and this may well increase the mobility of the drops too. One may presume

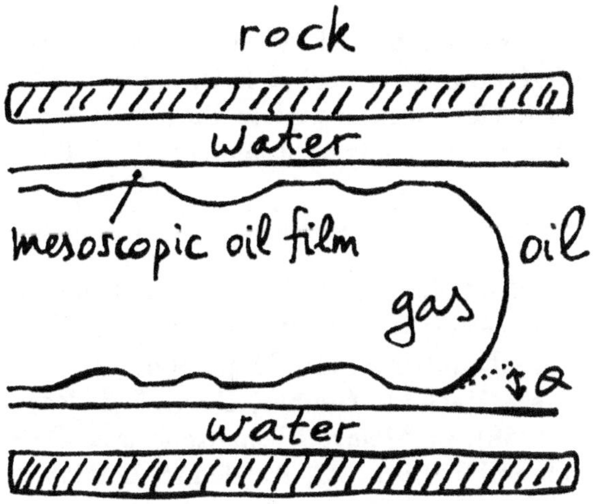

Figure 4. Frustrated complete wetting of oil on water.

that the amount of retention on the part of lingering drops is reduced relative to the partial wetting case. Moreover, relative to the complete wetting case, it is not clear *a priori* which state is the most effective for oil recovery. Are mobile drops more efficient for transporting oil than a wetting film?

Along the lines of [1] and other studies of displacement in uniformly wetted porous media [10] one may propose new experiments using a reservoir filled with sand, imbibed with hexane, brine and air. In view of the large temperature window, about $30K$, between T_{1w} and T_{cw} found experimentally [6], it would be a challenge to investigate frustrated complete wetting in this more complex environment, in the presence of sand as an additional substrate under the presumably macroscopic water layer. It remains to be seen whether three instead of two regimes of oil recovery efficiencies exist, corresponding to the three qualitatively different equilibrium wetting states. And, from a practical point of view, it will be most interesting to verify which state is the most conducive to oil flow.

Acknowledgements

We thank Ralf Blossey and Harvey Dobbs for a critical reading of the manuscript. J.O.I. is a Research Director of the Fund for Scientific Research of Flanders, and has been supported by the Belgian Concerted Action Programme (GOA) and Interuniversity Poles of Attraction Programme

(IUAP). K.R. has been supported by the Wetenschappelijke Onderzoeksgemeenschap (WOG) during a visit to K.U.Leuven. LPS de l'ENS is URA 1306 du CNRS, associée aux universités Paris 6 et Paris 7. This research has been supported by the contract Joule JOF3-CT95-0013.

References

1. O. Visika and J.-M. Lombard, Society of Petroleum Engineers (SPE) Reservoir Engineering (1996) February, 54.
2. J.W. Cahn, J. Chem. Phys. (1977) 66, 3667.
3. For reviews, see P.-G. de Gennes, Rev. Mod. Phys. (1985) 57, 827; S. Dietrich (1988), in "*Phase Transitions and Critical Phenomena*", eds. C. Domb and J. Lebowitz, Academic, London, Vol.12, p.1; for an introduction, see, e.g., J.O. Indekeu, Acta Physica Polonica (1995) 26, 1065-1100.
4. D. Bonn, H. Kellay, and G.H. Wegdam, Phys. Rev. Lett. (1992), 69, 1975.
5. K. Ragil, J. Meunier, D. Broseta, J.O. Indekeu and D. Bonn, Phys. Rev. Lett. (1996) 77, 1532.
6. N. Shahidzadeh, D. Bonn, K. Ragil, D. Broseta, and J. Meunier, Phys. Rev. Lett. (1998) 80, 3992.
7. K. Ragil, D. Bonn, D. Broseta, and J. Meunier, J. Chem. Phys. (1996) 105, 5160.
8. J.O. Indekeu, K. Ragil, D. Bonn, D. Broseta and J. Meunier (1998), J. Stat. Phys., in print.
9. S. Dietrich and M. Napiórkowski, Phys. Rev. A (1991) 43, 1861.
10. N.R. Morrow and F.G. McCaffery (1978), in "*Wetting, Spreading and Adhesion*", ed. J.F. Padday, Academic, London, p.289.

COMPLEX FLUIDS

ATOMISTIC COMPUTER SIMULATION STUDIES OF SURFACTANT SYSTEMS

S. BANDYOPADHYAY, M. TAREK AND M. L. KLEIN

Center for Molecular Modeling and
Department of Chemistry,
University of Pennsylvania,
Philadelphia, PA 19104-6323

Abstract. Computer simulation has emerged as a powerful probe of the behavior of surfactant systems. The past few years have seen several novel applications, which have given important insights into the nature of the water-surfactant interface as well as the behavior of surfactants at different liquid-vapor, liquid-liquid and liquid-solid interfaces. These lectures focus on surfactants and lipids in aqueous environments.

1. Introduction

The study of surfactants has a long history. Detailed understanding of such systems at a microscopic level has important implications for a wide range of scientific and technological research areas such as detergency, lubrication, molecular self-assembly, ion-transfer, drug delivery, etc. With the advent of powerful computers and sophisticated methodologies computer simulation has become an important alternative technique to investigate surfactant behaviors at interfaces. In this regard, computer simulation studies of water and aqueous-phase binary mixtures continue to be an area of interest, due to the universality of water as a polar solvent and it's important role in biological systems. The focus herein is the nature of the surfactant-water interface, and bilayer systems. Although simulation studies are often motivated by contemporary experiments, we focus mostly on molecular dynamics (MD) simulations. The readers interested in further details of the simulations and experiments are urged to investigate the cited references.

C. Caccamo et al. (eds.), New Approaches to Problems in Liquid State Theory, 347–358.

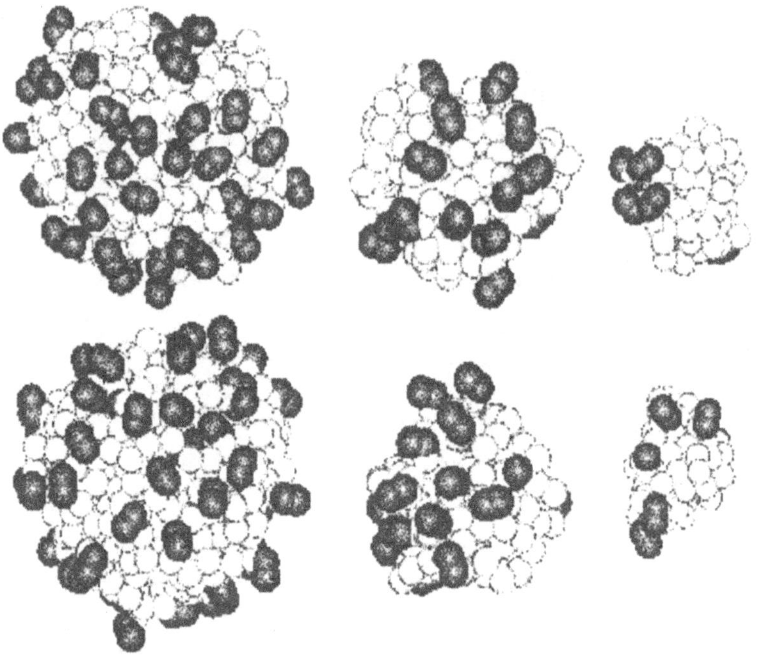

Figure 1. Snapshots of the water/ethanol simulation systems: large (left), medium (center), and small (right) droplet, respectively. From top to bottom, the snapshots are 100 ps apart and are taken from the last 200 ps of the MD trajectories. The atoms are drawn as spheres with their corresponding van der Waals radii and the carbon atoms are highlighted.

2. Surface properties of water and water/alcohol mixtures

The surface properties of water [1,2], small water droplets, and water/alcohol mixtures are relevant to the understanding of nucleation phenomena in liquid-phase binary mixtures. Tarek and Klein [3] have reported an MD study of the shape and surface structure properties of water/ethanol droplets as a function of the droplet size. The classic approach is based on the capillary approximation, where the droplets are usually described as spherical objects with a definite composition and their thermodynamic properties are used to derive the nucleation rate. Tarek and Klein studied three different droplet sizes of approximately 8, 11 and 16Å in diameter, with 12% ethanol mole fraction. From MD trajectories that spanned over a nanosecond, it was found that in each of the clusters ethanol aggregated at the droplet surface. In addition, the ratio of ethanol molecules at the surface to the number in the bulk increased as the size of the cluster decreased. Their analysis showed that significant shape fluctuations occur irrespective of the cluster size. The droplets were no longer spherical, as shown in the snapshots displayed in Figure 1. Moreover, the clusters deviate severely from sphericity as size decreases, with characteristic shape fluctuations ranging from 30 to 120 ps, as found for spherical micelles in solutions. These calculations reveal aspects of the structure and dynamics of small clusters of binary mixtures and also suggest possible reasons for the discrepency between nucleation rates predicted by existing theories [4] and the experiment [5]. First, the theoretical models do not allow for complete segregation of surfactant at the interface. Second, the thermodynamic description does not take into account the contribution of shape fluctuations to the surface free energy. It is likely, as in the case of micellar aggregates in solutions, these fluctuations have to be taken into account in estimating the free energy required to form a mixed critical nucleus from the vapor phase. Laaksonen *et al.* [6] have reported MD simulations at room temperature for water-methanol binary mixtures over a wide range of compositions. They have studied in detail the effects of the alcohol on water structure and water on methanol structure. Large structural changes were observed. In methanol-rich solution the water-water correlations were found to be very pronounced, while methanol retained most of its pure liquid structure. In water-rich solution, a high degree of ordering was observed with characteristic tetrahedral arrangements of water molecules around the hydroxyl group of the methanol. Recently, small angle neutron scattering (SANS) has been used to study the properties of nanodroplet aerosols [7].

3. Surfactant-water interfaces

The molecular level study of the phenomena at the surfactant-water interface is an active field in the area of research with significant contributions from both experiment [8,9] and theory [10]. The increasing power of computers, and the development of sophisticated simulation methodologies [11] over last few years, means that these techniques can now play a crucial role in understanding the interfacial properties of surfactants.

Shelley *et al.* [12] have performed MD simulations using both nonpolarizable and polarizable models to study the structure and electrostatics of the surfactant-water interface. In particular, they have studied the sodium octanoate-water system in the lyotropic liquid crystalline mesophase E, where the surfactants form very long columnar micelles, which pack in a two-dimensional hexagonal patern. Their simulations were carried out in the canonical (NVT) ensemble at room temperature and the systems consisted of a single columnar micelle containing 54 octanoate ions, 54 Na^+ counterions, and 503 water molecules. These simulations predicted

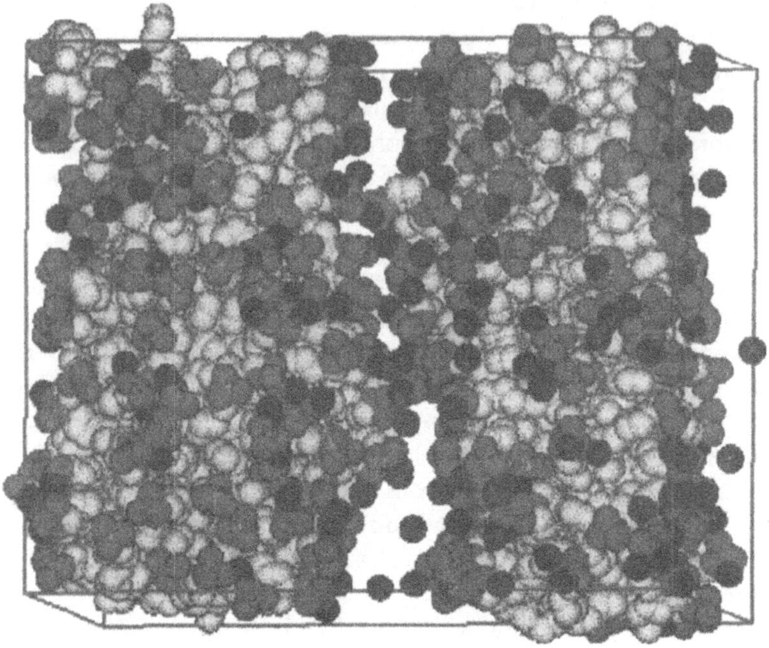

Figure 2. Snapshot of the final configuration of two cylindrical SDS micelles taken from an MD simulation of SDS micelle in mesophase E [13]. The surfactant head groups are highlighted and the counterions are shown as black spheres. Water molecules have been omitted for visual clarity and the simulation cell boundaries have been outlined.

that the interfacial region has a narrow water-hydrocarbon interface, about 4 Å thick, and the Na^+ ions and carboxylate head groups make up a characteristic electrostatic double layer around these micelles. It was shown that the contributions of the individual components to the overall potential drop are large (about 7 V). However, due to the collective nature of dielectric screening, significant cancellation of the individual contributions occur to give a net potential drop of less than 0.3 V across the interface. The structural properties of another important and commonly studied surfactant sodium dodecylsulfate (SDS) in E mesophase has also been studied recently [13]. This simulation was carried out on a large system, which consisted of two columnar micelles, each containing 128 dodecylsulfate anions, plus 256 Na^+ counterions, and 4350 water molecules. A 260 ps trajectory was generated at constant temperature (T=333 K) and constant pressure (P=0 atm) using a new advanced molecular dynamics package, PINY-MD [14]. Figure 2 displays a configuration of the system taken from this simulation.

The simulation study revealed that there is a small but distinct deviation from two-dimensional hexagonal symmetry of micelles in mesophase E, in agreement with experiment [15]. It was only possible to demonstrate such a distortion due to the presence of two distinct cylindrical micelles in the simulation system.

4. Surfactants at liquid-liquid and liquid-vapor interfaces

Proper understanding of the behavior of surfactants at liquid (in particular at water) interfaces is crucial to get an insight into processes like lubrication, detergent action, molecular self-assembly etc [16]. Recently, there have been several experimental studies of monolayers at the water-vapor interface. For example, vibrational sum frequency generation [17] (VSFG) and neutron reflection studies [18,19] are carried out on monolayers of SDS and also on other types of surfactants. There are also a number of MD studies reported over last few years on different surfactants at water-vapor interfaces. Recently, Berkowitz and coworkers [20] have reported MD simulations to study the effect of the interface on the properties of SDS. They carried out two simulations, one at the water-vapor interface and the other at the water-CCl_4 interface. From the simulation they observed a significant difference in the configurational properties of the surfactant at the two interfaces. The preferred orientation of the surfactant molecules at low surface coverage was different in the two systems. The water-vapor interface provides an attractive surface for the hydrocarbon tail of the surfactant and therefore it prefers a bent conformation. This allows the head group region to insert into the aqueous phase and achieve maximum solvation while the hydrocarbon tail lies down at the interface maximizing its van der Waals interaction with the surface water. In contrast to this, the van der Waals interaction between the SDS and the solvent is rather uniform, which was indicated by the fact that on average the chains prefer to remain straight at the water-CCl_4 interface with an inclination of about 40° from the surface normal. Villalba *et al.* [21] have studied interfacial behavior of a model heptane-water system in presence of nonylphenol triethoxylated surfactants. They reported the dependence of the superficial energy and entropy of the system as a function of surfactant concentration using RUV-2 model [22] of ternary oil/water/surfactant systems. They found that the surface free energy changes linearly with temperature but shows a minimum with respect to the concentration of the surfactant.

5. Solid-liquid interfaces

Understanding the adsorption mechanism of surfactant molecules at the solid-liquid interface is a key issue towards modeling technological processes that use surfactant molecules, such as detergency, lubrication, water purification, etc. The self-assembly of surfactant molecules in bulk solutions have been well-studied. However, less is known about how the aggregation process of surfactant molecules are influenced by the presence of a solid boundary surface. Recent studies of self-assembled structures formed by different types of surfactants at solid-liquid interfaces using atomic force microscope (AFM) have revealed that many of the novel aggregate geometries observed in bulk solution also occur near an interface [23-26]. The AFM images show interfacial aggregates with a high degree of curvature and periodic structure, in contrast to flat monolayers and bilayers. The surface has strong influence on the organization of these surfactants. The interfacial structure of the cationic surfactant tetradecyltrimethylammonium bromide (C_{14}TAB) was hemi-cylinders on hydrophobic surfaces (such as graphite), full cylinders on mica and spheres on silica [24].

To date, no computer simulation study has attempted to use a fully atomistic model to study such interfacial aggregation, mainly due to the large time-scale of such processes. Recently, Bandyopadhyay *et al.* [27] have carried out a long MD simulation of 1 nanosecond duration to characterize the aggregate structure of surfactants at an hydrophobic surface-water interface. In particular, they have studied molecular organization of cetyltrimethylammonium bromide (C_{16}TAB) at the interface of a flat hydrophobic surface and water. Their simulation system consisted of two such surfaces separated by a distance of 70Å. One of the surfaces was covered with 94 erect surfactants initially arranged in monolayer, while the other surface was covered with two hemi-cylindrical micelles each containing 64 monomers (at a surface coverage of 45Å^2 per molecule). The surfactant layers were then separated by a water layer with appropriate dimensions. The simulation was carried out at constant volume and temperature (NVT). The initial configuration and the configuration of the system after 1 ns are shown in Figure 3. It was observed that the monolayer at the interface rearranged into roughly hemi-cylinder micelles, while the initial hemi-cylinders remain almost intact with slight deformation in shape. The simulation suggests that cationic surfactants like CTAB form stable hemi-cylindrical aggregates at hydrophobic surface-water interfaces, which is in agreement with AFM measurements [23].

Recently, Wijmans and Linse [28] have reported Monte Carlo simulations using a lattice model to study the adsorption of amphiphilic oligomeric surfactant molecules at hydrophobic interfaces. It was observed that when the surfactant solution is brought into contact with a hydrophobic surface, the surfactant tails, which are hydrophobic in nature get adsorbed by this surface with hydrophilic head groups sticking out into the bulk solution.

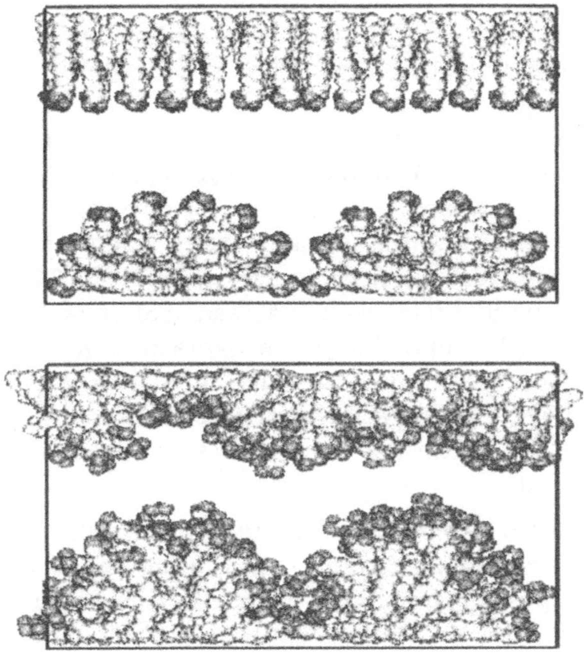

Figure 3. Profile snapshots of the starting configuration (top), and the configuration after 1 ns (bottom), obtained from an MD simulation of cetyltrimethylammonium bromide ($C_{16}TAB$) confined between two hydrophobic surfaces [27]. The surfactant head groups have been highlighted. Water molecules and counterions have been omitted for viewing clarity. The hydrophobic surfaces are at the top and bottom of the simulation cell, which is indicated by a box.

6. Lipid bilayers

X-ray and neutron diffraction plus deuterium nuclear magnetic resonance (^{2}H NMR) are powerful tools to obtain general structural information on biomembranes [29,30]. Experimental limitations alongwith the increase of computer CPU performance have encouraged implementation of novel computational methods in the study of biomembranes. Two recent reviews [31, 32] discussed the progress in this field in detail. We therefore restrict our discussion on publications that appeared last year.

Because of inadequate forcefields most of the previous MD simulations on lipids were restricted to those with saturated chains, namely dipalmitoylphosphatidylcholine (DPPC) and dimyristoylphosphatidylcholine (DMPC) [33, 34]. Feller *et al.* [33] have performed a series of MD simulations on fully hydrated liquid crystalline DPPC bilayers as a function of molecular surface area. Their results support the value of 62.9 Å^2/DPPC as obtained from X-ray data. Hydrogen bonding between water and phosphatidylcholine was studied by Gierula *et al.* [34] using MD simulation of a hydrated DMPC bilayer membrane in the liquid crystalline phase. They observed that on average there were 5.3 H-bonds formed per DMPC between water and DMPC oxygens, 4 of these involve non-ester phosphate oxygens, 0.9 carbonyl oxygens and the remaining involve ester phosphate oxygens. They also showed formation of many water bridges between DMPC molecules to generate clusters. Recently, Husslein *et al.* [35] have carried out a constant pressure and temperature MD simulation of diphytanolphosphatidylcholine (DPhPC) lipid bilayer in hydrated liquid crystal phase. This is an important lipid, which is widely used in the study of membrane channel activity. Their simulation results, (based on current generation force-field) for the electron density profiles agree reasonably well with X-ray diffraction data.

Only recently there have been attempts to simulate unsaturated lipid bilayers, which are essential structural components of biomembranes. Feller *et al.* [36] have proposed a potential energy function for unsaturated hydrocarbons and carried out MD simulations of a water/octene interface and a dioleoyl phosphatidylcholine (DOPC) bilayer at a low hydration of 5.35 waters/lipid. Comparison between DOPC and DPPC showed that even though the temperature, hydration level, and surface area/lipid for DOPC are lower than DPPC, the two systems are almost identically disordered in the bilayer interior. A strong interaction of the double bond in DOPC with water was indicated, which is responsible for membrane permeability and for stabilizing complex lipid-protein structures. Hyvonen *et al.* [37] have reported a long MD simulation to study the structure and dynamics of biologically important diunsaturated 1-palmitoyl-2-linoleoyl-sn-glycero-3-phosphatidylcholine (PLPC) bilayer. The structural properties of the phosphatidylcholine headgroup, the glycerol backbone, and the hydrating water were investigated. They have also studied in detail the structure of the double bond region and the effects of the diunsaturation on the bilayer interior.

7. Conclusions and outlook

The currently available simulation methodologies have the ability to offer visual images of surfactant interfaces. These images are a powerful complement to real experiments. In many of the examples cited above there is quantitative agreement with experiment, which gives added confidence to the simulators. The successes to date indicate that the use of atomistic computer simulations as a means of probing the behavior of surfactant systems will inevitably continue.

Acknowledgements

The authors acknowledge support from the National Institute of Health, the National Science Foundation and the Procter & Gamble company.

References

1. Zakharov, V.V., Brodskaya, E.N., and Laaksonen, A. (1997) Surface tension of water droplets: A molecular dynamics study of model and size dependencies, *J. Chem. Phys.* **107**, 10675-10683.
2. Alejandre, J., Tildesley, D.J., and Chapela, A. (1995) Molecular dynamics simulation of the orthobaric densities and surface-tension of water, *J. Chem. Phys.* **102**, 4574-4583.
3. Tarek, M., and Klein, M.L. (1997) Molecular dynamics study of two-component systems: The shape and surface structure of water/ethanol droplets, *J. Phys. Chem. A* **101**, 8639-8642.
4. Zeng, X.C., and Oxtoby, D.W. (1991) Binary homogeneous nucleation theory for the gas-liquid transition - a nonclassical approach, *J. Chem. Phys.* **95**, 5940-5947.
5. Strey, R., and Viisanen, Y. (1993) Measurement of the molecular content of binary nuclei - use of the nucleation rate surface for ethanol-hexanol, *J. Chem. Phys.* **99**, 4693-4704.
6. Laaksonen, A., Kusalik, P.G., and Svishchev, I.M. (1997) Three-dimensional structure in water-methanol mixtures, *J. Phys. Chem.* **101**, 5910-5918.
7. Wyslouzil, B.E., Cheung. J.L., Wilemski, G., and Strey, R. (1997) Small angle neutron scattering from nanodroplet aerosols, *Phys. Rev. Lett.* **79**, 431-434.
8. Naujok, R.R., Paul, H.J., and Corn, R.M. (1996) Optical second harmonic generation studies of azobenzene surfactant adsorption and photochemistry at the water/1,2-dichloroethane interface, *J. Phys. Chem.* **100**, 10497-10507.
9. Gragson, D.E., McCarty, B.M., and Richmond, G.L. (1997) Ordering of interfacial water molecules at the charged air/water interface observed by vibrational sum frequency generation, *J. Am. Chem. Soc.* **119**, 6144-6152.
10. Benjamin, I., (1995) Theory and computer-simulations of solvation and Chemical-reactions at liquid interfaces, *Acc. Chem. Res.* **28**, 233-239.
11. Martyna, G.J., Tuckerman, M.E., Tobias, D.J., and Klein, M.L. (1996) Explicit reversible integrators for extended systems dynamics, *Mol. Phys.* **87**, 1117-1157.

12. Shelley, J.C., Sprik, M., and Klein, M.L. (1997) Structure and electrostatics of the surfactant-water interface, *Progr. Colloid Polym. Sci.* **103**, 146-154.
13. Bandyopadhyay, S., Klein, M.L., Martyna, G.J., and Tarek, M. Molecular dynamics studies of the hexagonal mesophase of sodium dodecylsulfate in aqueous solution, *Mol. Phys.* (in press).
14. Martyna, G.J., Samuelson, S., Hughs, A., Tobias, D.J., Yarne, D., Klein, M.L., and Tuckerman, M.E. PINY-MD: A simulation package for next generation simulators, (in preparation).
15. Leigh, I.D., McDonald, M.P., Wood, R.M., Tiddy, G.J.T., and Trevethan, M.A. (1981) Structure of liquid-cryatalline phases formed by sodium dodecylsulfate and water as determined by optical microscopy, X-ray diffraction and nucler magnetic resonance spectroscopy, *J. Chem. Soc. Faraday Trans. 1* **77**, 2867-2876.
16. Gelbart, W.M., and Ben-Shaul, A. (1996) The "new" science of "complex fluids", *J. Phys. Chem.* **100**, 13169-13189.
17. Gragson, D.E., McCarty, B.M., and Richmond, G.L. (1996) Surfactant/water interactions at the air/water interface probed by vibrational sum frequency generation, *J. Phys. Chem.* **100**, 14272-14275.
18. Purcell, I.P., Thomas, R.K., Penfold, J., and Howe, A.M. (1995) Adsorption of SDS and PVP at the air-water interface, *Colloids Surf. A* **94**, 125-130.
19. Lu, J.R., Hromadova, M., Thomas, R.K., and Penfold, J. (1996) Structure of hydrocarbon chains in surfactant monolayers at the air/water interface: Neutron reflection from dodecyl trimethylammonium bromide, *J. Chem. Soc. Faraday Trans.* **92**, 403-408.
20. Schweighofer, K.J., Essmann, U., and Berkowitz, M. (1997) Simulation of sodium dodecylsulfate at the water-vapor and water-carbontetrachloride interfaces at low surface coverage, *J. Phys. Chem. B* **101**, 3793-3799.
21. Urbina-Villalba, G., Landrove, R.M., Guaregua, J.A. (1997) Molecular dynamics simulation of the interfacial behavior of a heptane/water system in the presence of nonylphenol tryethoxylated surfactants. 1. Surface energy, surface entropy, and interaction energies as a function of temperature and surfactant concentration, *Langmuir* **13**, 1644-1652.
22. Urbina-Villalba, G., and Reif, I. (1996) Simple theoretical framework for a systematic molecular dynamics study of the interfacial properties of oil/water systems in the presence of surfactant molecules, *Colloids Surf. A* **106**, 175-190.
23. Manne, S., Cleveland, J.P., Gaub, H.E., Stucky, G.D., and Hansma, P.K. (1994) Direct visualization of surfactant hemimicelles by force microscopy of the electrical double layer, *Langmuir* **10**, 4409-4413.

24. Manne, S., and Gaub, H.E. (1995) Molecular organization of surfactants at solid-liquid interfaces, *Science* **270**, 1480-1482.
25. Jaschke, M., Butt, H.J., Gaub, H.E., and Manne, S. (1997) Surfactant aggregates at a metal surface, *Langmuir* **13**, 1381-1384.
26. Patrick, H.N., Warr, G.G., Manne, S., and Aksay, I.A. (1997) Self-assembly structures of nonionic surfactants at graphite/solution interfaces, *Langmuir* **13**, 4349-4356.
27. Bandyopadhyay, S, Shelley, J.C., Tarek, M., Moore, P.B., and Klein, M.L. Surfactant aggregation at a hydrophobic surface, *J. Phys. Chem. B* (in press).
28. Wijmans, C.M., and Linse, P. (1997) Monte Carlo simulations of the adsorption of amphiphilic oligomers at hydrophobic interfaces, *J. Chem. Phys.* **106**, 328-338.
29. McCabe, M.A., Griffith, G.L., Ehringer, W.D., Stillwell, W., and Wassall, S.R. (1994) ^{2}H NMR studies of isomeric $\omega 3$ and $\omega 6$ polyunsaturated phospholipid membranes, *Biochemistry* **33**, 7203-7210.
30. Holte, L.L., Peter, S.A., Sinnwell, T.M., and Gawrisch, K. (1995) ^{2}H nuclear magnetic resonance order parameter profiles suggest a change of molecular shape for phosphatidylcholines containing a polyunsaturated acyl chain, *Biophys. J.* **68**, 2396-2403.
31. Merz, K., and Roux, B. (eds), *Biological membranes: a molecular perspective from computation and experiment*, Birkhauser, Boston 1996.
32. Tobias, D.J., Tu, K., and Klein, M.L. (1997) Atomic-scale molecular dynamics simulations of lipid membranes, *Curr. Opin. Colloid Interface Sci.* **2**, 115-126.
33. Feller, S.E., Venable, R.M., and Pastor, R.W. (1997) Computer simulation of a DPPC phospholipid bilayer: Structural changes as a function of molecular surface area, *Langmuir* **13**, 6555-6561.
34. Gierula, P.M., Takaoka, Y., Miyagawa, H., Kitamura, K., and Kusumi A. (1997) Hydrogen bonding of water to phosphatidylcholine in the membrane as studied by a molecular dynamics simulation: Location, geometry, and lipid-lipid bridging via hydrogen-bonded water, *J. Phys. Chem. A* **101**, 3677-3691.
35. Husslein, T., Newns, D.N., Pattnaik, P.C., Zhong, Q., Moore, P.B., and Klein, M.L. Constant pressure and temperature molecular dynamics simulation of the hydrated diphytanolphosphatidylcholine lipid bilayer, *J. Chem. Phys.* (in press).
36. Feller, S.E., Yin, D., Pastor, R.W., and MacKerell, A.D. (1997) Molecular dynamics simulation of unsaturated lipid bilayers at low hydration: Parameterization and comparison with diffraction studies, *Biophys. J.* **73**, 2269-2279.

37. Hyvonen, M.T., Rantala, T.T., and Korpela, M.A. (1997) Structure and dynamic properties of diunsaturated 1-palmitoyl-2-linoleoyl-sn-glycero-3-phosphatidylcholine lipid bilayer from molecular dynamics simulation, *Biophys. J.* **73**, 2907-2923.

ON EFFECTIVE INTERACTIONS IN COLLOIDAL PHYSICS

T. BIBEN
Laboratoire de Physique de l'ENS-Lyon (URA 1325 CNRS)
46 Allée d'Italie, 69364 Lyon Cedex 07, France

1. Introduction

Colloidal dispersions are made of "large" particles suspended in a solvent of "small" molecules (usually water). A particle can be considered as a colloid if its individual dynamics is essentially Brownian. Therefore, a colloidal particle must be large compared to the size of a solvent molecule, but not too massive in order to prevent fast sedimentation. Experimentally the size of these particles can range from $2nm$ to $1\mu m$ [1]. Although these boundaries are only indicative, as consequence it is possible to prepare mixtures of large and small colloidal particles with a diameter ratio of some few hundred, which is significantly different from usual atomic fluids. Another important, and more fundamental difference, is polydispersity in size and shape: in a colloidal suspension particles are not identical, and the width of the size distribution depends on the preparation process (some few percent in model systems).

From the point of view of statistical mechanics, and if for simplicity we assume monodispersity, a colloidal suspension can be modelled by a binary mixture of large and small particles, standing for the colloidal particles and the molecules of the solvent, respectively. Although the formalism for binary mixtures is well known [2], the problem is complicated here due to the different time and length scales involved. Some attempts have been made to apply liquid theory to the full mixture in order to describe the structure of colloidal suspensions using Integral Equation theory [3, 4, 5, 6], or to investigate relaxation processes using kinetic theory [7]. However, the lack of simulation data, due to the so-called ergodicity problem [8], rends predictions of these theories very difficult to control.

For these reasons, a simplified model, which assumes water as a continuum, is usually prefered. If we consider static properties such as the pair structure, the problem can be reduced to a single component system in-

C. Caccamo et al. (eds.), New Approaches to Problems in Liquid State Theory, 359–378.

teracting through an effective static potential. Hydrodynamic forces must be included when investigating dynamical properties. Statistical mechanics can be applied provided the effective interaction between the colloidal particles is known. Many studies have been devoted to the determination of these effective interactions, and the most successful theory is due to Derjaguin, Landau, Verwey and Overbeck (DLVO) [9]. The DLVO theory leads to an effective pair interaction between the colloidal particles, which includes steric (i.e. core repulsions between colloids) and electrostatic effects like the Van der Waals attraction, due to a difference in polarizability between the colloidal particles and the solvent, and the screening of the structural charges of the colloids by the co- and counter-ions of the solvent. This theory is still the most successful for colloidal particles. However, at low ionic strength, colloidal systems exhibit phase transitions suggesting existence of a long range attraction between colloidal particles [10, 11]. This attraction is not consistent with the repulsive nature of the DLVO potential and some authors have suggested that the solution to this problem could be in the density dependence of the effective potential [12]. This raises the problem of the thermodynamic ensemble in which the effective interaction is defined. Shall we consider the solvent as a reservoir fixing the density or the chemical potential of the ionic species ?

To obtain an answer to this question it is useful to reconsider the derivation of the effective interaction presented above from a microscopic point of view. Moreover, this microscopic model, accounting for the discrete nature of the solvent, enables us to discuss a feature that is not included in the DLVO theory: the layering of the solvent molecules around the colloidal particles and the consequences for the effective force. As we shall see, although this force occurs at very short distances (some few diameters of the solvent molecules), its strength can be large enough to induce structural changes.

2. Microscopic description

The model we consider here is a binary mixture of small and large particles, the solvent molecules and the colloidal particles (or alternatively a mixture of small and large colloidal particles). This model can easily be extended to an arbitrary number of components, to account for polydispersity or to treat more complicated solvents. For simplicity, the bare interactions between particles are assumed to be described by pair potentials. Effective interactions between colloidal particles are obtained formally by integrating over the degrees of freedom of the solvent molecules. The degrees of freedom (positions and momenta) of the colloidal particles will be denoted by Γ_c, and the degrees of freedom of the solvent molecules by Γ_s.

If V^{cc}, V^{ss} and V^{cs} denote the interaction energies between the N_c colloidal particles, the N_s solvent molecules, and the crossed interactions, respectively, and if K^c and K^s represent the kinetic energies of the colloids and of the solvent molecules, respectively, then the Hamiltonian of the system reads:

$$H(\Gamma_c, \Gamma_s) = K^c(\Gamma_c) + K^s(\Gamma_s) + V^{cc}(\Gamma_c) + V^{ss}(\Gamma_s) + V^{cs}(\Gamma_c, \Gamma_s) \quad (1)$$

Following the usual statistical mechanics prescription, macroscopic quantities like the pair structure can be obtained by averaging some microscopic quantities (or observables) over the possible configurations of the particles. The probability distribution of these configurations depends on the external constraints. If the quantities we consider do not depend on the degrees of freedom of the solvent molecules, then the probability distribution can be integrated with respect to the degrees of freedom of these molecules and expressed in terms of an effective potential. A very convenient statistical ensemble for the calculation of the phase diagram is the grand-canonical ensemble. In this ensemble, the system can exchange particles with a reservoir fixing the chemical potential. But to go further, two limiting cases must be considered:

2.1. NEUTRAL FLUIDS

If the particles are neutral, the number of solvent molecules and the number of colloidal particles can be varied independently. Thus, in the grand-canonical ensemble, the two chemical potentials μ_s and μ_c are independent variables. The resulting expression for the average of an observable $A(\Gamma_c)$ is [2]:

$$\begin{aligned} < A >= \quad & \frac{1}{\Xi} \sum_{N_c=0}^{+\infty} \frac{h^{-3N_c}}{N_c!} \sum_{N_s=0}^{+\infty} \frac{h^{-3N_s}}{N_s!} \\ & \int\int d\Gamma_c d\Gamma_s A(\Gamma_c) \exp\left\{ -\beta \Big[H(\Gamma_c, \Gamma_s) - \mu_c N_c - \mu_s N_s \Big] \right\} \quad (2) \end{aligned}$$

where Ξ is the partition function (normalization of the probability) and h is Planck's constant. β is the inverse temperature $1/k_B T$. Extracting from the exponential all the contributions that do not depend on Γ_s and N_s, equation (2) can be rewritten:

$$\begin{aligned} < A >= \quad & \frac{1}{\Xi} \sum_{N_c=0}^{+\infty} \frac{h^{-3N_c}}{N_c!} \\ & \int d\Gamma_c A(\Gamma_c) \exp\left\{ -\beta \Big[H^c(\Gamma_c) - \mu_c N_c \Big] \right\} \Xi^s[V^{cs}(\Gamma_c, \Gamma_s)] \quad (3) \end{aligned}$$

where $H^c(\Gamma_c) = K^c(\Gamma_c) + V^{cc}(\Gamma_c)$ is the bare Hamiltonian of the colloidal particles and $\Xi^s[V^{cs}(\Gamma_c, \Gamma_s)]$ is the partition function of the solvent placed in the external potential of the colloidal particles in configuration Γ_c. By definition:

$$\Xi^s = \sum_{N_s=0}^{+\infty} \frac{h^{-3N_s}}{N_s!} \int d\Gamma_s \exp\left\{-\beta\left[K^s(\Gamma_s) + V^{ss}(\Gamma_s) + V^{cs}(\Gamma_c, \Gamma_s) - \mu_s N_s\right]\right\} \quad (4)$$

Ξ^s can be inserted in the exponential to give :

$$< A >= \frac{1}{\Xi} \sum_{N_c=0}^{+\infty} \frac{h^{-3N_c}}{N_c!} \int d\Gamma_c A(\Gamma_c) \exp\left\{-\beta\left[K^c(\Gamma_c) + \mathcal{V}^{eff}(\Gamma_c) - \mu_c N_c\right]\right\} \quad (5)$$

where the effective potential is defined by:

$$\mathcal{V}^{eff}(\Gamma_c) = V^{cc}(\Gamma_c) - k_B T \ln\left\{\Xi^s[V^{cs}(\Gamma_c, \Gamma_s)]\right\} \quad (6)$$

This last equation shows that in the grand canonical ensemble, the effective potential is the sum of the bare interactions between the colloidal particles and the grand potential of the solvent in the external field these generate. It must be noted that this effective potential is a function of N_c and of the chemical potential of the solvent. This result is not a surprise, but before discussing it, it is interesting to restart the derivation for the case of charged particles.

2.2. CHARGED FLUIDS

If colloidal particles carry an electrostatic charge, the solvent must be charged as well. As a result, the constraint of charge neutrality does not allow to fix the number of colloidal particles and the number of solvent molecules independently. Therefore, in the grand-canonical ensemble there is only one chemical potential. We can rewrite all the equations of the previous subsection by setting $\mu_s = 0$ and removing the summation over N_s. The final expression we obtain for the effective potential is:

$$\mathcal{V}^{eff}(\Gamma_c) = V^{cc}(\Gamma_c) - k_B T \ln\left\{Q^s[V^{cs}(\Gamma_c, \Gamma_s)]\right\} \quad (7)$$

where now Q^s is the *canonical* partition function of the solvent in the external potential of the colloidal particles. As a result, even if we fix the chemical potential of the colloidal particles, the effective potential is the sum of the bare interactions and the *Helmholtz* free energy of the solvent in the external field of the colloids. The effective potential is now a function of the *density* of the colloidal particles.

2.3. DISCUSSION

Since the effective potential is related to either the free energy or the grand potential of the solvent in the external field of the colloids, there is no reason to consider it pairwise additive. From the point of view of computer simulations, expressions (6) and (7) show that it is possible, in principle, to compute the structural properties of colloidal suspensions by combining Monte Carlo moves for the colloids with Density Functional Theory (DFT) for the solvent [13]. However, as we shall see, in some cases, such as structural forces, pairwise additivity can be a good approximation [14, 15]. Although, special care must be taken in such cases. If we consider an expansion of the effective potential in terms of a constant term, a single particle contribution, a pair contribution, and so on, pairwise additivity results from the truncation of this expansion to the third term:

$$\mathcal{V}^{eff}(\Gamma_c) = V_0 + \sum_{I=1}^{N_c} V_1(\mathbf{X}_I) + \sum_{I<J} V_2(\mathbf{X}_I, \mathbf{X}_J) + \mathcal{O}(\text{triplet}) \qquad (8)$$

where $\mathbf{X}_I$ and $\mathbf{X}_J$ are the positions of colloidal particles I and J. $\mathcal{O}(\text{triplet})$ represents negligible triplet contributions. When the system is translationally invariant, equation (8) can be rewritten :

$$\mathcal{V}^{eff}(\Gamma_c) = V_0 + N_c V_1 + \sum_{I<J} V_2(|\mathbf{X}_I - \mathbf{X}_J|) + \mathcal{O}(\text{triplet}) \qquad (9)$$

For neutral fluids, V_0, V_1 and V_2 are functions of the chemical potential of the solvent. As a result, V_0 can be removed from the potential and V_1 can be included in the chemical potential of the colloids. With this redefinition of the chemical potential of the colloids, i.e. $\mu_c^{eff} = \mu_c - V_1$, the pair potential V_2 can be used in standard liquid theory to investigate the phase diagram.

On the contrary, for charged fluids V_0, V_1 and V_2 are functions of the *density* of the colloidal particles. As a result none of these contributions can be removed from or included in an effective chemical potential. Worst, the effective potential is now *state-dependent.*

As mentioned previously, these two cases are limiting cases. In a real suspension there are usually extra ionic species (salt) and the situation is intermediate. Indeed, salt can be treated as a neutral system since its concentration and the concentration of the colloidal particles can be fixed independently. As a consequence, the effective potential exhibits a contribution where the salt is treated in the grand-canonical ensemble in spite of the counter-ions being treated in the canonical ensemble. The resulting effective potential is a function of the chemical potential of salt and of the density of the colloids. At large ionic strength, salt is the dominant ionic species, and as a result the effective potential depends mostly on the chemical potential of salt and not on the colloidal concentration. However, at very low ionic strength, the contribution of the counterions can become important, and it should be possible to observe a crossover to a regime where the effective potential becomes density dependent.

We will now focus on structural interactions.

3. Structural interactions

This type of interaction occurs when large (colloidal) particles are placed in a bath of small particles. The layering of the small particles around the colloids induces an oscillating force that can be either attractive or repulsive depending on the separation between the colloidal particles. This force is similar in nature to the structural force observed by Horn and Israelaschvili [16] when a thin liquid film is confined between two plates. This force is strongly attractive at short separations due to the "osmotic depletion effect". This effect has been introduced by Asakura and Oosawa in 1958 [17] to explain reversible flocculation observed in colloid-polymer mixtures. In these experiments, polymers play the role of the small particles. The depletion effect is due to volume exclusion between a large and a small particle. If we represent the small and the large particles by hard spheres of diameter σ_s and σ_l, respectively, the depletion effect occurs when two large particles are separated by a distance smaller than σ_s (the depletion well). In this case, the small particles are expelled from the region separating the large particles. The resulting reduction of the local pressure in this region induces a strong attraction between the colloids.

The problem is to know the consequences of this structural force on the phase diagram of the large (colloidal) particles. Most of the results presented below will be found in reference [15]. For this reason, we shall discuss the main features only.

3.1. THE MODEL

The simplest model we can imagine to investigate structural forces is that of a hard sphere binary mixture of small and large particles (diameters σ_s and σ_l respectively). However, as we have mentionned, the key ingredient for the depletion effect is the exclusion volume between a large and a small sphere. An exclusion volume between the small spheres is not a necessity (although it strongly affects the osmotic pressure and the oscillations outside the depletion well). We can then consider the generic class of non-additive hard sphere mixtures for which the minimum distance between a large and a small sphere is $(\sigma_s + \sigma_l)(1 + \Delta)/2$. Δ is the non-additivity parameter, and when $\Delta = 0$ the mixture reduces to the usual (additive) hard sphere model. The pair structure of this model have been investigated for nearly symmetrical systems ($\sigma_s/\sigma_l \simeq 1$) [18], but a similar study for very asymmetric mixtures ($\sigma_s/\sigma_l \ll 1$) still lacks. The model considered by Asakura and Oosawa [17] for colloid-polymer mixtures corresponds to the extreme case where $\sigma_s = 0$, polymers can freely overlap each other, and the exclusion volume between a colloidal particle and a polymer is fixed by $\Delta > 0$. The model we will consider in the following is the additive hard sphere mixture for which $\Delta = 0$ and σ_s is small but finite.

3.2. MICROSCOPIC THEORIES

In this part, the small and the large spheres are treated at the same level. A description of the structural and thermodynamical properties of the mixture is provided by the pair correlation functions of the fluid. These correlation functions provide informations on the local organization of the particles in the fluid. The pair distribution functions $g_{\alpha\beta}(r)$ give the density of particles of type "α" at a distance "r" from a central particle of type "β". In Fourier space, the correlation functions are related to the structure factors of the fluid, which represent density fluctuations at a given wavevector "k", and can be measured in diffraction experiments. Thermodynamical quantities like the compressibility are obtained in the $k \to 0$ limit. This last property allows us, in principle, to investigate the phase diagram of the fluid. The usual starting point for the theoretical calculation of pair-correlation functions in liquids is the well known Ornstein-Zernike equations [19]:

$$h_{\alpha\beta}(r) = c_{\alpha\beta}(r) + \sum_{\gamma} \rho_\gamma c_{\alpha\gamma}(r) * h_{\gamma\beta}(r) \tag{10}$$

where $h_{\alpha\beta}(r)$ is the pair correlation function for particles of species α and β respectively. This pair correlation function is related to the pair distribution function $g_{\alpha\beta}(r)$ through the simple equation $h_{\alpha\beta}(r) = g_{\alpha\beta}(r) - 1$. $c_{\alpha\beta}(r)$ is the Ornstein-Zernike direct correlation function, and ρ_α is the number density of "α" particles. "$*$" denotes a convolution. It is clear that equations (10) do not constitute a closed system, and a set of closure relations is necessary to compute both $g(r)$'s and $c(r)$'s. Every closure of Eqn. (10) leads to an integral equation for $g(r)$. The search for good approximate closure relations of Eqn. (10) has resulted in a large number of such integral equations. For hard-sphere mixtures, the first successful integral equation was the famous Percus-Yevick equation [20]. However, more recent integral equations, such as the ones that result from the closures proposed by Rogers and Young (RY) [21] and by Ballone, Pastore, Galli and Gazzillo (BPGG) [22], have proven to provide a more accurate estimate of the pair structure of a hard sphere fluid, in particular at high densities. Introducing the notation $\gamma_{\alpha\beta}(r) = h_{\alpha\beta}(r) - c_{\alpha\beta}(r)$, the PY, RY and BPGG closure relations take the form:

$$g_{\alpha\beta}(r) = \exp(-\beta U_{\alpha\beta}(r))\left[1 + \gamma_{\alpha\beta}(r)\right], \text{ (PY)} \tag{11}$$

where $U_{\alpha\beta}(r)$ is the pair potential between a particle of type "α" and a particle of type "β". The factor β appearing in the exponential is the inverse temperature $\beta = 1/k_B T$.

$$g_{\alpha\beta}(r) = \exp(-\beta U_{\alpha\beta}(r))\left[1 + \frac{\exp(\gamma_{\alpha\beta}(r) f_{\alpha\beta}(r)) - 1}{f_{\alpha\beta}(r)}\right], \text{ (RY)} \tag{12}$$

where the function $f_{\alpha\beta}(r) \equiv 1 - \exp(-\xi_{\alpha\beta} r)$ varies between 0 and 1. And finally:

$$g_{\alpha\beta}(r) = \exp(-\beta U_{\alpha\beta}(r)) \exp\{[1 + s\gamma_{\alpha\beta}(r)]^{1/s} - 1\}. \text{ (BPGG)} \tag{13}$$

In the latter two expressions, additional adjustable parameters occur, namely "$\xi_{\alpha\beta}$" for the RY closure and "s" for the BPGG approximation. However, these parameters can not be chosen at will. They take values that eliminate the so-called thermodynamical inconsistency that plagues many approximate integral equations, such as PY. A manifestation of this inconsistency is that there is a difference between the pressure as computed from the virial equation of state, and as computed from the compressibility relation. A detailed discussion of these integral equations applied to very asymmetric mixture of hard spheres can be found in reference [23], and partially in references [4, 6, 24, 25]. We briefly summarize the most important theoretical predictions here, namely the stickiness and the predicted phase diagram for a size ratio $y = \sigma_s/\sigma_l = 0.1$.

The first important result have been obtained within the PY approximation. In the "colloidal" limit, i.e. the packing fractions $\eta_s = \pi\rho_s\sigma_s^3/6$ and $\eta_l = \pi\rho_l\sigma_l^3/6$ are fixed and the size ratio $y = \sigma_s/\sigma_l$ goes to zero, it has been shown that a δ-peak appears in the pair distribution function of the large spheres [24], and that the fluid of large spheres reduces to the well known sticky-sphere model due to Baxter [25]. The Baxter's model is known to present a spinodal instability when stickiness of the particles

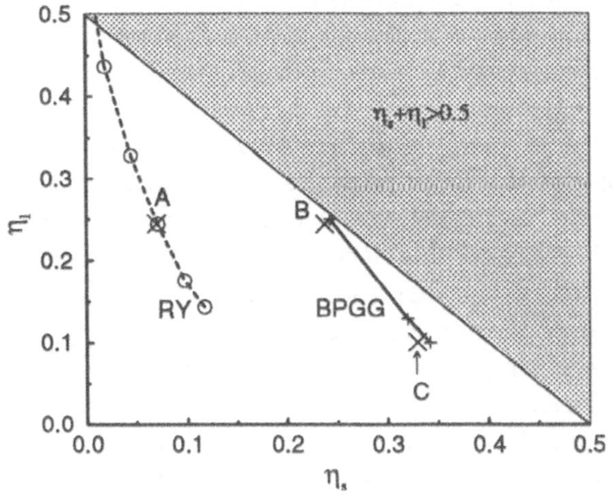

Figure 1. **Predicted phase diagram for a size ratio of 0.1. The dashed line is the predicted RY spinodal line and the full line corresponds to the region where the BPGG theory fails to converge. Below these two lines the system is stable (or metastable) whereas above them the mixture is expected to phase separate. We note the discrepancies between the two theories. The shaded region corresponds to a total packing fraction larger than 0.5 where solidification should occur. Simulation data will be presented for points A, B and C.**

becomes large. However, in the colloidal limit, the hard-sphere mixture we consider is always in the stable regime. Indeed, it is well known from the work of Lebowitz and Rowlinson [27, 28] that the PY approximation predicts a complete miscibility of the two components. However, the PY approximation is known to give non-physical results at small values of the size ratio y, for example the pair distribution function becomes negative at distances close to contact [23]. The self-consistent closures RY or BPGG show that PY strongly underestimates the stickiness. Moreover, these two integral equations exhibit singularities that are not present in the PY approximation. If we consider the concentration fluctuations, the RY closure predict a spinodal instability [4] and the BPGG closure a strong enhancement of the fluctuations. Interpretation of these singularities in terms of a phase diagram is always difficult since we know from the work of Belloni [29] that other integral equations, like HNC, can present spurious singularities. The results we obtained are presented in figure 1. In figure 1, points A, B and C represent the thermodynamical states we used to perform Monte Carlo (MC) simulations. From these results we can conclude that the self-consistent theories tend to favor the stickiness of the large spheres, and their tendency to phase separation. Such a phase transition is very interesting from the point of view of thermodynamics since temperature does not play any role. In usual transitions like the liquid-vapor transition, temperature controls the competition between energy and entropy. In this case, the competition involves the entropy of the large particles and the entropy of the small ones. The entropy of the small particles increases with the accessible volume left by the large spheres, and this accessible volume is maximized when the large spheres are sticked together and form a dense configuration. On the contrary, the entropy of the large spheres is maximized when they occupy all the possible space. Examples of purely entropic phase transitions have been obtained by considering small and large parallel hard cubes [30], system for which the depletion effect is strongly enhanced, or lattice models [31, 32]. However, for large and small sphere mixtures, the competition looks to be more subtle, and theories are very sensitive to approximations. Unfortunatly, MC simulations are very difficult to perform in the region of the phase diagram where integral equations predict a spinodal instability. One possible solution is to consider the effective model for which the degrees of freedom of the small spheres have been integrated out.

3.3. THE MEAN FORCE

In the first section we have derived generic expressions for the effective potential. Although these expressions are formally exact, the grand potential of the small particles in the external field of the large spheres is not very easy to handle. More interesting is the mean force acting on a large particle. The mean force is the gradient of the effective potential with respect to the coordinates of the large particle. If we extract the bare interaction between the large particles, the mean force induced by the small component on particle "K" can be written [15]:

$$\mathcal{F}_K(\{\mathbf{X}_I\}) = \int d\mathbf{r}\ \nabla U_{sl}(r)\ \rho^{(1)}(\mathbf{r}|\{X_I\})\ , \tag{14}$$

where $U_{sl}(r)$ is the bare pair potential between a large and a small sphere, $\{\mathbf{X}_I\}$ denotes the positions of the large spheres and $\rho^{(1)}(\mathbf{r}|\{X_I\})$ is the density of the small particles at position $\mathbf{r}$ in the field of the large particles. For hard spheres, this expression reduces to [14]:

$$\beta\mathcal{F}_K(\{X_I\}) = \xi^2 \int d\hat{\sigma}\rho^{(1)}(-\xi^+\hat{\sigma}|\{\mathbf{X}_I\})\hat{\sigma} \tag{15}$$

where $\xi = (\sigma_s + \sigma_l)/2$ is the radius of the pair excluded volume and $\hat{\sigma}$ is a unit vector pointing towards the center of the large sphere. The integral is

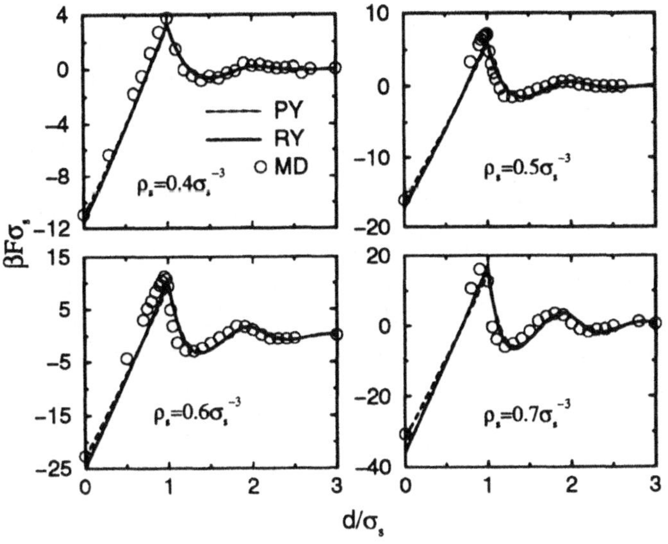

Figure 2. Effective force between two colloids. MD simulations versus the superposition approximation combined with the PY and RY theories for $g_{sl}(r)$ (see eq. (16)). d is the separation between the surfaces of the colloids, and a positive value of the force corresponds to a repulsion. The force is measured at fixed chemical potential μ_s. ρ_s corresponds to the density of the small spheres in the reservoir. These figures illustrate the very short range of the force $\sim 3\sigma_s$.

then calculated just outside the surface of the excluded volume associated to colloidal particle "K". The interpretation of Eqn. 15 is very simple: the mean force acting on colloidal particle "K" is simply the surface integral of the normal component of the instantaneous local pressure at the surface of this particle. Of course, the above "derivation" is static in nature and says nothing about the time scale on which this mean force is a meaningful concept. It is therefore interesting to note that Eqn. 15 has also been derived from a multiple time scale analysis of kinetic theory in reference [7].

3.4. PAIRWISE ADDITIVITY OF THE MEAN FORCE

Like the effective potential, the mean force depends on the positions of the large particles, and on the chemical potential of the small particles. A great simplification occurs when the mean force can be written as a sum of pair interactions. In this case, we can define an effective pair potential that depends only on the chemical potential of the small spheres. When a phase separation occurs, this dependence ensures the effective potential to be the same in both phases. A simple test of pairwise additivity can be performed by considering three colloidal particles in a bath of small spheres. The test consists in a comparison between the force felt by one of the colloids surrounded by its two neighbors to the sum of the forces induced by each neighbor independently. Thus, we have to compute $\mathcal{F}_1(\mathbf{X}_1, \mathbf{X}_2, \mathbf{X}_3) - \mathcal{F}_1(\mathbf{X}_1, \mathbf{X}_2) - \mathcal{F}_1(\mathbf{X}_1, \mathbf{X}_3)$, where "1" denotes the central colloid. Expression (15) relates the force felt by one of the colloids to the distribution of the small spheres at contact. Although integral equations provide good estimates of $g_{sl}(r)$, the distribution of the small spheres around a single large spheres, the theories presented above do not give informations on the distribution of the small particles around two or three large spheres. Some extensions have been studied to provide these distribution functions [14], but we will use here a very simple ansatz: the superposition approximation. This approximation assumes that the distribution of the small particles around two colloids can be obtained by multiplying the distributions around a single colloid. The resulting density profile for two colloidal particles reads:

$$\rho^{(1)}(\mathbf{x}|\mathbf{X}_1, \mathbf{X}_2) \simeq \rho_s g_{sl}(\mathbf{x} - \mathbf{X}_1) g_{sl}(\mathbf{x} - \mathbf{X}_2) \tag{16}$$

and for three colloids:

$$\rho^{(1)}(\mathbf{x}|\mathbf{X}_1, \mathbf{X}_2, \mathbf{X}_3) \simeq \rho_s g_{sl}(\mathbf{x} - \mathbf{X}_1) g_{sl}(\mathbf{x} - \mathbf{X}_2) g_{sl}(\mathbf{x} - \mathbf{X}_3) \tag{17}$$

Such an approximation is expected to be very crude from the point of view of $\rho^{(1)}$. Indeed, at short distances, where the layering of the small particles is the most pronounced, we expect the superposition approximation to overestimate the degree of order of the small particles. An eventual breakdown of pairwise additivity would come precisely from the layering of the small particles, this drawback of the superposition approximation rends the test even more severe. The worst situation for pairwise additivity is when the three colloids are sticked together. To test pairwise additivity of the force, we have considered the special case where the central colloidal particle is sticked to its two neighbors, and the angle between the neighbors goes from 180 degrees (the three particles are aligned) to 60 degrees which correspond to close packing. The largest observed deviation to pairwise additivity is only 0.5% of the force! From this we conclude that pairwise additivity is not only a good approximation, but an excellent one. Moreover, measurements of the effective force between two colloids by Molecular Dynamics (MD) simulations showed that the two particle superposition approximation give amazingly good results [15]. The same conclusion has also been obtained by Attard [14] on the basis of a more sophisticated theory. From the measurement of the effective force between two colloids, we can extract the effective pair potential by integration:

$$\beta V(r_s) = \int_{r_s}^{+\infty} \beta \mathcal{F}(r_s') \sigma_s dr_s' \tag{18}$$

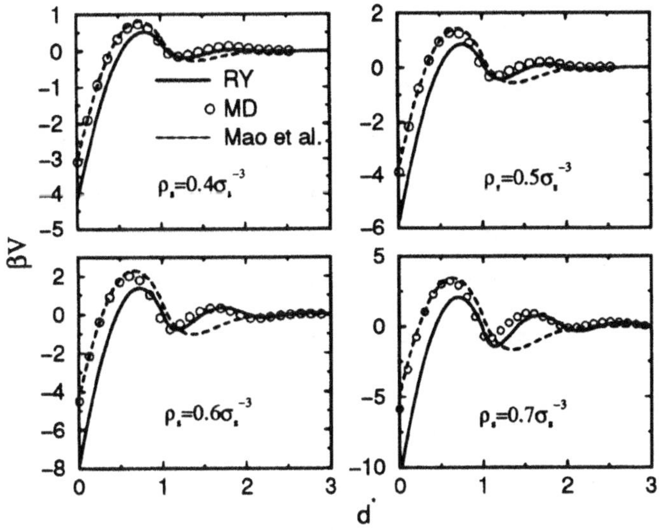

Figure 3. Effective potential between two colloidal particle. MD simulations versus the superposition approximation, and the virial theory presented by Mao et al. in reference [33].

The pair potential we can extract from equation (18) is defined up to a constant. This constant can be included in the constant term V_1 that appears in equation (9) and then in the effective chemical potential of the large spheres also. Data for the effective potential is presented in figure 3. It is interesting to compare this data obtained from MD simulations, and integral equations using the superposition approximation to other theories based on virial expansions of the potential [33] (see also reference [34]).

3.5. TOWARDS THE PHASE DIAGRAM

Once the effective potential is known, the pair structure of the large spheres can be computed in a MC simulation.

State point A of the phase diagram presented in figure 1 is an interesting point since at such values of η_l and η_s it is possible to compare microscopic approaches to the effective potential approach. Predictions of integral equations can be compared to MC simulations of the full binary mixture and to MC simulations of the effective fluid using the effective potential. In all simulation data of the effective model presented here, the number of colloidal particles in the simulation box is 2700. We observe in figures 4 and 5 an excellent agreement between all these approaches. The RY predictions

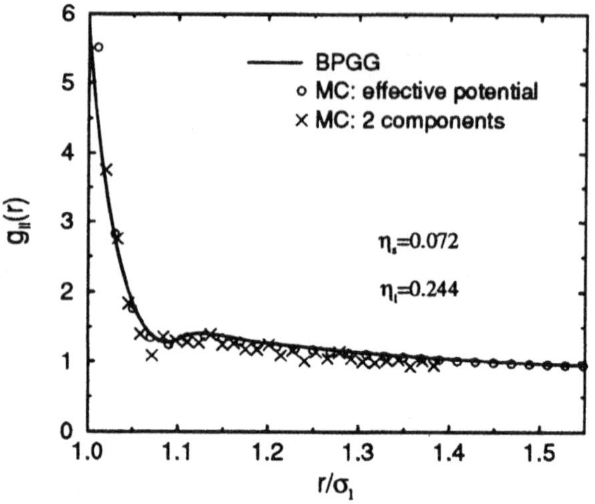

Figure 4. Pair distribution function of the colloidal particles at state point A. The BPGG theory is compared to simulations of the full binary mixture and of the effective model.

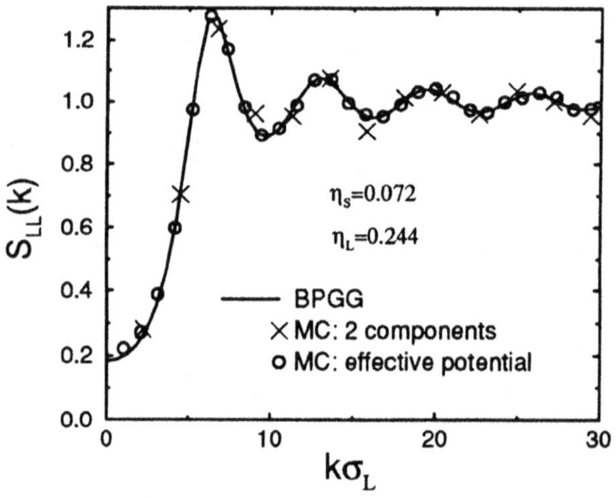

Figure 5. Structure factor of the colloidal particles at state point A. Same comment as in figure 4.

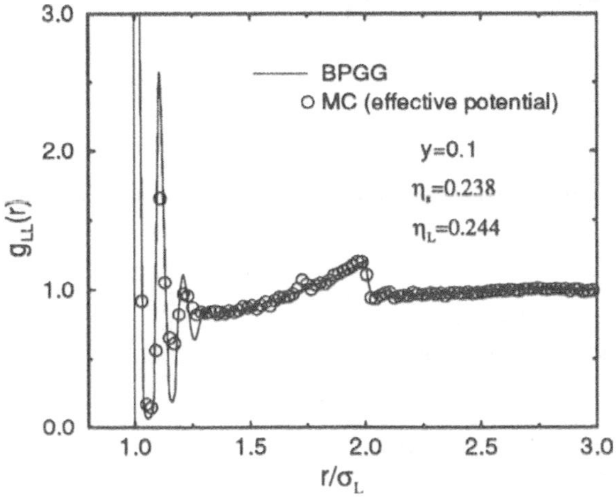

Figure 6. **Pair distribution function of the colloidal particles at state point B. We note the good agreement between the BPGG theory and MC data for the effective fluid.**

have not been plotted since state point A lies on the spinodal line predicted by this theory. As no singular behavior is observed in figure 4 and 5 we conclude that the RY theory overestimates the instability of the hard sphere mixture.

State point B is quite interesting since BPGG theory predicts important concentration fluctuations in this region. Unfortunately no simulation data for the full mixture is available in this region. Therefore, in figures 6 and 7 we compare the predictions of the BPGG theory with the effective model only. First of all, we can see that in figure 6 the pair distribution function is qualitatively different from the result presented in figure 4, for state point A. The sharp peak at contact suggests strong bonds between the colloidal particles. The contact peak is characteristic of pair formation, and the plateau between $1 < r/\sigma_l < 2$ indicates triplet formation. There is, however, no sign of long range order. Indeed, a comparison of the structure factors (figure 7) indicates that the concentration fluctuations at small k (i.e. the osmotic compressibility) are relatively large for both theory and simulation. It must be noted that due to the large packing fractions we can expect the solid phase to be in the neighborhood of point B. However, in the BPGG theory, solid phases are not taken into account since the system is assumed to be homogeneous from start. Thus, the enhancement of fluctuations observed in the BPGG structure factors cannot be due to

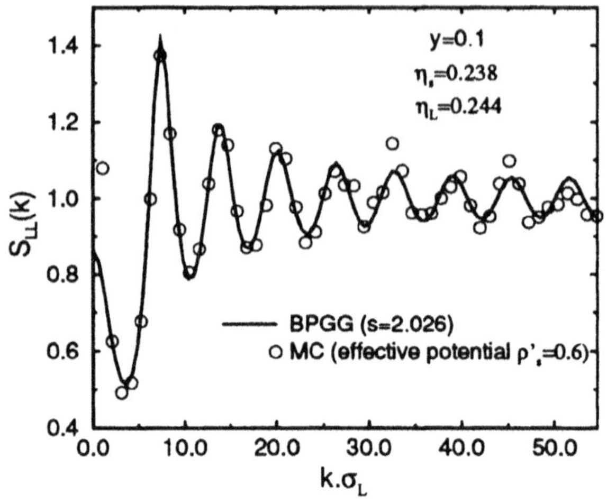

Figure 7. Structure factor of the colloidal particles at state point B. Same comment as in figure 6.

the proximity of a crystalline phase.

State point C is also in the region where BPGG predicts important fluctuations, although simulation data now exhibits quite a different behavior. First of all, the pair distribution function presented in figure 8 is no more stationary, and after 2×10^6 MC cycles (one cycle is one trial move for each particle), which is quite a long run, the contact value reaches values as large as 400! (the BPGG prediction is 90). This contact peak is very narrow (only 5% of the diameter of the *small* particles). The number of nearest neighbors is proportional to the integral of this peak, and it remains of the order of 4, as we can see in figure 9. This figure represents the evolution of the histogram of the number of particles contained in the contact peak during the simulation. This histogram is relatively broad, showing that the system is quite disordered. It is interesting to note the emergence of a peak in figure 8 at $r \simeq 1.75\sigma_l$. This peak corresponds to a bipyramidal structure, and this type of structure has been observed in MC simulations of sticky spheres [35]. In fact, we observe here two types of processes. The fast process is the formation of a huge percolating cluster containing half of the particles, as we can see from a cluster analysis in figure 10. The slow process is the structural reorganization of the cluster, with the emergence of local bipyramidal structures. This reorganization of the cluster is very slow since it requires bond breaking. The final equilibrium state is still un-

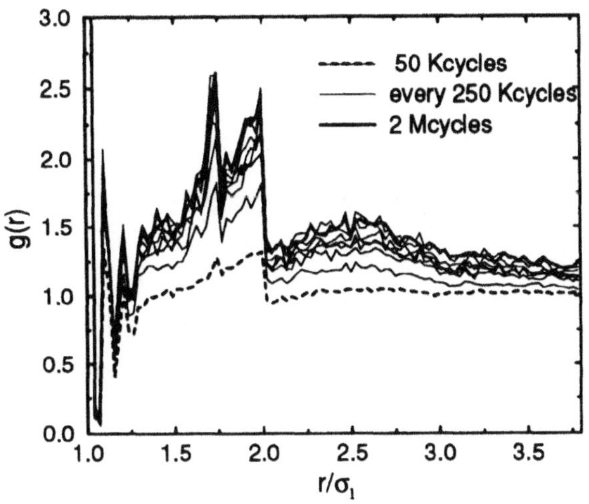

Figure 8. Pair distribution function of the colloidal particles at state point C. Even after 2×10^6 MC cycles (see the text), the pair structure has not reach equilibrium.

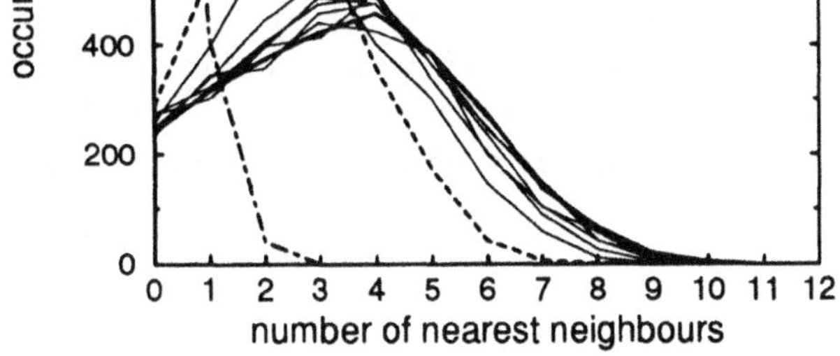

Figure 9. Histogram of the number of neighbors contained in the contact peak of the pair distribution function.

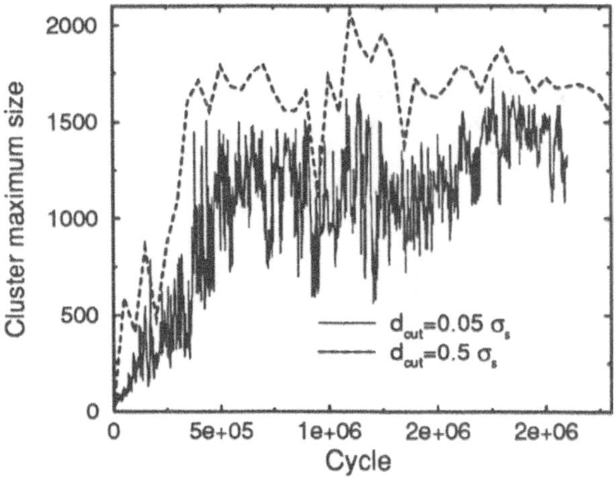

Figure 10. Evolution of the number of particles contained in the largest cluster during the simulation. d_{cut} is the cutoff used to identify connected particles. Since the contact peak $(g_{ll}(\sigma_l))$ is very narrow, the results presented here are relatively insensitive to the value of d_{cut}.

known, but recent simulations tend to favor the formation of a crystalline phase [36, 37].

From this study we conclude that the depletion effect is able to induce an instability of the fluid phase, which confirms the predictions of the integral equation theory. However, instead of a fluid-fluid phase separation, we observe the formation of a gel. Although MC simulations do not give access to the actual dynamics of the system, it gives us strong indications that a dynamical slowing down should be observed in the neighborhood of the transition. Finally, following the conclusion in reference [36], it is tempting to conjecture that the observed gel-like system is the result of a competition between a fluid-fluid phase separation and a fluid-solid phase transition.

4. Conclusion

In the first part we have discussed the definition of the effective potential induced between the colloidal particles from the point of view of statistical mechanics. From this study we may conclude that extra constraints, like charge neutrality, can strongly affect the definition of the effective potential. Moreover, the density dependence resulting from the constraint of charge neutrality not only affects the pair potential V_2, when pairwise additivity is a good approximation, but also V_0 and V_1 (see equation (9)). These contributions appear in the effective potential and cannot be removed in general. Although the introduction of effective interactions strongly simplifies the problem, the effective potential must be manipulated with extreme care. In the very important exemple of structural forces, the effective potential can be used to investigate the phase diagram. This reduction of the problem enabled us to confirm the existence of a spinodal instability of the fluid phase. Moreover, a very interesting gel-like structure has been observed, but the actual equilibrium phase is still unknown. Although hard spheres constitutes the simplest fluid we can imagine, it remains one of the hardest systems to study.

References

1. Pusey, P.N., (1991) in "Liquids, Freezing and the Glass Transition", Les Houches 1989 edited by J.P. Hansen, D. Levesque and J. Zinn-Justin, North Holland, Amsterdam.
2. Hansen, J.P. and Mc Donald,I.R.,(1986) "Theory of Simple Liquids", 2nd edition, Academic Press, London.
3. Attard, P., and Patey, G.N., (1990) *J. Chem. Phys.* **92**, 4970.
4. Biben, T., and Hansen, J.P., (1991) *Phys. Rev. Lett.*, **66**, pp. 2215.
5. Biben, T., and Hansen, J.P., (1991) *J. Phys. Cond. Matter*, **42** v 3, pp. F65.
6. Caccamo, C., and Pellicane, G., (1997) *Physica*, **A 235**, pp. 149.
7. Piasecki, J., Bocquet, L., and Hansen, J.P., (1995) *Physica*, **A 218**, pp. 125.
8. Fries, P. and Hansen, J.P., (1983) *Molec. Phys.*, **48**, pp. 891.
9. Verwey, E.J.W. and Overbeek, J.T.G., (1948) "Theory of the Stability of Lyophobic Colloids", Elsevier, Amsterdam.
10. Ito, K., Hiroshi H. and Ise N., (1994) *Science* **263**, pp. 66; Tata, B.V.R., Yamahara, E., Rajamani, P.V., and Ise, N., (1997) *Phys. Rev. Lett.* **78**, pp. 2660.
11. Tata, B.V.R., Rajalakshmi, M., and Arora, A.K., (1992) *Phys. Rev. Lett.* **69**, pp. 3778; Palberg, T., and Würth, M., (1994) *Phys. Rev. Lett.* **72**, pp. 786.
12. Van Roij, R., and Hansen, J.P., (1997) *Phys. Rev. Lett.* **79**, pp. 3082.
13. Löwen, H., Hansen, J.P., and Madden, P., (1992) *Phys. Rev. Lett.* **68**, pp. 1081,
14. Attard, P., (1989) *J. Chem. Phys.* **91**, pp. 3072 and pp. 3083.
15. Biben, T., Bladon, P., and Frenkel, D., (1996) *J. Phys. Cond. Matter* **8**, pp. 10799.
16. Horn, R. G., and Israelachvili, J. N., (1981) *J. Chem. Phys.* **75**, pp. 1400.
17. Azakura, S., and Oosawa, F., (1958) *J. Pol. Sci.* **33**, 183.
18. Gazzillo, D., (1991) *J. Chem. Phys.* **95**, pp. 4565, and references therein.
19. Ornstein, L.S., and Zernike, (1914) *F., Prok. Akad. Sci.* (Amsterdam) **17**, pp. 793.
20. Percus, S.K., and Yevick, G.J., (1958) *Phys. Rev.* **110**, pp. 1.
21. Rogers, F.J., and Young, D.A., (1984) *Phys. Rev.* **A 30**, pp. 999.
22. P. Ballone, P., Pastore, G., Galli, G., and Gazzillo, D., (1986) *Molec. Phys.* **59**, pp. 275.
23. Biben, T., (1993) PHD thesis, Universté Claude Bernard-Lyon I, France.
24. Biben, T., and Hansen, J.P., (1990) *Europhys. Lett.* **12** pp. 347.
25. Heno, Y., and Regnaut, C., (1991) *J.Chem.Phys* **95**, pp. 9204.
26. Baxter, R.J., (1968) *J. Chem. Phys.* **49**, pp. 2770.
27. Lebowitz, J.L., (1964) *Phys. Rev.* **A 133**, pp. 895.
28. Lebowitz, J.L., and Rowlinson, J.S., (1964) *J. Chem. Phys.* bf 41, pp. 133.
29. Belloni, L., private communication.
30. Dijkstra, M., Frenkel, D., and Hansen, J.P., (1994) *J. Chem. Phys.* **101**, 3179;

Dijkstra, M., and Frenkel, D., (1994) *Phys. Rev. Lett.* **72**, 298.
31. Frenkel, D., and Louis, A., (1992) *Phys. Rev. Lett.* **68**, pp. 3363.
32. Van Duijneveldt, J.S., and Lekkerkerker, H.N.W., (1993) *Phys. Rev. Lett.* **71**, pp. 4264.
33. Mao, Y., Cates, M.E., and Lekkerkerker, H.N.W., (1995) *Physica* **A 222**, pp. 10.
34. Götzelmann, B., Evans, R., and Dietrich, S., to appear in *Phys. Rev.* **E**.
35. Kranendonk, W.G.T., and Frenkel, D., (1988) *Mol.Phys.* **64**, pp. 403.
36. Dijkstra, M., Van Roij, R., and Evans, R., (1998) preprint.
37. Buhot, A., and Krauth, W., (1998) to appear in *Phys. Rev. Lett.*

THEORY AND SIMULATIONS OF POLYMERS AT SURFACES

A. YETHIRAJ
Department of Chemistry, University of Wisconsin, Madison, WI 53706

Abstract. This chapter discusses the application of liquid state methods such as integral equations, density functional theory, and computer simulation to investigate the properties of polymer melts and blends at surfaces. The focus is on three problems: I) the density profiles of polymers at solid surfaces, II) the nature and mechanism for the segregation to a surface of one component from a blend of polymers, and III) the non-equilibrium behaviour of confined liquids undergoing shear flow. Monte Carlo and molecular dynamics simulations and a density functional theory that address these problems are described.

1. Introduction

The behaviour of polymers near surfaces is important in many technological applications including polymer processing, lubrication, and coatings. In most of these applications the properties of the liquid within a few angstroms of the surface are crucial to the success of the application. It is difficult to experimentally obtain properties with such a fine resolution, and the theoretical study of these systems using liquid state methods and computer simulation has been popular and important.

In this chapter, we describe research in our group [1, 2, 3, 4, 5] on three different problems in the general area of polymers at surfaces. The first problem concerns the density profiles of polymers at solid surfaces, for which we have performed computer simulations [1] and developed an accurate density functional theory [2, 3]. The second problem concerns the segregation to the surface of one component from a binary blend (or mixture) of polymers for which we have done extensive Monte Carlo simulations [4]. The third problem concerns the behaviour of polymers under shear flow, for

C. Caccamo et al. (eds.), New Approaches to Problems in Liquid State Theory, 379–392.

which we have performed molecular dynamics simulations [5]. These three areas are introduced separately below.

The simplest problem concerning the surface behaviour of polymers is the structure of the liquid at a smooth solid surface. Computer simulations of polymers at surfaces have been popular [6, 7, 8] but, because they are computationally intensive, have been restricted to rather simple models of polymers. Liquid state theories for this problem include density functional theories [9, 10, 11, 12] and integral equations [13]. Integral equations (based on the growing adsorbent model) are reasonably accurate for the density profiles of hard chains at hard walls but fail at low densities and for stiff chains [14]. The density functional theory described in this work, on the other hand, is very accurate; about as accurate as similar theories for simple liquids. The disadvantage is that it requires a single chain simulation as part of the iterative procedure to obtain the density profile.

The segregation to the surface of one component from a mixture of two or more polymers is of technological importance. In the last decade there has been considerable attention focused on segregation caused by differences in the topology of the molecules of the polymer blend [15, 16]. These experiments involve polyolefin materials where the elementary chemical units are similar (i.e. CH_X groups) but differences arise from the architecture of the molecules. An example is a blend of polyethylene (PE), which is a linear chain, and polyethylenepropylene (PEP) which has a single CH_3 side group hanging off every fourth backbone carbon atom. Experiments on blends of polyolefins shows that the more branched component segregates to the surface, independent of the type of surface (air, silicon, etched silicon). We present simulation results to investigate the mechanism for this segregation.

Although the majority of theoretical work has been on equilibrium systems some of the important practical applications are in systems far from equilibrium. We will discuss an algorithm we have developed for the molecular dynamics simulations of polymers undergoing shear flow. The main technical issue is the proper method of thermostatting the fluid in order to mimic experimental systems. We find that the density profiles are relatively insensitive to shear, even though there is some chain stretching and alignment with the flow. There is considerable slip at the surfaces and a large temperature rise in the fluid at typical simulation conditions.

The rest of this chapter is organized as follows: in section 2 we discuss the density profiles of polymers at surfaces, in section 3 we discuss the surface segregation from polymer blends, in section 4 we discuss the behaviour of polymer melts under shear flow, and in section 5 we present some conclusions.

2. Density Profiles of Polymers at Surfaces

2.1. DENSITY FUNCTIONAL THEORY

In density functional theory one starts with an approximation for the grand free energy, Ω, as a functional of the density profile. If $\mathbf{R}$ denotes the positions of all the N monomers on a polymer molecule and $\rho_M(\mathbf{R})$ is the molecular density as a function of these positions, then at equilibrium

$$\frac{\delta\Omega}{\delta\rho_M(\mathbf{R})} = 0, \tag{1}$$

and this condition determines the density profile. Generally the free energy functional is divided into an ideal and excess part where the former is known exactly and approximations are sought for the latter. The major difference between the density functional theory of simple liquids and that of polymers is that in polymers, because of the intramolecular potential of interaction, the ideal gas functional is not analytically tractable. This makes the implementation of the theory more difficult.

The functional Ω is related to the Helmholtz free energy functional, $F[\rho_M]$ via a Legendre transform:

$$\Omega[\rho_M(\mathbf{R})] = F[\rho_M(\mathbf{R})] + \int [\Phi(\mathbf{R}) - \mu]\,\rho_M(\mathbf{R})d\mathbf{R}, \tag{2}$$

where μ is the chemical potential and $\Phi(\mathbf{R})$ is the external field. The functional $F[\rho_M]$ is generally expressed as the sum of an ideal (F^{id}) and excess (F^{ex}) part,

$$F[\rho_M(\mathbf{R})] = F^{id}[\rho_M(\mathbf{R})] + F^{ex}[\rho_M(\mathbf{R})]. \tag{3}$$

The ideal functional is known exactly:

$$F^{id}[\rho_M(\mathbf{R})] = k_BT\int d\mathbf{R}\rho_M(\mathbf{R})\,[\ln\rho_M(\mathbf{R}) - 1] + \int d\mathbf{R}V(\mathbf{R})\rho_M(\mathbf{R}), \tag{4}$$

where k_B is Boltzmann's constant, T is the temperature, and $V(\mathbf{R})$ describes all the intramolecular interactions (including bonding and long-range excluded volume). The main approximation in any density functional theory is the expression for the excess free energy functional.

In general the free energy functional can be decomposed as

$$F[\rho_M(\mathbf{R})] = F^{id}[\rho_M(\mathbf{R})] + F^{ex}[\rho(\mathbf{r})], \tag{5}$$

where

$$\rho(\mathbf{r}) = \int d\mathbf{R}\sum_{i=1}^{N}\delta(\mathbf{r} - \mathbf{r}_i)\rho_M(\mathbf{R}) \tag{6}$$

is the average density of sites at position $\mathbf{r}$, $\mathbf{r}_i$ are the positions of the beads ($\mathbf{R} = \{\mathbf{r}_i\}$).

With such a decomposition one can employ the weighted density approximation for F^{ex} [11]:

$$F^{ex}[\rho(\mathbf{r})] = \int \rho(\mathbf{r}) f(\bar{\rho}) d\mathbf{r} \tag{7}$$

where $f(\bar{\rho})$ is the excess (over ideal gas) free energy per site of the bulk fluid evaluated at a site density $\bar{\rho}(\mathbf{r})$,

$$\bar{\rho}(\mathbf{r}) = \int \rho(\mathbf{r}') w(|\mathbf{r} - \mathbf{r}'|) d\mathbf{r}' \tag{8}$$

is the weighted density, and $w(r)$ is the weighting function, normalized so that $\int w(\mathbf{r}) d\mathbf{r} = 1$. The density functional theory is completely specified by the functions $f(\rho)$ and $w(r)$.

The density functional theory has the structure of a self-consistent field theory where the density profile is obtained from a simulation of a single chain in the field due to the rest of the fluid and the surface. A formal minimization of Ω gives

$$\frac{\delta \Omega}{\delta \rho_M(\mathbf{R})} = \frac{\delta F}{\delta \rho_M(\mathbf{R})} + \Phi(\mathbf{R}) - \mu = 0, \tag{9}$$

which may be written as

$$\rho_M(\mathbf{R}) = \exp\left[-\beta V(\mathbf{R}) + \beta\mu - \beta\Phi(\mathbf{r}) + \beta \sum_{i=1}^{N} \lambda(\mathbf{r}_i)\right] \tag{10}$$

where $\beta = 1/k_B T$ and $\lambda(\mathbf{r})$ is the self-consistent field given by

$$\lambda(\mathbf{r}) = \frac{\delta F^{ex}}{\delta \rho(\mathbf{r})} = f(\bar{\rho}(\mathbf{r})) + \int d\mathbf{r}' \rho(\mathbf{r}') w(|\mathbf{r} - \mathbf{r}'|) f'(\bar{\rho}(\mathbf{r}')) \tag{11}$$

where $f' = df/d\rho$.

The approximation for F^{ex} reduces to choices for the function $f(\rho)$ and the weighting function $w(r)$. $f(\rho)$ may be obtained from an equation of state, of which several are available for polymers. The equations we use are listed in reference [3]. We use two different choices for the weighting function which are described below.

For hard sphere liquids, the simplest choice for $w(r)$, suggested by Stell [17] and Nordholm et al. [18], is

$$w(r) = \frac{3}{4\pi\sigma^3} \Theta(\sigma - r) \tag{12}$$

where $\Theta(r)$ is the Heaviside step function, and σ is the hard sphere diameter. A more sophisticated choice of $w(r)$, suggested by Curtin and Ashcroft (CA) [19], is

$$-k_B T\hat{c}(k) = 2f'(\rho)\hat{w}(k) + \rho f''(\rho)\hat{w}^2(k) + 2\rho f'(\rho)\hat{w}(k)\frac{\partial \hat{w}(k)}{\partial \rho}, \tag{13}$$

where primes denote derivatives with respect to density, carets denote Fourier transforms, and $\hat{c}(k)$ is the direct correlation function obtained using the PRISM theory [20].

The implementation of the density functional theory entails the iterative solution of one equation

$$\rho(\mathbf{r}) = \int d\mathbf{R}\left[\sum_{i=1}^{N}\delta(\mathbf{r}-\mathbf{r}_i)\right]\exp\left[-\beta V(\mathbf{R}) + \beta\mu - \beta\Phi(\mathbf{r}) + \beta\sum_{i=1}^{N}\lambda(\mathbf{r}_i)\right] \tag{14}$$

with $\lambda(\mathbf{r})$ given by Equation (11). A Newton-Raphson procedure (described previously [2]) is used; the complexity of $V(\mathbf{R})$ necessitates a single chain simulation at each iteration.

2.2. COMPARISON TO MONTE CARLO SIMULATIONS

The density profiles of hard chain polymers at surfaces are governed by a competition between packing and conformational entropic effects. At low densities conformational entropic effects dominate, and since a single chain suffers a loss in conformational entropy near the surface it prefers the region away from the surface. At high densities, packing effects dominate and the chain density is enhanced near the surface. These trends are seen in figure 1 which compares predictions of the density functional theory to simulations of freely-jointed tangent sphere hard chains [1] with N=20 and for various densities. In the figure $\eta(z)$ is the volume fraction of sites as a function of distance from the surface, with the distance of closest approach at z=0. The theory is in excellent agreement with the simulations in most cases. The calculations depicted here use the simple weighting function but almost identical results (not discernible if plotted in figure 1) are obtained using the CA weighting function [3].

The fact that the theory is relatively insensitive to the choice of the weighting function may be understood from the following argument. The weighted density functional theory guarantees that the exact wall sum rule, $\rho(z=0) = \beta P$, is satisfied where the pressure P is input through the bulk equation of state. The density profiles are therefore constrained both at the surface and far away (where the fluid is bulk-like). At low densities, the density profile is monotonically increasing with distance and, as long as the

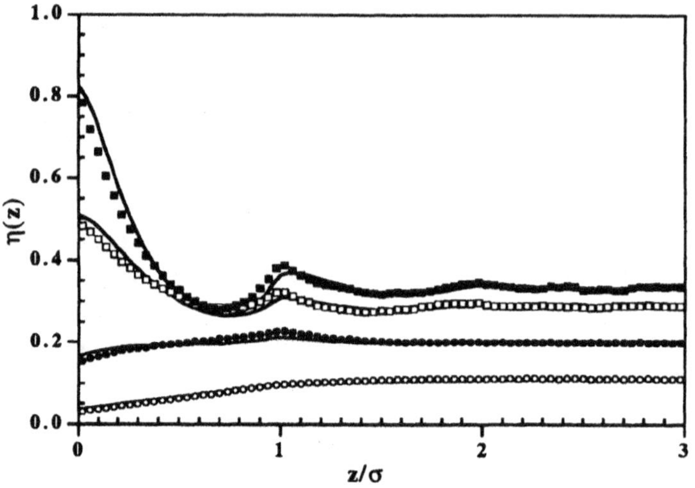

Figure 1. Comparison of density functional theory predictions (lines) to Monte Carlo simulations for the density profiles of freely jointed hard chains for N=20 and various densities. The bulk volume fraction can be read off from the value of the profile at large distances ($z \approx 3$).

range of $w(r)$ used is correct, the predictions are very accurate because the single chain entropic effects are well accounted for through the Monte Carlo simulation of a single molecule. Therefore discrepancies between theoretical predictions with different choices for $w(r)$ become apparent only at higher densities.

Figure 2 compares the density functional theory using the CA weighting function and the simple weighting function to Monte Carlo simulations of freely-jointed 3-mers [21] for η=0.4. In this case the theory with the simple weighting function has oscillations in the density profile that are out of phase with the simulation results. The theory with the CA weighting function is in quantitative agreement with the simulations. In fact it is about as accurate for hard chains as similar theories are for hard spheres. Similar results are seen for semiflexible polymers and for more realistic alkane-like models of polymers [2, 3].

3. Surface Segregation from Polymer Blends

Experiments have shown that when a binary blend of polymers where the molecules of one component are more branched than the other is placed at a surface, the more branched component preferentially segregates to the surface. Sikka et al. [16] correlated the surface segregation with the statistical segment length and argued that such a segregation was entropic in nature

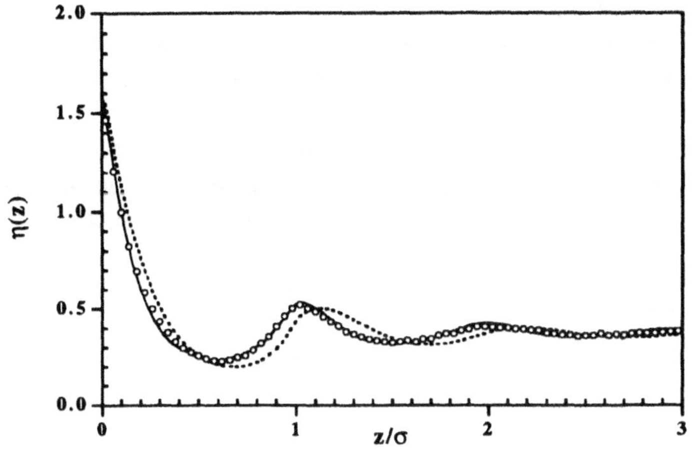

Figure 2. Comparison of density functional theory predictions with the Curtin-Ashcroft weighting function (—) and the simple weighting function (- - -) to Monte Carlo simulations for the density profiles of freely jointed hard trimers for η=0.4.

and caused by the fact that the branched polymers were conformationally smaller than the linear polymers. A theoretical analysis on a blend of stiff and flexible chains which included only entropic effects [22] concurred with the above picture.

From the results presented in the previous section one would expect entropic segregation to produce exactly the opposite of what is observed in experiment. In particular, since linear chains pack against the surface more efficiently than branched chains, one would expect that in high density melts the linear chains would segregate to the surface from purely entropic arguments. In fact, theory and simulations on blends of stiff and flexible chains show exactly this effect [23], i.e. in a blend of stiff and flexible chains, the stiffer chains are always preferred at the surface.

So what then is the mechanism for the segregation observed experimentally? Several possibilities present themselves. The branched chains contain a greater number of CH_3 groups, and these might have a greater affinity for the surface. Alternatively the interactions between two branched chains might be weaker than that between two linear chains. In this case there is an enthalpic advantage to placing linear chains amongst each other in the bulk where they can have more neighbours. This would cause a segregation of the branched chains to the surface. A related question is if, independent of the dispersion interactions, the difference in topology alone could cause a preferential segregation of one component to the surface.

We have investigated [4] the effect of topology on the surface segregation from polymer blends using Monte Carlo simulations on a simple model of polymer blends. The molecules of both components consist of N=19 freely-jointed tangent spheres, and differ only in how these spheres are connected together. In the branched chains the backbone consists of 13 spheres and every other backbone bead has a branch composed of one bead attached to it. The site-site intermolecular potential, $u_{ff}(r)$, is the same between any two beads in the system:

$$\begin{aligned}\beta u_{ff}(r) &= \infty, \quad r < \sigma \\ &= -\epsilon_{ff}\left[\sigma e^{-\kappa(r/\sigma-1)}/r - e^{-\kappa}/2\right], \quad \sigma < r < 2\sigma \\ &= 0, \quad r > 2\sigma \end{aligned} \tag{15}$$

where κ is the inverse range of the shifted and cut-off Yukawa potential (κ=2.5 in this work), the subscript ff denotes "fluid-fluid", and ϵ_{ff} is the dimensionless strength of the attraction. The cutoff at $r = 2\sigma$ is employed because it results in an enormous savings in computer time when link cell lists are used to calculate the energy during the simulation. The range of the potential is chosen to mimic a Lennard-Jones potential. The fluid is confined between two identical and parallel walls. The wall-fluid potential is $u_{wf}(z) = \phi(z) + \phi(H - z)$ where H is the separation of the surfaces, z is the perpendicular distance from one of the surfaces,

$$\begin{aligned}\beta\phi(z) &= \infty, z < 0, \\ &= -\epsilon_{wf}\left[e^{-\kappa z/\sigma} - e^{-\kappa}\right], 0 < z < \sigma, \\ &= 0, z > \sigma, \end{aligned} \tag{16}$$

and ϵ_{wf} is the dimensionless strength of the wall-fluid attraction.

This model is chosen because it allows one to vary the attractive forces and therefore isolate the effects of entropy and enthalpy. When the fluid-fluid attractions are turned off and wall-fluid attractions (identical for both species) are present, the *linear* component segregates to the surface. The magnitude of this segregation increases as the strength of the wall-fluid attraction is increased. This may be attributed to the fact that the linear chains can pack against the surface more efficiently than the branched chains. Wall-fluid attractions provide an additional enthalpic incentive for the linear chains to segregate to the surface. However, when fluid-fluid attractions (of comparable strength to the wall-fluid attractions) are turned on the *branched* polymers segregate to the surface. The reason for this is that the linear chains pack amongst each other more efficiently than the branched chains: In the presence of fluid-fluid attractions this provides an incentive for the linear chains to be in the bulk where their sites can have

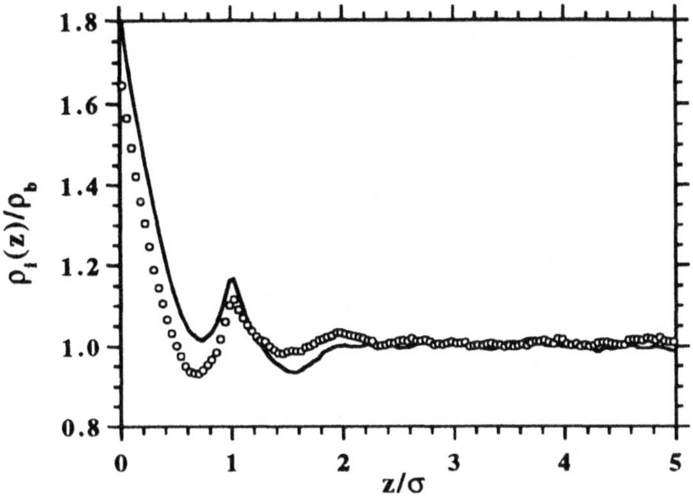

Figure 3. Density profiles (normalized to the value in the middle of the pore) of branched (o) and linear (—) polymers in an athermal blend at a hard wall.

more neighbours than near a surface. Consequently the segregation observed in the experiments [15, 16] is most likely an enthalpic effect caused by differences in packing. Since wall-fluid attractions favour the linear chains at the surface and fluid-fluid attractions favour the branched chains at the surface, the segregation is a strong function of the relative strengths of these attractions. This suggests that the behaviour observed in experiments is not universal and rather different results might be obtained with more strongly interacting surfaces.

Figure 3 depicts the density profiles of linear and branched chains (denoted $\rho_i(z)$) relative to the "bulk" value in the middle of the pore (denoted ρ_b) for the athermal blend (ϵ_{ff}=0) and for hard walls (ϵ_{wf}=0). In this purely entropic case, the linear polymers segregate to the surface although the segregation is very small and probably not measurable in an experiment.

A segregation of branched molecules to the surface occurs when fluid-fluid attractions are present. Figure 4 depicts the density profiles of branched and linear polymers relative to the bulk value for ϵ_{ff}=1 and ϵ_{wf}=0. In this case the branched polymers are preferentially segregated to the surface. This is because it is energetically favourable to place the linear polymers away from the surface since these molecules can pack amongst each other better than the branched molecules.

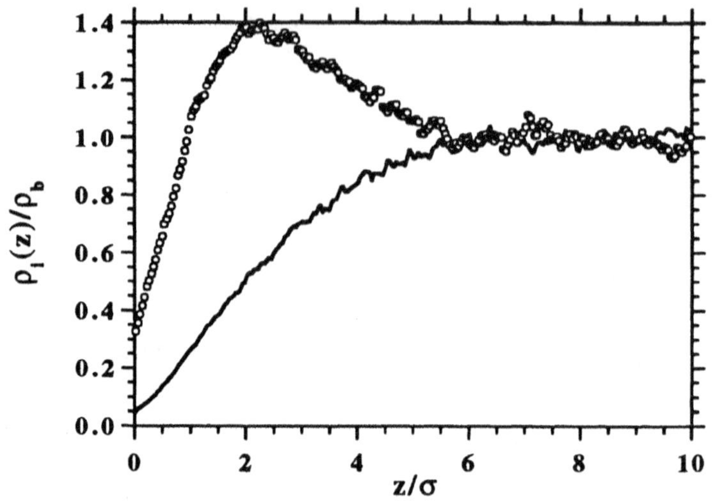

Figure 4. Density profiles (normalized to the value in the middle of the pore) of branched (o) and linear (—) polymers in an attractive blend (ϵ_{ff}=1) at a hard wall.

4. Polymers under Shear Flow

The behavior of confined polymer melts in shear flow is important in polymer extrusion operations when the surface finish of the extrudate is of significance. Many defects such as sharkskin patterns are observed, the molecular origin of which are unclear. In addition to an understanding of flow induced defects, molecular simulation can provide insight into the boundary conditions at the surface (i.e. slip or stick) and into the numerical values of the transport coefficients of confined liquids. We have developed simulation algorithms that can shed light on these issues.

Although there have been many simulations of confined polymers under shear [24, 25], most of these do not mimic any experimental system. In particular, it has been popular to impose a constant temperature across the entire film by coupling each atom in the fluid to a heat bath [24, 25]. This is only possible in a real fluid if the thermal conductivity of the fluid is infinite. Our approach has been to use walls that are composed of atoms attached with stiff springs to a plane and maintain these atoms at a constant temperature by coupling them to a heat bath. Shear is imposed by moving the plane containing the wall atoms, and the liquid temperature can rise due to viscous heating effects. By analyzing simulation results in the framework of continuum mechanics we can extract both the boundary conditions as well as the viscosity and thermal conductivity.

4.1. SIMULATION METHOD

The simulated system consists of the sheared fluid confined between two atomistic walls. The atoms that comprise the walls and the beads on the polymer chains interact via a continuous repulsive Lennard-Jones site site potential, $u(r)$, given by

$$u(r) = 4\epsilon \left[\left(\frac{\sigma}{r}\right)^{12} - \left(\frac{\sigma}{r}\right)^{6} + \frac{1}{4} \right] \theta\left(r_c - r\right) \tag{17}$$

where σ and ϵ are the Lennard-Jones parameters, and $r_c = 2^{1/6}\sigma$. We use σ, ϵ, and $\tau = (m\sigma^2/\epsilon)^{1/2}$ to set length, energy, and time scales, respectively (m is the mass of an atom). The walls consist of a single layer of atoms attached to face centered cubic (FCC) lattice sites by stiff harmonic springs. The polymer molecules are composed of beads connected with finitely extendable non-linear elastic (FENE) springs.

The two walls are moved with the same velocity but in opposite directions along the x-axis. The wall temperature is kept constant by coupling the motion of the wall atoms to a heat bath maintained at the desired temperature. The trajectories of the wall atoms are calculated by a stochastic generalization of the velocity Verlet algorithm [26]. The trajectories of the fluid atoms are calculated using a velocity Verlet algorithm. The results reported here correspond to $T^* = 4.0$, $\rho\sigma^3 = 0.8$, and N = 20.

4.2. DENSITY, VELOCITY, AND TEMPERATURE PROFILES

As expected, we observe a layering of the polymers at the surface similar to what was seen earlier for a fluid at a smooth wall. Interestingly, this layering is insensitive to the shear rate, even at large dimensionless shear rates of γ=0.2. Figure 5 compares the site density profile for a 20-mer chain fluid at nominal shear rates of 0 (equilibrium) and 0.2. The density profiles under shear and at equilibrium are identical within the statistical uncertainty of the data. This is the case even for much higher shear rates (γ=1) which are several orders of magnitude higher than what can be achieved in an experiment.

For no-slip boundary conditions and Couette flow, continuum mechanics predicts a linear velocity profile between the surfaces. Figure 6 depicts the velocity profile of a polymer melt under shear. There is large slip at the walls but otherwise the velocity profiles are roughly linear (the dashed line is a linear fit to the velocity profile). The slip at the surface causes the actual shear rate to be much smaller than the nominal shear rate. This must be accounted for in calculating the transport properties from a simulation.

The slip behaviour is also manifested in the temperature profile. Figure 7 shows that the temperature profile in the polymer film is roughly

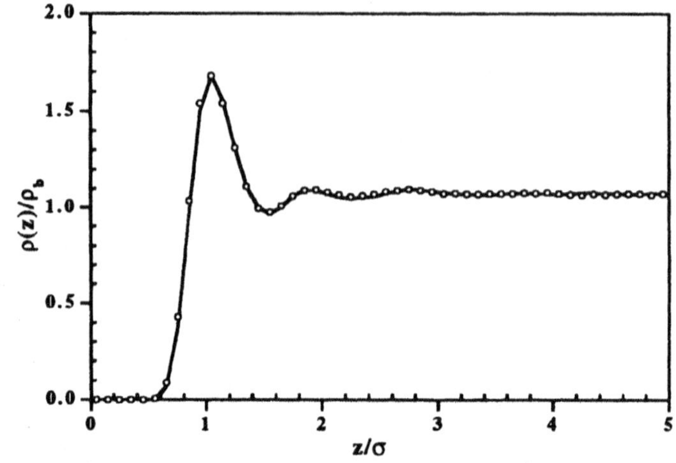

Figure 5. Density profiles of N=20 chains for T^*=4.0 and $\rho\sigma^3$=0.8 for γ=0.0 (line) and 0.2 (o). Note that the density profile is insensitive to the shear rate.

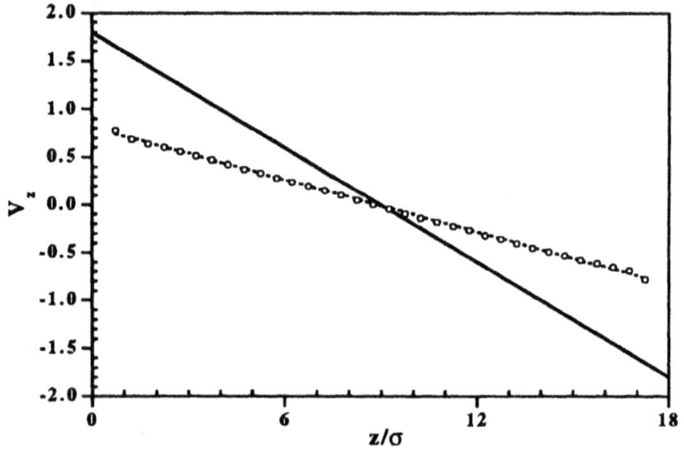

Figure 6. Velocity profile of N=20 chains for T^*=4.0 and $\rho\sigma^3$=0.8 for γ=0.2. The solid line is the continuum mechanics prediction with no slip. The dashed line is a linear fit to the simulation results. With the exception of the slip at the walls, the simulated profile is in good agreement with that expected from continuum mechanics.

parabolic, as expected from continuum mechanics, except for the discontinuity in temperature at the surfaces. The rise in temperature is quite large, reflecting the high viscosity of polymeric liquids.

These results show that at the shear rates typically used in molecular

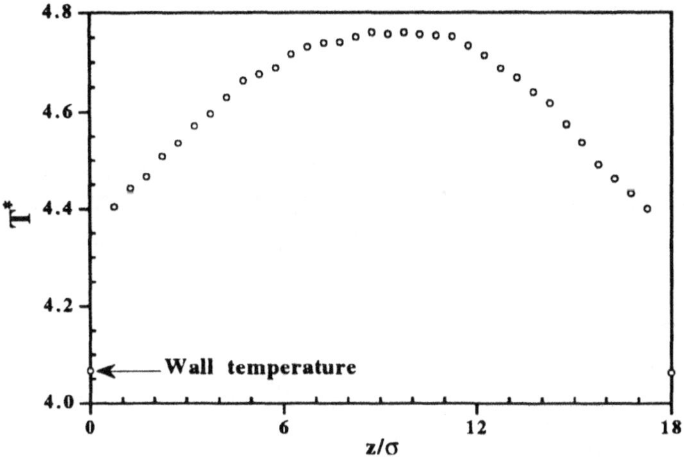

Figure 7. Temperature profile of $N=20$ chains for $T^*=4.0$ and $\rho\sigma^3=0.8$ for $\gamma=0.2$.

simulations, viscous heating can lead to a large temperature rise in the sheared fluid. For polymers this temperature rise can be significant. These findings indicate that caution should be exercised in interpreting the results of isothermal non-equilibrium molecular dynamics simulations of fluids at high shear rates.

Other results from these simulations are summarized below. The magnitude of slip increases with increasing shear rate, molecular weight, and degree of confinement. The chains also stretch and align in the flow direction although the former is not a large effect (about 10% at the highest shear rates).

5. Summary and Conclusions

The surface behaviour of polymers is of considerable practical and scientific importance. In this chapter we have shown that the techniques of computer simulation and density functional theory can be effectively used to tackle polymeric problems. The implementation of these approaches is more computationally demanding than in simple liquids but the availability of fast computers makes this less of a problem these days.

For example, we have described a density functional theory for polymers that is extremely accurate for the density profiles of polymers at surfaces, about as accurate as similar theories are for hard spheres at surfaces. More complex phenomena such as the surface segregation from polymer blends have also been investigated using judiciously chosen models. The flow behaviour of polymers can also be addressed using molecular dynamics simulations.

Acknowledgments

This work was supported by the Sloan Foundation and the National Science Foundation (through grant number CHE 9732604 to A.Y. and a computer grant number CHE 9522057 to the Department of Chemistry).

References

1. A. Yethiraj, J. Chem. Phys. **101**, 2489 (1994).
2. A. Yethiraj and C. E. Woodward, J. Chem. Phys. **102**, 5499 (1995).
3. A. Yethiraj, J. Chem. Phys. (in press).
4. A. Yethiraj, Phys. Rev. Lett. **74**, 2018 (1995).
5. R. Khare, J. J. de Pablo, and A. Yethiraj, Macromolecules **29**, 7910 (1996).
6. R. Dickman and C. K. Hall, J. Chem. Phys. **89**, 3168 (1988).
7. S. K. Kumar, M. Vacatello, and D. Y. Yoon, J. Chem. Phys. **89**, 5209 (1988).
8. A. Yethiraj and C. K. Hall, Macromolecules **23**, 1635 (1990).
9. D. Chandler, J. D. McCoy, and S. J. Singer, J. Chem. Phys. **85**, 5971 (1986).
10. W. E. McMullen and K. F. Freed, J. Chem. Phys. **92**, 1413 (1990).
11. C. E. Woodward, J. Chem. Phys. **94**, 3183 (1991).
12. E. Kierlik and M. L. Rosinberg, J. Chem. Phys. **100**, 1716 (1994).
13. A. Yethiraj and C. K. Hall, J. Chem. Phys. **95**, 3749 (1991).
14. A. Yethiraj, ACS Symposium Series **629**, 274 (1996).
15. U. Steiner, J. Klein, E. Eiser, A. Budkowski, and L. J. Fetters, Science **258**, 1126 (1992).
16. M. Sikka, N. Singh, A. Karim, and F. S. Bates, Phys. Rev. Lett. **70**, 307 (1993).
17. G. Stell, in "The Equilibrium Theory of Classical Fluids" (H. L. Frisch and J. L. Lebowitz, eds.) W. A. Benjamin, New York, 1964.
18. S. Nordholm, M. Johnson, and B. C. Freasier, Aust. J. Chem. **33**, 2139 (1980).
19. W. A. Curtin and N. W. Ashcroft, Phys. Rev. A **32**, 2909 (1985).
20. J. G. Curro and K. S. Schweizer, J. Chem. Phys. **87**, 1842 (1987).
21. S. Phan, M. L. Rosinberg, E. Kierlik, A. Yethiraj, and R. Dickman, J. Chem. Phys. **102**, 2141 (1995).
22. G. H. Fredrickson and J. P. Donley, J. Chem. Phys. **97**, 8941 (1992).
23. A. Yethiraj, S. K. Kumar, A. Hariharan, and K. S. Schweizer, J. Chem. Phys. **100**, 4691 (1994).
24. E. Manias, G. Hadziioannou, I. Bitsanis, and G. Ten Brinke, Europhys Lett. **24**, 99 (1993).
25. P. A. Thompson, G. S. Grest, and M. O. Robbins, Phys. Rev. Lett. **68**, 3448 (1992).
26. M. P. Allen and D. J. Tildesley, *Computer Simulation of Liquids* Oxford University Press, Oxford, U.K. 1992.

STATIC STRUCTURE AND CHAIN CONFORMATIONS IN POLYELECTROLYTE SOLUTIONS

A. YETHIRAJ
Department of Chemistry, University of Wisconsin, Madison, WI 53706

Abstract. A microscopic approach is described for the static structure and conformational properties of polyelectrolyte solutions. The theory predicts liquid like structure on long lengthscales in dilute solutions, which diminishes at concentrations higher than the overlap threshold concentration, only to reappear in concentrated solutions. The predictions for the static structure factor are in good agreement with scattering experiments on solutions of rigid tobacco mosaic virus particles if the charge on the particles is treated as adjustable. The theory predicts that $k_{max} \sim \rho^{\nu}$, where k_{max} is the position of the maximum in the structure factor and ρ is the concentration with ν=1/3 and 1/2 in dilute and semi-dilute solutions, respectively, in agreement with experiment. The theory is in good agreement with computer simulations of flexible polyelectrolytes, both for the conformational properties and the static structure factor. It is argued that a reversible binding of counterions to the polyions must be invoked in order to understand the experimental results in the framework of this theory.

1. Introduction

Polyelectrolyte solutions are of considerable biological and technological importance. For example, the properties of RNA and DNA, and their interactions with proteins and other cellular matter are of fundamental importance, and the applications of synthetic polyelectrolytes in gels, surfactants, and resins, are of industrial importance. Despite considerable study, however, these systems are still believed to be one of the least understood substances in soft matter [1]. The reason for this is two-fold. One reason is that there has been little synergy between experiment and theory. On one hand theoretical study is complicated by the presence of long-range

C. Caccamo et al. (eds.), *New Approaches to Problems in Liquid State Theory*, 393–405.

interactions and chain connectivity, and most theories have focused on isolated polyelectrolyte molecules. On the other hand, experiments in dilute solution are notoriously difficult, and most experiments have focused on semi-dilute solutions, where different chains overlap considerably. A second reason is that different experiments on apparently similar systems often show radically different behaviour, and this has caused some confusion in the community.

In this chapter we describe recent work [2, 3, 4, 5] in our group on the theory of polyelectrolyte solutions. The key difference between our approach and that of previous workers is that we use the standard methods of liquid state theory, e.g. integral equations, and can thus investigate many molecule systems ranging from dilute to concentrated polyelectrolyte solutions rather than just single chains. This approach also allows us to go beyond the standard random phase approximation, which is popular in polymer physics but has been shown to be inappropriate for charged polymers. This chapter focuses on the static structure and conformational properties of polyelectrolyte solutions with a goal of understanding a series of recent experiments that are described below.

The static structure of polyelectrolyte solutions is of current interest. Recent light scattering and neutron scattering experiments [6, 7] on solutions of salt free Tobacco Mosaic Virus (TMV) particles show a characteristic peak in the structure factor. This peak broadens and moves towards higher wavevectors as the concentration is increased. At some (low) concentrations a second peak in the structure factor is also observed. If the position of the maximum is denoted k_{max}, then $k_{max} \sim \rho^{\nu}$ where ρ is the TMV concentration, with $\nu \approx 1/3$ and $1/2$ in the dilute and semi-dilute regimes, respectively. The data strongly suggests a liquid-like ordering of the TMV particles on long length scales.

Light scattering and neutron scattering experiments have also been performed on polyelectrolyte solutions composed of *flexible* chains such as sodium polyacrylic acid (NaPAA) [8] and sodium polystyrene sulphonate (NaPSS) [9]. The qualitative behaviour of the structure factor is similar to that described above for TMV.

The determination of chain conformations is experimentally more challenging because it requires contrast matching (in neutron scattering) via isotopic substitution. There are only a few such studies; these are on different systems and report conflicting results. There have been some computer simulations of model systems [10] which show that in salt free solutions the size of the chains decreases rapidly as the concentration is increased. These calculations provide benchmarks against which theories can be tested.

The integral equation theory described in this chapter is successful in reproducing all the qualitative behaviour observed in computer simulations and experiment suggesting that the phenomena are adequately described if the electrostatic interactions are accounted for at a level possible with standard liquid state methods. An important conclusion is that most experiments are not in the asymptotic large chain length regime and therefore are not expected to show universal behaviour. We also find that it is necessary to invoke a binding of counterions to the polyions in order to fit the data.

The rest of this chapter is organized as follows. Section 2 describes the theoretical methods, section 3 presents results for rod-like TMV molecules, and section 4 presents results for flexible polymers. Some conclusions are presented in section 5.

2. Theoretical Methods

2.1. MOLECULAR MODELS

Slightly different models are used in the study of rigid and flexible polyelectrolyte molecules, although in both cases the molecules are modeled as a collection of N charged interaction sites with a charge qe (e is the charge on an electron and q can take on fractional values) on each site. The solvent is treated as a dielectric continuum and the small ions are included through an effective potential between sites on the polymer molecules.

In the model for the rigid TMV molecules, the sites are placed in a linear arrangement (with length L) and a hard sphere plus screened Coulomb interaction is used between all sites, i.e. the site-site interaction is

$$\begin{aligned} \beta v(r) &= \infty, \quad r < \sigma \\ &= \Gamma \frac{e^{-\kappa r}}{r}, \quad r > \sigma \end{aligned} \tag{1}$$

where σ is the hard sphere diameter, $\Gamma = q^2 l_B$, $l_B \equiv e^2/\epsilon k_B T$ is the Bjerrum length, ϵ is the dielectric constant of the solvent, k_B is Boltzmann's constant, T is the temperature, $\beta = 1/k_B T$, and κ is the inverse Debye screening length given by $\kappa^2 = 4\pi\Gamma\rho$ in salt-free solutions.

In the model for the flexible polymers there is an intramolecular contribution, $\mathcal{H}_i$, to the Hamiltonian in addition to the site-site screened Coulomb interaction. In this work $\mathcal{H}_i$ is given (using continuum notation) by,

$$\mathcal{H}_i = \frac{3}{2l} \int_0^L \left(\frac{\partial \mathbf{r}(s)}{\partial s} \right)^2 ds + \frac{3l}{8} \int_0^L \left(\frac{\partial^2 \mathbf{r}(s)}{\partial s^2} \right)^2 ds \tag{2}$$

where s is the arc length of the chain measured from one end, $\mathbf{r}(s)$ is the position of point s, l is the persistence length, and L is the length of the chain. The first term comes from harmonic bonding springs between adjacent sites, and the second term comes from an energy cost for bending a polymer molecule.

2.2. POLYMER REFERENCE INTERACTION SITE MODEL THEORY AND CLOSURE APPROXIMATIONS

The Polymer Reference Interaction Site Model (PRISM) theory, which been successfully applied to neutral polymers [11], consists of an equation that relates the total correlation function, $h(r)(\equiv g(r)-1$ where $g(r)$ is the site-site pair correlation function), to the single chain structure factor $\hat{\omega}(k)$, and the site-site direct correlation function, $c(r)$, via a nonlinear integral equation which is given (in Fourier space) by,

$$\hat{h}(k) = \hat{\omega}(k)\hat{c}(k)\hat{\omega}(k) + \rho\hat{\omega}(k)\hat{c}(k)\hat{h}(k) \tag{3}$$

where carets denote Fourier transforms and k is the momentum transfer variable. In the implementation of the theory it is assumed that $\hat{\omega}(k)$ is known. This indeed is the case for rigid molecules, but not for flexible chains. In the latter case $\hat{\omega}(k)$ must be determined in a self-consistent fashion as described in the next section. The static structure factor, $\hat{S}(k)$, is given by $\hat{S}(k) = \hat{\omega}(k) + \rho\hat{h}(k)$.

Several closure relations for the PRISM equation have been suggested. For models which include a hard core, we have the exact core condition $g(r) = 0$ for $r < \sigma$, and we seek a relation between $h(r)$ and $c(r)$ outside the hard core. Closures can be obtained via analogy with the theory of simple liquids. For example, the Percus-Yevick (PY) closure is given by,

$$c(r) = \left[1 - e^{-\beta v(r)}\right] g(r), \tag{4}$$

which for hard sphere polymers reduces to $c(r) = 0$ for $r > \sigma$.

For charged chains we use the reference Laria Wu Chandler (R-LWC) closure [12, 13]:

$$\begin{aligned} \omega * c * \omega(r) &= \omega * c_0 * \omega(r) + \omega * (-\beta v) * \omega(r) \\ &+ h(r) - h_0(r) - \ln \frac{g(r)}{g_0(r)}, \quad r > \sigma, \end{aligned} \tag{5}$$

where c_0 and h_0 are the correlation functions for an uncharged hard chain "reference" system at the same density ρ, and asterisks denote convolution integrals.

A rather stringent test of the theory is to compare it to simulations at infinite dilution ($\rho \rightarrow 0$) where the PRISM theory is the least accurate.

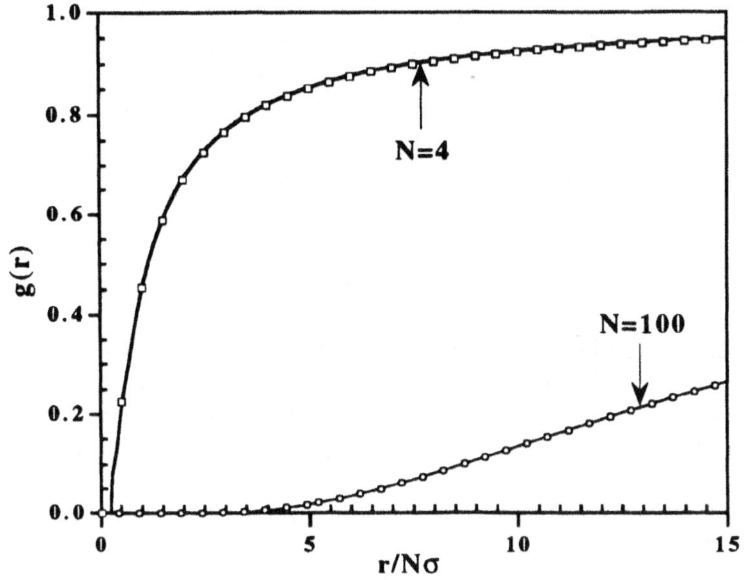

Figure 1. Comparison of $g(r)$ at infinite dilution predicted by the PRISM theory with the R-LWC closure to Monte Carlo simulations (circles) for Γ=0.2, κ=0, and N=4 and 100.

Figure 1 compares theoretical predictions for the pair correlation function to simulations results for Γ=0.2, κ=0, and N=4 and 100. At short distances the value of $g(r)$ is very small because of the strong repulsion between the molecules. $g(r)$ monotonically increases as the separation is increased and asymptotically goes to 1 for large r. The PRISM theory with R-LWC closure is accurate at all distances for both short and long chains. In fact, it is more accurate for charged rods than the PRISM theory with PY closure is for *hard* rods. For large r, $g(r) \sim \exp(-\Gamma N^2/r)$ similar to that between two spheres each with a charge of Nqe. This limiting behaviour is predicted correctly by the R-LWC closure.

2.3. SELF-CONSISTENT DETERMINATION OF INTRAMOLECULAR AND INTERMOLECULAR CORRELATION FUNCTIONS

For non-rigid molecules, within the PRISM formalism, the intramolecular and intermolecular correlation functions must be determined in a self-consistent fashion. The general procedure is to replace the many chain system by a single chain whose sites interact via the sum of the bare interaction and a pairwise decomposable medium-induced interaction due to the other molecules in solution. This medium-induced interaction, $W(r)$, is obtained in a self-consistent fashion as follows. Starting with and initial guess for $W(r)$, $\hat{\omega}(k)$ is obtained using an approximation perturbation theory. The next guess for $W(r)$ is then obtained from the PRISM theory using the $\hat{\omega}(k)$ just calculated. This procedure is described below.

The effective single chain Hamiltonian, $\mathcal{H}$ is

$$\beta\mathcal{H} = \beta\mathcal{H}_{sing} + \sum_{\alpha<\gamma} \beta W(\mid \mathbf{r}_\alpha - \mathbf{r}_\gamma \mid) \tag{6}$$

where $\mathcal{H}_{sing}$ is the Hamiltonian of an isolated chain, and $\mathbf{r}_\alpha$ is the position of site α on the chain. The solvation free energy has been assumed to be pairwise decomposable, and independent of site identity. The medium-induced potential, in the hypernetted chain approximation, is given by [14, 15]

$$\beta\hat{W}(k) = -\rho\hat{c}(k)\hat{S}(k)\hat{c}(k). \tag{7}$$

The calculation of $\hat{\omega}(k)$ for a given $W(r)$ is done using an approximate perturbation theory, the starting point for which is the construction of the Hamiltonian, $\mathcal{H}$, of the full interacting system and a trial Hamiltonian, $\mathcal{H}_1$, about which a perturbation series in the mean square end-to-end distance can be developed. The Hamiltonian is given by

$$\begin{aligned} \beta\mathcal{H} &= \frac{3}{2l}\int_0^L \left(\frac{\partial \mathbf{r}(s)}{\partial s}\right)^2 ds + \frac{3l}{8}\int_0^L \left(\frac{\partial^2 \mathbf{r}(s)}{\partial s^2}\right)^2 ds \\ &+ \frac{1}{2}\int_0^L ds \int_0^L ds' V[\mathbf{r}(s) - \mathbf{r}(s')] \end{aligned} \tag{8}$$

where $V(r) = \beta W(r) + \Gamma e^{-\kappa r}/r$ is the sum of the bare and the medium induced interaction between beads. A suitable reference system is,

$$\beta\mathcal{H}_1 = \frac{3}{2l_1}\int_0^L \left(\frac{\partial \mathbf{r}(s)}{\partial s}\right)^2 ds + \frac{3l_1}{8}\int_0^L \left(\frac{\partial^2 \mathbf{r}(s)}{\partial s^2}\right)^2 ds, \tag{9}$$

i.e. $\mathcal{H}_i$ with l replaced by an effective persistence length, l_1.

The Edwards-Singh approach [16] is to expand $< R^2 >$ in a perturbation series about $< R^2 >_1$ and choose the parameter l_1 in an optimal fashion. If the first non-zero term in the series is annihilated, then

$$< R^2 >_1 < \mathcal{H}_1 - \mathcal{H} >_1 - < R^2(\mathcal{H}_1 - \mathcal{H}) >_1 = 0, \tag{10}$$

and this results in a single nonlinear equation for l_1.

Once the optimum l_1 has been determined, the mean square end-to-end distance is given by

$$< R^2 >= Ll_1 - \frac{l_1^2}{2}(1 - e^{-2L/l_1}) \tag{11}$$

and the single chain structure factor is

$$\hat{\omega}(k) = \frac{2}{L}\int_0^L ds(L-s)\exp\left\{-\frac{k^2}{6}\left[l_1 s + \frac{l_1^2}{2}\left(e^{-2s/l_1} - 1\right)\right]\right\}. \tag{12}$$

3. Structure of Tobacco Mosaic Virus Solutions

Scattering experiments have addressed two specific issues: the scaling of the position of the peak, k_{max}, in $\hat{S}(k)$ with concentration, and the detailed features in $\hat{S}(k)$. We find that the scaling behaviour is not very sensitive to the choice of model parameters but the height and width of $\hat{S}(k)$ are. We first present results for the scaling behaviour and then present results for the absolute intensity.

Figure 2 compares theoretical predictions for k_{max} as a function of the reduced concentration to experimental data for TMV and fd-virus solutions ($\rho^* = N/L^3$ in the theory). The dashed line is the theoretical prediction and the solid lines are least squares fit to the theory in the high and low concentration regimes with slopes of 1/3 and 1/2. The theory is in very good agreement with the experiments. Note that there are no adjustable parameters in this comparison.

A quantitative comparison between theory and experiment requires the charge on the TMV molecules. Since this is not available from any measurement, it is treated as adjustable and obtained by requiring the height of the first peak in the structure factor to match the experimental value at each concentration. Figure 3 compares theoretical predictions to experimental data [7] for various concentrations below the overlap threshold concentration. In all cases a good fit is obtained.

The charge obtained from the fit to experimental data decreases as the concentration is increased. If this is due to a chemical equilibrium between polyions and counterions the equilibrium constant can be determined from a simple model [19]. If the counterions and binding sites of TMV are denoted H^+ and $site^-$, and the occupied site is denoted $H - site$, then the chemical equilibrium relation may be written as

$$H^+ + site^- = H - site, \tag{13}$$

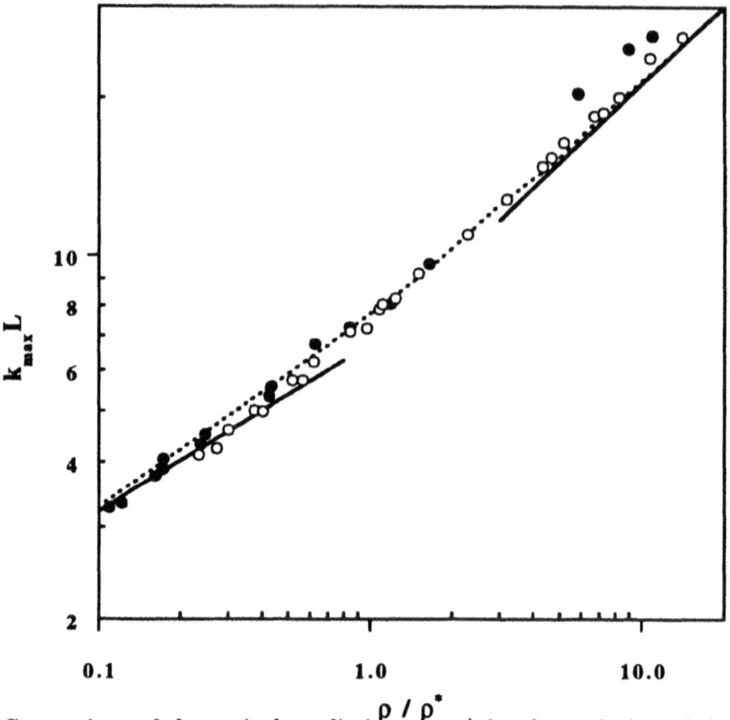

Figure 2. Comparison of theoretical predictions (- - -) for the variation of the position of the first peak in the structure factor with concentration to experiments on tobacco mosaic virus [7] (•) and fd-virus [17] (o). The solid lines are fits to the theory in the low concentration and high concentration regimes with slopes of 1/3 and 1/2. There are no adjustable parameters in this comparison.

with an association constant K defined as

$$K = \frac{[H - site]}{[H^+][site^-]}. \tag{14}$$

In the above the brackets denote concentrations, and activity coefficients have been ignored. The additional constraints for the chemical equilibrium are:(1) mass conservation $[H^+] + [H - site] = SC_{TMV}$ where C_{TMV} is the TMV concentration and S is the total number of binding sites on a TMV molecule; and (2) electroneutrality $[H^+] = [site^-]$ in salt-free solutions. If Q_{TMV} is the exposed charge on the TMV particles (which we obtain by fitting theory to experiment for the structure factors) then $[H^+] = [site^-] = Q_{TMV}C_{TMV}$. An analysis of the fit charge using this method gives S=271 and $K = 2.04 \times 10^5$ inverse molar, respectively, for dilute solutions, which are consistent with other estimates in the literature.

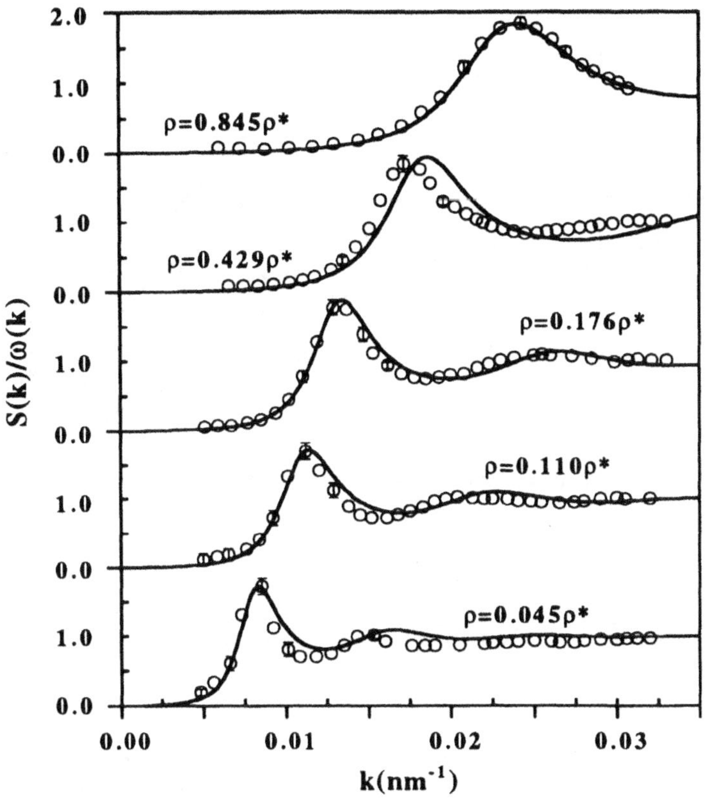

Figure 3. Comparison of theory for $\hat{S}(k)$ to TMV experiments [7] for various concentrations (as marked).

4. Solutions of Flexible Polyelectrolytes

The theory predicts the expected exponents for the scaling of the root mean square end-to-end distance $R \equiv \sqrt{< R^2 >}$ with degree of polymerization N. For uncharged chains $R \sim N^{3/5}$ and $R \sim N^{1/2}$ in dilute and concentrated solutions, respectively. For isolated chains, $R \sim N$ and $R \sim N^{3/5}$ in salt-free and high salt solutions, respectively. In very concentrated solutions, $R \sim N^{1/2}$ independent of the salt concentration. These scaling exponents are identical to those obtained by Muthukumar [18].

Figure 4 compares predictions of the theory (solid lines) to the computer simulations [10], for R in salt-free solutions. At low concentrations the chains are expanded, but not quite rod-like. As the concentration is increased the chain size gradually decreases. For all chain lengths, the theory

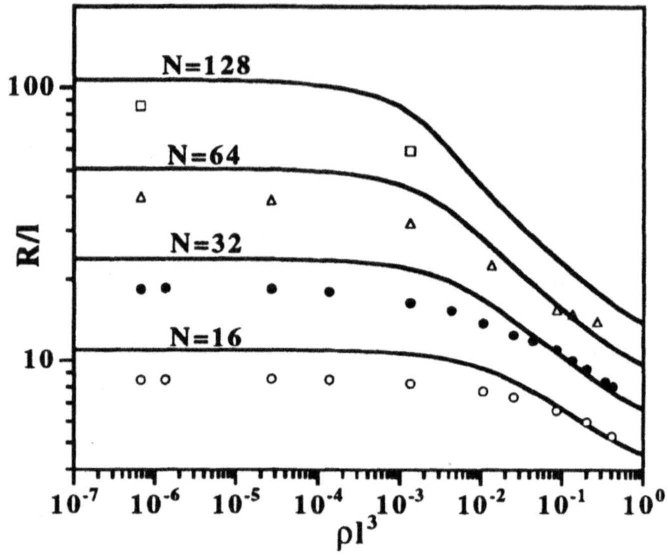

Figure 4. Comparison of theoretical predictions (lines) to molecular dynamics simulations [10] (symbols) for the root-mean-square end-to-end distance in salt-free solutions for N ranging from 16 to 128 (as marked).

is in good agreement with the simulations with no adjustable parameters. In particular the predictions of the theory do not deteriorate as the chain length is increased.

The theory is also in good agreement with simulation results for the static structure factor in salt-free solutions both for the position of the peak and for the intensity. Figure 5 compares results for the static structure factor in salt-free solutions to computer simulations. At low concentrations there is a prominent peak in the structure factor. As the concentration is increased this peak broadens and moves to higher wavevectors. The agreement between theory and simulation is good at all concentrations except ρl^3=0.0106 where the theory does not reproduce the position of the peak very accurately.

The predictions of the theory for the variation of k_{max} with concentration are in good agreement with computer simulations (not shown) although the asymptotic regime is not reached for the short chains used in the simulations. It is only for very long chains that the semi-dilute scaling is observed in the theory. Figures 6 (a) and (b) depict theoretical predictions for the scaling of k_{max} with ρ for N=200 and 10^4, respectively. For N=200 a crossover is observed where the exponent changes from 1/3 in the dilute regime to something higher, but never attains the asymptotic semi-dilute

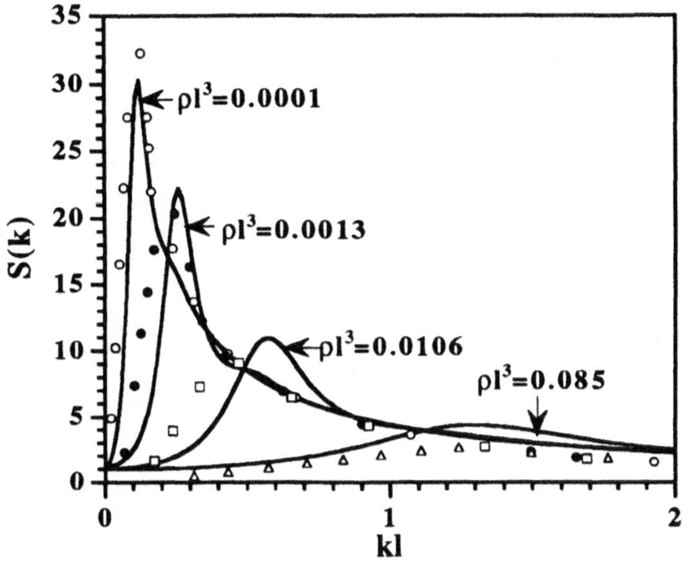

Figure 5. Comparison of predictions of the perturbation theory (solid lines) to molecular dynamics simulations [10] (symbols) for the static structure factor in salt-free solutions for N=32.

value of 1/2. For $N = 10^4$ the semi-dilute regime is much larger and the asymptotic exponent of 1/2 is clearly seen over several decades of concentration. This slow approach to the asymptotic behaviour is expected from experimental results [9].

5. Summary

The static structure and conformational properties of polyelectrolyte solutions have been the focus of considerable experimental study. In this chapter we have shown that the experimental results can be understood using standard liquid state approaches such as integral equations. In particular, good results are obtained with the self-consistent PRISM approach both for the conformational properties and the static structure factor.

Acknowledgments

I thank Dr. C.-Y. Shew for useful discussions. This work was supported by the Sloan Foundation and the National Science Foundation (through grant number CHE 9732604 to A.Y. and a computer grant number CHE 9522057 to the Department of Chemistry).

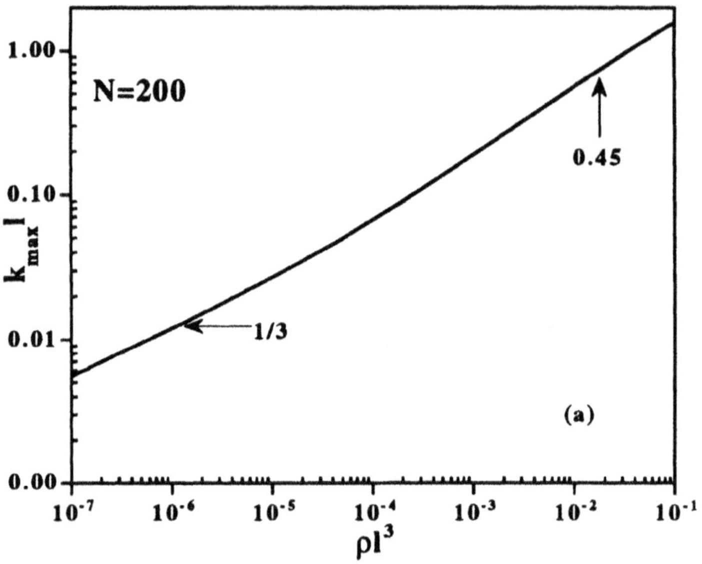

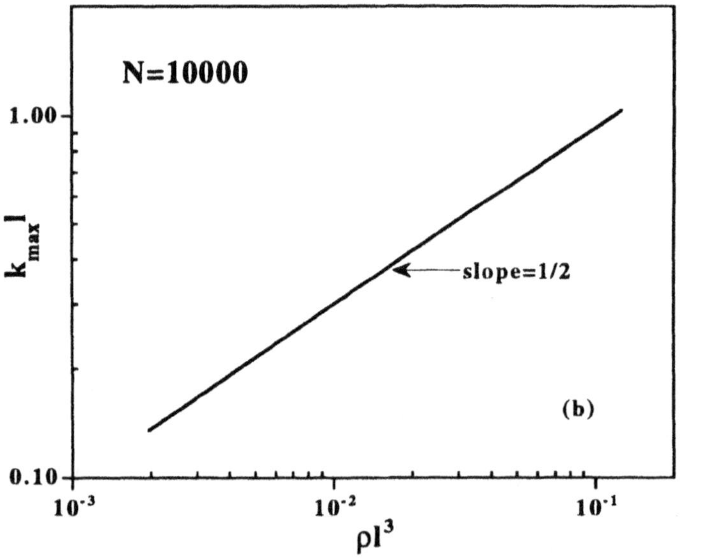

Figure 6. Theoretical predictions for the variation of the position of the peak in the structure factor with concentration for (a) $N = 200$, and (b) $N = 10^4$. The slopes of the lines are marked.

References

1. S. Forster and M. Schmidt, *Advances in Polymer Science* **120**, 50 (1995).
2. A. Yethiraj and C.-Y. Shew, *Phys. Rev. Lett.* **77**, 3937 (1996).
3. C.-Y. Shew and A. Yethiraj, *J. Chem. Phys.* **106**, 5706 (1997).
4. A. Yethiraj, *Phys. Rev. Lett.* **78**, 3789 (1997).
5. A. Yethiraj, *J. Chem. Phys.* **108**, 1184 (1998).
6. E. E. Maier, S. F. Schulz, and R. Weber, Macromolecules **21**, 1544 (1988).
7. E. E. Maier, R. Krause, M. Deggelmann, M. Hagenbüchle, R. Weber, and S. Fraden, *Macromolecules* **25**, 1125 (1992).
8. J. Yamanaka, H. Matsusoka, H. Kitano, N. Ise, T. Yamaguchi, S. Saeki, and M. Tsubokawa, *Langmuir* **7**, 1928 (1991).
9. W. Essafi, F. Lafuma, and C. E. Williams, *J. Phys. II France* **5**, 1269 (1995).
10. M. Stevens and K. Kremer, *J. Chem. Phys.***103**, 1669 (1995).
11. K. S. Schweizer and J. G. Curro, *J. Chem. Phys.***87**, 1842 (1987).
12. D. Laria, D. Wu, and D. Chandler, *J. Chem. Phys.* **95**, 4444 (1991).
13. K. S. Schweizer and A. Yethiraj, *J. Chem. Phys.* **98**, 9053 (1993).
14. D. Chandler, Y. Singh, and D. M. Richardson, *J. Chem. Phys.* **81**, 1975 (1984).
15. C. Grayce, A. Yethiraj, and K. S. Schweizer, *J. Chem. Phys.* **100**, 6857 (1994).
16. S. F. Edward, and P. Singh, *J. Chem. Soc. Faraday Trans. II*, **75**, 1001 (1979).
17. R. Weber, private communication. The data appear in Ph.D dissertations of E. E. Maier and M. Hagenbüchle, University Konstanz, Germany.
18. M. Muthukumar, J. Chem. Phys. **105**, 5183 (1996).
19. C.-Y. Shew and A. Yethiraj, J. Chem. Phys. (submitted).

THEORIES OF HYDROPHOBIC EFFECTS AND THE DESCRIPTION OF FREE VOLUME IN COMPLEX LIQUIDS

L. R. PRATT, S. GARDE AND G. HUMMER
Theoretical Division, Los Alamos National Laboratory
Los Alamos, New Mexico 87545 USA

Abstract. Recent progress on molecular theories of hydration of nonpolar solutes in aqueous solution has led to new ways of thinking about the old issue of free, or "available", volume in liquids. This article surveys the principal new results with particular attention to general issues of packing in liquids.

1. Introduction

Aqueous solutions of colloidal solutes are preeminent examples of complex liquids. In such settings, attention is often directed towards the issues of macromolecular structure, aggregation, and dynamics. However, one aspect of these problems is stubbornly associated with the small size scale of molecules, in particular the small size of a water molecule. That problem is the molecular understanding and description of hydration effects – hydrophobic effects – that stabilize membranes, micelles, folded proteins, and aggregates of such structures. This presentation will focus on that basic, molecular scale issue underlying aqueous solutions of interest both to biophysics and material science of colloids.

Because hydrophobic effects are so broadly discussed, it might be surprising to notice that workers on molecular theories of hydrophobic effects have not achieved good agreement on that molecular theory, at least when we proceed beyond the primitive stage of the principles of statistical mechanics and specification of the intermolecular interactions involved. The empirical fact is that different researchers hold different opinions on the correctness of several available theories, each of which agrees with a set of experimental data at least roughly. A sampling of different perspectives is

C. Caccamo et al. (eds.), New Approaches to Problems in Liquid State Theory, 407–420.

available from the references [1-33]; this collection is not intended to be complete, however.

Furthermore, the simplest extensions of our experimental information can spark new debates about our understanding of hydrophobic effects. A current example, the effect of pressure on hydrophobic stabilization of folded proteins, is discussed below.

In consequence of such observations, the theory surveyed here was designed for maximal simplicity on the fundamentals of statistical mechanics and on the physical assumptions applied to the particular problems. The theory that was developed is relevant to the traditional packing problems of theories of liquids and we emphasize that connection in the discussion here.

A central problem for the theory of hydrophobic effects is the old, but imperfectly solved, theoretical problem of finding space for a solute in liquid solvents. As is well recognized, liquids are dense, disordered materials. It is this combination of attributes that makes these problems difficult. The description of available volume in hard core model liquids is central to the modern understanding of the van der Waals equation of state and is basic to such ordering phase transitions as the hard sphere freezing and the liquid crystal phase transitions [34]. In the specific motivating case considered here, the particular molecular structuring characteristic of liquid water is expected to be important. So we must preserve the fidelity of the description of the structure of liquid water, in addition to tackling the problem of packing in dense, disordered materials.

This lecture develops a new conceptualization of this old problem and new techniques for predicting the fractional free volume accessible to hard core molecules in condensed phases. This new approach is based upon an information theory perspective that has general applicability and was initially explicitly heuristic. Information was sought on a condensed medium of interest and on that basis a prediction of the fractional free volume accessible to a hard model solute was made. For some important cases considered so far, the detail of information required for accurate, interesting predictions has been surprisingly modest.

2. Available Volume Statistics

Consider the solubility of inert gases in aqueous solutions. The medium is liquid water and the solute is idealized as a hard object, perfectly repelling the center (oxygen atom) of each water molecule. For such models the interaction part of the chemical potential of the solute is obtained as

$$\beta\Delta\mu = -\ln p_0 \ , \tag{1}$$

with p_0 the probability that the hard solute could be inserted into the system without overlap of van der Waals volume of the solvent; $1/\beta = k_B T$. This is a specialization of Widom's formula [35, 36]

$$\exp\{-\beta\Delta\mu\} = \langle\exp\{-\beta\Delta U\}\rangle_0 . \tag{2}$$

ΔU is the change in the solute-solvent interaction potential energy upon placement of the solute in an arbitrary position in the solvent and the average indicated by $\langle\ldots\rangle_0$ is over the thermal motion of the solvent unaffected by the solute. The solute is a test particle for this calculation. For the hard core model being considered, ΔU is either *zero* or *infinity*, so the average sought involves a random variable with value either *one* or *zero*; the averaging collects the fraction of solute placements that would be allowed. If presented with a thermal configuration of a large volume of solvent, we might estimate these quantities by performing many trial placements of the solute throughout the solvent and determining the fraction of those trial placements that would be allowed. This estimates V_{free}/V, the fractional free volume accessible to the solute. Thus, Eq. (1) is a free volume formula [37], exact for the model being considered.

The operation of these formulae can be viewed alternatively: Imagine identifying a molecular scale volume at an arbitrary position in the liquid system by (1) hypothetical placement of the solute and (2) determination of those positions of water oxygen atoms that would be excluded due to solute-solvent interactions. We will call this volume the observation volume. With such a molecular scale volume defined we could keep track, say during a simulation calculation, of the probabilities p_n that $n = 0, 1, \ldots$ oxygen atom occupants are observed. As the notation suggests, p_0 is the probability that no occupants are observed in the molecular volume.

Our strategy for predicting of p_0 will be to model the distribution p_n and to extract the extreme value p_0. This is a primitive approach to theories of $\beta\Delta\mu$ and solubilities of inert gases water. Both more and less subtle theoretical works on these topics have been long available. "Less subtle" here means simulation calculations, techniques more straightforwardly useful than many "more subtle" approaches. The "more subtle" means here that further statistical quantities have been introduced for spherical solutes with the intention that they might facilitate more expansive approximate theories. These include

$$d_1(\lambda) = -\frac{dp_0(\lambda)}{d\lambda} , \tag{3}$$

$$G(\lambda) = \left(\frac{-1}{4\pi\lambda^2\rho}\right)\frac{d\ln p_0(\lambda)}{d\lambda} . \tag{4}$$

λ is the radius of a center-to-center exclusion sphere. These quantities have been useful in suggesting physical theories because they have interpretations that are appreciated physically. $d_1(\lambda)$, is the distribution function of distances λ from an arbitrary point in the liquid to the *nearest* solvent center. $4\pi\lambda^2\rho G(\lambda)$ [Eq. (4)] is, in view of Eq. (1), the derivative with respect to exclusion radius of the hydration free energy due to intermolecular interactions, in thermal energy units k_BT. It gives the compressive force exerted by the solvent on the hard spherical solute. In addition, $4\pi\lambda^2\rho G(\lambda)d\lambda$ is the expected number of solvent centers in a shell of radius λ and width $d\lambda$ outside a hard sphere that excludes solvent centers from a ball of radius λ.

3. Simulation Results for Liquid Water

Some of the simulation work has determined the quantities of Eqs. (3) and (4) for molecular liquids represented realistically at the current state-of-the-art [14, 16]. Thus we know that $d_1(\lambda)$ for liquid water and for liquid n-hexane are both unimodal with maximum displaced by a distance less than 0.1Å; the maximum occurs at slightly smaller distances for liquid n-hexane than for liquid water [14, 16]. This difference in the most probable cavity size between liquid n-hexane and liquid water is not large. The difference in the most probable cavity size between liquid water and a reference random medium, with sites of the same radius and distributed randomly at the water density, is greater than the difference between n-hexane and water. So the differences observed between the two molecular liquids considered are not *purely* reflections of molecular size and density. This comparison addresses the idea that the low solubility of inert gases in liquid water might be due to the small size of the water molecule and the possibility that "interstitial" cavities would, on this basis, be smaller in water than in the organic liquid [3, 38]. The fact that these differences between water and n-hexane are slight is associated with the fact that the basic units considered in n-hexane are the methyl and methylene groups. These are not so different in size from a water molecule. It should be noted also that on a packing fraction basis, typical organic liquids are denser than liquid water [6].

The notable distinction between the results for liquid water and liquid n-hexane is that the distribution $d_1(\lambda)$ is narrower for liquid water. This suggests that the liquid water phase is less flexible than the liquid n-hexane phase in opening cavities of substantial size.

Simulation calculations have also produced $G(\lambda)$ for $0< \lambda <3.0$Å, approximately. This size range covers the simplest atomic solutes He and Ne but not much more. This does, however, permit comparison between water and organic solvents, and it permits comparison of available theories with the simulation data. Thus for the range 2.0Å< λ <3.0Å, $G(\lambda)$ for liquid water is approximately two-times larger than for n-hexane. Water exerts a higher compressive force on the surface of an inert solute than do typical organic liquids; water squeezes-out hydrophobic solutes [39].

The checking of theories against the available simulation data $G(\lambda)$ for water has also been revealing [14, 16]. We now know that the predictions of the scaled particle model [40, 41] are significantly below the numerically exact results for $G(\lambda)$. The Pratt-Chandler (PC) integral equation theory [42-44] predicts results for $G(\lambda)$ that are significantly too large. The more pragmatic revised scaled particle model due to Stillinger [2] typically predicts $G(\lambda)$ between those two theories and with some empiricism about interpolation junctions can describe the available simulation data satisfactorily [14, 16]. For sizes λ ≫3.0Å the available computer simulation data are less extensive, the theories less convincing, and the checking has been pursued less vigorously. See, however, the recent results of Reference [45].

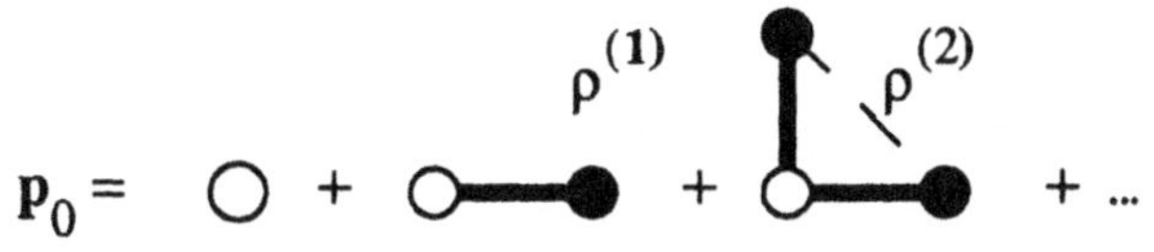

$$= \quad 1 \quad - \quad \langle n\rangle_0 \quad + \quad (1/2)\langle n(n-1)\rangle_0 \quad + \quad \ldots$$

Figure 1. Mayer-Montroll expansion for the insertion probability p_0. The standard notation follows [50, 51], *e.g.*, the solid bonds indicate Mayer f-functions. The second line gives the evaluations for the diagrams shown in the case of a hard core solute. n is the number of solvent centers in the observation volume.

4. Information Model

We now return to the theoretical program of predicting p_0. What are the standard theoretical tools for this? The most immediate guiding theory is the 'inclusion-exclusion' development [46] of Eq. (2) [47, 48, 49]:

$$p_0 = 1 + \sum_{m=1}^{\infty} \frac{(-1)^m}{m!} \int_v d\mathbf{r}_1 \int_v d\mathbf{r}_2 \cdots \int_v d\mathbf{r}_m \rho^{(m)}(\mathbf{r}_1, \mathbf{r}_2, \ldots, \mathbf{r}_m) \ , \quad (5)$$

where $\rho^{(m)}$ is the m-body joint density for solvent centers. This is depicted in a standard way in Figure 1. These are standard combinatorial results, frequently seen in forms such as [46]

$$p_0 = 1 - \langle n\rangle_0 + \sum_{m=2}^{\infty} \frac{(-1)^m}{m!} \langle n(n-1)\cdots(n-m+1)\rangle_0 \ . \quad (6)$$

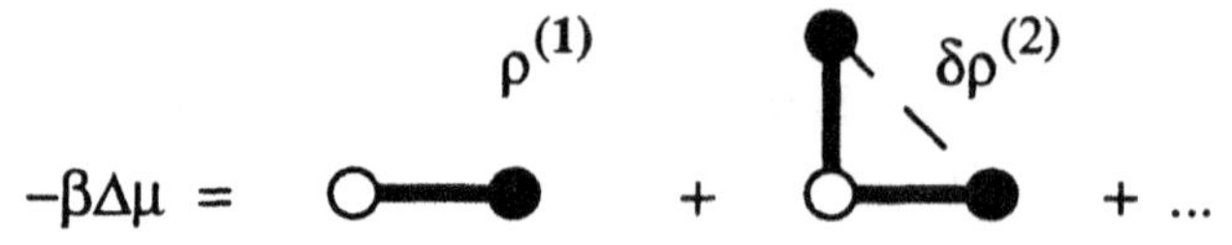

$$= \quad -\langle n\rangle_0 \quad + \ (1/2)\{\langle n(n-1)\rangle_0 - \langle n\rangle_0^{\,2}\} + \ldots$$

Figure 2. 'Virial' expansion, notation as in Figure 1. Virial typically connotes a low density expansion but here we are assuming full knowledge of the medium correlation functions in the absence of the solute; $\delta\rho^{(2)}(1,2) = \rho^{(2)}(1,2) - \rho^{(2)}(1)\rho^{(1)}(2)$. Thus from the perspective of a density expansion this organization shuffles the contributions to the virial coefficients. In fact, here contributions are ordered according to the number of bonds attached to the root point. Successive contributions have a structure that may be derived from that of familiar cumulants with the formal replacement $\langle n^k\rangle_0 \to \langle n(n-1)\cdots(n-k+1)\rangle_0$; then p_0 can be formally expressed as $\langle e^{-n}\rangle_0$. See [52, 53, 54]. Table 1 gives formulae for contributions through 5th order.

Here the random variable n is the number of solvent centers within the observation volume and, *e.g.*, $\langle n\rangle_0$ is the expected number of centers within the observation volume. Several important points can be made from Figure 1. The first is that the primordial available volume model is obtained from the first two terms shown $\beta\Delta\mu \approx -\ln[1-\langle n\rangle_0]$. The second point is direct and basic: p_0 is naturally expressed in terms of occupancy moments, indeed binomial moments here. The sum truncates sharply for cases where a finite maximum number of particles can be present in the observation volume. The sum can be of practical value in the case where only the first nontrivial term is retained. For large solute volumes or solvent densities, this sum is not directly useful; we seek a way to exploit the same information but in a more broadly useful form.

The next most immediate theoretical guidance comes from the 'virial' expansion depicted in Figure 2. This can be considered a resummation of the series Figure 1 and is better in the sense that such a truncation cannot produce a negative probability as truncation of Figure 1 can do. However, when that trouble is avoided Figure 2 can still be less compact and truncations can be less accurate. For example, single term approximation to Figure 1 can be satisfactory and then Figure 2 with a single term is likely to be less so.

Ultimately these considerations avoid the issue that we have only limited information and we want to make the best prediction of p_0 that we can. When the problem is stated this way what to do next is clear: we model the probabilities p_n on an information theory basis. We consider a relative

TABLE 1. Successive contributions to the series $\beta\Delta\mu = \sum_1^\infty a_j$ of Figure 2. $\binom{n}{j} \equiv n(n-1)\cdots(n-j+1)/j!$ is the binomial coefficient.

j	a_j
1	$+\langle\binom{n}{1}\rangle_0$
2	$-\langle\binom{n}{2}\rangle_0 + \langle\binom{n}{1}\rangle_0{}^2/2$
3	$+\langle\binom{n}{3}\rangle_0 - \langle\binom{n}{1}\rangle_0\langle\binom{n}{2}\rangle_0 + \langle\binom{n}{1}\rangle_0{}^3/3$
4	$-\langle\binom{n}{4}\rangle_0 + \langle\binom{n}{1}\rangle_0\langle\binom{n}{3}\rangle_0 + \langle\binom{n}{2}\rangle_0{}^2/2 - \langle\binom{n}{1}\rangle_0{}^2\langle\binom{n}{2}\rangle_0 + \langle\binom{n}{1}\rangle_0{}^4/4$
5	$+\langle\binom{n}{5}\rangle_0 - \langle\binom{n}{1}\rangle_0\langle\binom{n}{4}\rangle_0 + \langle\binom{n}{1}\rangle_0{}^2\langle\binom{n}{3}\rangle_0$ $-\langle\binom{n}{2}\rangle_0\langle\binom{n}{3}\rangle_0 + \langle\binom{n}{1}\rangle_0\langle\binom{n}{2}\rangle_0{}^2 - \langle\binom{n}{1}\rangle_0{}^3\langle\binom{n}{2}\rangle_0 + \langle\binom{n}{1}\rangle_0{}^5/5$

or cross information entropy [55],

$$\eta(\{p_n\}) = -\sum_{n=0}^{\infty} p_n \ln\left(\frac{p_n}{\hat{p}_n}\right) , \tag{7}$$

where $\hat{p}_n$ represents a "default model" chosen heuristically. The moments that enter into the series Figure 1, obtained from simulations if necessary, are the information typically used. We then maximize this information entropy subject to the constraints that the probabilities reproduce the available information. The formal maximization of this entropy gives probabilities

$$p_j \propto \hat{p}_j \exp\left(-\sum_{k=1}^{k_{max}} \zeta_k \binom{j}{k}\right) \tag{8}$$

where the ζ_k are Lagrange multipliers to be adjusted so that the probabilities finally reproduce the information given initially. The machinery for doing this can be developed straightforwardly. For example, the normalization of the probabilities can be deferred at intermediate stages of the calculation. Then the final thermodynamic result can be given in terms of the required normalization factor

$$\beta\Delta\mu = \ln \sum_{n=0} \frac{\hat{p}_n}{\hat{p}_0} \exp\left(-\sum_{k=1}^{k_{max}} \zeta_k \binom{n}{k}\right) . \tag{9}$$

This is suggestive of the calculation of a partition function for a modest-sized set of states with effective interactions. The interesting questions then involve the predictions extracted from the p_j for properties other than the given information. In our case, the property of first interest is $\beta\Delta\mu$.

We can make the comforting observation that use of only $\langle n\rangle_0$ and the natural default model $\hat{p}_j \propto 1/j!$ produces the Poisson distribution as expected; $\zeta_1 = -\ln\langle n\rangle_0$ and $\beta\Delta\mu = \langle n\rangle_0$ as would be found by retaining only the first term shown in Figure 2. Note that evaluation of the second term there with the Poisson distribution gives zero as it should [53]; $\langle \delta n^2 \rangle_0 = \langle n \rangle_0$ for the Poisson distribution.

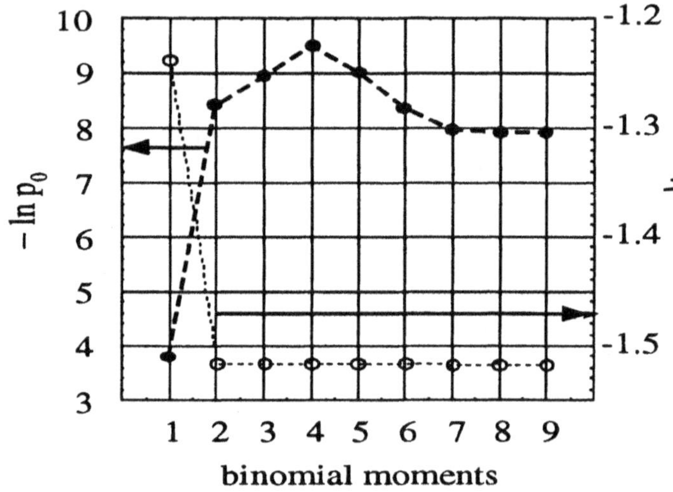

Figure 3. Convergence of predictions for hydration free energy and information entropy with numbers of binomial moments employed; $\hat{p}_j \propto 1/j!$. The moment information was obtained from computer simulation of liquid water [58]. λ=3.0Å. The filled (open) circles use the left (right) scale.

Figure 3 shows how this prediction scheme works-out when the solvent is computer simulated liquid water and the solute is a hard sphere of a size appropriate for comparison with a Ne atom. The immediate point is that a model based upon the two moments that might be obtained from experiment, $\langle n \rangle_0$ and $\langle n(n-1) \rangle_0$, fortuitously provides the most satisfactory simple prediction of p_0. The Poisson (one moment) model is not satisfactory. Inclusion of moments higher than the second is not advantageous unless several higher moments are available. The behavior of the information entropy η suggests that the initial two moments do a good job of describing the distribution and the subsequent higher moments are 'uninformative.'

These points are further remarkable because recent analyses [56, 57] have underscored the fact that Percus-Yevick approximate integral equations can be derived on the assumption that solvent density fluctuations are distributed according to a Gaussian probability functional. The results of the present investigations, including particularly [56-59], give further insight and support to those ideas. The PC approximate integral equation theory of hydrophobic effects [42], at its inception a Percus-Yevick analogue, can be given a similar basis [56]. The PC theory is thereby given a better foundation than was available at its genesis.

Note that present two-moment model is not precisely an assumption of a Gaussian probability functional for a density field. The occupancies here are required to be nonnegative integers. That this be true of *all* subvolumes of the observation volume is an important restriction.

TABLE 2. Approximate evaluation of contributions to the series Figure 2 for the circumstances of Figure 3, λ=3.0Å. The final row gives the result obtained directly from simulation [58]. The two right-most columns elaborate a direct idea for reorganization of this series, motivated by the form of the single-term result $\beta\Delta\mu \approx -\ln[1-\langle n\rangle_0]$. Though this resummation improves the prediction after each term, after five terms the result is still not as successful as the two-moment information theory shown in Figure 3.

j	$\langle\binom{n}{j}\rangle_0$	a_j	$\sum_1^j a_k$	$r_j \equiv \frac{a_j/a_{j-1}}{(1-1/j)}$	$-\ln[1-r_j]+\sum_1^j[a_k - {r_j}^k/k]$
1	3.77	3.77	3.77	-	-
2	5.75	1.36	5.13	0.72	5.42
3	4.57	0.75	5.88	0.83	6.29
4	2.04	0.50	6.38	0.89	6.91
5	0.51	0.36	6.74	0.90	7.21
∞	-	-	7.93	-	-

That specifically binomial moments are involved in the series above emphasizes the point that the occupancies must be nonnegative integers. Further perspective on the convergence issue is obtained by examination of successive contributions to the series Figure 2. Numerical results are shown in Table 2. The first and second order terms make a significant contribution but by themselves are not close to the full answer. The values of the additional terms do not establish a rapid convergence to the known full answer. However, we can view the information theory model as a technique for reorganization of the series.

4.1. ENTROPY CONVERGENCE

When the typical occupation numbers n are large the granularity of the distribution p_n is expected to be less significant, at least near the center of the distribution and when viewed on a coarse enough scale. In such circumstances the predictions of two-moment information theory models are not significantly different than those of the PC theory. It is remarkable that a simple calculation along these lines gives a convincing explanation of the puzzling and contentious issue of "entropy convergence" in hydrophobic hydration [60].

The phenomenon to be explained is the following: entropies of transfer of non-polar molecules from gas phase or a non-polar solvent into water converge at a temperature of about 400 K to approximately zero entropy change. Similar behavior was also seen in the microcalorimetry experiments on unfolding of several globular proteins. This behavior is insensitive to the particular hydrophobic solute molecule. Since the entropy is a temperature derivative of a hydration free energy, the convergence temperature identifies a region where graphs of hydration free energy versus temperature are extremal, in fact, maximal. Below that region the hydrophobic hydration free energy increases with temperature but above that region the hydrophobic hydration free energy decreases as the temperature is raised.

The two-moment information theory model above was applied to this problem for hard sphere solutes in water with the heuristic modification that a flat default model was used; $\hat{p}_j \propto$ constant for $j \leq j_{max}$ and zero otherwise. This latter adjustment was found empirically to give slightly better hydration free energies. The results of the model and simulation calculations accurately agreed on the temperature dependence of the hydration free energies. To analyze this agreement the information theory model was simplified to a continuous Gaussian distribution that then gives

$$\Delta\mu \approx \frac{k_{\rm B}T}{2}\left\{\frac{\langle n\rangle_0{}^2}{\langle\delta n^2\rangle_0} + \ln[2\pi\langle\delta n^2\rangle_0]\right\} \tag{10}$$

$$= T\rho_{sat}(T)^2\{k_{\rm B}v^2/2\langle\delta n^2\rangle_0\} + T\{k_{\rm B}\ \ln(2\pi\langle\delta n^2\rangle_0]/2\}\ . \tag{11}$$

v is the observation volume and $\rho_{sat}(T)$ is the liquid density along the vapor saturation curve so that $\langle n\rangle_0 = \rho_{sat}(T)v$. $\langle\delta n^2\rangle_0$ was found to be insensitive to temperature for the relevant conditions. Further, the first term of Eq. (11) is larger than the second. Thus the non-monotonic behavior of the free energy with temperature and the entropy convergence is a consequence of the non-monotonic variation of $T\rho_{sat}(T)^2$ with temperature. The only molecular parameter to complicate matters is the volume v and with this formula v does not affect the entropy convergence temperature [60]. Thus the temperature of entropy convergence is about the same for a wide family of solutes.

The physical point is: the entropy convergence phenomenon occurs for water because of the low and temperature insensitive values of $\langle\delta n^2\rangle_0$. In fact, the isothermal compressibility of water at low pressure has a minimum value at T=319 K. That temperature differs substantially from the observed entropy convergence temperatures but it is not necessary that these temperatures be approximately equal, just that $\langle\delta n^2\rangle_0$ be insensitive to temperature in the region $d[T\rho_{sat}(T)^2]/dT \approx 0$.

The technical point of Eq. (10) is: this formula is simple and effective but how it is obtained from the series Figure 2 is not simple. Part of the complication is that two additional twists have been interjected, the flat default model and the continuous approximation.

4.2. PRESSURE DENATURATION OF PROTEINS

It is a common view, based upon our current understanding, that hydrophobic effects provide a nonspecific, cohesive stabilization of compact protein structures. However, it has been argued [62] that our current understanding of hydrophobic effects is not consistent with the experimental facts of pressure denaturation of globular proteins. The information theory model of the previous section was applied also to study hydration free energies and potentials of mean force (pmfs) for two and three hydrophobic spherical solutes in water as a function of pressure [63]. As is well known, those pmfs exhibit contact and solvent-separated minima corresponding, respectively, to cases where the hydrophobic spheres contact each other or where a water molecule intervenes. It was found that increasing pressure shifted the free energy balance of those two cases towards the solvent-separated circumstance. This suggested an intercalation mechanism for pressure denaturation: as the pressure of the liquid is raised, water molecules are forced into protein structure. A similar point of view can be taken of the formation of clathrate hydrates at elevated pressures: at low pressure hydrophobic effects lead to close contacts and to clustering of hydrocarbon gases dissolved in water. Pressure increases stabilize the crystalline phase that eliminates close solute contacts. If attention is focused on the hydrocarbon material this behavior might seem counter-intuitive because the hydrocarbon material seems to expand. But the thermodynamic principle is that increasing pressure stabilizes the phase of lower volume. Therefore, we conclude that the system may be packed more efficiently and have a lower total volume when water molecules are intercalated into the hydrocarbon clusters. These topics will surely be the subject of further research.

5. Concluding Comments

Identification of some generalizations and future directions for these theories will provide concluding comments. Firstly, we note that the generalization of these ideas to treat continuous, rather than only hard core repulsive, solute-solvent interactions is known [64].

Secondly, we note the importance in the context of the aqueous solutions of our restriction here to small molecule solutes. It is well recognized that treatment of larger solutes requires consideration of the multiphasic character of these solutions on large length scales [2, 58, 63]. For large enough hard sphere solutes dissolved in water close to phase coexistence, the possibility that the solvent will pull away from the solute surface requires specific attention. Further subtleties arise when the solute-solvent interactions are not just repulsive but include attractive interactions too [65, 66]. These issues will surely be the subject of further research in the area of hydrophobic effects. It seems likely that an appropriately designed default model should be able to describe such effects in a physical manner.

Thirdly, we note that these approaches provide some unanticipated answers [64] to questions such as "How is water different from hydrocarbon liquids as a solvent for nonpolar solutes?" The importance of the low and temperature insensitive values of the isothermal compressibility of liquid water is noteworthy. However, such answers are not in the format that is most often intended when such questions are asked. Most often such questions solicit information about particular patterns of solvent structure in the neighborhood of a hydrophobic solutes. Some groundwork has been laid for consideration of those detailed structural issues in a format consistent with the discussion of this paper [67]. Pursuit of answers about the detailed structural issues and their relevance to hydrophobic effects will surely be the subject of future research.

Finally, we note again the relevance of these ideas to the classic problems of packing in liquids. The importance of these issues is reflected in the significance of the hard sphere fluid system to our understanding of liquids. Hard core model systems may not be directly realistic. But it continues to surprise that when attention is directed to new physical problems, *e.g.* the thermodynamics and structure of glasses or folded proteins, understanding of basic packing problems is again requested. Such problems surface quite broadly [66-72]. It would be interesting to see the ordering phase transitions associated with packing problems, crystallization and liquid crystallization, analyzed on these bases. An initial step along such lines for the hard sphere fluid has been taken [75] but more work is deserved.

We hasten to add that the results so far have not superseded previous theoretical results. But this new approach offers the possibility of better, more physical understandings of packing problems in the equilibrium statistical mechanics of non-crystalline materials and the previous theories of them. This approach *has* achieved new understanding for the problems of primitive hydrophobic effects. By exploiting information external to conventional theories, even simulation data, and by proposing a pattern for utilizing that information, these approaches begin to respond to Andersen's [76] request for a 'theory of theories.'

Acknowledgement

This work was supported by the LDRD program at Los Alamos.

References

1. A. Ben-Naim and H. L. Friedman. *J. Phys. Chem.* 71:448 449, 1967.
2. F. H. Stillinger. *J. Soln. Chem.*, 2:141 158, 1973.
3. B. Lee. *Biopolymers*, 24:813 823, 1985.
4. A. Ben-Naim. *J. Chem. Phys.*, 90:7412 7425, 1989.
5. P. L. Privalov and S. J. Gill. *Pure and Applied Chemistry*, 61:1097 1104, 1989.
6. A. Pohorille and L. R. Pratt. *J. Amer. Chem. Soc.*, 112:5066 5074, 1990.
7. K. A. Dill. *Science*, 250:297, 1990.
8. P. L. Privalov, S. J. Gill, and K. P. Murphy. *Science*, 250:297 298, 1990.
9. K. P. Murphy, P. L. Privalov, and S. J. Gill. *Science*, 247:559 561, 1990.
10. N. Muller. *Acc. Chem. Res.*, 23:23 28, 1990.
11. L. R. Pratt. Oil and water don't mix. In *CLS Division 1991 Annual Review*, 5285 Port Royal Rd., Springfield, VA 22161, 1991. National Technical Information Service U. S. Department of Commerce. LA-UR-91-1783.
12. K. A. Sharp, A. Nichols, R. F. Fine, and B. Honig. *Science*, 252:106 109, 1991.
13. K. A. Sharp, A. Nicholls, R. Friedman, and B. Honig. *Biochem.*, 30:9686 9697, 1991.
14. L. R. Pratt and A. Pohorille. *Proc. Natl. Acad. Sci. USA*, 89:2995, 1992.
15. T. Lazaridis and M. E. Paulaitis. *J. Phys. Chem.*, 96:3847 3855, 1992.
16. L. R. Pratt and A. Pohorille. Proceedings of the EBSA 1992 International Workshop on Water-Biomolecule Interactions. Societa Italiana di Fisica, Bologna, 1993.
17. B. Lee. *Protein Science*, 2:733 738, 1993.
18. D. van Belle and S. J. Wodak. *J. Am. Chem. Soc.*, 115:647 652, 1993.
19. A. Ben-Naim and R. M. Mazo. *J. Phys. Chem.*, 97:10829 10834, 1993.
20. A. Holtzer. *Biopolymers*, 34, 1994.
21. L. X. Dang. *J. Chem. Phys.*, 100:9032 9034, 1994.
22. D. van Belle, M. Prevost, G. Lippens, and S. J. Wodak. *ACS Symposium Series*, 568:318 334, 1994.
23. B. Madan and B. Lee. *Biophysical Chemistry*, 51:279 289, 1994.
24. D. Sitkoff, K. A. Sharp, and B. Honig. *Biophysical Chemistry*, 51:397 409, 1994.
25. A. Holtzer. *Biopolymers*, 35:595 602, 1995.
26. S. K. Kumar, I. Szleifer, K. Sharp, P. J. Rossky, R. Friedman, and B. Honig. *J. Phys. Chem.*, 99:8382 8391, 1995.
27. D. Horvath, D. van Belle, G. Lippens, and S. J. Wodak. *J. Chem. Phys.*, 104:6679 6695, 1996.
28. T. Headgordon. *Proc. Natl. Acad. Sci. USA*, 92:8308 8312, 1995.
29. A. D. J. Haymet, K. A. T. Silverstein, and K. A. Dill. *Faraday Discussions*, 103:117 124, 1996.
30. S. Garde, G. Hummer, and M. E. Paulaitis. *Faraday Discussions*, 125 139, 1996.
31. K. A. Sharp, S. Kumar, P. J. Rossky, R. A. Friedman, and B. Honig. *J. Phys. Chem.*, 100:14166 14177, 1996.
32. M. Prevost, I. T. Oliveira, J. P. Kocher, and S. J. Wodak. *J. Phys. Chem.*, 100:2738 2743, 1996.
33. P. M. Wiggins. *Physica A*, 238:113 128, 1997.
34. D. Chandler, J. D. Weeks, and H. C. Andersen. *Science*, 220:787 794, 1983.
35. B. Widom. *J. Chem. Phys.*, 39:2808 2812, 1963.
36. B. Widom. *J. Phys. Chem.*, 86:869 872, 1982.
37. H. Reiss. *J. Phys. Chem.*, 96:4736 4747, 1992.
38. B. Lee. *Biopolymers*, 31(8):993 1008, 1991.
39. F. M. Richards. *Scientific American*, 264(1):54, 1991.
40. R. A. Pierotti. *J. Phys. Chem.*, 67:1840 1845, 1963.
41. R. A. Pierotti. *Chem. Rev.*, 76:717 726, 1976.
42. L. R. Pratt and D. Chandler. *J. Chem. Phys.*, 67:3683 3704, 1977.
43. L. R. Pratt. *Ann. Rev. Phys. Chem.*, 36:433, 1985.

44. L. R. Pratt and D. Chandler. *Methods in Enzymology*, 127:48, 1985.
45. G. Hummer and S. Garde. *Phys. Rev. Letts.*, 80:4193 4196, 1998.
46. J. Riordan. *An Introduction to Combinatorial Analysis*. Princeton University Press, Princeton, NJ, 1978.
47. J. E. Mayer and E. Montroll. *J. Chem. Phys.*, 9:2 16, 1941.
48. H. Reiss, H. L. Frisch, and J. L. Lebowitz. *J. Chem. Phys.*, 31:369 380, 1959.
49. N. G. van Kampen. *Stochastic Processes in Physics and Chemistry*. North-Holland, New York, 2nd edition, 1992. Section II.4.
50. H. C. Andersen. Volume 5 of *Modern Theoretical Chemistry, Statistical Mechanics, Part A: Equilibrium Techniques*. Plenum, New York, 1977.
51. J.-P. Hansen and I. R. McDonald. *Theory of Simple Liquids*. Academic Press, San Diego, CA, 2nd edition, 1991. Chapter 4.
52. R. Kubo. *J. Phys. Soc. Japan*, 17:1110 1120, 1962.
53. P. Carruthers. *Phys. Rev. A*, 43:2632 2639, 1991.
54. R. L. Graham, D. E. Knuth, and O. Patashnik. *Concrete Mathematics*. Addison-Wesley, Reading, MA, 2ndmerely edition, 1994. Section 2.6.
55. J. E. Shore and R. W. Johnson. *IEEE Transactions on Information Theory*, 26(1):26 37, 1980.
56. D. Chandler. *Phys. Rev. E*, 48:2898 2905, 1993.
57. J. K. Percus. *J. de Phys. IV*, 3:49 57, 1993.
58. G. Hummer, S. Garde, A. E. Garcia, A. Pohorille, and L. R. Pratt. *Proc. Nat. Acad. Sci. USA*, 93:8951, 1996.
59. B. J. Berne. *Proc. Nat. Acad. Sci. USA*, 93:8880, 1996.
60. S. Garde, G. Hummer, A. E. Garcia, M. E. Paulaitis, and L. R. Pratt. *Phys. Rev. Letts.*, 77:4966 4968, 1996.
61. L. R. Pratt, *Hydrophobic Effects*, in *The Encyclopedia of Computational Chemistry*, Schleyer, P. v. R.; Allinger, N. L.; Clark, T.; Gasteiger, J.; Kollman, P. A.; Schaefer III, H. F.: Schreiner, P. R., Eds.; John Wiley & Sons, Chichester, 1998.
62. W. Kauzmann. *Nature (London)*, 325:763 764, 1987.
63. G. Hummer, S. Garde, A. E. Garcia, M. E. Paulaitis, and L. R. Pratt. *Proc. Natl. Acad. Sci. USA*, 95:1552 1555, 1998.
64. G. Hummer, S. Garde, A. E. García, M. E. Paulaitis, and L. R. Pratt. Hydrophobic Effects on a Molecular Scale. preprint 1998. LA-UR-98-2758, http://xxx.lanl.gov/abs/physics/9807001.
65. J. D. Weeks, R. L. B. Selinger, and J. Q. Broughton. *Phys. Rev. Letts.*, 75:2694 2697, 1995.
66. J. D. Weeks, K. Vollmayr, and K. Katsov. *Physica A*, 244:461 475, 1997.
67. L. R. Pratt and R. A. LaViolette. Quasi-chemical theories of associated liquids. *Molec. Phys.*, 1998. (in press).
68. M. D. Eldridge, P. A. Madden, and D. Frenkel. *Nature*, 365:35 37, 1993.
69. D. Frenkel. *Physics World*, 6:24 25, 1993.
70. D. Frenkel. *J. Phys.: Condens. Matter*, 6:A71 A78, 1994.
71. M. Dijkstra, D. Frenkel, and J. P. Hansen. *J. Chem. Phys.*, 101:3179 3189, 1994.
72. M. Dijkstra and D. Frenkel. *Phys. Rev. Letts.*, 72:298 300, 1994.
73. D. Frenkel. *Future Generation Computer Systems*, 10:207 212, 1994.
74. M. Dijkstra and R. Vanroij. *Phys. Rev. E*, 56:5594 5602, 1997.
75. G. E. Crooks and D. Chandler. *Phys. Rev. E*, 56:4217 4221, 1997.
76. H. C. Andersen. *Ann. Rev. Phys. Chem.*, 26:145 166, 1975.

THEORETICAL AND SIMULATION STUDY OF MODEL PROTEIN SOLUTIONS

C. CACCAMO, D. COSTA AND G. PELLICANE
Istituto Nazionale per la Fisica della Materia (INFM) and Dipartimento di Fisica, Università di Messina Contrada Papardo, C.P. 50, 98166 Messina, Italy

1. Introduction

Phase equilibria in protein solutions are of great interest since on their knowledge rests the possibility to govern the process of protein crystallization. As is well known, such a control is far from having been achieved since the experimental protocols for the growth of crystals pure enough, and of sufficiently large size so to allow the determination of the protein structure, are not well defined.

Such a status of the art constitutes a great stimulus for both theoretical and simulation studies of model fluids which can mimic protein solutions. The aim of such investigations is to obtain as much information as possible on phase transition conditions and in particular, to ascertain which characteristics of the potential play a substantial role in determining phase equilibria. Now, very recent experimental [1] and simulation [2] studies of protein solutions have shown that the most favourable conditions for promoting a controlled nucleation process, and thus for getting a regular crystal growth, occur when the fluid-fluid critical point falls just below the sublimation curve. This last aspect seems of peculiar interest since it is known [3, 4, 5] that in one component fluids the liquid-vapor binodal flattens when the range of the potential decreases till, eventually, it shifts below the sublimation curve and the liquid phase becomes metastable.

One of the purpose of this study is to extend the investigations hitherto performed on one component systems to the multicomponent case, so that solvent effects on the diluted protein molecules can be explicitly taken into account. We shall consider to such an aim hard-core Yukawa mixtures (HCYM) in which hard sphere particles interact through an attractive Yukawa tail.

C. Caccamo et al. (eds.), New Approaches to Problems in Liquid State Theory, 421–436.

Albeit such a system constitutes a rather crude modelization of a protein solution, the corresponding potential has a simple analytic form which allows one to tune the depth of the attractive well at short range and the decay of the interactions at long range, features which are known to be most effective [5] in determining the phase portrait of the fluid.

We shall try to determine the phase diagram of HCYM both by theoretical means and simulation, by extending the results we have recently obtained for these same systems in the symmetric case [6]. Specifically, we shall make use of the modified hypernetted-chain [7] approximation, which is known to be both accurate and thermodynamically selfconsistent. We note that the determination of the phase diagram of a fluid mixture in the framework of an iterationally solvable theory, in which thermodynamic consistency is also to be imposed, is by itself a rather demanding task for which quite few results are available (see [8] for a review). In this respect the study at issue, though stimulated by an experimental problem of great interest, has a more general applicability which seems of interest in the scientific scenario of this NATO-ASI.

We also perform a series of computer simulation investigations of the phase diagram of HCYM by employing the Gibbs Ensemble Monte Carlo [9] and the Semigrand-canonical Monte Carlo [10, 11] methods. The adoption of two simulation strategies reveals crucial in order to get a complete representation of the phase diagram for at least one type of the HCYM investigated. Also, simulation acts as the necessary benchmark of the theoretical predictions, and is the only information source in those thermodynamic regions in which the solution algorithm of the modified hypernetted-chain fails to converge.

We shall introduce in sect. 2 the model potential, the theory and the simulation approaches. We then report and discuss the results for the phase diagram of various mixtures in sec. 3. A few concluding remarks follow in sect. 4.

2. Model, theory and simulation approach

We consider a binary fluid mixture in which particles interact through the potential

$$\begin{aligned} v_{ij}(r) &= \infty & r < \sigma_{ij} \\ v_{ij}(r) &= -K_{ij}\exp[-z_{ij}(r-\sigma_{ij})]/r & r \geq \sigma_{ij} \end{aligned} \tag{1}$$

where i and j are indices of the particle species, $\sigma_{ii} \equiv \sigma_i$ is the hard sphere diameter of the i-th species, and $\sigma_{12} = (\sigma_1 + \sigma_2)/2$. Most of the results we shall report below are obtained for a completely symmetric configuration, in which the two species have equal diameter and equal like interaction amplitudes, namely $\sigma_1 = \sigma_2$ and $K_{11} = K_{22}$. The inverse decay length z_{ij} is also held fixed and the mixture is solely characterized by the ratio $K_{12}/K_{11} = \nu$ of the unlike to like interaction strength. Asymmetric mixtures will be characterized by the size ratio $\alpha = \sigma_2/\sigma_1$ of the hard core diameters as well as by different interaction strength parameters K_{ij} and z_{ij}.

In what follows we shall measure distances in σ_1 units, and energies in K_{11} units. We also introduce the reduced temperature $T^* = k_B T/K_{11}$, the reduced density $\rho^* = (\rho_1 + \rho_2\alpha^3)\sigma_1^3$, and the concentration $\chi_i = \rho_i/\rho$, where ρ_i and ρ are the i-th and total number density of particles, respectively.

We now briefly outline our theoretical and simulation approaches. In the modified hypernetted-chain (MHNC) approximation, a closure to the Ornstein-Zernike equation

$$h_{ij}(r) = c_{ij}(r) + \sum_k \rho_k \int c_{ik}(|\mathbf{r} - \mathbf{r}'|) h_{kj}(\mathbf{r}') d\mathbf{r}' \tag{2}$$

(where $h_{ij}(r) = g_{ij}(r) - 1$ and $c_{ij}(r)$ are the pair and direct correlation function, respectively), is provided by the cluster expansion expression for the radial distribution funtion (rdf) $g_{ij}(\mathrm{r})$:

$$g_{ij}(r) = \exp[-\beta v_{ij}(r) + h_{ij}(r) - c_{ij}(r) + E_{ij}(r)] \quad . \tag{3}$$

We calculate the bridge function $E_{ij}(r)$ in eq. (3) by using its Percus-Yevick theory [12] counterpart for hard sphere mixtures, $E_{ij}^{HS}(r;\sigma_{ij}^*)$. The hard sphere diameters σ_{ij}^* are used as adjustable parameters in order to impose the thermodynamic consistency of the theory. The latter amounts to the request that the osmotic compressibilities obtained from the virial equation of state (EOS), are equal to those calculated from fluctuation theory, namely that

$$\left(\beta \frac{\partial P}{\partial \rho_i}\right)_{T,\rho_{j(j\neq i)}} = 1 - \sum_j \rho_j \tilde{c}_{ij}(q=0) \tag{4}$$

where $\tilde{c}_{ij}(q = 0)$ is the Fourier transform of $c_{ij}(r)$, estimated in the long wavelength limit.

The solution of the MHNC is obtained through iterational procedures which are well established in the literature (see [8]).

The liquid-vapor binodal line is then constructed by calculating the chemical potential from an integral with respect to ρ of the EOS, performed along isothermal paths so to obtain the excess free energy of the system. An isothermal-isochore path is adopted in order to reach the liquid side of the binodal without crossing the unstable region below the critical point. Along the isochore portion of the path, contributions to the excess free energy are obtained by integrating the excess energy with respect to the inverse temperature. Note that thermodynamic consistency is ensured all along the integration process.

Gibbs Ensemble Monte Carlo (GEMC) simulations at constant volume are mostly used in order to study the liquid-vapor coexistence curve. Typical runs were performed with $N = 400$ total number of particles, enclosed in two cubic boxes with periodic boundary conditions.

The GEMC procedure implies the swapping of particles between the two simulation boxes and, at relatively high densities, such moves are frequently rejected thus causing ergodic sampling problems. In order to tackle the description of liquid-liquid type equilibrium we then use the Semigrand-canonical Monte Carlo (SGCMC) approach in which one performs a single simulation of one of the coexisting boxes. Within this framework, it is not necessary to perform volume changes and the equilibrium between the coexisting phases is achieved by means of particle identity interchanges, rather than through direct particle transfers. A "reduced" version of the SGCMC, suited to deal with symmetric mixtures [11], was applied in this work.

Similarly to the GEMC case, SGCMC simulations are performed with $N = 400$ particles, enclosed in a cubic box with periodic boundary conditions at a fixed number density.

We also try to locate the onset of freezing in our HCYM. To such an aim, we adopt an approach proposed by other authors [13] in which one monitors the behavior of the multiparticle residual entropy defined as

$$\Delta s \equiv s_{ex} - s_2 \quad , \tag{5}$$

where s_{ex} is the the excess entropy of the system per particle in units of the Boltzmann constant, and the two-particle term

$$s_2 = -\frac{1}{2}\rho \sum_{i,j} \int \{g_{ij}(r) \ln[g_{ij}(r)] - g_{ij}(r) + 1\} dr \quad . \tag{6}$$

As it has been verified and thoroughly discussed in several recent studies [4, 13, 14], the vanishing of Δs acts as quite a sensitive indicator of the freezing transition in several one component fluids, and preliminary evidence is available that the same condition can also be associated to phase separation in two component mixtures [15].

3. Results and discussion

3.1. SYMMETRIC MIXTURES

With reference to our previous work [6], we first consider the case $\sigma_1 = \sigma_2$ and $K_{11} = K_{22}$. We also fix a unique value for the potential decay rate, $z_{ij} = 2.45$.

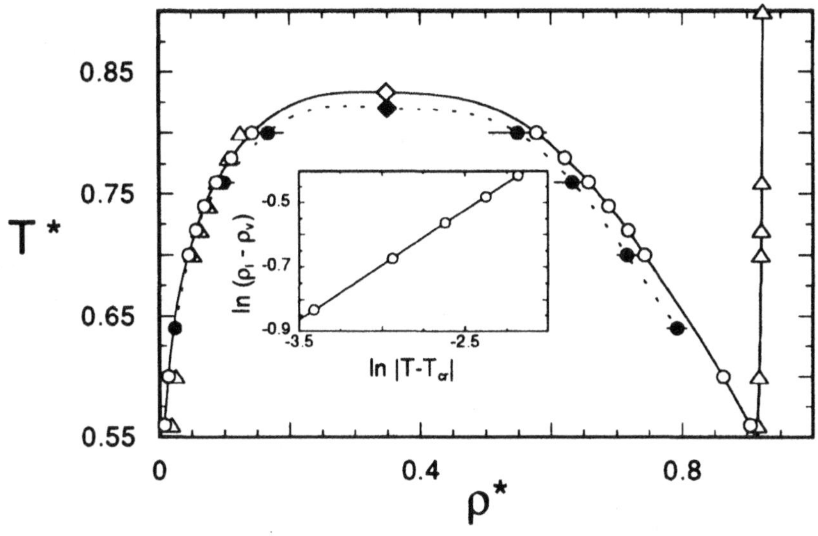

Figure 1. Phase diagram of the mixture with $\nu = 0.9$. Filled and open circles: GEMC and MHNC predictions for the binodal points, respectively. The interpolated estimate of the critical point is reported as a filled and a open diamond, respectively, for the two cases. Statistical uncertainties in the simulation are shown as horizontal bars. Open triangles correspond to the $\Delta s = 0$ locus. The drawn lines are guides for the eye. Note that the MHNC does not converge close to the critical point ($0.80 < T^* < 0.88$). In the inset: power law fit (7) of the last five binodal points up to $T^* = 0.8$.

3.1.1. *Liquid-vapor and liquid-liquid equilibrium*

GEMC and MHNC calculations of the coexistence lines for two mixtures with ν=0.9 and ν=0.7, are reported in figs. 1 and 2, respectively.

As it is possible to observe, liquid-vapor phase separation takes place only for not too high densities. It also appears that when ν decreases the binodal curve shifts downwards with a marked decrease of the critical temperature.

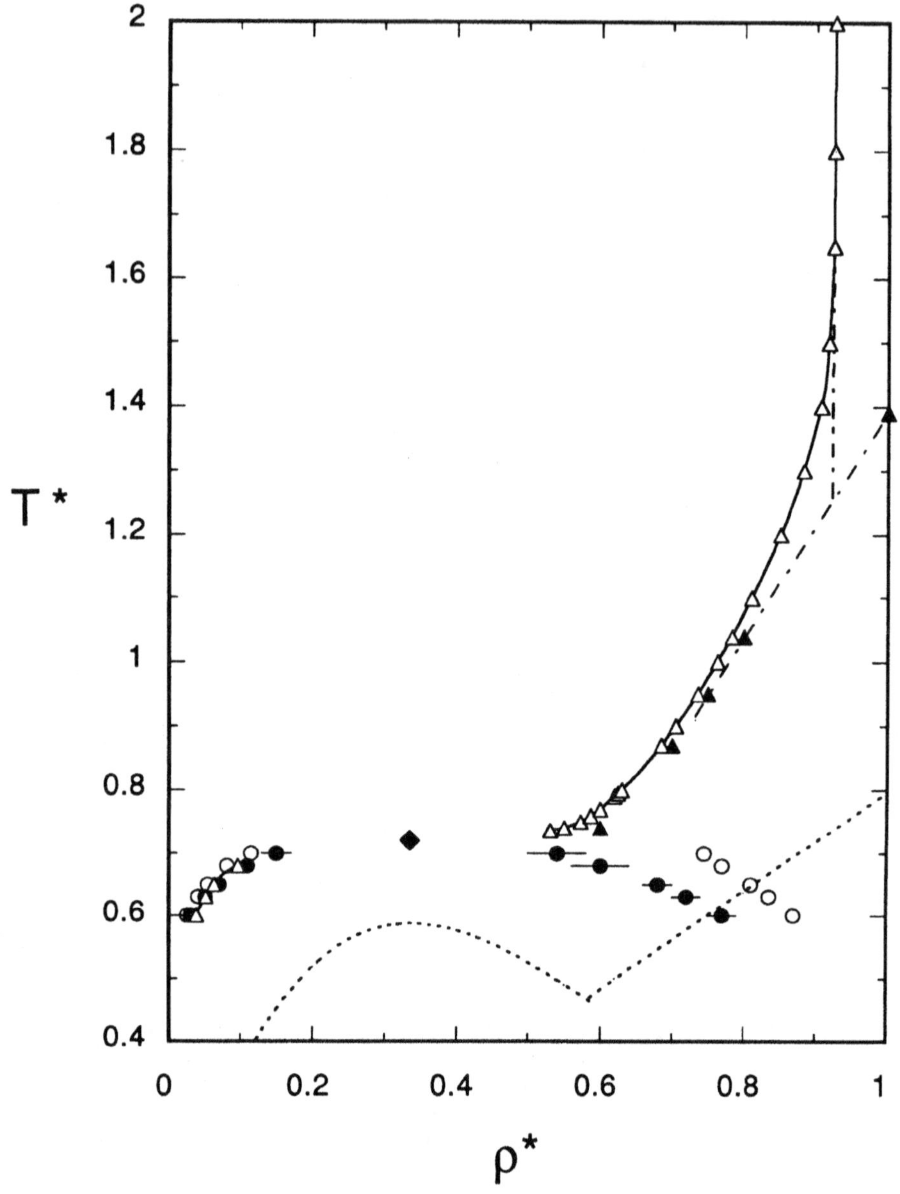

Figure 2. Phase diagram of the mixture with $\nu = 0.7$. Filled and open circles: GEMC and MHNC binodal line, respectively. Filled diamond: GEMC liquid-vapor critical point (interpolated). Filled triangles: SGCMC consolute critical points; open triangles: MHNC estimate of the $\Delta s = 0$ locus. Continuous and dash-dotted lines are guides for the eye. Dotted line: MSA predictions for the liquid-vapor and liquid-liquid spinodal (see ref. [16]).

At $\nu = 0.9$ the MHNC binodal compares quite satisfactorily with the simulation results. A similar good agrement is also visible at $\nu = 0.7$ for the gas-branch of the binodal; the liquid branch however, turns out to be shifted to densities higher than the GEMC results. Such a discrepancy is quite probably due to the fact that, since liquid-liquid phase separation takes place in the high density regime (see below), we should actually investigate in that region the coexistence between *two* liquids and a vapor phase. In this respect both the MHNC and the GEMC estimates of the liquid-branch at ν=0.7 might deserve further assessment.

The GEMC critical points in figures 1 and 2 were obtained by fitting the binodal in terms of the scaling law for the densities

$$\rho_l - \rho_g \propto |T - T_{cr}|^\beta \tag{7}$$

with the effective non-classical critical exponent $\beta = 0.32$, and by making use of the law of rectilinear diameters.

The MHNC liquid-vapor critical point at $\nu = 0.9$ is determined by fitting, through the same scaling law (7), a set of of binodal points close to the highest temperature that we could explore before the solution algorithm of the MHNC ceases to converge. The critical temperature and density so estimated, $T^*_{cr} = 0.83$ and $\rho^*_{cr} = 0.35$, are fairly close to the GEMC values, $T^*_{cr} = 0.82$ and $\rho^*_{cr} = 0.35$, respectively.

Unfortunately, a MHNC estimate of the critical parameters at $\nu = 0.7$ turns out to be not feasible since the solution algorithm of the theory ceases to work too much before the critical region is approached.

We now turn to discuss liquid-liquid equilibrium.
SGCMC results are reported in fig. 3 in the form of temperature-concentration coexistence curves at fixed density.

One can see a systematic increase of the critical temperature with the density; the former is obtained by interpolating the simulation points through the scaling law

$$\Delta\chi \propto |T - T_{cr}|^\beta, \tag{8}$$

where $\Delta\chi$ is the difference in concentration between the two coexisting phases at the given temperature T. The critical concentration is always 0.5 for symmetry reasons.

We also plot, in fig. 2, the line of consolute critical points obtained from the coexistence lines reported in fig. 3. A very short extrapolation of this line intersects the binodal line at $\rho^*_c \simeq 0.57$ and $T^*_c \simeq 0.7$. These density and temperature coordinates mark the "crossover" from the liquid-vapor to the liquid-liquid equilibrium regime (this can also be verified through specific

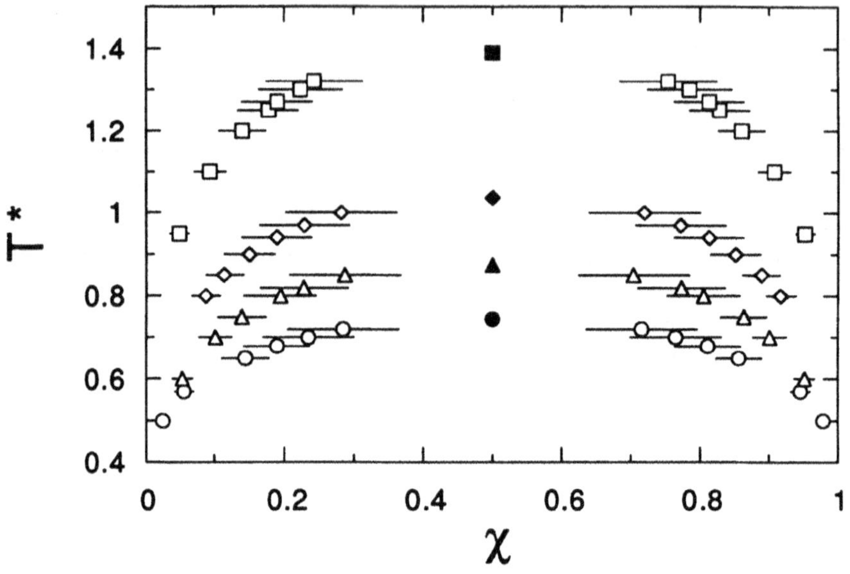

Figure 3. SGCMC results for liquid-liquid equilibrium at fixed overall density of the system. Circles: $\rho^* = 0.60$, triangles: $\rho^* = 0.70$, diamonds: $\rho^* = 0.80$, squares: $\rho^* = 1.00$. The filled markers refer to the estimated critical parameters. Statistical uncertainties are shown as horizontal bars.

GEMC simulations at $\rho^* \simeq 0.6$ [6], not reported here). With reference to previous studies on Lennard-Jones mixtures [17], we can identify the intersection point between the consolute critical point line and the binodal as the critical end point of our HCYM.

It is now interesting to observe the behavior of the liquid -vapor coexistence curve in connection with liquid-liquid equilibrium. In fact, the trend of the binodal to flatten when ν decreases from 0.9 to 0.7, allows one to predict that for sufficiently low ν the curve will become so flat that the "homogeneous" liquid phase will completely disappear; thus, when unlike interactions become sufficiently weak, the system might exhibit a continuous transformation from the vapor phase to liquid-liquid coexistence (see fig. 2).

It is worth to consider this behavior in comparison with what happens in the pure hard-core Yukawa fluid [3]; in that case, the liquid phase can become metastable with respect to the sublimation line, and the system can perform a direct transition from the vapor to the solid phase, provided the potential becomes sufficiently short ranged and deeply attractive. In the present HCYM case, the liquid phase can become metastable, and the system can perform a direct transformation from the vapor to liquid -liquid coexistence, provided the relative strength of the unlike to like interaction becomes sufficiently small. Thus, in mixtures of hard-core Yukawa fluids we can have two different mechanisms of metastability of the (homogeneous) liquid phase.

The picture that emerges from this first set of results is that like interactions, which we have chosen as dominant on their unlike counterpart, favour self-aggregation processes and these, in turn, act as the driving mechanism for liquid-liquid phase separation. On the other hand, self-aggregation is enhanced at high densities since particles get closer and closer to explore the most attractive part of the Yukawa potential; more thermal energy will be then necessary in order to settle a homogeneous phase, and this explains the increase of the critical consolute temperature when ρ^* increases (see fig. 3). By converse, at low densities, the internal energy cannot balance the entropic contributions to the total free energy associated with self-aggregation, and a minimization of the thermodynamic potential is obtained by crossing to liquid-vapor equilibrium.

Previous studies [18] of the dependence of phase stability on the magnitude of ν confirm such a picture.

3.1.2. $\Delta s = 0$ *loci and freezing*

In figs. 1 and 2 we report the density-temperature *loci* for which $\Delta s = 0$, according to MHNC calculations.

We first discuss the $\nu = 0.7$ case. It appears that, in the density region $0.6 \leq \rho^* \leq 0.8$ the $\Delta s = 0$ line is quite well coincident with the line of critical consolution points as determined from the simulation (see fig. 2).

We substantiate this observation by calculating within the MHNC the $q \to 0$ limit of the concentration-concentration structure factor, $S_{cc}(0)$, by means of the formula

$$S_{cc}(q) = \chi_1\chi_2\left[\chi_2 S_{11}(q) + \chi_1 S_{22}(q) - \sqrt{\chi_1\chi_2} S_{12}(q)\right], \tag{9}$$

where $S_{ij}(q)$ are the partial structure factors.

$S_{cc}(0)$ is quite large along the $\Delta s = 0$ locus, in the density range $0.6 \leq \rho^* \leq 0.8$, as visible in fig. 4; thus, we have an indication that the system is in the neighbourhood of liquid-liquid phase separation. This is also confirmed by the high values attained by $S_{cc}(0)$ along an isochore path at $\rho^* = 0.8$, as also shown in fig. 4. Note that a true divergence of $S_{cc}(0)$ is expected to occur on the spinodal line, which however we do not calculate in the MHNC.

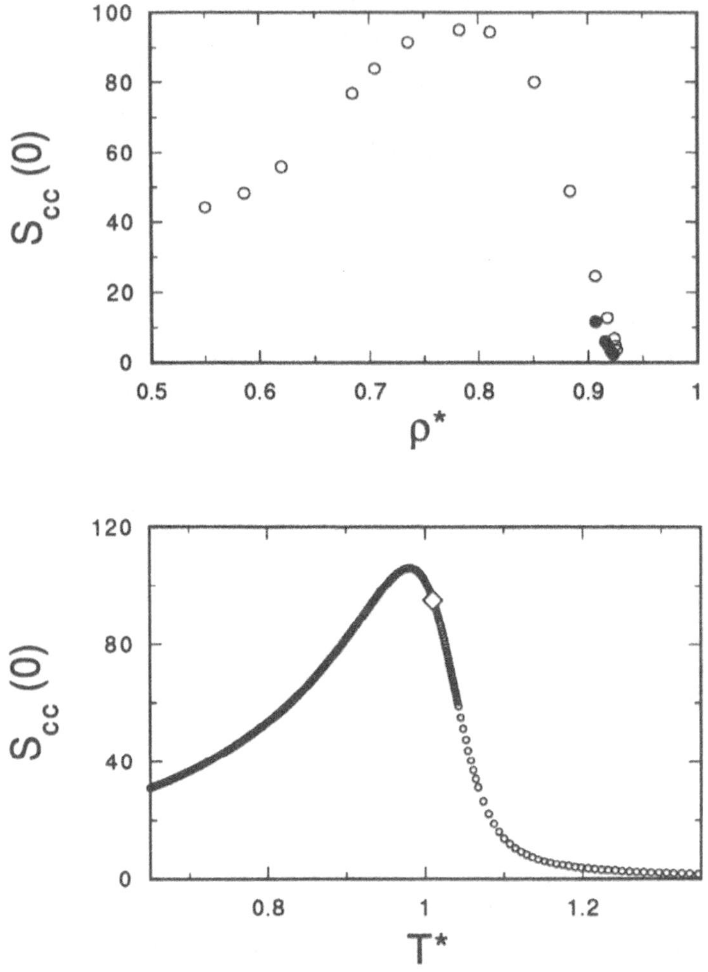

Figure 4. Top: $S_{cc}(0)$ (see text) as a function of the reduced density along the $\Delta s = 0$ *loci*, at $\nu = 0.7$ (open circles), and $\nu = 0.9$ (filled circles). Note the (relative) reduction of concentration fluctuations at $\nu = 0.9$ in the high density regime. Bottom: $S_{cc}(0)$ calculated along the isochore path $\rho^* = 0.8$ as a function of the temperature. The diamond on the curve corresponds to the point where $\Delta s = 0$.

For very high densities ($\rho^* > 0.8$), the $\Delta s = 0$ line becomes almost vertical (see fig. 2). Consistently with the results of previous studies, we interpret this portion of the $\Delta s = 0$ line as the freezing line of the mixture.

Unfortunately, at variance of the one component hard-core Yukawa fluid, we don't have simulation results for the freezing line of HCYM to compare with.

Note that, similarly to the $\nu = 0.7$ case, in the $\nu = 0.9$ case the $\Delta s = 0$ *locus* appears practically vertical, but at very high density ($\rho^* \simeq 0.92$). We again interpret such a *locus* as the freezing line; this intersects in a fairly sharp manner the binodal line, thus identifying the triple point for which we estimate $T_{tr} = 0.56$ and $\rho^*_{tr} = 0.92$.

3.2. ASYMMETRIC MIXTURES

We consider various cases of mixtures for which the size ratio α (see sect.2) is equal to 2, 3 and 4, respectively. We also choose the potential parameters in such a way that the biggest sized particles experience an interaction potential deeper and shorter ranged than that experienced by the smaller particles. For instance, for the mixture in which $\alpha = 2$, we set, as shown in

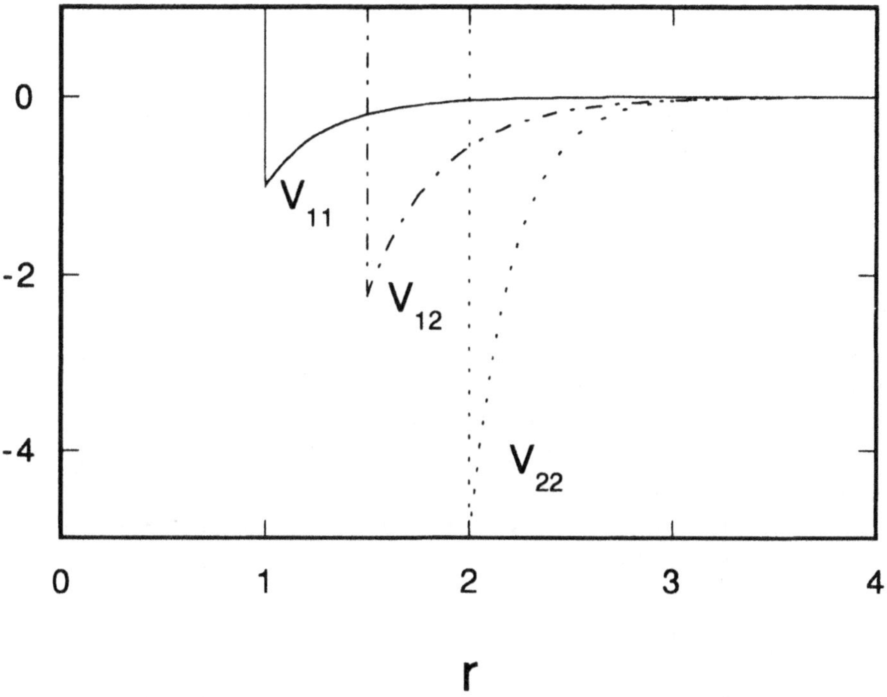

Figure 5. Interaction potential for an asymmetric HCYM with $\alpha = 2$. Distances and potential energies are in σ_1 and K_{11} units, respectively.

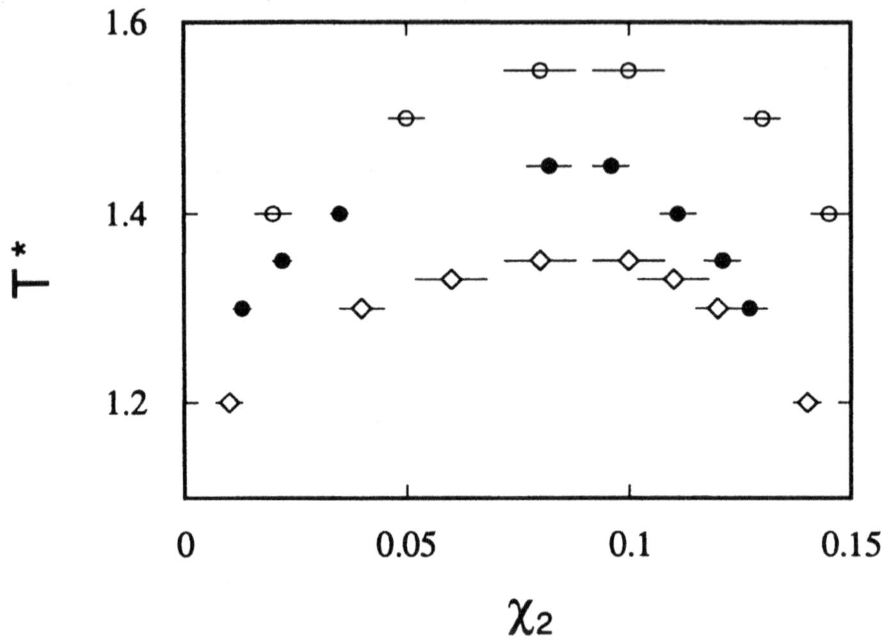

Figure 6. GEMC results for the phase diagram of different asymmetric HCYM in which the overall $\rho^* = 0.1$. Open diamonds: $\alpha = 2$; dots: $\alpha = 3$; Open circles: $\alpha = 4$.

fig. 5, $K_{22} = 5K_{11}$ and $z_{22} \simeq 2z_{11}$; K_{12} is then fixed in terms of a geometric mean of the like amplitudes.

The phase diagram is then investigated by means of GEMC simulations at fixed overall density $\rho^* \simeq 0.1$. Note that the concentration of particles 2 never exceeds 15% of the total number of particles in the mixtures. These calculations yield the fluid-fluid equilibrium curves shown in fig. 6.

One feature that emerges is that both coexisting phases contain a small percentage of big sized particles. The effect of the size difference also appears as a substantial one with a marked increase of the critical temperature with the size ratio. However, the equilibrium curve remains confined in a relatively narrow temperature range.

It is also interesting to look at the internal structure of the coexisting phases. This is shown in fig. 7 where the evolution in temperature of the radial distribution functions inside the two simulation boxes of the GEMC, is visible. We report on the left $g_{22}(r)$ for the phase less populated of particles 2, corresponding to the low density branches of the coexistence curves shown in fig. 6.

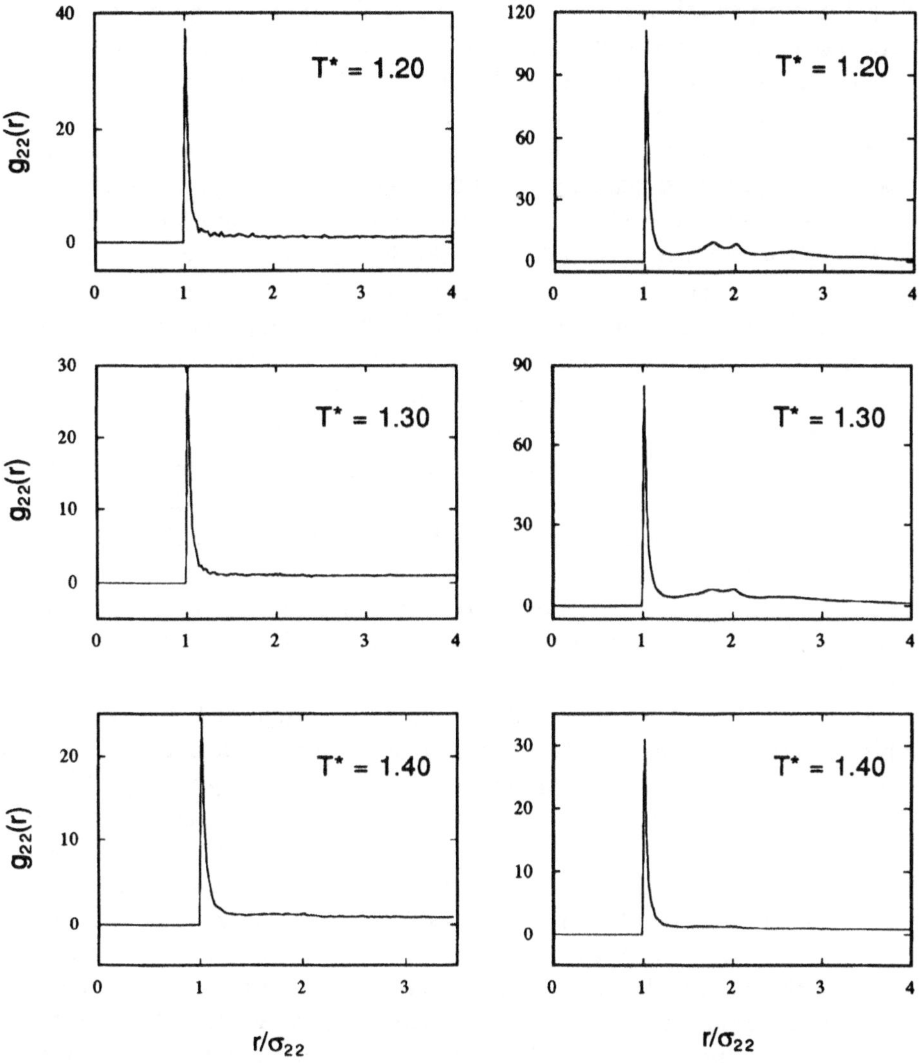

Figure 7. GEMC radial distribution function at different temperatures T^* for the coexisting phases of a HCYM with $\alpha = 2$ (see the phase diagram of fig. 6). Left: phase with smaller χ_2; right: phase with greater χ_2.

We can see that the rdf doesn't show any significant structure at any temperature except for the very high "contact" value, this last feature indicating the probable formation of dimers. By converse, $g_{22}(r)$ for the more concentrated 2-particle phase, shown on the right of fig. 7, develops (at subcritical temperatures) marked structures (note the scale) at the second and third n.n. distances. Given the low overall packing fraction at which these simulations were performed, we are prompted to conclude that in this case the bigger sized particles tend to aggregate in ordered clusters.

It is possible that, although we have performed quite long GEMC simulations, the "styckyness" of the potential causes some ergodic sampling problems, and thus that the system is not fully equilibrated. The results however show a coherent variation with the potential parameters; this seems to support their overall reliability. Moreover, we have performed GEMC simulations by doubling the total number of particles, and we have obtained results practically identical to those here shown.

We have not attempted to perform simulations at densities ρ^* significantly higher than 0.1. Actually, in the presence of considerable asymmetries, it becomes difficult to perform the swapping of particles, and the box volume modifications proper of the GEMC method. Such steps are in fact safe only at relatively low density. However, for most protein solutions the packing fraction is of the order we have considered, so that the information we get seems in any case useful.

In particular, the results indicate that the range of temperatures over which the mixture segregates, expands with the increasing size asymmetries. For the actual size asymmetries encountered in typical protein solution, where α is at least of the order of 10, we should expect fluid/fluid phase coexistence even when protein are rather diluted and at relatively high temperatures. Such a result might have also been anticipated on qualitative grounds. The interesting point is however that there is a substantial effect exerted on the phase coexistence line by the strength of the interaction between the large and the small spheres (i.e. the solvent-protein interaction in our modelization). We have seen in fact that the decrease of the (relative) intensity of such cross-interaction can result in a dramatic flattening of the liquid-vapor binodal.

4. Concluding remarks

We have calculated the phase diagram of two-component hard-core Yukawa mixtures (HCYM), with the aim to model protein solutions and to predict phase separation processes that can occur in those fluids, especially in connection with the problem of protein crystallization.

We have taken advantage of the combined use of a refined theory as the MHNC, and of two simulation methodologies, namely the GEMC technique and the SGCMC technique, in order to obtain as much information as possible on the phase portrait of the model mixtures considered. The freezing conditions have also been explored thanks to a freezing criterion previously developed by other authors.

Although the size asymmetries we could consider in this study are still pretty far from the actual ones occuring in real protein solutions, the combined study of the fully symmetric HCYM and of moderately asymmetric mixtures, has allowed us to unravel some of the basic mechanisms that are likely to be determinant in fixing the relative positions of the coexistence lines in the phase portrait of protein solutions, and hence the conditions for protein crystallization.

The emerging feature is that the size asymmetry expands the region of phase separation by increasing the critical point temperature of the fluid-fluid coexistence line; however, if the solvent-solute interaction strength becomes relatively small with respect to like interactions, the fluid-fluid curve might considerably flatten. This evidence adds to the one already gathered from previous studies on one component systems about the flattening of the fluid-fluid coexistence curve, which occurs when the range of the interparticle potential decreases. Our approach yields a first, albeit rough, quantitative estimate of the relative importance of these two processes in order to determine the overall phase diagram of a mixture.

In order to improve our description we have in progress attempts to introduce in the MHNC a set of more accurate bridge functions or, alternatively, to use some else thermodynamically selfconsistent theory. On the simulation side the difficulties seem more severe, and one should probably resort to one of the most recent simulation techniques by which particle moves that are usually forbidden because of spatial hindrance, can be performed with a properly tailored phase space sampling algorithm.

References

1. M. Muschol and F. Rosenberger, J. Chem. Phys. **107**, 1953 (1997).
2. P. R. ten Wolde and D. Frenkel, Science **277**, 1975 (1997).
3. M. H. J. Hagen and D. Frenkel, J. Chem. Phys. **101**, 4093 (1994).
4. E. Lomba and N. E. Almarza, J. Chem. Phys. **100**, 8367 (1994).
5. C.F.Tejero, A.Daanoun, H.N.W.Lekkerkerker and M.Baus, Phys.Rev.Lett. **73**, 752 (1994);
6. C. Caccamo, D. Costa and G. Pellicane, J. Chem. Phys. **109**, 4498 (1998).
7. Y. Rosenfeld and N. W. Ashcroft, Phys. Rev. A **20**, 1208 (1979).
8. C. Caccamo, Phys. Rep. **274**, 1 (1996).
9. A. Z. Panagiotopoulos, Mol. Phys. **71**, 813 (1987); A. Z. Panagiotopoulos, N. Quirke, M. Stapleton, and D. J. Tildesley, Mol. Phys. **63**, 527 (1988).
10. D. A. Kofke and E. D. Glandt, Mol. Phys. **64**, 1105 (1988).
11. E. de Miguel, E. M. del Rio, and M. M. Telo da Gama, J. Chem. Phys. **103**, 6188 (1995).
12. J. L. Lebowitz, Phys. Rev. **133**, A895 (1964).

13. P. V. Giaquinta and G. Giunta, Physica A **187**, 145 (1992); P. V. Giaquinta, G. Giunta, and S. Prestipino Giarritta, Phys. Rev. A **45**, 6966 (1992).
14. C. Caccamo, G. Giunta, and G. Malescio, J. Chem. Phys. **103**, 6188 (1995).
15. F.Saija and P.V.Giaquinta, J.Phys.Condens.Matter **8**, 8137 (1996).
16. C. Caccamo and G. Giunta, Mol. Phys. **78**, 83 (1993).
17. N. B. Wilding, Phys. Rev. Lett. **78**, 1488 (1997); Phys. Rev. E **55**, 6624 (1997).
18. C. Caccamo, G. Giunta, and C. Hoheisel, Phys. Lett. A **158**, 325 (1991).

QUANTUM FLUIDS

AB INITIO LIQUIDS: SIMULATING LIQUIDS BASED ON FIRST PRINCIPLES

DOMINIK MARX
Max–Planck–Institut für Festkörperforschung
Heisenbergstrasse 1, 70569 Stuttgart, Germany

Abstract. The basic ideas underlying the Car–Parrinello *ab initio* approach to molecular dynamics simulations are introduced. The central aspect of this technique is the first principles determination of the interactions from concurrent electronic structure calculations. Thus the usual step of introducing parameterized interaction potentials in order perform computer simulations of liquids is avoided. The currently most successful *ab initio* schemes treat the electrons within Hohenberg–Kohn–Sham density functional theory where generalized gradient approximations to the exchange–correlation energy are used. One example of such a first principles simulation of a liquid, the hydration of Be^{2+} in water, is presented. Secondly, an extension of the "traditional" Car–Parrinello *ab initio* molecular dynamics scheme to include quantum-mechanical nuclei is introduced. Within the Born–Oppenheimer separation of nuclei and electrons the nuclear degrees of freedom are quantized using Feynman's path integral formulation of quantum statistical mechanics, whereas the electrons are still represented using density functional theory. This *ab initio* path integral technique opens the way to study quantum effects in chemically complex systems such as, e.g., an excess proton in water.

1. Why First Principles Simulations?

Over the years, computer simulations have become a well–established third pillar in modern physics and chemistry, thereby complementing both experiment and analytical theory. In the traditional approach to atomistic computer simulations of liquids, a first and crucial step is to provide "a model" that describes the interactions between the atoms or molecules [1, 2, 3, 4]. In the case of molecules, there exist also "internal degrees of freedom", such

C. Caccamo et al. (eds.), New Approaches to Problems in Liquid State Theory, 439–457.

as vibrations and rotations in addition to the translations, and sometimes also several electronic or magnetic states. Often these internal degrees of freedom are frozen in some combination of states, i.e., the molecules are approximated as rigid non–spherical objects. Very large molecules, which are at the heart of biophysics and biochemistry, have the possibility to fold in several ways, which leads to many different conformers. A host of different functional forms were introduced over the years to describe these interactions between and within molecules [5], such as Morse, Lennard–Jones, Buckingham potentials and many more in addition to the electrostatic Coulomb interactions. The specific parameterization of these functions is either based on empirical input, on separate *ab initio* quantum–chemical calculations, or on a combination of both. Having devised such an interaction potential, standard molecular dynamics or Monte Carlo simulation procedures [2, 3, 4] are used to generate the desired statistical–mechanical ensemble which allows one to compute observables.

Although extremly powerful and successful, this traditional approach has well–known difficulties and also limitations. Selecting, representing, and fitting the interatomic potential functions [6] is a non–trivial task, in particular when it comes to include non–pairwise interactions. Electronic polarization phenomena, which are associated with charge transfer due to solvation, are intrinsically many–body interactions; see e.g. Refs. [7, 8] for recent progress in taking this into account in the case of water models. Along the same lines, it was explicitly shown that three–body interactions supplementing the usual "Coulomb plus two–body potential" approximation are crucial in order to simulate correctly the hydration shell of simple but highly charged ions such as Be^{2+} in liquid water [9, 10]. This particular example will be discussed in Sect. 2.2.

In the case of liquid solutions it is also of major concern to treat the interactions between the solvent and the solute at about the same level of accuracy, which is of particular importance if hydrogen–bonded interactions are present. However, probably the most severe obstacle for simulating complex molecular assemblies such as liquids *via* simple potential models are situations where chemical transformations or reactions between different species occur. The simplest reaction in water, proton diffusion, is already very difficult to deal with based on using parameterized (model) potentials.

In principle, it is easy to get rid of all these problems by solving, at each molecular dynamics (or Monte Carlo) step, the electronic Schrödinger equation for a given configuration of the nuclei, instead of relying on (model) potentials with an *a priori* fixed parameterization. In practice, this is a formidable task and clever, i.e., computationally tractable, ways of dealing with this problem have to the found. A very successful approach to first principles simulations was introduced by Car and Parrinello [11]. They devised a fictitious dynamics approach in the framework of extended system molecular dynamics [2, 3] for solving the electronic structure problem "on the fly" using density functional theory for the electronic degrees of freedom. The essential features of Car–Parrinello molecular dynamics will be introduced in Sect. 2.1 and one particular example of its use in the world of liquids is given in Sect. 2.2. A more comprehensive overview about using *ab initio* molecular dynamics to study liquids and solutions is given in Ref. [12].

In standard Car–Parrinello approach to molecular dynamics [11], the nuclei are treated as classical point particles, which is a satisfactory approximation in many circumstances. There are, however, situations where this is no longer true, in particular, if motion of light hydrogen nuclei is involved. Recently, we have overcome this limitation by combining the Car–Parrinello approach with a Feynman path integral formulation which allows one to take efficiently into account the quantum nature of the nuclei [13, 14, 15]. This *ab initio* path integral technique is introduced in Sect. 3.1 and preliminary results of its application to study the excess proton in water are given in Sect. 3.2.

2. Ab Initio Molecular Dynamics

2.1. METHODOLOGY

Here we can only provide a sketch of the *ab initio* molecular dynamics philosophy and we refer the interested reader to Refs. [16, 17, 18, 19, 20, 21, 22] for more detailed discussions.

Let us assume at the outset that the Born–Oppenheimer approximation holds, i.e., the $I = 1, \ldots, N$ nuclei with cartesian positions $\{\mathbf{R}_I\}$ move at any instant of time on the energy surface $E_0(\{\mathbf{R}_I\})$ that is provided by the lowest-energy solution Ψ_0 of the electronic structure problem. The electronic ground state is used because interactions with higher-lying electronic states are assumed to be of no relevance; note that we use the notions nuclei, ions, and atoms interchangeably for simplicity. In addition, we shall assume here that the nuclei can be faithfully approximated as classical point-particles, i.e., nuclear quantum dispersion and tunneling effects are excluded from our present treatment.

The conceptually simplest approach to first principles or *ab initio* molecular dynamics is what is sometimes called the "Born–Oppenheimer dynamics". Let $\mathcal{H}(\{\mathbf{R}_I\})$ be the exact non-relativistic many-body electronic Hamiltonian at a given configuration of the nuclei $\{\mathbf{R}_I\}$. The total energy of this configuration is given by

$$E_0(\{\mathbf{R}_I\}) = \langle \Psi_0 | \mathcal{H}(\{\mathbf{R}_I\}) | \Psi_0 \rangle \ , \tag{1}$$

where $\Psi_0(\{\mathbf{R}_I\})$ is the exact ground–state eigenfunction of the Hamiltonian, i.e., $\mathcal{H}\Psi_0 = E_0\Psi_0$. The recipe to perform *ab initio* molecules dynamics then consists in obtaining the forces acting on the nuclei as the gradient of this potential energy, *via* the Hellmann–Feynman theorem,

$$\frac{\partial E_0(\{\mathbf{R}_I\})}{\partial \mathbf{R}_I} = \left\langle \Psi_0 \left| \frac{\partial \mathcal{H}(\{\mathbf{R}_I\})}{\partial \mathbf{R}_I} \right| \Psi_0 \right\rangle \ , \tag{2}$$

which in turn are used in conjunction with Newton's equations of motion,

$$M_I \ddot{\mathbf{R}}_I = -\frac{\partial E_0(\{\mathbf{R}_I\})}{\partial \mathbf{R}_I} \ , \tag{3}$$

to obtain an operable molecular dynamics scheme. At this stage, standard algorithms to integrate these equations of motion and thus to generate a trajectory in phase space can be used [2, 3, 4]. Assuming that ergodic sampling is achieved, this trajectory can be used to compute ensemble averages in the microcanonical ensemble from time averages. Let us stress that of course the standard restrictions of simulations apply also in the present case, i.e., only those processes can be investigated that have typical correlation lengths and relaxation times that are much smaller than the linear dimension of the simulation supercell and the simulation run time, respectively.

The essential feature of this straightforward "Born–Oppenheimer molecular dynamics scheme" (1)–(3) is that the electronic problem has to be solved at every timestep of the molecular dynamics run, using self–consistent diagonalization or minimization techniques to obtain the Hellmann–Feynman forces (2) as the necessary input to solve Newton's equations of motion (3). Is this tractable in practice? The maximum timestep that can be used is dictated by the period of the highest frequency phonon or vibration that occurs in the system. Thus, typical timesteps are about 0.1–0.5 fs, which means that of the order of 10 000 or more timesteps are needed to generate a trajectory of 1 ps or more. At this stage it becomes evident that a simple–minded implementation of the Born–Oppenheimer procedure based on solving 10 000 times and more the electronic structure problem from scratch is computationally intractable for useful applications. However, the wavefunction Ψ_0 is not expected to change much on the relevant timescale of fractions of femtoseconds which separates a given configuration in the molecular dynamics trajectory from the next one to be generated. Thus, every successful scheme has to make use of this fact in order to render *ab initio* molecular dynamics feasible in practice. A clever and elegant way of using the information of the wavefunction at a given timestep in order to compute the forces for the next timestep was devised by Car and Parrinello [11]. In addition, Car and Parrinello picked a very efficient scheme (in the sense of accuracy *vs.* computational effort) to take care on the electronic structure part of *ab initio* molecular dynamics, namely Hohenberg–Kohn–Sham density functional theory [23, 24].

Electronic density functional theory is reviewed in great detail in many places, e.g. in Refs. [25, 26, 27], and we collect here only the most important aspects for its use in conjunction with molecular dynamics. For a given nuclear configuration $\{\mathbf{R}_I\}$ the total energy of the system is the minimum of a functional [23, 24],

$$\begin{aligned} E^{\mathrm{KS}}[\{\phi_i\},\{\mathbf{R}_I\}] &= \sum_i^{\mathrm{occ}} f_i \left\langle \phi_i \left| -\frac{1}{2}\nabla^2 \right| \phi_i \right\rangle + \int d\mathbf{r}\ V_{\mathrm{ext}}(\mathbf{r})\rho(\mathbf{r}) \\ &+ \frac{1}{2}\int d\mathbf{r}d\mathbf{r}'\ \frac{\rho(\mathbf{r})\rho(\mathbf{r}')}{\mid \mathbf{r}-\mathbf{r}' \mid} + E_{\mathrm{xc}}[\rho]\ , \end{aligned} \tag{4}$$

of the electronic density,

$$\rho(\mathbf{r}) = \sum_i^{\mathrm{occ}} f_i \mid \phi_i(\mathbf{r}) \mid^2\ , \tag{5}$$

where the Kohn–Sham orbitals $\{\phi_i\}$ are a set of orthonormal auxiliary functions $\langle \phi_i \mid \phi_j \rangle = \delta_{ij}$ and

$$V_{\mathrm{ext}}(\mathbf{r}) = -\sum_{I,i} \frac{Z_I}{|\mathbf{R}_I - \mathbf{r}_i|} + \sum_{I<J} \frac{Z_I Z_J}{|\mathbf{R}_I - \mathbf{R}_J|} \tag{6}$$

is the external potential due to the Coulomb interactions between nuclei and electrons as well as between the nuclei themselves; f_i denote the occupation number of the orbitals and the summation includes only the occupied states.

The (unknown) exchange–correlation functional $E_{\mathrm{xc}}[\rho]$ contains all the intricacies of the interacting many–electron system. The simplest approximation to this functional is obtained by assuming that a non–uniform system such as a molecule or solid has locally the same exchange and correlation energy as a uniform electron gas of the same density

$$E_{\mathrm{xc}}^{\mathrm{LDA}}[\rho] = \int d\mathbf{r}\ \rho(\mathbf{r})\ \varepsilon_{\mathrm{xc}}^{\mathrm{LDA}}(\rho(\mathbf{r}))\ , \tag{7}$$

which defines the local density approximation LDA [24]. This approximation is known to be accurate in many cases, in particular for structures with strong covalent bonds. However, it is clearly not satisfactory for describing weaker interactions such as hydrogen bonds, due to the well–known overbinding defect of LDA, which leads to large binding energies and short bonds. But hydrogen bonds are crucial interactions in a host of different polar protic liquids, including, preeminently, water. Major improvements upon LDA are obtained by including in the exchange–correlation kernel ε not only the density itself, but also its local variation in space, i.e., its gradient

$$E_{\text{xc}}^{\text{GGA}}[\rho] \;=\; \int d\mathbf{r}\;\rho(\mathbf{r})\;\varepsilon_{\text{xc}}^{\text{GGA}}(\rho(\mathbf{r});\nabla\rho(\mathbf{r}))\;, \tag{8}$$

which defines the vast class of the so–called generalized gradient approximations GGA, see, e.g., Ref. [28]. Such approximations are demonstrated [29] to lead in general to an improvement over LDA, which is particularly important for hydrogen–bonded systems.

The minimization of the Kohn–Sham functional (4),

$$E_0^{\text{KS}}[\rho_0] = \min_{\{\phi_i\}} E^{\text{KS}}[\{\phi_i\}]\;, \tag{9}$$

with the orthogonality constraint on the orbitals applied, leads to the Kohn–Sham equations,

$$\mathcal{H}^{\text{KS}}\phi_i(\mathbf{r}) \;=\; \sum_j \Lambda_{ij}\phi_j(\mathbf{r})\;, \tag{10}$$

with the density (or orbital–dependent) Kohn–Sham Hamiltonian,

$$\mathcal{H}^{\text{KS}} = -\frac{1}{2}\nabla^2 + V_{\text{ext}}(\mathbf{r}) + \int d\mathbf{r}'\frac{\rho(\mathbf{r}')}{|\,\mathbf{r}-\mathbf{r}'\,|} + \frac{\delta E_{\text{xc}}[\rho]}{\delta\rho(\mathbf{r})}\;, \tag{11}$$

which have to be solved self–consistently. The set of orbitals converged to self–consistency yields the Hamiltonian $\mathcal{H}_0^{\text{KS}}$ and the ground–state electron density $\{\phi_i\} \to \rho_0$ *via* (5). The set $\{\Lambda_{ij}\}$ are Lagrange multipliers that enforce the orthogonality constraint and yield the Kohn–Sham orbital energies $\epsilon_i = \Lambda_{ii}$ after a unitary transformation to the diagonal representation.

The crucial idea underlying Car–Parrinello *ab initio* molecular dynamics consists in associating a fictitious Newtonian dynamics to the electronic orbitals in the framework of extended systems dynamics [2, 3], similar in spirit to the Andersen or Nosé–Hoover formulations of thermostats. The Car–Parrinello Lagrangian of the nuclei and the orbitals, which are treated as the extended system variables, reads

$$\begin{aligned}\mathcal{L}^{\text{CP}} \;=\;& \sum_i^{\text{occ}} \mu\left\langle\dot{\phi}_i\,\middle|\,\dot{\phi}_i\right\rangle + \sum_{I=1}^{N}\frac{1}{2}M_I\dot{\mathbf{R}}_I^2 \\ & - E^{\text{KS}}[\{\phi_i\},\{\mathbf{R}_I\}] + \sum_{ij}\Lambda_{ij}\left(\langle\phi_i\,|\,\phi_j\rangle - \delta_{ij}\right)\;,\end{aligned} \tag{12}$$

where the mutal orthogonality of the orbitals is imposed as usual by Lagrange multipliers. This leads to the following coupled set of classical Euler–Lagrange equations

$$M_I \ddot{\mathbf{R}}_I = -\frac{\partial E^{\mathrm{KS}}[\{\phi_i\},\{\mathbf{R}_I\}]}{\partial \mathbf{R}_I} \tag{13}$$

$$\begin{aligned} \mu \ddot{\phi}_i(\mathbf{r}) &= -\frac{\delta E^{\mathrm{KS}}[\{\phi_i\},\{\mathbf{R}_I\}]}{\delta \phi_i^\star} + \sum_{ij} \Lambda_{ij}\phi_j \\ &= -\hat{H}^{\mathrm{KS}}\phi_i + \sum_{ij}\Lambda_{ij}\phi_j \end{aligned} \tag{14}$$

for nuclei and electrons, where μ is the fictitious mass assigned to the orbital degrees of freedom. The idea is that a molecular dynamics simulation which was started with a minimized functional E_0^{KS}, i.e., with a set of wave functions ϕ_i that yield the ground–state density ρ_0 at the initial nuclear configuration $\{\mathbf{R}_I(t=0)\}$, will dynamically preserve the minimum property of the Kohn–Sham functional also in future configurations $\{\mathbf{R}_I(t \gg 0)\}$ generated by integrating (13)–(14). This is achieved if the fictitious kinetic energy $\sum_i \mu \langle \dot{\phi}_i \mid \dot{\phi}_i \rangle$ of the electrons stays negligibly small during the run. In this sense, the magnitude of this kinetic energy, or the corresponding "temperature" of the electrons, is a measure for how close the instantaneous value of the Kohn–Sham functional $E^{\mathrm{KS}}[\{\mathbf{R}_I(t)\}]$ comes to the minimized functional $E_0^{\mathrm{KS}}[\{\mathbf{R}_I(t)\}]$ which corresponds to the true electronic ground state at time t. Thus, the Car–Parrinello equations of motion (13)–(14) propagate the nuclei at the desired true physical temperature, whereas the orbital degrees of freedom are kept at a very low fictitious electronic temperature to guarantee accurate electronic ground–state forces.

In the limit of a vanishingly small fictitious electronic kinetic energy the Car–Parrinello equations of motion reduce to

$$M_I \ddot{\mathbf{R}}_I = -\frac{\partial E_0^{\mathrm{KS}}}{\partial \mathbf{R}_I} \tag{15}$$

$$0 = -\hat{H}_0^{\mathrm{KS}}\phi_i + \sum_{ij}\Lambda_{ij}\phi_j \tag{16}$$

with $\{\phi_i\} \to \rho_0$. As a result, a Car–Parrinello trajectory which was generated with a sufficiently small fictitious electronic kinetic energy over the entire run *via* (13)–(14) is expected to approximate very closely the Born–Oppenheimer trajectory as obtained by directly minimizing the functional E^{KS} at each molecular dynamics step according to the prescription (15)–(16).

At this stage, we are prepared to discuss applications of this *ab initio* machinery and we refer to the already cited review articles for further technical information concerning various approximations to the (unknown) density functional, basis set choices, pseudo potential schemes and the like.

2.2. APPLICATIONS: SOLVATION OF SIMPLE IONS IN WATER

The *ab initio* simulation approach is particularly well suited to study structure and dynamics of complex molecular liquids. In such cases, it is not at all trivial to devise reliable and accurate potentials. This problem is much less severe for simple monatomic liquids, e.g., noble gases such as fluid Argon or Neon. Since quite recently, extremly accurate potentials can be used in simulations of such fluids [30, 31, 32] in order to approach "chemical accuracy", i.e., relevant energy differences have to be accurate within less than the thermal energy at room temperature (corresponding to the order of 300 K which is roughly 1 kcal/mol, 0.01 eV, or 0.001 a.u.).

Arguably the simplest but most important complex liquid is water and, more generally, aqueous systems. Even the structure of pure bulk water at ambient conditions as determined from experiments is still subject of changes, re–analysises and debates, see, e.g., Refs. [33, 34]. *Ab initio* Car–Parrinello investigations of aqueous systems started in the early nineties with water clusters [35, 36] and ice [37], where is was clearly demonstrated that the local density approximation LDA [38] (7) is insufficient to account for the energetically subtle but crucial hydrogen–bonding effects; note that the bond energy of a hydrogen bond in the isolated water dimer $(H_2O)_2$ amounts to only about 0.2 eV, compared to nearly 5 eV for a covalent OH bond in the H_2O molecule. Major progress was made prossible by using gradient–corrected density functionals, such as various types of generalized gradient approximations GGA (8). In particular the semilocal correction to the electronic exchange energy according to Becke [39] was shown to be very effective in describing hydrogen bonding with a reasonable accuracy. This allowed for first principles investigations of pure bulk water at ambient [40] and supercritical [41] conditions, and more recently for the calculation of the infrared spectrum of bulk water [42]. A detailed assessment of the various approximations involved in such simulations of liquid water, in particular concerning the choice of the (approximate) density functional, is presented in Ref. [43].

The literature on *ab initio* simulations of aqueous solutions is much more scarce and includes investigations of excess H^+ and OH^- ions in water [44], of very concentrated HCl acid [45], of the solvation of Be^{2+} in water [46, 47, 10], of the addition of water to SO_3 which yields H_2SO_4 acid [48], and of an aqueous glucose solution [49]. An earlier review on *ab initio* molecular dynamics simulations of liquids and solutions was given by Sprik [12].

In the following we will focus on the investigation of the hydration behavior of Be^{2+} in water [46, 47, 10]. This small and doubly charged cation is particularly challenging for computer simulations because the quality of pairwise additive potentials, as used in many simulations of liquids, is expected to degrade with increasing charge density on the complexated ion. The many–body effects have their origin in the redistribution of the electronic charge density upon hydration and the significant polarization of the water molecules in the first hydration shell. Comparing the results from pioneering molecular dynamics simulations relying on the pair potential approximation to describe the cation–water interactions [50] with the results of simulations employing an explicit three–body Be^{2+}–$(H_2O)_2$ potential, it has been shown [51, 9] that the latter is crucial in order to obtain the correct tetrahedral symmetry of the hydration shell and the corresponding hydration number of four instead of six for Be^{2+} in water. We note here that the strategy of representing the many–body interactions by an *effective* pair potential was shown to be unsuccessful for the beryllium ion [51].

We refer to the original publications for the full details of this particular investigation [46, 47, 10]. Let us only mention that a single Be^{2+} ion was immersed in 31 H_2O water molecules contained in a cubic simulation box. The box length was close to 10 Å with periodic boundary conditions applied; the negative counter charge of $-2e$ was treated as a homogeneous background. The GGA density functional was obtained by adding Becke's exchange gradient–correction [39] to the Perdew–Zunger parameterization [38] of the local density approximation. The *ab initio* molecular dynamics simulation was performed at ambient conditions and statistical averages were obtained from a 1 ps run after careful equilibration.

We started the *ab initio* simulation from a configuration in which the beryllium ion was surrounded by six water molecules forming an approximately octahedral coordination shell. Two of the six water molecules were expelled simultaneously from the first hydration shell on a timescale of only 0.2–0.3 ps. The result was a fourfold coordinated situation that was stable without significant alterations until the end of the run. The resulting final radial distribution functions for the beryllium–water subsystem are presented in Fig. 1. The first peak of the BeO distribution has an average value of 1.66 Å with an r.m.s. width of ± 0.06 Å and its maximum is located at 1.65 Å. The value obtained from an X–ray diffraction study [50] of a fairly concentrated (5.3 molal) aqueous $BeCl_2$ solution is 1.67 Å. The first and second peaks of the BeO radial distribution function are clearly separated so that its integral reaches, between these peaks, a pronounced plateau with a value of exactly four. The BeH distribution function has its maximum around 2.37 Å, with an average value and r.m.s.

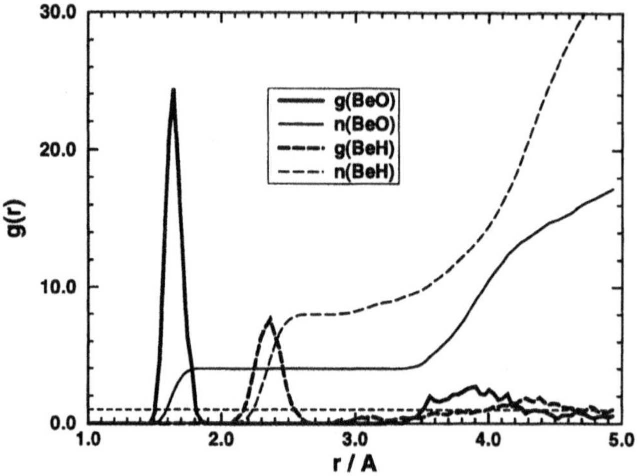

Figure 1. Beryllium–oxygen (solid lines) and beryllium–hydrogen (dashed lines) partial radial distribution functions $g(r)$ (bold lines) together with their running coordination numbers $n(r)$ (light lines) at ambient conditions. The horizontal dotted line marks the value unity.

width of 2.36±0.09 Å and integrates to eight at about 2.7 Å. This behavior of the beryllium–water radial distribution functions gives clear evidence that the beryllium di–cation is fourfold coordinated in liquid water, which is in line with the results from simulations relying on a three–body potential to describe the H_2O–Be–OH_2 interactions in the solvation shell [51, 9]. The most recent estimate from ^{18}O–NMR spectroscopy [52] is 4.1±0.2 from a $Be(ClO_4)_2$ solution at 295 K and the fits of the X–ray diffraction data [50] also give evidence for a coordination number of 4 instead of 6 at 298 K.

The geometrical arrangement of the water molecules in the first hydration shell can be characterized by two angles. The most probable angle between two oxygen atoms of hydration water molecules and the central beryllium ion $\angle_{\mathrm{O-Be-O}}$ is about 110°, which is close to 109.5° of the ideal tetrahedron. Secondly, the distribution function of the angular orientation of the (geometric) water dipole vector relative to the BeO axis $\angle_{\mathrm{Be-O-Dipole}}$ has its maximum at 180°. Thus, the first hydration shell of Be^{2+} as described by the surrounding oxygen atoms is in an average sense of tetrahedral symmetry with only small amplitude distortions due to thermal motion in the liquid.

One of the most striking results of our simulation is the change in the intramolecular geometry of the solvation molecules. In Table 1 we contrast the geometrical parameters of the water molecules in the first hydration shell with those of the water molecules that are further apart. While the latter molecules are very similar to those of pure bulk water obtained with

TABLE 1. Intramolecular water geometry of a single water molecule in its optimized static structure ("isolated H_2O"), of the four water molecules in the first hydration shell around the Be^{2+} ion in solution ("hydration water"), and of those 27 molecules that do not participate in the first hydration shell ("non–hydration water") at ambient conditions. Reported is the average r_{OH}^{av} and most probable r_{OH}^{max} OH bond length in Å, as well as the average $\angle_{HOH}^{max}$ and most probable $\angle_{HOH}^{max}$ HOH bond angle in degrees; the r.m.s. width of the peaks is given after the ± signs

	isolated H_2O	non–hydration water	hydration water
r_{OH}^{av}	0.996	1.008±0.028	1.028±0.033
r_{OH}^{max}	0.996	1.004	1.017
$\angle_{HOH}^{av}$	104.7	105.8±5.1	108.5±5.2
$\angle_{HOH}^{max}$	104.7	104.8	108.7

the same methodology [40], the solvation water molecules are significantly distorted. Both their HOH angle and OH distance increase significantly compared to pure bulk water. In particular the OH distance is larger than any of the OH distances found for isolated $[Be(H_2O)_n]^{2+}$ $n = 1, \ldots, 4$ clusters [46]. It is interesting to compare in more detail our findings in the bulk with the results of our calculations on isolated clusters [46]. These cluster calculations [46] revealed an increase of the HOH angle from 104.7° for an isolated water molecule to 108.5° for $n = 1$, which remains close to that value up to the $n = 4$ tetrahedral cluster and also in solution. The OH distance, however, does change a lot as a function of the size of the coordination shell. The cluster calculations show an initial increase of this bond length from 0.996 Å for an isolated water molecule to 1.020 Å for the $n = 1$ cluster, followed by a continuous decrease to 0.999 Å after adding up to $n = 4$ molecules to the beryllium ion. This means that a straightforward extrapolation from these data to predict the water geometry in solution is bound to fail. The same is true for the BeO distance which increases monotonically from 1.503 Å for $n = 1$ to 1.685 Å for the tetrahedral $n = 4$ cluster, a trend clearly not in line with the value 1.66 Å obtained in the bulk from the radial distribution function in Fig. 1.

These effects on the structure of the four closest water molecules of Be^{2+} are thus clearly due to the (hydrogen–bonded) interactions of the first solvation shell with the surrounding bulk water. In addition, it was found [10] that the substantial increase of the dipole moment found for the water molecules in the first solvation shell results from a subtle interplay of the nuclear and electronic contributions. In order to find results similar to those in our calculations it is essential to include at least the second solvation shell in the description. This underlines the importance of fully accounting for an embedding of the first hydration shell in the hydrogen–bonded network of liquid water *via* hydrogen bonds in order to simulate solutes in liquid water. It has already been attempted to generalize and improve the reaction field method by adding short–range corrections to the dielectric continuum theory to account for exactly these hydrogen–bonding effects [53]. However, this approach is unsatisfactory or at least very difficult as stressed by the authors themselves. A different route was taken in the framework of the "hydrated ion concept" where the central ion plus its first hydration shell, i.e., the hydrated ion, are treated as a unit which in turn is embedded in the atomistic solvent, see Ref. [54] and references therein. What is common to all these approaches is, however, that the solvent–solvent and solvent–solute interactions are not treated on the same level of accuracy. This aspect is, on the other hand, automatically guaranteed in the *ab initio* molecular dynamcis treatment.

3. Ab Initio Path Integrals

3.1. METHODOLOGY

Up to this point we focussed exclusively on those situations where the nuclei could be successfully approximated as classical particles. There are, however, many situations where quantum dispersion broadening and tunneling effects play an important role and cannot be neglected if the simulation aims at being realistic. A novel method, *ab initio* path integrals, was devised [13] to cope with such situations. The central idea is to quantize the nuclei using Feynman's path integral representation of quantum–statistical mechanics and at the same time to include the electronic degrees of freedom *à la* Car–Parrinello "on the fly".

It is well known that the partition function of a N–body quantum system can be mapped onto a $N \times P$–body classical system. This system is isomorphic to N interacting harmonic ring–polymers composed of P monomers each, and sometimes the polymer is called necklace and the monomers are called beads or replica; see Refs. [55, 56] for introductions into numerical path integration techniques based on model potentials. As a result of the simplest Trotter factorization the canonical partition function

$$\mathcal{Z} = \mathrm{Tr}\,\exp\left[-\beta\hat{\mathcal{H}}\right] \tag{17}$$

of N distinguishable nuclei moving in a given external electronic ground–state potential $E_0(\{\mathbf{R}_I\})$ can be exactly written as

$$\mathcal{Z} = \lim_{P\to\infty} \prod_{I=1}^{N} \prod_{s=1}^{P} \left[A_P \int d\mathbf{R}_I^{(s)}\right] \times$$

$$\exp\left[-\beta\sum_{s=1}^{P}\left(\sum_{I=1}^{N}\frac{1}{2}M_I\omega_P^2\left(\mathbf{R}_I^{(s)}-\mathbf{R}_I^{(s+1)}\right)^2+\frac{1}{P}E_0(\{\mathbf{R}_I^{(s)}\})\right)\right] \quad (18)$$

with $A_P = \left(M_I P/2\pi\hbar^2\beta\right)^{3/2}$, $\omega_P^2 = P/\beta^2\hbar^2$ and $\beta = 1/k_{\rm B}T$ after invoking the Born–Oppenheimer approximation. Traditionally, such quantum partition functions for finite discretizations P are sampled by generating a stochastic Markov trajectories with Monte Carlo techniques, which is the essence of all path integral Monte Carlo methods [55, 56]. Based on arguments similar to those discussed in Sect. 2.1, we opted instead for the generation of the trajectory by deterministic molecular dynamics propagation in order to make optimum use of previous information on the wavefunction. The Lagrangian,

$$\begin{aligned}
\mathcal{L} \;=\; & \frac{1}{P}\sum_{s=1}^{P}\left\{\sum_{i}^{\rm occ}\mu\left\langle\dot{\phi}_i^{(s)}\middle|\dot{\phi}_i^{(s)}\right\rangle - E^{\rm KS}[\{\phi_i\}^{(s)},\{\mathbf{R}_I\}^{(s)}]\right. \\
& \left. +\sum_{ij}\Lambda_{ij}^{(s)}\left(\left\langle\phi_i^{(s)}\middle|\phi_j^{(s)}\right\rangle-\delta_{ij}\right)\right\} \\
& +\sum_{s=1}^{P}\left\{\sum_{I=1}^{N}\frac{1}{2}M_I'\left(\dot{\mathbf{R}}_I^{(s)}\right)^2-\sum_{I=1}^{N}\frac{1}{2}M_I\omega_P^2\left(\mathbf{R}_I^{(s)}-\mathbf{R}_I^{(s+1)}\right)^2\right\}
\end{aligned} \quad ,(19)$$

introduced in Ref. [13] is the quantum–generalization of the "classical" Car–Parrinello Lagrangian (12), and properly reduces to it in the limit $P \to 1$. One can see that this Lagrangian yields P coupled Car–Parrinello–type equations of motion, which have a harmonic coupling

$$-M_I\omega_P^2\left(2\mathbf{R}_I^{(s)}-\mathbf{R}_I^{(s+1)}-\mathbf{R}_I^{(s-1)}\right) \quad (20)$$

between a given bead s and its two neighbors $s-1$ and $s+1$ along the necklace.

A few remarks are now in order. The equations of motion derived from the Lagrangian (19) have to be thermostatted [13, 14, 15] by the so–called Nosé–Hoover chain thermostats [57] in order to assure an ergodic sampling of the path integral [58]. In addition, the electronic degrees of freedom have to be thermostatted as well and *separately* [13, 14] at a much lower "temperature" so that the wavefunction is always sufficiently close to the electronic ground–state. The efficiency of the molecular dynamics sampling of the path integral's configuration space is greatly improved by using mode–transformed coordinates, such as the staging [58] or normal–mode variables as discussed in Ref. [15]. Finally, we mention that this approach is ideally suited for parallel implementation since the *electronic structure problem* of a given bead s is independent from *all* the other beads.

3.2. APPLICATIONS: PROTON TRANSFER IN WATER

After applying the *ab initio* path integral methodology from Sect. 3.1 to isolated molecules and clusters in the gas phase [59, 60, 61, 62, 63, 64], first investigations of bulk systems such as solid hydrogen under high pressure [65] and phase transitions in H_2O ice [66] followed. Currently, we are investigating the behavior of an excess proton in water, a system that was already studied by the standard Car–Parrinello techniques from Sect. 2.1 using heavy water D_2O and classical nuclei, see Ref. [44]. The present fully quantum–mechanical simulations are performed at ambient conditions using 32 H_2O molecules with one excess proton added and a path integral discretization of 8 replicas. Based on the results of a comparative investigation of the performance of several density functionals for liquid water [43], the so–called BLYP density functional [39, 67] is used here to describe the electronic structure.

One of the major issues to be clarified in the context of proton transfer in water is the molecular structure of the charge defect. Considering the energy released upon binding one proton to a single water molecule (i.e., the proton affinity of the H_2O molecule of 165 kcal/mol corresponding to roughly 8 eV or 80 000 K), it is immediately clear that the bare proton H^+ cannot exist in water as such and is expected to form a quite strongly bound complex $H^+\cdot(H_2O)_n$ with surrounding water molecules. However, since decades evidence for different numbers n are derived based on different types of experiments. Guided primarily by thermodynamic measurement, the Eigen school has proposed [68], that the excess proton forms an $H_9O_4{}^+$ complex ($n = 4$), where the H_3O^+ core is particularly strongly hydrogen–bonded to three H_2O molecules. Zundel and coworkers [69], on the other hand, have interpreted a host of infrared spectra as proving the existence of an $H_5O_2{}^+$ complex ($n = 2$), where the excess proton is shared between two water molecules giving rise to an easily polarizable and rapidly fluctuating proton. Finally, Giguère [70] interprets various experiments as being consistent with H_3O^+ as the important entity ($n = 1$).

In our fully quantum–mechanical *ab initio* path integral simulations we confirm that both $H_5O_2{}^+$ and $H_9O_4{}^+$ complexes are present in the liquid as can be seen by inspecting the trajectory. In order to investigate separately the structure of these complexes, we introduced a criterion based on the asymmetric stretch coordinate $\delta = R_{O_aH^\star} - R_{O_bH^\star}$ of the shared proton in the hydrogen–bonded atom triple $O_aH^\star O_b$. To be more precise: the purely geometric assignment of one $H_3O_a^+$ unit is in most cases unambiguous and O_b and $H^\star$ are those oxygen and hydrogen atoms that form together with O_a the shortest δ coordinate in this particular configuration. This $O_aH^\star O_b$ triple is what we call the "most active hydrogen bond". For $|\delta| \approx 0$ the

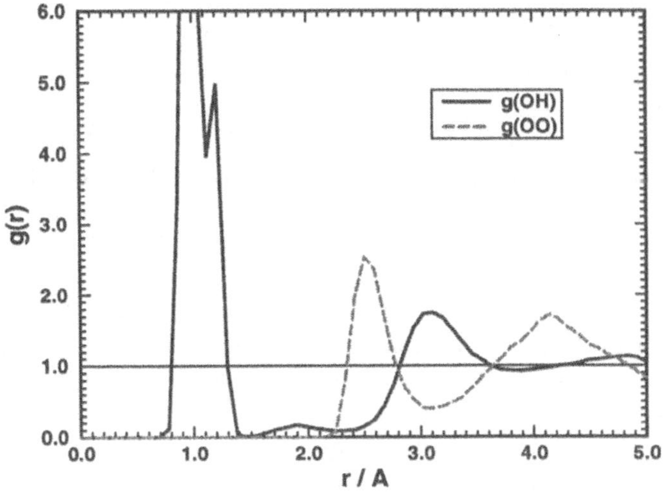

Figure 2. Partial radial distribution functions $g_{XY}(r)$ for $H_5O_2^+$-like complexes from the fully quantum-mechanical simulations of the excess proton in water at ambient conditions for oxygen-hydrogen (solid lines) and oxygen-oxygen (dashed lines) pairs; only those XY pairs are included where at least one of the two atoms belongs to the $O_aH^*O_b$ triple with $\delta \leq 0.1$, see text. The horizontal dotted line marks the value unity.

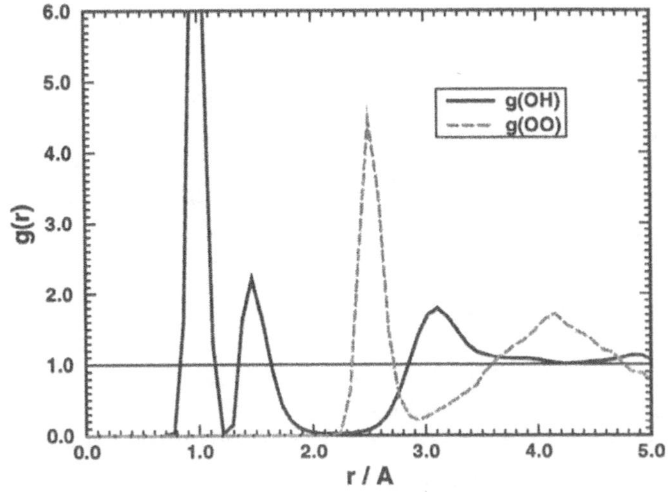

Figure 3. Partial radial distribution functions $g_{XY}(r)$ for $H_9O_4^+$-like complexes from the fully quantum-mechanical simulations of the excess proton in water at ambient conditions for oxygen-hydrogen (solid lines) and oxygen-oxygen (dashed lines) pairs; only those XY pairs are included where at least one of the two atoms belongs to the $H_3O_a^+$ complex and $\delta > 0.3$, see text. The horizontal dotted line marks the value unity.

proton resides approximately midway between two water molecules in that triple, whereas it is much closer to one particular water molecule for $|\delta| \gg 0$.

In Fig. 2 we present quantum–statistical averages of partial radial distribution functions $g_{XY}(r)$ for $|\delta| \leq 0.1$ Å, and the $|\delta| > 0.3$ Å case is shown in Fig. 3. For small $|\delta|$ values (Fig. 2) the *pure intra*complex OO radial distribution function (not shown here) peaks near $r \approx 2.4$ Å, compared to about 2.8 Å for bulk water. The OH pair distribution function $g_{OH}(r)$ determined from only the two water molecules in $H_5O_2^+$ peaks at 0.95 Å, which is the same value as determined for water molecules in the bulk. The secondary feature in the first OH peak close to 1.25 Å is a clear signature of the shared proton that resides approximately midway between O_a and O_b, and only a bump is left from the standard *inter*molecular OH peak at 1.8 Å as found in bulk water. This conforms with Zundel's view of a proton that is shared between two essentially unperturbed water molecules like $H_2O\cdots H^+\cdots OH_2$. For large $|\delta|$ values (Fig. 3) the first OO peak is located near $r \approx 2.5$ Å and integrates to 3.0 at its first minimum. The OH distribution function $g_{OH}(r)$ peaks in this case close to 1.0 Å instead around 0.95 Å for bulk water or in the $H_5O_2^+$ complex. These findings are in line with Eigen's view of an $H_3O^+\cdot(H_2O)_3$ complex with an essentially unperturbed H_3O^+ core molecular ion and three strongly bound H_2O molecules. In addition, the three water molecules coordinating the central H_3O^+ ion in the $H_9O_4^+$ complex are found to be approximately fourfold coordinated, and thus approach the average coordination expected in bulk water.

However, histograms of the probability of occurrence of configurations with a certain value of the δ coordinate turn out to be very flat and rather unstructured. This means that a clear–cut distinction between these $H_5O_2^+$ and $H_9O_4^+$ complexes cannot be made, and that *both* complexes are expected to be crucial in order to fully understand proton migration in water.

4. Where Do We Stand?

In this lecture course the basic ideas underlying both Car–Parrinello *ab initio* molecular dynamics [11] and *ab initio* path integrals [13] were introduced and discussed in Sect. 2.1 and Sect. 3.1, respectively. Both methods constitute a combination of previously separate methodologies, i.e., electronic structure calculations and molecular dynamics or path integral techniques, respectively. The crucial point is that these unified *ab initio* approaches no longer rely on predefined interaction potentials as an input for the simulation, but rather use electronic structure calculations to obtain the potentials or forces "on the fly" as the simulation proceeds. Thus, the approximations underlying the determination of the interactions are conceptually shifted onto another level, i.e., that of the approximations inherent to the particular electronic structure method chosen, such as Hohenberg–Kohn–Sham density functional theory within the generalized gradient approximation GGA as used here. Furthermore, it was demonstrated in Sect. 2.2 and Sect. 3.2 that first principles investigations of liquids, even including nuclear quantum effects such as proton dispersion and tunneling, are now possible. Finally, we mention that pioneering investigations of quite complex chemical reactions in solution are already initiated [48, 71, 72] and hold great promise for extending the realm of *ab initio* molecular dynamics [11].

Acknowledgements

It is a great pleasure to thank Ettore Fois, Jürg Hutter, Michiel Sprik, Mark Tuckerman, and in particular Michele Parrinello for fruitful and enjoyable collaborations on the topics covered in these lecture notes.

References

1. G. Ciccotti, D. Frenkel, and I. R. McDonald, *Simulation of Liquids and Solids* (North-Holland, Amsterdam 1987).
2. M. P. Allen and D. J. Tildesley, *Computer Simulation of Liquids* (Clarendon Press, Oxford 1987, reprinted 1990).
3. R. Haberlandt, S. Fritzsche, G. Peinel, and K. Heinzinger *Molekulardynamik – Grundlagen und Anwendungen* (Vieweg Verlag, Braunschweig 1995).
4. K. Binder and G. Ciccotti (Eds.), *Monte Carlo and Molecular Dynamics of Condensed Matter Systems* (Italien Physical Society SIF, Bologna 1996).
5. C. G. Gray and K. E. Gubbins, *Theory of Molecular Fluids Vol. 1* (Oxford, Clarendon 1984).
6. G. C. Schatz, Rev. Mod. Phys. **61**, 669 (1991).
7. M. Sprik, J. Chem. Phys. **95**, 6762 (1991).
8. D. Borgis and A. Staib, Chem. Phys. Lett. **238**, 187 (1995).
9. M. M. Probst, E. Spohr, K. Heinzinger, and P. Bopp, Molec. Sim. **7**, 43 (1991).
10. D. Marx, M. Sprik, and M. Parrinello, Chem. Phys. Lett. **272**, 360 (1997).
11. R. Car and M. Parrinello, Phys. Rev. Lett. **55**, 2471 (1985).
12. M. Sprik, J. Phys.: Condens. Matter **8**, 9405 (1996).
13. D. Marx and M. Parrinello, Z. Phys. B (Rapid Note) **95**, 143 (1994); note that a misprinted sign in the equation was corrected in Ref. [14].
14. D. Marx and M. Parrinello, J. Chem. Phys. **104**, 4077 (1996).
15. M. E. Tuckerman, D. Marx, M. L. Klein, and M. Parrinello, J. Chem. Phys. **104**, 5579 (1996).
16. D. K. Remler and P. A. Madden, Mol. Phys. **70**, 921 (1990).
17. G. Galli and M. Parrinello, in *Computer Simulations in Materials Science* p. 282, edited by M. Meyer and V. Pontikis (Kluwer, Dordrecht 1991).
18. G. Pastore, E. Smargiassi, and F. Buda, Phys. Rev. A **44**, 6334 (1991).
19. M. C. Payne, M. P. Teter, D. C. Allan, T. A. Arias, and J. D. Joannopoulos, Rev. Mod. Phys. **64**, 1045 (1992).
20. G. Galli and A. Pasquarello, in *Computer Simulation in Chemical Physics* p. 261, edited by M. P. Allen and D. J. Tildesley (Kluwer, Dordrecht 1993).
21. R. Car, in *Monte Carlo and Molecular Dynamics of Condensed Matter Systems* p. 601, edited by K. Binder and G. Ciccotti (Italien Physical Society SIF, Bologna 1996).

22. M. Parrinello, Solid State Commun. **102**, 107 (1997).
23. P. Hohenberg and W. Kohn, Phys. Rev. **136**, B864 (1964).
24. W. Kohn and L. J. Sham, Phys. Rev. **140**, A1133 (1965).
25. R. M. Dreizler and E. K. U. Gross, *Density-Functional Theory* (Springer, Berlin 1990).
26. R. G. Parr and W. Yang, *Density-Functional Theory of Atoms and Molecules* (Oxford University Press, Oxford 1989).
27. R. O. Jones and O. Gunnarsson, Rev. Mod. Phys. **61**, 689 (1989).
28. J. P. Perdew, J. A. Chevary, S. K. Vosko, K. A. Jackson, M. R. Pederson, D. J. Singh, and C. Fiolhais, Phys. Rev. B **46**, 6671 (1992).
29. B. G. Johnson, P. M. W. Gill, and J. A. Pople, J. Chem. Phys. **98**, 5612 (1993).
30. E. Ermakova, J. Solca, H. Huber, and D. Marx, Chem. Phys. Lett. **246**, 204 (1995).
31. J. Solca, A. J. Dyson, G. Steinebrunner, and H. Huber, Chem. Phys. **224**, 253 (1997).
32. B. Kirchner, E. Ermakova, J. Solca, and H. Huber, Chem. Eur. J. **4**, 383 (1998).
33. A. K. Soper, J. Phys.: Condens. Matter **9**, 2717 (1997).
34. A. K. Soper, F. Bruni, and M. A. Ricci, J. Chem. Phys. **106**, 247 (1997).
35. K. Laasonen, F. Csajka, and M. Parrinello, Chem. Phys. Lett. **194**, 172 (1992).
36. K. Laasonen, M. Parrinello, R. Car, Ch. Lee, and D. Vanderbilt, Chem. Phys. Lett. **207**, 208 (1993).
37. C. Lee, D. Vanderbilt, K. Laasonen, R. Car, and M. Parrinello, Phys. Rev. Lett. **69**, 462 (1992); Phys. Rev. B **47**, 4863 (1993).
38. J. P. Perdew and A. Zunger, Phys. Rev. B **23**, 5048 (1981); this LDA parameterization is based on the Quantum Monte Carlo data of D. M. Ceperley and B. J. Alder, Phys. Rev. Lett. **45**, 566 (1980).
39. A. Becke, Phys. Rev. A **38**, 3098 (1988).
40. K. Laasonen, M. Sprik, M. Parrinello, and R. Car, J. Chem. Phys. **99**, 9080 (1993).
41. E. S. Fois, M. Sprik, and M. Parrinello, Chem. Phys. Lett. **223**, 411 (1994).
42. P. L. Silvestrelli, M. Bernasconi, and M. Parrinello, Chem. Phys. Lett. **277**, 478 (1997).
43. M. Sprik, J. Hutter, and M. Parrinello, J. Chem. Phys. **105**, 1142 (1996).
44. M. Tuckerman, K. Laasonen, M. Sprik, and M. Parrinello, J. Phys.: Condens. Matter **6**, A93 (1994); J. Phys. Chem. **99**, 5749 (1995); J. Chem. Phys. **103**, 150 (1995).
45. K. Laasonen and M. L. Klein, J. Amer. Chem. Soc. **116**, 11620 (1994); J. Phys. Chem. **98**, 10079 (1994).
46. D. Marx, J. Hutter, and M. Parrinello; Chem. Phys. Lett. **241**, 457 (1995).
47. D. Marx, E. Fois, and M. Parrinello, Int. J. Quantum Chem. **57**, 655 (1996).
48. E. J. Meijer and M. Sprik, J. Phys. Chem. A **102**, 2893 (1998).
49. C. Molteni and M. Parrinello, J. Am. Chem. Soc. **120**, 2168 (1998); see also Chem. Phys. Lett. **275**, 409 (1998).
50. T. Yamaguchi, H. Ohtaki, E. Spohr, G. Pálinkás, K. Heinzinger, and M. M. Probst, Z. Naturforsch. **41a**, 1175 (1986).
51. M. M. Probst, E. Spohr, and K. Heinzinger, Chem. Phys. Lett. **161**, 405 (1989).
52. P.-A. Pittet, G. Elbaze, L. Helm, and A. E. Merbach, Inorg. Chem. **29**, 1936 (1990).
53. B. Marten, K. Kim, C. Cortis, R. A. Friesner, R. B. Murphy, M. N. Ringnalda, D. Sitkoff, and B. Honig, J. Phys. Chem. **100**, 11775 (1996).
54. R. R. Pappalardo and E. Sánchez Marcos, J. Phys. Chem. **97**, 4500 (1993); E. Sánchez Marcos, J. M. Martínez, and R. R. Pappalardo, J. Chem. Phys. **105**, 5968 (1996); J. M. Martínez, R. R. Pappalardo, E. Sánchez Marcos, K. Refson, S. Díaz-Moreno, and D. Muñoz-Páez, J. Phys. Chem. B **102**, 3272 (1998).
55. M. J. Gillan, in *Computer Modelling of Fluids, Polymers, and Solids*, edited by C. R. A. Catlow, S. C. Parker, and M. P. Allen (Kluwer, Dordrecht 1990).
56. D. M. Ceperley, Rev. Mod. Phys. **67**, 279 (1995).
57. G. J. Martyna, M. L. Klein, and M. Tuckerman, J. Chem. Phys. **97**, 2635 (1992).

58. M. E. Tuckerman, B. J. Berne, G. J. Martyna, and M. L. Klein, J. Chem. Phys. **99**, 2796 (1993).
59. D. Marx and M. Parrinello, Nature (London) **375**, 216 (1995).
60. D. Marx and A. Savin, Angew. Chem. Int. Ed. Engl. **36**, 2077 (1997).
61. D. Marx and M. Parrinello, Science **271**, 179 (1996).
62. M. E. Tuckerman, D. Marx, M. L. Klein, and M. Parrinello, Science **275**, 817 (1997).
63. I. Štich, D. Marx, M. Parrinello, and K. Terakura, Phys. Rev. Lett. **78**, 3669 (1997); J. Chem. Phys. **107**, 9482 (1997).
64. R. Rousseau and D. Marx, Phys. Rev. Lett. **80**, 2574 (1998).
65. S. Biermann, D. Hohl, and D. Marx, J. Low Temp. Phys. **110**, 97 (1998).
66. M. Benoit, D. Marx, and M. Parrinello, Nature (London) **392**, 258 (1998); see also a *News and Views* comment by J. Teixeira, Nature (London) **392**, 232 (1998); M. Benoit, D. Marx, and M. Parrinello, Comp. Mat. Sci. **10**, 88 (1998); D. Marx, in *Computer Simulation of Rare Events and the Dynamics of Classical and Quantum Condensed Phase Systems* (Enrico Fermi Summer School, Lerici 1997) B. J. Berne, G. Ciccotti, und D. F. Coker (Eds.), in press.
67. C. Lee, W. Yang, and R. G. Parr, Phys. Rev. B **37**, 785 (1988).
68. E. Wicke, M. Eigen and Th. Ackermann, Z. physik. Chem. (N.F.) **1**, 340–364 (1954); for a review see M. Eigen, Angew. Chem. internat. Edit. **3**, 1 (1964).
69. G. Zundel and H. Metzger, Z. physik. Chemie (N.F.) **58**, 225 (1968); for a review see G. Zundel, in The Hydrogen Bond — Recent developments in theory and experiments. II. Structure and spectroscopy p. 683–766 edited by P. Schuster, G. Zundel, C. Sandorfy (North–Holland Publishing Company, Amsterdam 1976).
70. P. A. Giguère, J. Chem. Educ. **56**, 571 (1979).
71. A. Curioni, M. Sprik, W. Andreoni, H. Schiffer, J. Hutter, and M. Parrinello, J. Am. Chem. Soc. **119**, 7218 (1997).
72. M. Mohr, H. Zipse, D. Marx, and M. Parrinello, in preparation.

PROGRESS IN THE THEORY OF INHOMOGENEOUS ^{4}HE

L. REATTO
INFM and Dipartimento di Fisica,
Università degli Studi di Milano,
Via Celoria 16 I-20133 Milano, Italy

1. Introduction

^{4}He in the liquid state is a prototypal quantum fluid of strongly interacting particles. At the low energy of interest these atoms can be considered as structurless Bosons and the interatomic interaction can be represented in good approximation by well characterized pair forces. The basic microscopic theoretical tools are, on one hand the variational theory and, on the other hand, "exact" simulation methods. A rather complete description of the ground state properties of bulk ^{4}He has been achieved. "Exact" simulation methods[1], i.e. Green's Function Monte Carlo (GFMC) and Diffusion Monte Carlo (DMC), give the equation of state (EOS) with such an accuracy that these methods can be used to test the accuracy of the interatomic potential which is used as an input.

The variational method has reached a rather satisfactory level of accuracy, the ground state energy being reproduced by the most advanced theory[2] to within about 3% over the full density range of the liquid state. Great advances have been obtained lately also for the excited states. Two kinds of excitations are fundamental to superfluid ^{4}He. One represents eigenstates of momentum, the phonon-maxon-roton branch of excitations which is most directly probed experimentally by inelastic neutron scattering. The other kind of excitations is vorticity. Exact simulation methods have difficulties in treating such excited states due to a sign problem but some interesting progress has recently been obtained[3]. The variational theory is able now[4, 5] to reproduce the features of the large wave vector excitations, maxons and rotons, rather accurately, the deviation from experiment being at the level of 5%. The results came after many years of effort when the typical deviations were at the level of 20-30%.

C. Caccamo et al. (eds.), New Approaches to Problems in Liquid State Theory, 459–479.

Lately much attention has been given to inhomogeneous states of ^{4}He, like a thin film of ^{4}He adsorbed on a substrate, the free surface of the liquid or a cluster of atoms. These represent an area in which significant experimental advances have recently been achieved.

It is well known that pair correlations have a basic role in liquid ^{4}He but in order to achieve a quantitative description three-body correlations (triplet terms) are needed in the wave function (wf). A different strategy has been followed with the introduction[6] of the so called Shadow Wave Function (SWF). Here correlations beyond the pair level are introduced via the use of subsidiary variables so that correlations among the ^{4}He atoms are produced both by direct terms and indirectly via the coupling to these subsidiary variables, the shadows, which are integrated over. This appears to be a very efficient way of introducing many-body couplings, also beyond the pair and triplet terms, which become important as the freezing transition is approached[2, 7]. In addition, the SWF strategy turns out to be very useful also for the excited states[4] since it allows for backflow effect in a natural way. More recently the SWF approach has been extended also to treat bound states of the atoms in which the self–binding is achieved just via interparticle correlations[8].

The outline of the chapter is as follows. First we review the ground state of ^{4}He. In Sect. 3 we discuss the roton excitations and what we learn by considering excitations in the form of wave packets. As example of a strongly inhomogeneous state we discuss in Sect. 4 a free slab of ^{4}He and the liquid-vacuum interface. The case of a doped cluster is considered in Sect. 5. Some conclusions are presented in Sect 6.

2. Ground state of bulk ^{4}He

The simplest non trivial wf for the ground state of N bosons is a Jastrow form:

$$\Psi_J(\vec{r}_1....\vec{r}_N) = \prod_{i<j} e^{-\frac{1}{2}u(r_{ij})} \tag{1}$$

where $u(r_{ij})$, called the pseudopotential, is a real function of the pair distance $r_{ij} = |\vec{r}_i - \vec{r}_j|$ between atoms. Eq.(1) incorporates the basic features of the ground state of bosons: it is symmetric under exchange of particles, it is real with no nodes and easily incorporates the core condition with $u(r)$ becoming large and positive where the pair potential $v(r)$ is strongly repulsive. The pseudopotential $u(r)$ is a variational function which is determined by minimization of the expectation value of the Hamiltonian. This expectation reads

$$E_J = \langle \Psi_J | H | \Psi_J \rangle / \langle \Psi_J \mid \Psi_J \rangle = \frac{N}{2}\rho \int d^3 r g(r) \left[\frac{\hbar^2}{4m} \nabla^2 u(r) + v(r) \right] \quad (2)$$

in the case in which the interatomic potential is of the pair form. $g(r)$ is the radial distribution function

$$g(r) = \langle \delta(\vec{r}_1) \delta(\vec{r} - \vec{r}_2) \rangle / \rho^2 \quad , \quad \rho = N/V. \quad (3)$$

This function is of interest by itself because its Fourier transform is directly related to the static structure factor $S(q)$ which can be measured by X–rays or by neutron scattering.

The Jastrow function is already of some value for liquid ^{4}He but it is not enough for a quantitative theory. A more general form is the so called Feenberg form[9]

$$\Psi(\vec{r}_1, .., \vec{r}_N) =$$

$$= \exp\left\{ -\frac{1}{2} \sum_i u^{(1)}(\vec{r}_i) - \frac{1}{2} \sum_{i<j} u(\vec{r}_i, \vec{r}_j) - \frac{1}{2} \sum_{i<j<l} u^{(3)}(\vec{r}_i, \vec{r}_j, \vec{r}_l) + ... \right\} \quad . \quad (4)$$

The one–body term $u^{(1)}(\vec{r})$ is supposed to be present only when the system is dishomogenous like in the solid phase or in presence of an external potential like that due to a substrate. Notice the profound difference between these two cases. If there is an external potential the hamiltonian is not translationally invariant so that it is obvious that the one–body terms $u^{(1)}$ is present. On the other hand a crystalline solid state is a manifestation of spontaneous translational broken symmetry and the presence of the one–body terms is not obvious.

In case of a bulk liquid $u^{(1)}$ is not present, the pair term depends only on the pair distance, $u^{(2)}(\vec{r}_i, \vec{r}_j) = u(r_{ij})$, and similarly the higher order terms depend only on relative distances.

The level beyond the Jastrow one includes the triplet contributions, Ψ_{J+T}. This approximation is excellent at the equilibrium density[7] but a rapidly increasing deviation of the energy from experiment as freezing is approached indicates that correlations beyond the triplet level become relevant. This is shown in Fig.1. These results are based on an expansion of $u^{(2)}$ and $u^{(3)}$ on a basis set and the resulting minimization problem is solved by a reweighting MC technique[7].

It is useful to notice a formal analogy between our quantum problem and a classical fluid. The probability density of finding the N atoms in a

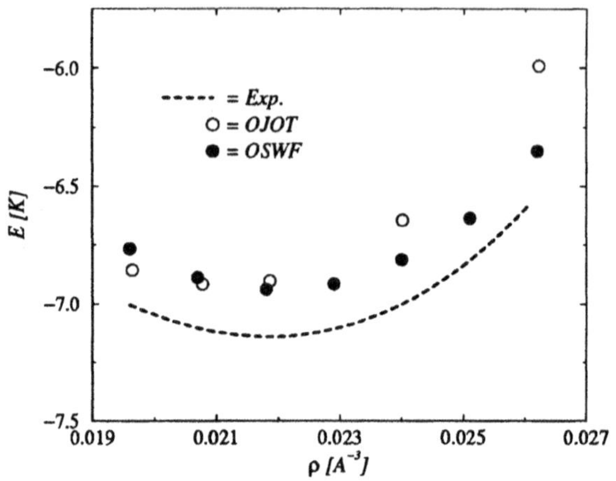

Figure 1. Equation of state of ^{4}He at $T = 0$ in the liquid region

configuration $R \equiv (\vec{r}_1, ...\vec{r}_N)$ is just the square of the wave function. Within the variational ansatz (1) this is formally equal to the probability density in configuration space of classical particles interacting with a pair interaction $v^*(r)$ at temperature T^*such that $v^*(r)/K_B T^* = u(r)$. Similarly with wf (4) one has equivalence with a classical system in presence of an external potential and of a three-body interaction, in addition to the two-body one. This equivalence explains why similar techniques, like MC or the method of integral equation for $g(r)$, find use in the quantum problem.

In the SWF method a set $S = (\vec{s}_1, ...\vec{s}_N)$ of N subsidiary variables is introduced[6]. The position of each atom is coupled to the position of one of the subsidiary variables and these are integrated over. Therefore the unnormalized wf reads

$$\Psi(R) = \int dSF(R, S) \tag{5}$$

and the correlating factor $F(R, S)$ is assumed to contain only pair terms, i.e.

$$F(R, S) = \Phi_p(R)\,\Theta(R, S)\,\Phi_S(S) \tag{6}$$

where Φ_p and Φ_S are Jastrow functions with pseudopotentials $u_p(r)$ and $u_s(r)$, respectively. The kernel Θ has the form $\exp[-\sum_i u_{sp}(|\vec{r}_i - \vec{s}_i|)]$ .The three pseudopotentials u_p, u_s and $u_{ps}$are variational functions. A basis set expansion of these functions and a MC minimization give[2] an improved

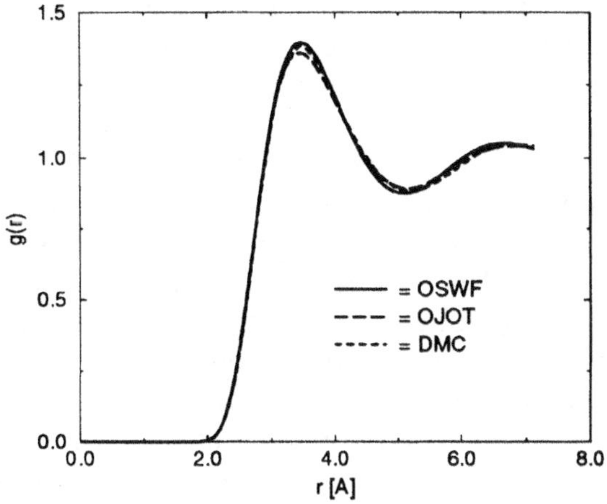

Figure 2. Comparison between the OSWF, OJOT and DMC results for the pair distribution function at ρ_{eq} in the liquid

energy with respect to a J + T wf in particular at large densities and this indicates that close to freezing correlations beyond the pair and triplet levels are important. In fact the SWF can be considered as a way to incorporate interparticle correlations to all order, as it can be seen by performing a cluster expansion of $\ln \Psi(r)$. In addition a physical motivation can be given for the shadow variables as a way of representing the quantum delocalization of a particle. Not only the energy is given accurately by the SWF but also other quantities like $g(r)$ (see Fig.2) and the BE condensation fraction which turns out to be $n_0 = 8.1\%$ at ρ_{eq}. This has to be compared with $n_0 = 7.2\%$ given by DMC. Another important point is that the same SWF (5) describes a solid when the density is above $\rho \cong 0.028$ Å^{-3}. The crystalline order is obtained as a spontaneous broken symmetry[10], at variance with other theories in which the crystalline order has to be introduced explicitly in the wf by one–body terms. This capability of the SWF of describing on the same basis either a liquid or a solid depending on density is a unique property of SWF. This has had important implications as it opened the way to a microscopic study of disorder phenomena in a quantum solid like the liquid-solid interface[11] or a vacancy[12] in solid ^{4}He.

3. Excited states

Many properties of superfluid ^{4}He are determined by the low-lying excited states which at low temperatures can be considered as independent excitations. These are eigenstates of momentum $\vec{p} = \hbar\vec{k}$ and the excitation spectrum ε_k is central to the theory of the system. For instance the fact that the quantity ε_k/k is bounded below by a non-zero value is at the basis of superfluidity. At small k the excitations are phonons and this is common to most of the condensed matter. What is specific to ^{4}He is that there are no other excitations of very low energy. If there is a consensus on the nature of phonons, i.e. longitudinal density fluctuations, the situation is different with the so called rotons, the excitations characterized by a minimum of ε_k at finite k_0. These are excitations rather unique to ^{4}He with special properties: they carry a large momentum but at $k = k_0$ they have zero group velocity, they have a positive group velocity for $k > k_0$ but this is negative for $k < k_0$ so that the momentum is in the opposite direction to the motion. Finally at low T rotons have been observed[13] to propagate over macroscopic distances and the theoretical indication is that they would propagate with no measurable damping in the $T \to 0K$ limit.

The computation of ε_k has been a long standing problem of many-body physics, started with the work of Feynman[14]. This is based on the simple ansatz for a single excited state.

$$\Psi_{\vec{k}}^F(R) = F_{\vec{k}}\Psi_0, \quad F_{\vec{k}} = \rho_{\vec{k}}/\sqrt{NS_0(k)} \tag{7}$$

where $\Psi_0(R)$ is the normalized ground state, $\rho_{\vec{k}}$ is the density fluctuation

$$\rho_{\vec{k}} = \sum_j e^{i\vec{k}\cdot\vec{r}_j} \quad , \tag{8}$$

the denominator in eq.(7) for $F_{\vec{k}}$ is the appropriate normalization factor and $S_0(k)$ is the static structure factor for the ground state

$$S_0(k) = \left\langle \Psi_0 \left| \rho_{\vec{k}}\rho_{-\vec{k}} \right| \Psi_0 \right\rangle / N. \tag{9}$$

The excitation spectrum has a very simple expression

$$\varepsilon_F(k) = \hbar^2 k^2 / 2mS_0(k) \tag{10}$$

and it has the qualitative features as observed in ^{4}He. Keeping in mind that at $T = 0 \quad S_0(k) \underset{q\to 0}{\sim} \hbar\, k/2mc$, c being the sound velocity, ε_k has the expected linear behavior at small k and $\varepsilon_F(k)$ has a minimum roughly in correspondence with the main maximum of $S(k)$. This description is all too simple in the large k region, the roton energy being of order of $20K$ at the equilibrium density, more than twice of the experimental value $8.61K$. The origin of the discrepancy was immediately recognized as due to the absence of correlated motion between atoms in the wf (7) and the next step was taken by Feynman and Cohen[15] with the introduction of the so called backflow contribution. In quantum mechanics the argument of the exponential function in $F_{\vec{k}}$ has the role of the velocity potential in classical hydrodynamics so in order to account for the backflow the position $\vec{r}_j$ of the $j-th$ atom in $\rho_{\vec{k}}$ is replaced by $\vec{r}_j + A\sum_{l(\neq j)}(\vec{r}_j - \vec{r}_\ell)/|\vec{r}_j - \vec{r}_\ell|^3$. In classical hydrodynamics the coefficient A is equal to $(4\rho)^{-1}$ but here A is taken as a variational parameter. The computation with such wf was not easily implemented so what has always been used is the expression linearized in A, i.e. the Feynman factor $F_{\vec{k}}$ in (7) is replaced by

$$F_{\vec{k}}^{FC}(R) = \sum_j e^{i\vec{k}\cdot\vec{r}_j}\left(1 + A\sum_{\ell(\neq j)}\frac{\vec{r}_j - \vec{r}_\ell}{|\vec{r}_j - \vec{r}_\ell|^3}\right) \tag{11}$$

Notice that this can be rewritten in the more general form

$$F_{\vec{k}}^{FC}(R) = \rho_{\vec{k}} + \sum A_{\vec{k},\vec{k}'}\rho_{\vec{k}-\vec{k}'}\rho_{\vec{k}'} \quad , \tag{12}$$

and the hydrodynamic form (11) corresponds to $A_{\vec{k},\vec{k}'} = 4\pi A\frac{\vec{k}'\cdot\vec{k}}{k^2}$. Therefore the backflow can be considered as a coupling between two phonon excitations of wave vector $\vec{k}-\vec{k}'$ and $\vec{k}'$. The FC wf gives an improved roton energy but its value of about $14K$ is still much higher than experiment. Another significant quantity is the strength Z_q of the single excitation contribution to the dynamical structure factor $S(k,\omega)$. At small T $S(k,\omega)$ contains a very sharp peak so that one writes $S(k,\omega) = Z_k\,\delta(\omega - \varepsilon_k/\hbar) + S_m(k,\omega)$ where $S_m(k,\omega)$ represents any broad contribution of $S(k,\omega)$. Since the integral of $S(k,\omega)$ over all ω gives $S(k)$, it is clear that

$$f_k = Z_k/S(k) \tag{13}$$

represents the relative efficiency for exciting a single excitation by a neutron compared with all possible processes. With the Feynman wf $Z_k = S(k)$ so that $f_k = 1$ at all k. The FC wf gives a reduced value of f_k at large k but still strongly overestimates its value compared with experiment.

The FC theory has been improved by considering additional processes which lead to contribution in the wf of higher order in the $\rho_{\vec{k}}$ variable and terms of cubic and of fourth order have been taken into account. Sophisticated many-body techniques[16] have been used but notwithstanding efforts over some decades the roton energy has remained above $10K$, about 20% above experiment and an even larger deviation is found at densities above equilibrium.

A new attack[17] to this problem has been given by SWF. The basic observation is the following. Suppose that the density fluctuation is written not in terms of the positions of the atoms but in terms of the shadow variables so that we write

$$\Psi_{\vec{k}}(R) = \int dSF(R,S)\sigma_{\vec{k}} \quad , \quad \sigma_{\vec{k}} = \sum_j e^{i\vec{k}\cdot\vec{s}_j}. \tag{14}$$

One can prove that this is an exact eigenstate of momentum with eigenvalue $\hbar\vec{k}$ so that this is a suitable trial wf. It is not difficult to show that this Feynman-like form of wf actually contains backflow when looked from the point of view of the variables $\{\vec{r}_j\}$. In fact the relative motion of $\vec{r}_i$ with respect to $\vec{s}_i$ modifies the simple density fluctuation of the s variables. What is also significant is that this induced backflow is in the exponential form. This implies that the wf (14) implicitly contains backflow contributions to all order in the $\rho_{\vec{k}}$ variables. The actual computation of the spectrum ε_k gave an excellent energy, about $10K$ at ρ_{eq}. It is remarkable that such simple wf with no variational parameter for the excited state gives an energy at the level of the most sophisticated many-body theories.

An improvement over the form (14) is obtained if $\sigma_{\vec{k}}$ is modified by an explicit backflow contribution as follows[4]

$$\Psi_{\vec{q}}(R) = \int dSF(R,S)\tilde{\sigma}_{\vec{k}},$$

$$\tilde{\sigma}_{\vec{k}} = \sum_j \exp\left\{ i\vec{k}\cdot\left[\vec{s}_j + \sum_{l(\neq j)} (\vec{s}_j - \vec{s}_\ell)\lambda(|\vec{s}_j - \vec{s}_\ell|)\right]\right\} \tag{15}$$

where the backflow function λ is assumed of a short range form with some variational parameters. These parameters are determined from minimization of the total energy of the excited state

$$E_k = \left\langle \Psi_{\vec{k}} |H| \Psi_{\vec{k}} \right\rangle / \left\langle \Psi_{\vec{k}} \mid \Psi_{\vec{k}} \right\rangle. \tag{16}$$

From this the excitation energy is obtained as $\varepsilon_k = E_k - E_0$, where E_0 is the corresponding energy of the ground state, i.e. (16) with $\vec{k} = 0$. Few observations are in order. The first is that E_k is a lower bound to the total energy of the excited state but the variational bound is lost for ε_k because ε_k is obtained as the difference of two approximate expectation values. Notice that this is true also for all the other computations of ε_k. The second observation is that the explicit backflow in (15) is in the exponential form and no expansion is performed. Since averages are computed by MC there is no reason to make such an expansion. From the computational point of view to extract ε_k is very demanding since this quantity is obtained as the difference of two extensive quantities. Since the binding energy of ^{4}He is about $-7K$ and the roton energy about $9K$, in a computation with about 100 ^{4}He atoms the statistical error in the computation of E_k must

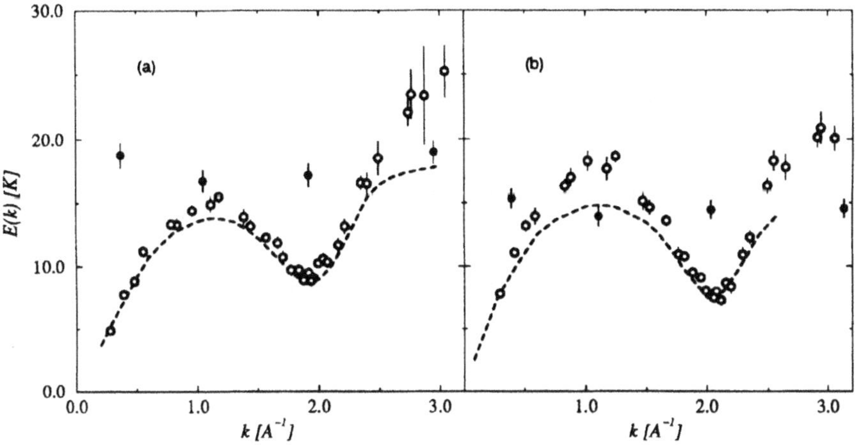

Figure 3. (a): Excitation spectrum computed with a fully optimized excited state SWF (open circles) and excitation spectrum for a double roton excitation (filled circles) at equilibrium density. (b): the same of (a) at freezing density.

be below few parts in 10^4 in order to get the roton energy within an accuracy of 5%. Last observation is that we can compute $E_{\vec{k}}$ only for the wave vectors which are compatible with the periodic boundary condition, i.e. $\vec{k} = \left(\frac{2\pi}{L}n_x, \frac{2\pi}{L}n_y, \frac{2\pi}{L}n_z\right)$ with $n_{x,y,z}$ integers and L being the side of the simulation box.

The results[2] for ε_k at ρ_{eq} and at the freezing density are shown in Fig.3. There is an overall quantitative agreement with experiment at both densities and the typical deviation is 5%. Only two regions of discrepancy are present: at large k beyond the roton minimum and in the maxon region for the higher density. In both cases ε_k is above twice the roton energy and as a consequence the single excited state gets mixed with a double roton excitation which has lower energy. This is not included in the wf (15). Also the result for the efficiency f_k of the single excitation in $S(k,\omega)$ is in good agreement with experiment and we conclude that now we have a quantitative theory of the excitations in superfluid ^{4}He at low temperature.

The previous computation, however, does not give much insight on what is a roton excitation and which model is more appropriate. We have studied[18] Gaussian–like wave packet built with the wf (16). In this way the system is no more homogeneous and we have computed the local density $\langle\rho(\vec{r})\rangle_{\vec{k}_0,\vec{r}_0,\delta}$ and the local momentum $\left\langle\vec{P}(\vec{r})\right\rangle_{\vec{k}_0,\vec{r}_0,\delta}$. $\vec{k}_0$ represents the average momentum of the packet and δ its width in wave–vector and $\vec{r}_0$ is the

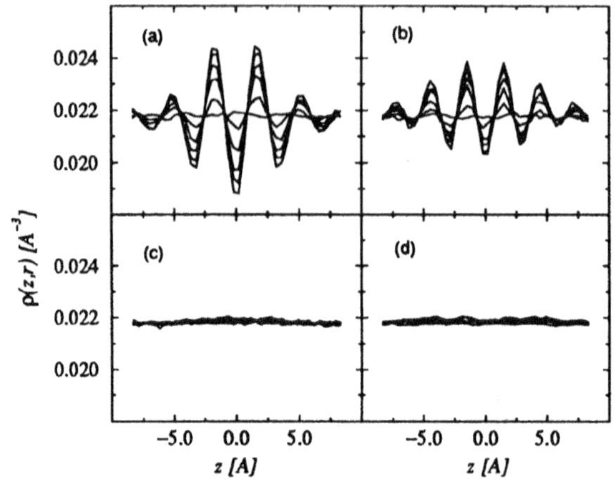

Figure 4. Local density $\langle\rho(z,r_\perp)\rangle$ ($r_\perp = (x^2+y^2)^{1/2}$) for a R^- (a), roton (c), R^+ (b) and distinguished particle (d) wave-packet at ρ_{eq}, plotted for $r_\perp \simeq 0.7$, 2.1, 3.15, 4.3, 5.95 and 8.15 Å. The density fluctuations decrease monotonically with $r_\perp$ in (a) and (b).

position of the center of the packet. One of the models of rotons describes the excitation as a quasiparticle. In order to test this we have also constructed a similar wave–packet for a distinguishable ^{4}He atom. This means that we selected one of the ^{4}He atoms and we have given the momentum only to this atom. In terms of the wf this is equal to (15) but now $\tilde{\sigma}_{\vec{k}}$ in place of the summation over j contains just one term, the one which corresponds to the distinguished atom.

With these wave packets we have computed the local density and the local momentum. For a general wave vector k_0 the local density has a modulation in space. The roton wave vector k_R is special: when $k_0 = k_R$ all modulations disappear (see Fig.4). Notice that this is true only for the optimized backflow otherwise there is some residual modulation. In the case of the distinguishable particle the local density is unaffected by the excitation for all wave vectors. This shows that in this respect there is a close resemblance between a quasiparticle and a roton but only exactly at k_R, not for other k.

The local momentum is shown in Fig.5. At $k_0 = k_R$ one notices the clear signature of backflow with the presence of a component of $\langle\vec{p}\rangle$ perpendicular to $\vec{k}$ which is reminiscent of a smoke ring. This backflow pattern turns out to be essentially equal to that of the distinguishable excitation. From $\langle\vec{p}\rangle$ one can compute the vorticity which is maximum along a circle normal to $\vec{k}_0$ and of radius of ~ 3.4Å (Fig.6). The circulation around this line is only about 2% of the quantum of circulation. The pattern of $\langle\vec{p}\rangle$ changes when

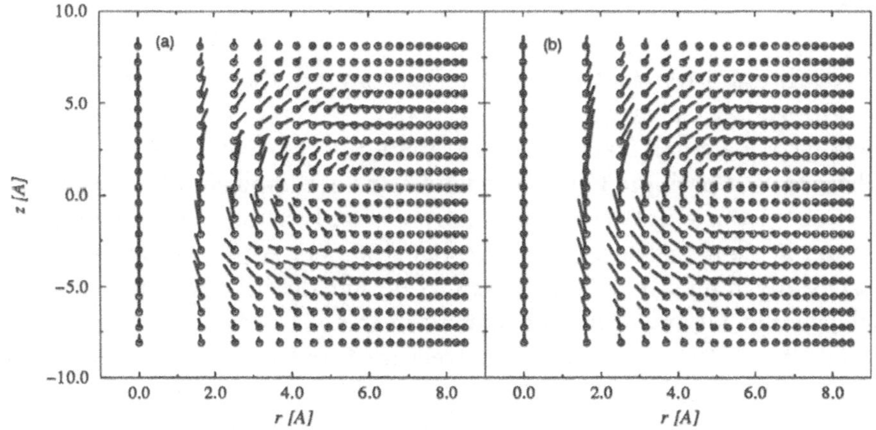

Figure 5. The vectors represent the local momentum $\langle \vec{p}(z, r_\perp) \rangle$ computed for the R^- (a) and roton (b) wave-packets. The open circles represent the starting point of the vectors. The center of the packet is at $z = 0$ and $r = 0$.

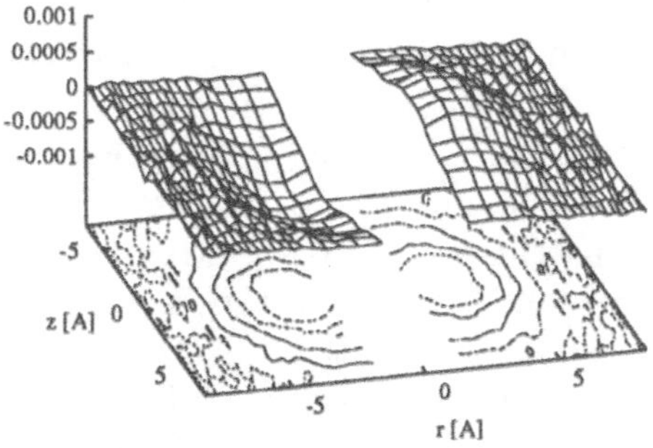

Figure 6. Vertical axis represents $\vec{\nabla} \times \langle \vec{v}(\vec{r}) \rangle$ in $h/m\AA^2$ units at ρ_{eq} for a roton wave-packet.

$k_0 \neq k_R$, the backflow pattern being modified by a modulation related to the value of k_0.

We conclude that exactly at k_R the excitation is very similar to a quasi particle excitation but this is no more true as we move from k_R. The backflow has a ring–like structure as in a smoke ring, but it is due to interparticle correlations and it is not a quantized vortex ring which represents a singularity in the order parameter.

4. Self-bound states

The ground state of N atoms of ^{4}He is a bound state. This means that if the atoms have a large volume V available they will form a droplet occupying only a small fraction of the volume V. The center of mass can be anywhere in V if periodic boundary conditions are applied. This means that we have coexistence between a dense droplet and vacuum. The correlations which are present in the Jastrow or Jastrow plus triplet wf for bulk ^{4}He are essentially repulsive so that these wf do not describe a droplet but the particles would fill any volume allowed to them. In order to have a droplet one would need strong attractive correlations and it turns out that these have to be so strong that the particles freeze. SWF contains some attractive correlations but these are not strong enough to give binding.

In the standard treatment of a self-bound state a one-body term is introduced in the wf as in Eq.(4). The one-body pseudopotential $u^{(1)}(\vec{r})$ modulates the local density and in this way a cluster[19], a slab[20] or a system filling half space[21] can be represented. The variational principle allows the determination of the form of $u^{(1)}(\vec{r})$. Such a wf has the unphysical feature that it does not allow the motion of the center of mass. This defect is overcome if the coordinates in the one–body factors are interpreted as $\vec{r}_i - \vec{r}_{cm}$ where $\vec{r}_{cm} = \frac{1}{N}\sum \vec{r}_i$ is the center of mass. Notice that this changes what appears as a one-body term in the wf in a N-body terms since $u^{(1)}$ depends on the coordinates of all the particles.

The previous approach produces reasonable results for the energy and the density profile of self-bound states of ^{4}He. This wf has the feature to have two-and perhaps three-body correlating factors and then N-body terms and in these last ones the information on the overall size of the system has to be present because this is the range of $u^{(1)}$. These features are unappealing and we have been searching for a wf in which only local correlations are present. We have succeeded[8] by the following extension of the SWF:

$$\Psi_{glue}(R) = \int dS F(R,S) L(S) \quad , \tag{17}$$

$$L(S) = \prod_{i=1}^{N} \exp\left[-D\frac{(\hat{n}_i - 1)^2}{\hat{n}_i}\right] . \tag{18}$$

$\hat{n}_i$ represents a suitable normalized local density operator for the shadow variables around the $i-th$ shadow and we have taken

$$\hat{n}_i = A^{-1} \sum_{j(\neq i)} \ell(|\vec{s}_i - \vec{s}_j|) \quad , \quad \ell(s) = e^{-\mu s^2} . \tag{19}$$

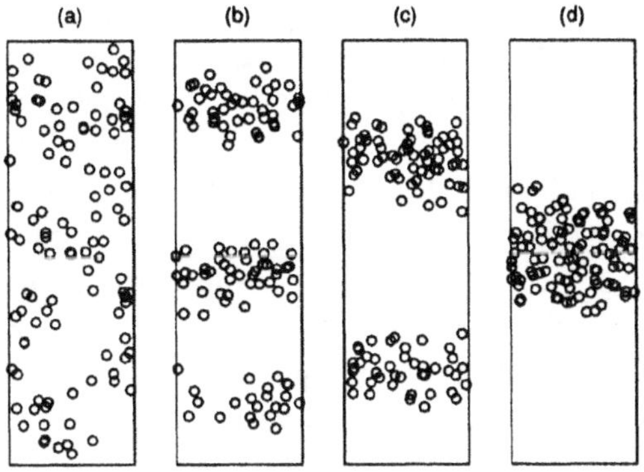

Figure 7. Projection of the positions of the particles on the $x - z$ plane after that the z side of the simulation box has been expanded. The panels correspond to configurations after 250 (a), 7500 (b), 10^5 (c) and 2×10^5 (d) Monte Carlo steps.

The factor $F(R, S)$ is equal to Eq.(6) as in the standard SWF. D, A and μ are variational parameters.

We have called this wf a glue-SWF because the factor L has the role of a "spring" which tries to keep $\widehat{n}_i$ not too far from one. A key role is due to the presence of $\widehat{n}_i$ in the denominator in the argument of the exponential in (18): no single particle can "evaporate" from the fluid otherwise the wf would drop to zero. Indeed, we have verified by MC computations that the wf (17) describes a self-bound state. For instance, we have performed the following simulation experiment. First we run a standard simulation of N atoms in a box of volume $V = L^3$ with $N/V = \rho_{eq}$. At a certain point we expand by a large factor, 5 for instance, the box in the direction z and at the same time all z components of $\vec{r}_i$. Thus the average density drops by a factor of 5 so that $\widehat{n}_i$ is well below 1. The evolution of the system is such that the particles rapidly coalesce leaving most of the box void of particles and a well defined slab of material is formed (see Fig.7). If the box is expanded in all three directions the stable state turns out to be a cluster. The resulting energy and density profile are roughly similar to those obtained with a standard wf. For instance, for a system of 108 ^{4}He atoms (surface area Å=290.6 Å^2 and slab thicknesses 18.6 Å) we get an energy per particle $E = -\ 5.38 \pm 0.03K$. The previous best variational result[20] for the same geometry is $E = -5.31 \pm 0.03K$. From the results for slabs of different thickness we extract a surface tension $\sigma = 0.302 \pm 0.014\ K/\text{Å}^2$. The experimental value is $\sigma = 0.274K/\text{Å}^2$. Similarly good results have been obtained for clusters. We conclude that the glue-SWF is able to describe self-bound states of ^{4}He. It is interesting to observe that the wf (17) does not have a classical analogue. We already discussed that a standard wf (1) gives a probability density in configuration space formally equal to that of a fictitious classical fluid. Similarly a SWF has an analogue in terms of a fluid of flexible triatomic molecules. This is not true for the glue-SWF because there is no "potential" that is able to reproduce the effect of the local density operator $\widehat{n}_i$ at the denominator in the exponential function in (18). This corresponds to the property of the glue-SWF that the liquid coexists with vacuum, i.e. it has zero vapor pressure, whereas any classical fluid with finite range forces has a finite vapor pressure.

5. Density fluctuations and Bose–Einstein condensate at the free surface of ^{4}He

The zero point motion of the low energy excitations in a quantum fluid determines the small k behavior of the static structure factor. For instance the zero point motion of phonons in bulk ^{4}He gives the characteristic behavior[14] $S(k) = \hbar k/2mc$ at small k where c is the sound velocity. The same behaviour is present in a two dimensional system. In ^{4}He with a free surface in addition to phonons another branch of excitations appears, the ripplons. They propagate close to the liquid-vacuum interface with the energy

$$\varepsilon_r\left(k_{||}\right) = \hbar\sqrt{\frac{\sigma}{m\rho_{eq}}}k_{||}^{3/2} \quad . \tag{20}$$

$\vec{k}_{||}$ is the wave vector parallel to the surface, σ the surface tension and ρ_{eq} the density well inside the liquid. Let us define a z-dependent structure factor

$$S\left(k_{||}, z, z'\right) = \left\langle \frac{A^{-2}}{\rho(z)\rho(z')} \sum_m e^{i\vec{k}_{||}\cdot\vec{r}_m} \delta\left(\vec{r}_m \cdot \hat{\vec{j}} - z\right) \sum_\ell e^{i\vec{k}_{||}\cdot\vec{r}_\ell} \delta\left(\vec{r}_\ell \cdot \hat{\vec{j}} - z'\right)\right\rangle \tag{21}$$

where $\hat{\vec{j}}$ is the unit vector in the z direction normal to the surface and A is the cross section of the slab. $S(k_{||}, z, z')$ measures the density fluctuations of wave vector $k_{||}$ between particles, one at height z and one at height z'. The diagonal part $S(k_{||}, z, z)$ gives the static structure factor for particles within a layer characterized by the coordinate z. It has been shown[22] that $S(k_{||}, z, z')$ has the following behaviour

$$S\left(k_{\|}, z, z'\right) = 2\sqrt{\frac{\hbar^2}{m\sigma\rho_{eq}}\frac{d\sqrt{\rho_1(z)}}{dz}\frac{d\sqrt{\rho_1(z')}}{dz'}}k_{\|}^{-1/2} \tag{22}$$

at small $k_{\|}$. The $k_{\|}^{-1/2}$ divergence at $k_{\|} = 0$ is a manifestation of large density fluctuations due to ripplons. Notice that these fluctuations are present only when z and z' are in the interfacial region where $d\rho_1(z)/dz$ is non zero. As we move from the interface to the interior the $k_{\|}^{-1/2}$ divergence crosses over to the linear behaviour in $k_{\|}$ typical of the bulk so that $S(k_{\|}, z, z)$ does not diverge but it vanishes at $k_{\|} = 0$.

We have computed[23] the z-dependent structure factor (21) for a slab of ^4He described by a standard wf Eq.(1) and by the glue-SWF. In the case of a standard wf with one-body terms and bulk like pair terms we do not find any enhancement of the density fluctuations in the surface region as it is shown in Fig.8a where $S(k_{\|}, z, z)$ is plotted: as z moves from the interior of the slab to the surface $S(k_{\|}, z, z)$ evolves from a bulk like behaviour to that of a very diluted system for which $S(k) = 1$. On the contrary in the case of the glue-SWF $S(k_{\|}, z, z)$ becomes large at small $k_{\|}$ when z is such that $d\rho_1(z)/dz$ is large (Fig. 8.b). Actually the computed $S(k_{\|}, z, z')$ turns out to agree quantitatively with the value obtained from the ripplon formula (22) in which the values are used for σ and $\rho_1(z)$ as given by the same wf. We conclude that the glue-SWF contains the basic phenomena of ripplon fluctuations whereas the standard wf does not[24]. Notice that in our computation there is a lower cut off to $k_{\|}$ due to the finite size of the system so that we cannot describe the $k_{\|}^{-1/2}$ singularity but only the enhancement of the density fluctuations at small but finite $k_{\|}$. In addition, to recover the $k_{\|}^{-1/2}$ singularity the pseudopotentials should have a suitable long range contribution which is absent in the present parametrization of the glue–SWF.

In a slab geometry also the momentum distribution for $\vec{k}$ parallel to the surface has a well defined meaning and one can consider a z dependent momentum distribution $n(k_{\|}, z)$. Of special interest is the Bose-Einstein condensate fraction $n_0(z)$, i.e. the fraction of particles with $k_{\|} = 0$, as function of z. With a standard wf it was found[25] that full condensation ($n_0 = 1$) is present in the outer surface region where the local density $\rho_1(z)$ is small. We have verified this by using a standard wf with one–body and bulk like Jastrow terms. This is shown in Fig. 9a where n_0 is plotted not as function of z but as function of $\rho_1(z)$. In bulk ^4He n_0 is about 10% but n_0 becomes 70% when $\rho_1(z)$ is about 10% of ρ_{eq} and it reaches full condensation when $\rho_1(z)$ is of order and below 1% of ρ_{eq}. The result is quite different with the glue-SWF as shown in Fig.9b: $n_0(z)$ rapidly rises

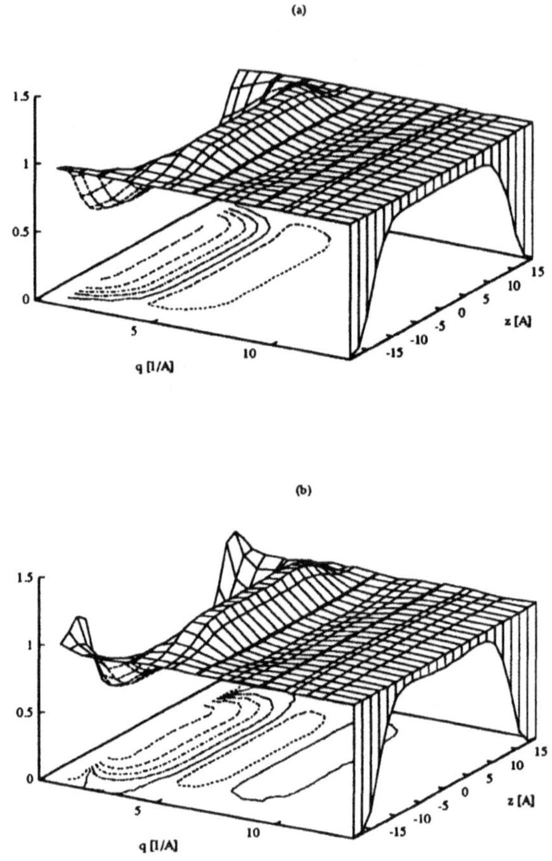

Figure 8. On the vertical axis the local static structure factor $S(q_{\parallel}, z, z)$ is plotted for a standard wf (a) and for a glue-SWF (b). The density profile $\rho(z)$ is also shown on the right hand side. Results for $N = 162$, slab thickness $27\mathring{A}$ and surface area $290.6\mathring{A}^2$.

to about 51% when $\rho_1(z)$ is about 10% of ρ_{eq} but further out in the tail region $n_0(z)$ decreases to small values. It is found that this depletion of the condensate has an associated promotion of particles at small momentum and this takes place in the surface region where the ripplon enhancement of $S(k_{\parallel}, z, z)$ is maximum. This indicates that ripplon fluctuations have a strong effect on the momentum distribution and that the tail region of the liquid-vacuum interface is very different from a rarefied system. In any case we find that there is an enhancement of the condensate in the surface region and it has been speculated[13] that this is relevant in experiments of quantum evaporation.

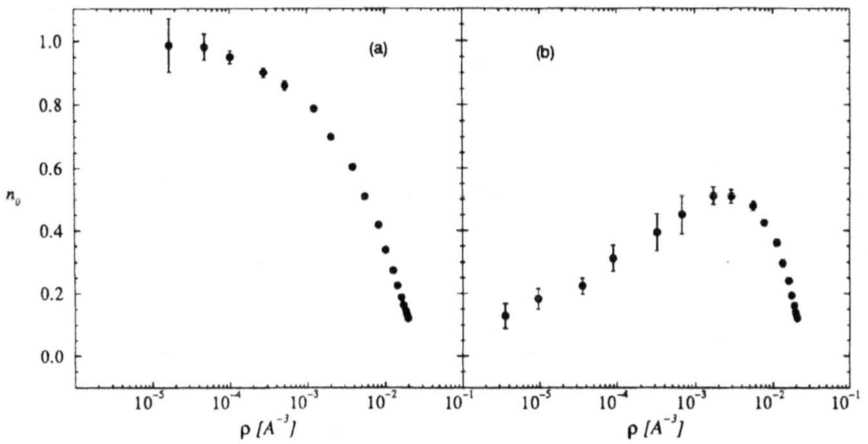

Figure 9. n_0 as function of the value of the local density of the slab for a standard wf (a) and for the glue-SWF (b).

6. Solid like order in doped ^{4}He clusters

Small clusters of ^{4}He have been the subject of much interest both experimentally and theoretically. The study of how the cooperative superfluid properties arise and approach the bulk behaviour as the number N of atoms of the cluster increases is of general interest. Evidence for superfluid behaviour and for roton excitations has been found experimentally[26]. As mentioned above the glue-SWF is able to describe in a satisfactory way also clusters of ^{4}He. Here I do not discuss the results for pure clusters but I consider only the case of a cluster with an impurity atom[27]. No atom or molecule are soluble in bulk superfluid ^{4}He but a number of atoms and molecules are found to dissolve inside a small cluster of ^{4}He atoms. For an impurity atom like Xe or for a molecule like SF_6 the attractive well of the interatomic potential between the impurity and one ^{4}He atom is much stronger than the one between ^{4}He atoms. Therefore we might expect that the ^{4}He atoms around the impurity pack themselves at much higher density than usual and a solid like order should be present. A SWF should be specially useful under these conditions given its flexibility to describe either a liquid or a solid. In our computation we have used the glue-SWF (17) with the optimized pseudopotentials[2] of the bulk system for the ^{4}He atoms. For the impurity-^{4}He pseudopotentials we have used the Mc Millan form $u_p(r) = (b/r)^5$ for the real variables, a Gaussian coupling $u_{PS}(r-s) = -C\,|\vec{r}-\vec{s}|^2$ between the impurity position and the position of its shadow and the rescaled interatomic potential $u_s(s) = \alpha v_{AZ}(\gamma s)$ as

N	glue-SWF	VMC[29]	DMC[29]
40	-17.85(3) [K]	-17.17(1) [K]	-19.41(2) [K]
70	-12.60(2) [K]	-11.84(1) [K]	-13.64(1) [K]
112	-9.76(2) [K]	-9.01(1) [K]	-10.39(2) [K]
240	-7.20(2) [K]	-6.44(1) [K]	-7.43(1) [K]

TABLE 1. Energy per particle of SF_6-^4He_N clusters.

pseudopotential between the shadow of the impurity and the shadow of a ^{4}He atom. v_{AZ} is the interatomic ^{4}He potential[28]. b, c, α, γ and the glue parameters for the impurity are the variational parameters for the doped clusters. Some of our results for the impurity SF_6 are shown in Table 1. One can see that there is a very large improvement over previous variational results[29] and the glue-SWF results are rather close to the "exact" results obtained by DMC. The local density $\rho_1(r)$ of the ^{4}He atoms around the impurity shows a well defined structure corresponding to two shells of ^{4}He atoms and as an example the result for a cluster of $N = 112$ atoms is shown in Fig.10a. A standard variational wf gives a much weaker local structure and the glue–SWF are in much better agreement with the density profile given by DMC. In order to study the presence of solid-like order around the impurity we have computed an angular correlation $P(\theta, \varphi)$ as follows. We take as origin of the coordinates the impurity and as polar axis the direction determined by the impurity and one of the ^{4}He atoms within a distance between 2.5Å and 5.6Å . This range corresponds to the first shell of neighbours. $P(\theta, \varphi)$ is the probability of finding a second ^{4}He atom within the same range of distance from the impurity at angles θ and φ with respect to the polar axis. Therefore $P(\theta, \varphi)$ is a three body correlation function. It is clear from Fig.10b that there is a well defined order which extends over the full range of θ and φ and one can count 20 atoms in the first shell. The angles for which $P(\theta, \varphi)$ has maxima correspond to an icosahedron. Well defined angular correlations are present also in $P(\theta, \varphi)$ computed for distances corresponding to the second shell of neighbours. These correlations extend only for $\theta \lesssim 90°$ so that the second shell does not have a true solid like order. Clusters of different sizes have essentially the same spatial order in the first shell of neighbours. It should be noticed that these ^{4}He atoms are rather mobile even if there is a very pronounced solid order: an atom which is in this first shell at a certain moment of the simulation, at a later stage of the simulation can be found anywhere in the cluster. It will be

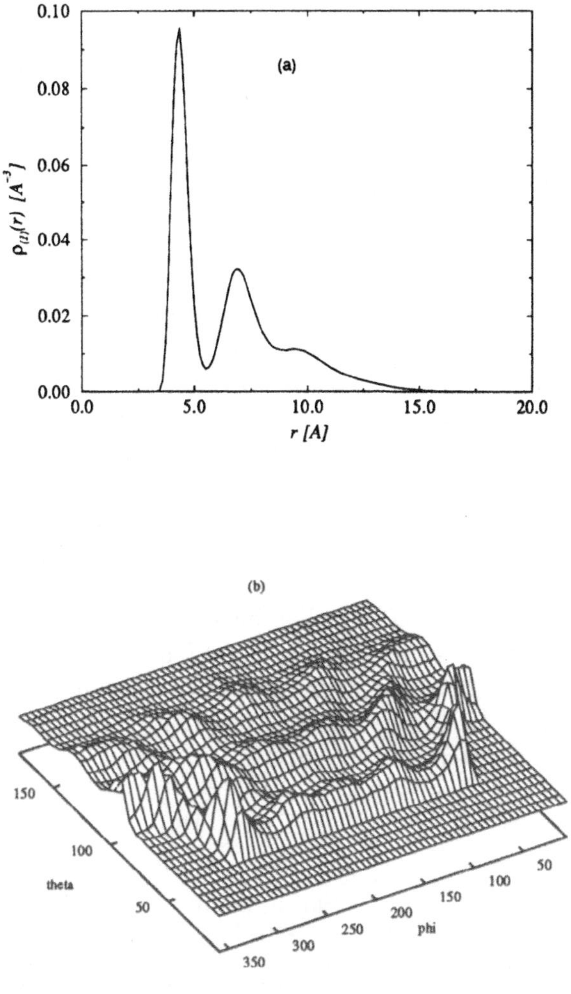

Figure 10. (a): Local density $\rho_1(r)$ of the ^{4}He atoms around a SF_6 molecule for a cluster of $N = 112$ atoms. (b): Angular correlation $P(\theta, \varphi)$ for the first shell of a cluster of $N = 240$ atoms.

very interesting to study the excitations of these doped clusters: we might expect the presence of roton like excitation with a decreased energy with respect to bulk ^{4}He.

7. Conclusion

The microscopic description of the condensed phases of ^{4}He has greatly improved and the ground state and excited states of the system can be considered as essentially established with only the interatomic potential as input. What is presently missing is a satisfactory theory of the system at temperatures of order of T_λ. Sometimes one finds the statement that we do not have a microscopic theory of ^{4}He. This is wrong unless by this one intends a *simple* model of the system, something like BCS theory for standard superconductors. Superfluid ^{4}He is a strongly interacting system and it is unlikely that a simple model will ever be able to give a comprehensive description of the system.

Much of the interest now is in the dishomogeneous states of ^{4}He. Here much has to be learned yet but some significant progress has been achieved. The energetics and the density profile of clusters and of the liquid - vacuum interface are under control. The full description of fluctuations, of correlations and of the off-diagonal density matrix is not complete but some advances have been obtained and the technique of shadow wave function appears to be very promising.

Acknowledgement

Most of the work presented here has been performed in collaboration with colleagues and former students. Their contribution and insight have been essential and let me thank D.E. Galli, G.L. Masserini, E. Cecchetti and R. Grisenti in Milano, S. Fantoni and S. Moroni in Trieste and G.V. Chester, M.H. Kalos and F. Pederiva at Cornell University.

References

1. For a general review, see: K.E. Schmidt and D.M. Ceperley, in "*Monte Carlo Methods in Condensed Matter Physics*", edited by K. Binder, Topics in Applied Physics, Vol. 71 (Springer, Berlin 1992).
2. S. Moroni, D.E. Galli, S. Fantoni and L. Reatto, *Phys. Rev.* **B58**, 909 (1998).
3. J. Boronat and J. Casulleras, *Europhys. Lett.* **38**, 291 (1998).
4. D.E. Galli, L. Reatto and S.A. Vitiello, *J. Low Temp. Phys.* **101**, 755 (1998).
5. K.E. Schmidt, M.H. Kalos, M.A. Lee, and G.V. Chester, *Phys. Rev. Lett.* **45**, 573 (1980).
6. S.A. Vitiello, K. Runge, and M.H. Kalos, *Phys. Rev. Lett.* **60**, 1970 (1988). For a review see: L. Reatto, in *Progress in Computational Physics of Matter*, eds. L. Reatto and F. Manghi (World Scientific, 1995), p.43.
7. S. Moroni, S. Fantoni and G. Senatore, *Phys. Rev.* **B52**, 13547 (1995).
8. D.E. Galli and L. Reatto, in: "*Monte Carlo and Molecular dynamics of Condensed Matter Systems*", K. Binder and G. Ciccotti (Eds.), SIF, Bologna, (1996).
9. E. Feenberg: "*Theory of Quantum Fluids*" (Academic, New York 1969).
10. F. Pederiva, A. Ferrante, S. Fantoni and L. Reatto, *Phys. Rev.* **B52**, 7564 (1995).
11. F. Pederiva, A. Ferrante, S. Fantoni and L. Reatto, *Phys. Rev. Lett.* **72**, 2589 (1994); Physica B**194-196**, 967 (1994).
12. F. Pederiva, G.V. Chester, S. Fantoni and L. Reatto, *Phys. Rev.* **B56**, 5909 (1997).
13. A.F.G. Wyatt, *Nature* **391**, 56 (1998).
14. R.P. Feynman, *Phys. Rev.* **94**, 262 (1954).

15. R.P. Feynman and M. Cohen, *Phys. Rev.* **155**, 1189 (1956).
16. D.K. Lee and F.J. Lee, *Phys. Rev.* **B11**, 4318 (1975). M. Manousakis and V.R. Pandharipande, *Phys. Rev.* **B30**, 5062 (1984).
17. W. Wu, S.A. Vitiello, L. Reatto and M.H. Kalos, *Phys. Rev. Lett.* **67**, 1446 (1991).
18. D.E. Galli, E. Cecchetti and L. Reatto, *Phys. Rev. Lett.* **77**, 5401 (1996).
19. V.R. Pandharipande, J.G. Zabolitzky, S.C. Pieper, R.B. Wiringa and U. Helmbrecht, *Phys. Rev. Lett.* **50**, 1676 (1983).
20. J.L. Valles and K.E. Schmdt, *Phys. Rev.* **B38**, 2879 (1988).
21. E. Krotscheck, *Phys. Rev.* **B32**, 5713 (1985).
22. E. Krotscheck, S. Stringari and J. Treiner, *Phys. Rev.* **B35**, 4754 (1987).
23. D.E. Galli and L. Reatto, *J. Low Temp. Phys.* (in the press).
24. A Feenberg wf gives the ripplon fluctuations only if the pair pseudopotential $u(\vec{r}_i, \vec{r}_j)$ depends not only on the pair distance r_{ij} but also on the distances z_i and z_j of the particles from the surfaces.
25. A. Griffin and S. Stringari, *Phys. Rev. Lett.* **76**, 259 (1996).
26. M. Hartmann, F. Mielke, J.P. Toennies, A.F. Vilesov and G.B. Benedek , *Phy. Rev. Lett.* **76**, 4560 (1996).
27. D.E. Galli, unpublished
28. R.A. Aziz, F.R.W. McCourt, C.C.K. Wong, *Mol. Phys.*, **61** (1987) 1487.
29. S.A. Chin and E. Krotscheck, in *"Recent Progress in Many-Body Theories"*, Vol.4, Edited by E. Schachinger *et al.*, Plenum Press, New York (1995).

TIGHT-BINDING LATTICE-GAS MODEL: ALKALI FLUIDS

M. REINALDO-FALAGAN AND P. TARAZONA
Departamento de Física Teórica de la Materia Condensada, Universidad Autónoma de Madrid, E-28049 Madrid, Spain

E. CHACON
Instituto de Ciencia de Materiales, Consejo Superior de Investigaciones Científicas, E-28049 Madrid, Spain

AND

J. P. HERNANDEZ
Department of Physics and Astronomy, University of North Carolina, Chapel Hill NC 27599-3255, USA

Abstract. A description of metal-atom fluids treats the ions using a lattice gas model and the electronic structure via a tight-binding description, with a single orbital per ion and hopping only to nearest neighbors. For monovalent atoms, the spectrum of electronic states is half filled; this provides the only cohesive energy in the model. Electrical neutrality of each cluster is imposed. The model results in an effective interaction between the ions which is not pair-additive and leads to strongly asymmetric liquid-vapor coexistence curves, similar to observations on the alkalis. The electrical conductivity is calculated with a Greenwood-Kubo formula. Our results identify the metal-nonmetal transition as percolative. It is located on the vapor side of the coexistence curve, in agreement with the experimental results. Recent advances replace the lattice gas by a continuous description of the ionic positions; also, the study of electronic correlation has been started, in the lattice-gas with an approximate Hubbard model for the electronic structure.

1. Introduction

The study of simple insulating liquids requires the potential between atom pairs as a function of their separation. It is obtainable from quantum me-

C. Caccamo et al. (eds.), New Approaches to Problems in Liquid State Theory, 481–490.

chanical calculations or a model fitted to the gas. The total interaction is taken to be pairwise additive and the interaction and the statistical problem are decoupled. The statistical mechanics of such liquids is treatable to obtain the thermodynamics and structures [1]. A striking deviation from this *simple liquid* behavior is that of liquid metals. There, the conduction electrons are delocalized and the system has to be treated as a mixture of (Boltzmann) ions and (Fermi-Dirac) electrons. For liquid metals, near the triple point, a double perturbation [2], of a *reference simple fluid* for the ions and *jellium* for the conduction electrons, is used successfully. The coexisting vapor is almost trivial due to its low density, but with a difference: the valence electrons are localized in neutral atoms or clusters. Near the critical point, and the metal-nonmetal transition, the interrelation between electronic delocalization and ionic structure becomes crucial and the low temperature approach fails.

Near-critical data for the alkalis [3, 4] motivate a search for theoretical improvements [5, 6]. Our main objective here is setting a minimal model, containing the main features of the coupling between electronic and ionic structures, to analyze the statistical problem. We have reported [7, 8] such calculations. Here, we sketch the qualitative effects of the present very simple model.

2. Tight-binding Lattice-gas Model

The first simplification consists in using a lattice-gas model to represent the ionic structure. Thus, the ions are constrained to partially occupy the sites of a chosen lattice. The critical behavior of such 'lattice-gas' models, with simple pair interactions, shows that they belong to the same universality class as simple fluids; so we may expect that the long-range fluctuations in the critical region are not qualitatively affected by this simplification.

The electronic properties of this model are described by the simplest one-electron tight-binding Hamiltonian for each ionic configuration, with N ions partially occupying the M sites of a BCC lattice. Thus, we postulate a single orbital at each occupied lattice site (no orbitals at the empty sites) and the electronic hopping is restricted to nearest-neighbor (nn) occupied sites. The one-electron energies, ϵ_n, and wavefunctions, $\phi_n(i)$ (the probability amplitudes that an electron occupies the orbital on the ion at site i), depend on the ionic configuration. They are the, n=1, N, solutions of the eigenvalue equation:

$$(\epsilon_a - \epsilon)\phi(i) - t\sum_{j}^{nn}\phi(j) = 0, \tag{1}$$

for $i = 1, N$ and with the sum over j extended to all the occupied sites which are nn to site i, either directly or through the periodic boundary conditions imposed. The site energy, ϵ_a, corresponds to that of the atomic orbital and t is the constant hopping parameter between occupied nn sites. The one-electron eigenstates are to be occupied, using Fermi statistics, with probability

$$f_n = [exp((\epsilon_n - \mu_e)/k_BT) + 1]^{-1}. \tag{2}$$

The electronic chemical potential μ_e is used to fix the total number of electrons to be equal to the total number of atoms. Here, the only energy associated with each ionic configuration is the electronic free energy:

$$U = 2\sum_{n=1}^{N} \epsilon_n f_n + k_BT[f_n ln f_n + (1 - f_n)ln(1 - f_n)]. \tag{3}$$

If orthonormal solutions to (1) are chosen, charge neutrality at each site is obtained.

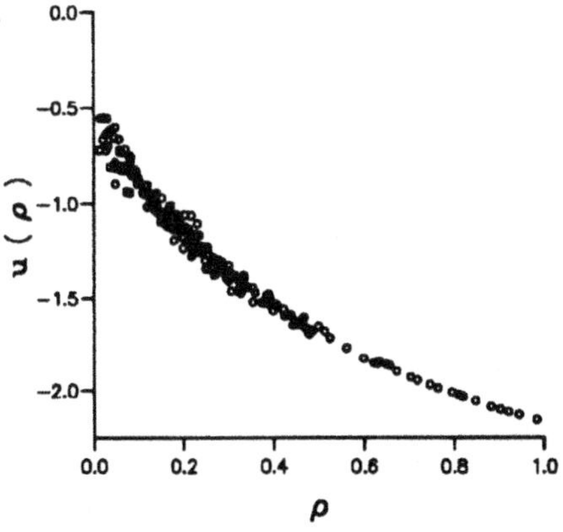

Figure 1. The electronic free energy per ion (in units of t) as a function of the mean fractional-occupation density of the BCC lattice, from a random-occupation ensemble and with electronic entropy corresponding to $k_BT/t = 0.4$.

Structural, thermodynamic, and electronic self consistency demand that the probability of finding any ionic configuration be proportional to its Boltzmann factor, $\exp(-(U - N\mu)/k_BT)$, in a grand-canonical ensemble with chemical potential μ. The self-consistency requirement is the same for very rarefied ($N << M$) configurations, whose electrons are localized in small clusters, or for very dense ($N \approx M$) ones, whose electrons are delocalized. In all cases, the preferential double occupancy of the lower energy states provides the cohesive energy of the system. Once the underlying lattice structure is chosen, the only parameters in the model are ϵ_a and t; we may absorb them by taking the energy origin to be $\epsilon_a = 0$ and using t as the energy scale.

3. Mean-field Approximation

As a beginning, we consider a macroscopic mean-field treatment ignoring electronic feedback to the structure and with random lattice occupancy. The lattice is occupied with mean density $\rho = N/M$. The electronic free energy in (3) is obtained, using exact diagonalizations, for a spectrum of such realizations with variable ρ. The energy per ion, $u = U/N$, as a function of ρ is presented in fig. 1 for the BCC lattice; a typical example, $k_BT/t = 0.4$, was chosen. The $u(\rho, T)$ have a substantial scatter around their mean, $\bar{u}(\rho, T)$, particularly at low ρ; the scatter is a realization dependence which should scale out with increasing lattice size (2000 sites are used here). Each isotherm of $\bar{u}(\rho, T)$ is a concave-up function which is strikingly different from the straight line obtained for nn pair-interacting systems.

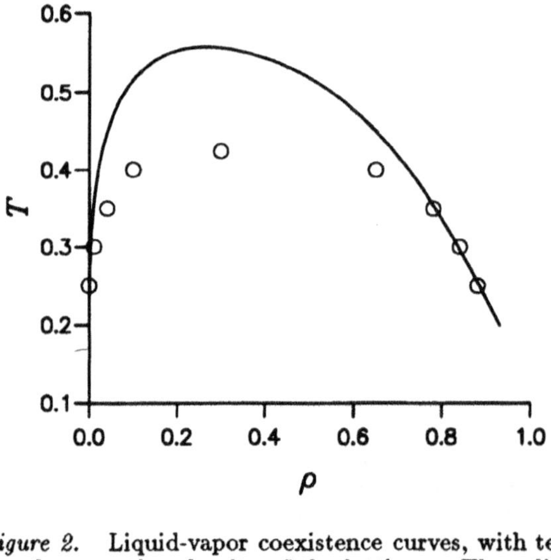

Figure 2. Liquid-vapor coexistence curves, with temperature in units of t/k_B, vs fractional-occupation density of the lattice, ρ. The solid line is the result of the mean-field calculation, eq. (4). The open circles are the self-consistent Monte Carlo results, using the coefficients in eq.(5).

The Helmholtz free energy per ion in this approximation,

$$F/N = \bar{u}(\rho, T) + k_BT \left[\ln(\rho) + \frac{(1-\rho)}{\rho} \ln(1-\rho) \right], \tag{4}$$

is used to obtain a liquid-vapor coexistence curve, shown as a solid line in fig. 2, with a critical temperature $k_B T_c^{mf}/t = 0.56$. The curve is asymmetric in density about the critical point, in contrast with the fully symmetric result of a lattice gas with nn pair interactions. In that case, the mean-field approximation overestimates the critical temperature by about 20%; we shall examine what results arise from fluctuations with our self-consistent model.

4. Monte Carlo Simulation

In our model, the electronic eigenvalues are obtained from (1) and occupied thermally to obtain (3). Self consistency is required since the eigenvalues depend on the ionic correlations through the occupied-site distribution which is itself a function of the Boltzmann factors, determined by (3).

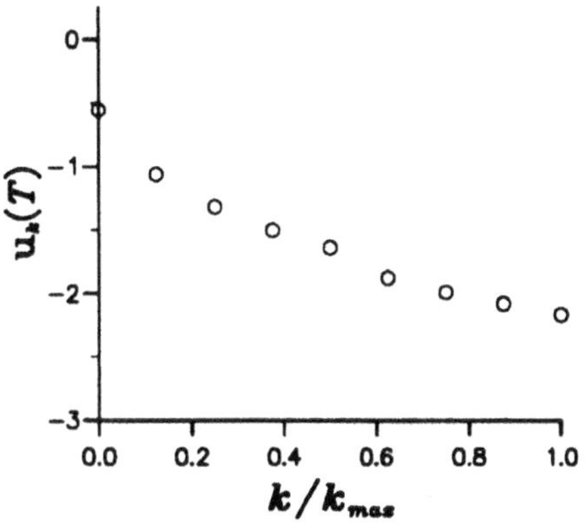

Figure 3. Self-consistent results at $k_B T/t = 0.4$ for the fitting parameters $u_k(T)$, giving the energy (in units of t) of a particle with k occupied nn sites, versus k normalized to the lattice coordination.

We fit [9, 10] the electronic free energy (3), for each T, a spectrum of densities, and representative ensembles of site occupations, as a function of the nn environments of the occupied sites. The fit uses least-squares, with

$$U(T) = \sum_{k=0}^{k_{max}} u_k(T) N_k, \tag{5}$$

where N_k are the number of ions with k occupied nn in each realization and k_{max} is the lattice coordination (8, for BCC). $k = 0$ implies isolated atoms; eq. (3) yields $u_0 = -2k_B T ln2$, an entropy overestimate by a factor of two. The remaining $u_k(T)$ are fitted.

We start, for each T, using (5) to fit exact diagonalizations for random occupations at each ρ. Then, using the $u_k(T)$, as site energies for particles with k occupied nn sites, grand-canonical Monte Carlo simulations are performed to obtain equilibrium configurations and ρ as a function of the chemical potential. In contrast to mean-field, the ion configurations are now correlated to the electronic energy by the local environments. We then use a spectrum of resulting configurations to repeat the process and iterate to self consistency at each T. Self-consistent parameters for (5) are shown in fig. 3. The phase coexistence curve and electronic properties are then obtained (see fig. 2).

5. Electrical Conductivity

In self-consistent calculations, the electronic structure will reflect the dramatic change in the ionic correlations accompanying the thermodynamic phase change; this should dominate the electronic properties. Here, ionic percolation is required for metallic conductivity. The Kirchoff's law model of Nield *et al* [11] is the simplest approach. It replaces pairs of occupied nn sites by a constant resistor. That 'bond resistor' is chosen to obtain the experimental conductivity in the dense liquid (near the triple point) and results can be compared to alkali data. The comparison is excellent and quite similar to that in [9].

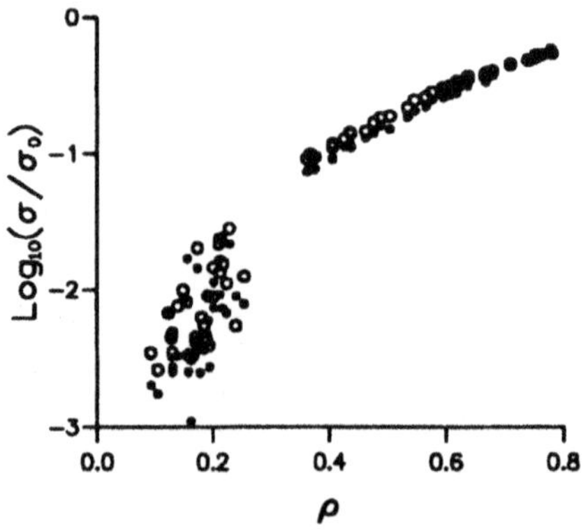

Figure 4. Electrical conductivity estimates σ, normalized to the conductivity at full lattice occupation σ_0, logarithmic scale vs density, for the isotherm $k_BT/t = 0.4$. The full circles are obtained using the Greenwood-Kubo quantum values of the bond resistors (eq. (6)); the open circles are the results of using a fixed bond resistor; both approaches then use the Kirchoff model for a spectrum of Monte Carlo configurations.

The above method neglects quantum effects which may preclude conduction even if the ionic structure percolates. We compare with a quantum estimate in which a loss of phase memory (due to scattering), after a nn hop, is assumed: a variant of the randomized phase model [12] in the Kubo-Greenwood formula [13], modified for our disordered ionic topology. In this alternative, the bond conductance is a sum over amplitudes, at nn sites, of equal energy (for elastic scattering) states:

$$1/R_{i,j} \propto \sum_{m \neq n} \frac{\partial f_n}{\partial \epsilon_n} |\phi_n(i)|^2 |\phi_m(j)|^2 \delta(\epsilon_n - \epsilon_m). \tag{6}$$

In our finite size system, the energy-conserving delta function will be replaced by a Gaussian of $0.16t$ variance (a value of little resulting importance, except near complete filling of the lattice). Energies and amplitudes are obtained from exact diagonalizations for typical equilibrium configurations. Macroscopic conductivities are obtained as in the Nield model but using (6). This approach caricatures the strong scattering expected in a hot disordered fluid. In our statistical sampling of ionic configurations, most configurations *do* give global resistances which are well described by the simplest version of the Nield model. The result of both procedures are in agreement, as is shown in fig. 4 for a near-critical isotherm.

6. Recent Advances

The replacement of the lattice gas by a continuum, with hopping depending on separation, and the inclusion, in the lattice gas, of Hubbard-type correlations between electrons are ongoing work. In both cases the philosophy is identical with the previously described model. The methods used are described below; the results will be published in the near future.

In the first case, we replace the lattice gas with a continuum, which demands a new definition for the hopping, in eq. (1); it will now depend exponentially on separation of the ion centers. The scaling parameter t is now the hopping for ions separated by a hard-sphere diameter. For ions whose centers are separated by x hard-sphere diameters, the hopping matrix element is taken to be the product of t with $exp[-3.89(x-1)]$. The constant, 3.89, has been chosen to yield a ratio between the dimer cohesion and that of the dense system which is intermediate between the experimental results for Rb and Cs; thus, it is a reasonable one for the alkalis. Also, we must redefine the right hand side of eq. (5), as we now have energetics which consist of hard-sphere repulsions plus effective attractions due to the valence-electron hopping, an effect which depends on the local environments around the ions. Such local environments will be specified by weighted sums (with the weight decreasing with distance) which measure the local density and its square. It is not enough to define the local density, as confusion will arise between environments with little occupation close to the chosen ion, but larger occupation at a larger distance, with other situations in which the population at closer separations is increased but decreased, in compensation, further away. Thus, for the right hand side of eq. (5), we use a sixth-degree polynomial in 'local density' plus a third-degree polynomial in 'local density squared'; the 'local density' (squared) being defined as the sum of the hopping-exponential dependence (its square), evaluated at the positions of ions in the environment of the one in question. The polynomial coefficients are the fitting parameters. Finally, in the interest of computational speed, the grand-canonical Monte Carlo simulations are replaced by so-called Gibbs ensemble simulations. Such simulations consist of two boxes in which a fixed number of atoms, in a fixed total volume, are shared, with a demand of equality of a specified temperature, as well as of unspecified pressure and chemical potential. The phase-coexistence densities are thus obtained from triple trials of: moving the position of ions (for equilibrium), changing ions between the boxes (to equate chemical potential), and increasing the volume of one box while reducing it in the other (to equate pressures).

In the alternative extension, in which electron correlations are to be included, we maintain the lattice gas for the ions and nn hopping. Then, the Gutzwiller [14] approximation is used to treat the Hubbard correlations among electrons, due to an onsite repulsive energy U. The free energy equivalent to eq. (3) now has an internal energy which is the average of the exact spectrum obtained from eq. (1), with a redefined thermal occupation (eq. (5) in ref. [15]), multiplied by a band narrowing factor q (it is dependent on the fractional double occupancy of the sites, d), plus the correlation energy Ud. The entropy effects are to be included in the average sense of ref. [15]. This d-dependent free energy is then minimized, for each neutral cluster, to obtain an average d and a minimized free energy; the latter then becomes the equivalent of eq. (3). The self-consistent process, starting from the new equivalent to eq. (5), then proceeds as before.

7. Summary

We have sketched a tight-binding lattice-gas model for metal-atom fluids, which can be solved for its thermodynamic, structural, and electronic properties. The model accounts for the inhomogeneity of statistical configurations driven by nonadditive interactions due to valence-electron delocalization. Monte Carlo simulations coupled to exact diagonalizations and site-energy fittings are used self consistently.

The qualitative consequence of the electronic feedback to the ionic structures, in our model, is the existence of a vapor-liquid phase transition. The resulting condensation in the model, driven by the electronic band energy, is qualitatively compared with alkali-fluid data. Despite the extreme simplicity of the model, the phase diagrams in fig. 5, our self-consistent results and experimental data for Cs (all in units of the critical temperatures and

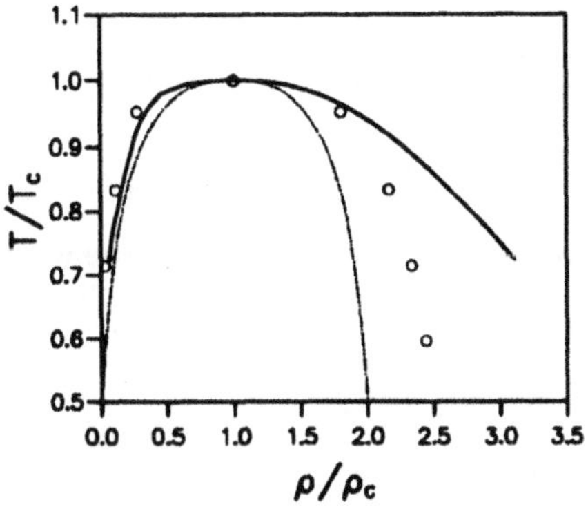

Figure 5. Reduced liquid-vapor coexistence curves, temperatures and densities scaled to their critical values. The full line is a fit to experimental data for Cs. The circles are the self-consistent Monte Carlo results from the present model. The dotted line is the result for a BCC lattice-gas model with additive nn pair interactions.

densities), show an appealing similarity. The asymmetry obtained contrasts with the coexistence curve results, also shown in fig. 5 and due to particle-hole symmetry, of lattice-gas models with additive nn pair interactions. Our results underestimate the density of the dense liquid at coexistence. This effect is a result of the reduced cohesion resulting from only using the electronic energy.

The electrical conductivity was estimated, using a quantum-resistor network and self-consistent ionic structures. The main reason that density reduction leads to nonmetallicity is the lack of percolation of the self-consistent ionic structure.

The model is a very simplified representation of a metal-atom fluid, as only electronic effects contribute to the cohesion. Nevertheless, the results show that the model has the correct basic ingredients: self consistency between the electronic and ionic structures allows a unified treatment of a spectrum of system properties and qualitative description of observations in the alkali fluids. Examination of the effects resulting from removing the lattice-gas approximation, replacing it with a continuum description, and from including Hubbard correlations among the electrons are in progress.

Acknowledgments

This work was partially supported by the DGICT (Spain)-grant PB94-005-C02 and by NATO-grant SA.5-2-05(CRG.940240). One of us (JPH) would like to thank IBERDROLA for partial support in the course of this work.

References

1. Hansen, J.P. and McDonald, I.R. (1986) *Theory of Simple Liquids*, Academic Press, New York.
2. Shimoji, M. (1977) *Liquid Metals*, Academic Press, New York.
3. Jüngst, S., Knuth, B. and Hensel F. (1985) Observation of Singular Diameter in the Coexistence Curves of Metals, *Phys. Rev. Lett.* **55**, 2160-2163.
4. Hensel, F., Jüngst, S., Noll, F. and Winter, R. (1985) Metal-Nonmetal Transition and the Critical Point Phase Transition in Fluid Cesium, in D. Adler and H. Fritzche (eds.), *Localization and Insulator-Metal Transitions*, Plenum Press, New York, pp. 109-117.
5. Stratt, R.M. (1990) The electronic structure of liquids, *Annu. Rev. Phys. Chem.* **41**, 175-205 .
6. Logan, D.E. and Wolynes, P.G. (1984) Self-consistent theory of localization in topologically disorder systems, *Phys. Rev. B* **29**, 6560-6562; (1986) Anderson localization in topologically disordered systems: The effects of band structure, *J. Chem. Phys.* **85**, 937-948; Logan, D.E. and Winn, M.D. (1988) The density of states of a spatially disordered tight-binding model, *J. Phys. C* **21**, 5773-5795.
7. Tarazona, P., Chacón, E. and Hernandez, J.P. (1996) Phase Diagram and fluid structure from a simple model of the alkalis, *J. Phys.: Condens. Matter* **8**, 9359-9362.
8. Reinaldo-Falagán, M., Tarazona, P., Chacón, E. and Hernandez, J.P. (1997) Thermodynamic and electronic properties of a tight-binding lattice-gas model, *J. Phys.: Condens. Matter* **9**, 9799-9810.
9. Tarazona, P., Chacón, E. and Hernandez, J.P. (1995) Simple model for the phase coexistence and electrical conductivity of alkali fluids, *Phys. Rev. Lett.* **74**, 142-145.
10. Chacón, E., Hernandez, J.P. and Tarazona, P. (1995) Theoretical models for the liquid-vapor and metal-nonmetal transitions of alkali fluids, *Phys. Rev. B* **52**, 9330-9341.
11. Nield, V.M., Howe, M.A. and McGreevy, R.L. (1991) The metal-nonmetal transition in expanded cesium, *J. Phys.: Condens. Matter* **3**, 7519-7525.
12. Hindley, N.K. (1970) Random phase model of amorphous semiconductors, *J. Non-Cryst. Sol.* **5**, 17-30.
13. Cusak, N.E. (1987) *The Physics of Structurally Disordered Matter*, Adam Higler, U. of Sussex, p. 192.
14. Gutzwiller, M.C. (1965) Correlation of electrons in a narrow s band, *Phys. Rev.* **137**, A1726-A1735.
15. Rice, T.M., Ueda, K., Ott, H.R. and Rudigier, H. (1985) Normal-sate properties of heavy-electron systems, *Phys. Rev. B* **31**, 594-596.

CURRENT-DENSITY FUNCTIONAL THEORY OF TIME-DEPENDENT LINEAR RESPONSE IN QUANTAL FLUIDS: RECENT PROGRESS

M. P. TOSI, M. L. CHIOFALO, A. MINGUZZI and R. NIFOSÌ
Istituto Nazionale di Fisica della Materia and Classe di Scienze
Scuola Normale Superiore, I-56126 Pisa, Italy

Abstract. Vignale and Kohn have recently formulated a local density approximation to the time-dependent linear response of an inhomogeneous electron system in terms of a vector potential for exchange and correlation. The vector potential depends on the induced current density through spectral kernels to be evaluated on the homogeneous electron gas. After a brief review of their theory, the case of inhomogeneous Bose superfluids is considered, with main focus on dynamic Kohn-Sham equations for the condensate in the linear response regime and on quantal generalized hydrodynamic equations in the weak inhomogeneity limit. We also present the results of calculations of the exchange-correlation spectra in both electron and superfluid boson systems.

1. Introduction

The evaluation of the dynamical properties of inhomogeneous quantal systems is a central problem in condensed matter physics. As an example we may recall the current efforts at elucidating the dynamical behaviour of dilute alkali vapours which have undergone Bose-Einstein condensation in magnetic traps. The experimental studies have concerned (i) the excitation of low-lying shape deformation modes in a regime where there is no detectable noncondensate fraction [1, 2] and the behaviour of mode frequencies and relaxation times as temperature is raised towards and above condensation [3]; (ii) the propagation of sound waves in the condensate and its thermal cloud [4, 5]; and (iii) antisymmetric oscillations of the condensate and the thermal cloud corresponding to second sound [5]. These experiments have stimulated a number of theoretical studies (see *e.g.* [6] and references given therein). The relationship between the dynamics of such confined fluids and that of a homogeneous Bose superfluid [7, 8] also is of great interest.

Local density approximations have been very useful in accounting for exchange and correlation (xc) in the ground state energy of inhomogeneous electron systems, and the search for dynamic extensions has attracted interest for some time (see *e.g.* [9]). A *scalar* time-dependent xc potential, which is a local functional of the time-dependent particle density, is useful in dealing with low-frequency phenomena. However, analysis of the constraints coming from basic conservation laws has shown that inconsistencies can arise and are associated with the non-existence of a gradient expansion at finite frequency ω for the xc potential in terms of the density alone (see *e.g.* [10]). Recently Vignale and Kohn [11, 12] have overcome these difficulties by resorting to a dynamic xc *vector* potential even in the case of a system subject to an external *scalar* potential. They got an

C. Caccamo et al. (eds.), New Approaches to Problems in Liquid State Theory, 491–502.

expression for the xc vector potential in the linear response regime in terms of kernels to be evaluated on the homogeneous electron gas at the local equilibrium density. Their expression becomes exact when the equilibrium density and the external potential are slowly varying in space, on length scales set by $1/k_F$ and v_F/ω where k_F and v_F are the local Fermi wave-number and velocity. The kernels are directly related to current-current response functions [13].

In § 2 below we give a brief review of the work of Vignale and Kohn and then proceed in § 3 to report on our work regarding an inhomogeneous superfluid. Again for the latter the validity of the results is restricted to weak inhomogeneity (slow variations over length scales set by the interatomic distance and by c/ω where c is the local speed of sound) in the linear response regime. In § 4 we report some examples of xc spectra for both the electron fluid and the Bose superfluid, which have been obtained in basically the same decoupling scheme for the current-current correlations [13].

2. Weakly Inhomogeneous Electron Fluid in the Normal State

According to the Runge-Gross theorem [10] the problem of a many-electron system in an external potential $V(\mathbf{r},t)$ can be mapped into that of non-interacting electrons in an effective potential $V(\mathbf{r},t)+V_{ex}(\mathbf{r},t)$ (the excess potential $V_{ex}(\mathbf{r},t)$ includes both the Hartree term and the xc term). In the so-called "adiabatic local density approximation" (ALDA) the dynamic xc potential is evaluated from the xc energy density of the homogeneous electron gas in the form

$$V_{xc}^{ALDA}(\mathbf{r},t)=d\varepsilon_{xc}^{h}(n(\mathbf{r},t))/dn \quad , \tag{1}$$

as in the corresponding static problem. The ALDA embodies the correct electron-gas compressibility, but does not account correctly for plasmon dipersion and omits plasmon damping as well as transverse-current fluctuations. In the context of linearized hydrodynamics for a monatomic fluid, the ALDA embodies the speed of isothermal first sound, but does not account for sound-wave damping nor does it allow for the transition from a collisional to a collisionless regime which may take place *e.g.* as temperature is varied.

As already remarked in § 1, Vignale and Kohn [11, 12] have proposed using a dynamic xc vector potential in the linear-response regime to overcome the limitations of the ALDA for the dynamics of an inhomogeneous electron system in the normal state. The induced current density is written as

$$\delta\mathbf{j}(\mathbf{r},\omega)=\int d\mathbf{r}'\overleftrightarrow{\chi}_{KS}(\mathbf{r},\mathbf{r}';\omega)[\mathbf{a}(\mathbf{r},\omega)+\mathbf{a}_{H}(\mathbf{r},\omega)+\mathbf{a}_{xc}(\mathbf{r},\omega)] \quad , \tag{2}$$

where $\overleftrightarrow{\chi}_{KS}(\mathbf{r},\mathbf{r}';\omega)$ is the Kohn-Sham response of a non-interacting reference system, $\mathbf{a}(\mathbf{r},\omega)$ is the applied potential, $\mathbf{a}_H(\mathbf{r},\omega)$ is the Hartree mean-field term and $\mathbf{a}_{xc}(\mathbf{r},\omega)$ is the xc vector potential. This is related to the induced current density by

$$\mathbf{a}_{xc}(\mathbf{r},\omega)=\int d\mathbf{r}'\,\overleftrightarrow{\mathrm{f}}_{xc}(\mathbf{r},\mathbf{r}';\omega)\delta\mathbf{j}(\mathbf{r}',\omega) \quad . \tag{3}$$

The kernels $\overleftrightarrow{\mathrm{f}}_{xc}(\mathbf{r},\mathbf{r}';\omega)$ entering Eq. (3) have some general exact properties, which are needed to relate them in the weak-inhomogeneity limit to analogous kernels for the homogeneous electron fluid taken at the local equilibrium electron density $n(\mathbf{r})$ [12]. These are Onsager's reciprocity relation, the zero-force and zero-torque theorems (in brief, the net force and the net torque exerted by the xc potential on the system must vanish) and a Ward identity embodying the compressibility sum rule for the homogeneous electron fluid. As the final result the dynamic correction to the ALDA xc vector potential is expressed in terms of the induced *velocity* $\mathbf{u}(\mathbf{r},\omega)=\delta\mathbf{j}(\mathbf{r},\omega)/n(\mathbf{r})$ and of the homogeneous xc current-density kernel $\overleftrightarrow{\mathrm{f}}_{xc}(\mathbf{k};\omega)$, evaluated at the local equilibrium density. The latter may be separated into the sum of a longitudinal and a transverse part:

$$\overleftrightarrow{\mathrm{f}}^{\mathrm{h}}_{xc}(\mathbf{k};\omega)=\omega^{-2}[f^{L}_{xc}(\omega)\mathbf{kk}+f^{T}_{xc}(\omega)(k^2\overleftrightarrow{\mathbf{I}}-\mathbf{kk})]\ . \tag{4}$$

Rather than reporting the complicated expression of the xc vector potential derived by Vignale and Kohn [11] in the local density approximation, we refer to the later work of Vignale *et al.* [14], who rewrite it in a physically transparent way through a visco-elastic stress tensor $\overleftrightarrow{\sigma}$. Defining

$$\sigma^{(xc)}_{ij}=\eta_{xc}(\omega)\Big(\frac{\partial u_i}{\partial r_j}+\frac{\partial u_j}{\partial r_i}-\frac{2}{3}\delta_{ij}\vec{\nabla}\cdot\mathbf{u}\Big)+\zeta_{xc}(\omega)\delta_{ij}\vec{\nabla}\cdot\mathbf{u} \tag{5}$$

where $\eta_{xc}(\omega)$ and $\zeta_{xc}(\omega)$ are complex viscosity coefficients given by

$$\eta_{xc}(\omega;n)=-(n^2/i\omega)f^{T}_{xc}(\omega;n) \tag{6}$$

and

$$\zeta_{xc}(\omega;n)=-(n^2/i\omega)[f^{L}_{xc}(\omega;n)-\frac{4}{3}f^{T}_{xc}(\omega;n)-d^2\varepsilon_{xc}(n)/dn^2]\ , \tag{7}$$

one finds that the divergence of $\overleftrightarrow{\sigma}^{(xc)}$ determines the dynamical xc vector potential. The connection of the theory with generalized hydrodynamics in the linear-response, long-wavelength regime becomes fully evident when one introduces a weak-inhomogeneity assumption also for the Kohn-Sham response matrix, by replacing it by its diamagnetic part $\overleftrightarrow{\chi}_{KS}(\mathbf{r},\mathbf{r}';\omega)\rightarrow\delta(\mathbf{r}-\mathbf{r}')n(\mathbf{r},\omega)/m$. The induced current is then given by

$$-im\omega\delta\mathbf{j}(\mathbf{r},\omega)=n(\mathbf{r})[-i\omega\mathbf{a}(\mathbf{r},\omega)-\vec{\nabla}\delta v_H(\mathbf{r},\omega)-\vec{\nabla}(\delta n(\mathbf{r},\omega)/n^2K_T)]+\vec{\nabla}\cdot\overleftrightarrow{\sigma}(\mathbf{r},\omega) \tag{8}$$

where $\delta v_H(\mathbf{r},\omega)$ is the Hartree potential associated with the long-range Coulomb interactions, K_T is the local electron-gas compressibility and the full stress tensor $\overleftrightarrow{\sigma}$ differs from the expression in Eq. (5) by the replacement $\eta_{xc}(\omega)\rightarrow\eta_{xc}(\omega)-p_0(n)/i\omega$, p_0 being the pressure of the non-interacting electron gas.

In summary, the corrections to the ALDA restoring force in the linear-response, weak-inhomogeneity regime merely involve the visco-elastic effects associated with frequency-dependent elastic constants and damping coefficients, which are related to the imaginary part and the real part of the two viscosity coefficients $\eta_{xc}(\omega)$ and $\zeta_{xc}(\omega)$, respectively. For longitudinal motions, in particular, the plasmon excitation arises from the Hartree term in Eq. (8) and the role of the visco-elastic function $\zeta_{xc}(\omega)+4\eta_{xc}(\omega)/3$ is (i) to shift the plasmon dispersion coefficient away from the low-frequency value determined by K_T in Eq. (8), and (ii) to introduce plasmon damping from decay into multiple electron-hole pairs.

Equation (8) relates the generalized viscosity coefficients $\eta_{xc}(\omega)$ and $\zeta_{xc}(\omega)$, or equivalently the dynamic local field factors $f^{L}_{xc}(\omega)$ and $f^{T}_{xc}(\omega)$ entering Eqs (6) and (7), to the current-current response function of the homogeneous electron gas, *via* a finite-frequency generalization of the Kubo relations [13]. This property will be used in § 4 for an evaluation of these xc spectra, based on the exact equation of motion for the current response and on the reduction to two-pair excitation processes through a decoupling approximation [13, 15]. We shall first turn, however, to a discussion of an inhomogeneous Bose superfluid.

3. Dynamic Density Functional Theory for a Bose Superfluid

In a superfluid of Bose particles the set of basic dynamic variables must include the superfluid velocity field $\mathbf{u}_s(\mathbf{r},t)$ in addition to the total (superfluid plus normal-fluid) current density. We follow Hohenberg and Martin [7] in introducing the transformation

$$\psi^{\dagger}(\mathbf{r}) = [\hat{n}_c(\mathbf{r})]^{1/2} \exp[-i\hat{\varphi}(\mathbf{r})] \tag{9}$$

for the field operator in terms of the condensate density and phase operators: then the superfluid velocity operator is $\hat{\mathbf{u}}_s(\mathrm{r}) = m^{-1}\vec{\nabla}\hat{\varphi}(\mathbf{r})$, in accord with the irrotational character taken by the superfluid velocity field in the absence of vortices.

The linear response of the condensate density and phase to a dynamic gauge-breaking external field can be treated by means of the equation of motion for the field operator and is found to have the Hohenberg-Kohn-Sham structure compatible with a mapping of the interacting system into a single-particle reference system [16]. The condensate self-energy $\sigma(1)$ is introduced by

$$\sigma(1) = \int d\bar{2} v(\mathbf{r}_1, \mathbf{r}_{\bar{2}}) < \psi^{\dagger}(\bar{2})\psi(\bar{2})\psi(1) > \tag{10}$$

where $1 = (\mathbf{r}_1, t_1)$, $v(\mathbf{r}_1, \mathbf{r}_{\bar{2}})$ is the interparticle pair potential and the symbol $< \cdots >$ denotes the expectation value of an operator on the equilibrium ensemble at given temperature. The Kohn-Sham response functions $\chi_{KS}(1,1')$ of the condensate are constructed from a reference Green's function $G_{KS}(1,1')$ which is given by

$$G_{KS}^{-1}(1,1') = [i\frac{\partial}{\partial t_1} + \frac{1}{2m}\nabla^2 - V(\mathbf{r}_1) + \mu - \frac{\sigma_{eq}(\mathbf{r}_1)}{< \psi(\mathbf{r}_1) >_{eq}}]\delta(1,1') \quad , \tag{11}$$

where $V(\mathbf{r}_1)$ is the confining potential and μ is the chemical potential. The two-by-two matrix $\chi(1,1')$ of the response functions of the condensate can then be cast into the form

$$\chi = \chi_{KS} + \chi_{KS} \otimes K \otimes \chi \tag{12}$$

where the symbol $\otimes$ denotes integration over intermediate variables and the kernels in the matrix K are obtained as functional derivatives of the real and imaginary parts of

$$\delta\sigma(1) = \sigma(1) - \sigma_{eq}(\mathrm{r}_1) < \psi(1) > / < \psi(\mathbf{r}_1) >_{eq} \tag{13}$$

with respect to the condensate density and phase.

We should emphasize that (i) the Kohn-Sham reference system that we are proposing for the superfluid contains the interactions through the equilibrium value of the condensate self-energy entering the RHS of Eq. (11); and (ii) we are *not* advocating a density functional approach to the evaluation of the equilibrium state of the superfluid. In particular, a thermodynamic treatment based on the Gross-Pitaevskii equation already is in quantitative agreement with experiment for confined vapours of alkali atoms which have undergone Bose-Einstein condensation [17].

A similar fully microscopic treatment of the current density induced in the inhomogeneous superfluid by a further external driving field leads to microscopic expressions for the remaining response functions (the current-current response and the current-condensate cross response) [16]. While these expressions are quite complicated in the general case, these calculations allow one to complete contact with the well known two-fluid model of Tisza and Landau. Firstly, the total current-current response function has the general structure already found by Hohenberg and Martin [7], who used it to show that in an appropriate limit this function gives the response of the total current density to an external field defining the normal fluid velocity $\mathbf{u}_n(\mathbf{r},t)$. Secondly, a Ward identity for the current response of the noncondensate at fixed condensate, which was first given by Huang and Klein [18] can be used to introduce a function $\rho_s(1,2)$ through the equation

$$\vec{\nabla}_1\rho_s(1,2)=\vec{\nabla}_1[n_c(1)\delta(1,2)]-2m\,\mathrm{Im}\Big[<\psi^\dagger(1)>\frac{\delta\sigma(1)}{\delta\mathbf{a}(2)}\Big|_{<\psi>}\Big] \quad . \tag{14}$$

Within our microscopic treatment Eq. (14) can be rewritten in the form

$$\vec{\nabla}_1\rho_s(1,2)=\vec{\nabla}_2\cdot\frac{\delta\mathbf{j}(1)}{\delta\mathbf{u}_s(2)}\Big|_{\mathbf{a}} \quad . \tag{15}$$

It follows from Eq. (15) that the function $\rho_s(1,2)$ reduces for the homogeneous fluid in the static limit to the hydrodynamic definition of superfluid density given by Hohenberg and Martin [7]. It can also be shown that in the same limit the definition of superfluid density given by Griffin [8] is recovered from Eq. (14).

On the above grounds we have developed a two-fluid model for the generalized hydrodynamics of the inhomogeneous superfluid in the linear-response, weak-inhomogeneity regime [16]. We propose that Eq. (14) may provide a reasonable definition of the equilibrium superfluid density $\rho_s(\mathbf{r})$ in the weakly inhomogeneous case, when we take its static limit and the $\mathbf{k}=0$ component of its Fourier transform with respect to $\mathbf{r}_1-\mathbf{r}_2$. The functional derivative entering the RHS of Eq. (14) is a five-point correlation function, to be evaluated in this case on the equilibrium state.

With the above definition of the superfluid density, we first proceed to extend Landau's hydrodynamic theory for the homogeneous superfluid to finite-frequency phenomena. We use for this purpose the well-known memory function formalism, as described *e.g.* in the book of Forster [19]. We assume isothermal conditions, *i.e.* we neglect the couplings between temperature and density fluctuations. The form of the generalized hydrodynamic equations is dictated by some general considerations: (i) invariance under a Galileian transformation and Onsager symmetry must hold; (ii) as a consequence of the zero-force and zero-torque theorems, the time derivative of the current density $\mathbf{j}$ is driven by the divergence of a symmetric tensor of the second rank; (iii) the time derivative of the superfluid velocity $\mathbf{u}_s$ is the gradient of a scalar, in view of its irrotational character below threshold for vortex generation; and (iv) the internal driving forces are determined by the divergence of the normal-fluid velocity $\mathbf{u}_n$ and of the interdiffusion current $\mathbf{j}_r\equiv\mathbf{j}-n\mathbf{u}_n=\rho_s(\mathbf{u}_s-\mathbf{u}_n)$.

The generalized hydrodynamic approach is easily extended to a weakly inhomogeneous superfluid in isothermal conditions [16]. The general properties that we have listed above to infer the structure of the generalized hydrodynamic equations in the homogeneous case remain valid. In addition, a Ward identity is essential in relating the effects of a weak inhomogeneity on the excess kernels to their density dependence: it allows the inhomogeneous kernels to be set equal to those of the homogeneous fluid at the local densities of superfluid and normal fluid. The resulting dynamical equations are

$$-im\omega\delta\mathbf{j}(\mathbf{r},\omega)=n(\mathbf{r})[-i\omega\mathbf{a}(\mathbf{r},\omega)-\vec{\nabla}(\delta n(\mathbf{r},\omega)/n^2K_T)]+\vec{\nabla}\cdot\vec{\vec{\sigma}}(\mathbf{r},\omega) \tag{16}$$

and

$$-im\omega\delta\mathbf{u}_s(\mathbf{r},\omega)=-i\omega\mathbf{a}_s(\mathbf{r},\omega)$$
$$-\vec{\nabla}[(n^2K_T)^{-1}\delta n(\mathbf{r},\omega)-(Ts/c_V)\delta s(\mathbf{r},\omega)]+\vec{\nabla}\cdot\vec{\vec{\sigma}}^{(s)}(\mathbf{r},\omega) \quad , \tag{17}$$

where s and c_V are the entropy and heat capacity per particle, and the stress tensors are given by

$$\sigma_{ij} = [\eta(\omega) - p_0(n)/i\omega](\frac{\partial u_{ni}}{\partial r_j} + \frac{\partial u_{nj}}{\partial r_i} - \frac{2}{3}\delta_{ij}\vec{\nabla}\cdot\mathbf{u}_n)$$

$$+ \delta_{ij}[\zeta_2(\omega)\vec{\nabla}\cdot\mathbf{u}_n + \zeta_1(\omega)\vec{\nabla}\cdot\mathbf{j}_r] \quad (18)$$

and

$$\sigma_{ij}^{(s)} = \delta_{ij}[\zeta_3(\omega)\vec{\nabla}\cdot\mathbf{j}_r + \zeta_4(\omega)\vec{\nabla}\cdot\mathbf{u}_n] \ , \quad (19)$$

$p_0(n)$ being the ideal-gas pressure. The ALDA contributions have been explicitly written in Eqs. (16) and (17) in terms of the total particle density fluctuations and of the entropy fluctuations, the latter being associated with fluctuations in the density of the non-condensate. The corresponding thermodynamic coefficients are to be evaluated on the homogeneous fluid at the local equilibrium density. Evidently, the ALDA driving forces are responsible for the first and second sound modes.

The visco-elastic coefficients in Eqs. (18) and (19) are related to a set of excess kernels according to the following equations:

$$\zeta_1(\omega) = -(i\omega)^{-1}[f_{j_L u_s}(\omega) - \partial p_{xc}/\partial n|_T] \quad , \quad (20)$$

$$\zeta_2(\omega) = -(i\omega)^{-1}[f_{j_L j_L}(\omega) - 4f_{j_T j_T}(\omega)/3 - n\partial p_{xc}/\partial n|_T] \quad , \quad (21)$$

$$\zeta_3(\omega) = -(i\omega)^{-1}[f_{u_s u_s}(\omega) - \partial\mu_{xc}/\partial n|_T - (Ts/c_V)\partial\mu_{xc}/\partial T|_n] \quad , \quad (22)$$

$$\zeta_4(\omega) = -(i\omega)^{-1}[f_{u_s j_L}(\omega) - \partial p_{xc}/\partial n|_T] \quad (23)$$

and

$$\eta(\omega) = -(i\omega)^{-1} f_{j_T j_T}(\omega) \quad , \quad (24)$$

with $\zeta_1 = \zeta_4$ from Onsager symmetry. As already remarked, the kernels in these equations are those of the homogeneous fluid at the local equilibrium densities of superfluid and normal fluid at temperature T. The xc pressure and chemical potential are similarly those of the homogeneous fluid at the local equilibrium density.

Within the memory function formalism, the visco-elastic spectra in Eqs. (18) and (19) can be calculated from the linear response functions of the homogeneous superfluid through finite-frequency forms of the Kubo relations. These will be used in § 4.2 for a Bose vapour.

4. Evaluation of dynamic kernels

4.1. THE ELECTRON FLUID IN THE NORMAL STATE

As we have already remarked at the end of § 2, Eq. (8) implies that the visco-elastic coefficients are related to the current-current response function of the homogeneous electron fluid by a finite-frequency extension of the Kubo relations. Precisely,

$$\mathrm{Im}\, f_{xc}^{L(T)}(\omega) = \lim_{k\to 0} \frac{m^2\omega^2}{n^2k^2}\, \mathrm{Im}\, \tilde{\chi}_{L(T)}(k,\omega) \quad , \quad (25)$$

where $\tilde{\chi}_{L(T)}(k,\omega)$ is the proper longitudinal (or transverse) current-current response [13]. This emphasizes that the plasmon does not contribute to the xc kernels.

Starting from the equation of motion for the current density fluctuation operator $\hat{j}_{\vec{q}i}$, the following exact long-wavelength result can be derived:

$$\mathrm{Im}\,\tilde{\chi}_{ij}(\vec{k},\omega) = -(m^2\omega^4)^{-1}\sum_{\vec{q},\vec{q}'}\sum_{s,s'}\Gamma_{is}(\vec{q},\vec{k})\Gamma_{js'}(\vec{q}',-\vec{k})$$

$$\times\,\mathrm{Im}\left\{i\int_0^\infty dt\,\exp[i(\omega+i\eta)t] < [\hat{j}_{\vec{q}s}\hat{\rho}_{-\vec{q}},\hat{j}_{\vec{q}'s'}\hat{\rho}_{-\vec{q}'}] > \right\} + o(k^2)\ , \quad (26)$$

$\hat{\rho}_{\vec{q}}$ being the density fluctuation operator and the detailed expression of the coefficients Γ being given in ref. [15]. Decoupling of the four-point correlation function entering the RHS of Eq. (26) into all possible products of two-point correlation functions leads to the following approximate result [13]:

$$\mathrm{Im} f_{xc}^{L(T)}(\omega) = -\int_0^\infty \frac{d\omega'}{\pi}\int \frac{d^3q}{(2\pi)^3}\frac{(4\pi e^2/n\omega')^2}{(\omega-\omega')^2}\mathrm{Im}\,\chi_L(q,\omega-\omega')$$

$$\times\left[a_{L(T)}\,\mathrm{Im}\,\chi_L(q,\omega') + b_{L(T)}(\omega'/\omega)^2\,\mathrm{Im}\,\chi_T(q,\omega')\right]\ . \quad (27)$$

Here, the a's and b's are numerical coefficients given by $a_L = 23a_T/16 = 23/30$ and $b_L = 4b_T/3 = 8/15$. Equation (27) only includes the contribution from excitations of two correlated electron-hole pairs. It can also be derived by means of second-order perturbation theory followed by approximate resummation to infinite order through renormalization of single-particle excitations from bare to screened [20]. For further details of this calculation, including the various possible choices for the response functions in the RHS of Eq. (26) and the inclusion of final-state exchange processes restoring the exactly known coefficients of the $\omega^{-3/2}$ decay of the spectra at high frequency [21], the reader is referred to the original papers [13, 15].

Equation (27) shows that both the longitudinal and the transverse spectrum involve an LL and an LT channel, the former becoming dominant at high frequency and the latter at low frequency (on a frequency scale set by the plasma frequency ω_{pl}). The values of the numerical coefficients in Eq. (27) are such that $\mathrm{Im}\,f_{xc}^T = (16/23)\,\mathrm{Im}\,f_{xc}^L$ at high frequency and $\mathrm{Im}\,f_{xc}^T = (3/4)\,\mathrm{Im}\,f_{xc}^L$ at low frequency: as a consequence, the two spectra have very similar shapes, the relation $\mathrm{Im}\,f_{xc}^T \cong 0.72\,\mathrm{Im}\,f_{xc}^L$ being approximately valid at all frequencies. This is shown in Figure 1, which reports the two spectra for the electron fluid at a value of the coupling strength $r_s = 3$ in comparison with an earlier attempt by Gross and Kohn [22] at estimating the longitudinal spectrum by a smooth interpolation between high and low frequency. In contrast, our microscopic calculations show a sharp spectral threshold at $2\,\omega_{pl}$, which is due to the opening of a two-plasmon channel in the two-pair excitation spectrum [23]. The large spectral strength of the plasmon excitation at long wavelengths, as compared to single-pair excitations, accumulates most of the oscillator strength for two-pair processes in the spectral region just above $2\,\omega_{pl}$.

The real part of the longitudinal and transverse kernels is obtained from the spectra illustrated in Figure 1 by means of Kramers-Kronig relations, *i.e. via* Hilbert transform and fulfilment of the third-moment sum rules. As a consequence of the spectral threshold at $2\,\omega_{pl}$, a sharp minimum is found in $\mathrm{Re}\,f_{xc}^{L(T)}$ at this frequency. Strong xc effects are present in $\mathrm{Re}\,f_{xc}^{L(T)}$ in the region of the plasma frequency, where these functions decrease with increasing ω towards their minimum at $2\,\omega_{pl}$. This theoretical result is in accord with the available data on the plasmon dispersion coefficient from Electron Energy Loss experiments on the alkali metals [24], although band-structure effects will also play a significant role in a fully quantitative comparison between theory and experiment.

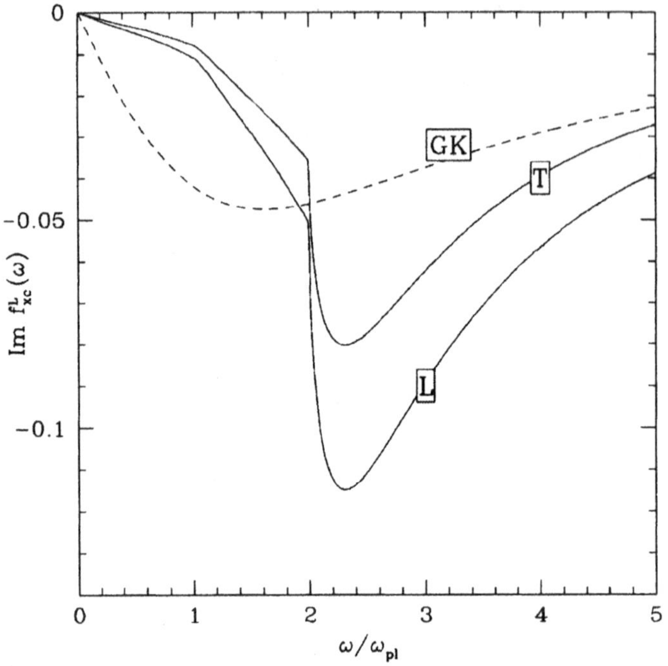

Figure 1. $\mathrm{Im} f_{xc}^{L,T}(\omega)$ for the electron gas at $r_s = 3$ in units of $2\omega_{pl} / n$, as functions of ω / ω_{pl}. The dashed line gives the Gross-Kohn interpolation scheme.

As a last point we briefly record the main results that have been obtained by this approach [15] for the elastic moduli and the viscosity coefficients of the fully degenerate electron gas in dimensionalities D = 3 and D = 2. These are: (i) the calculated electron-gas compressibility is in good agreement with the available Quantum Monte Carlo data on the xc energy; (ii) xc effects decrease the shear elastic constant for $r_s \geq 6$ in D = 3 and at all r_s in D = 2; (iii) the bulk viscosity vanishes identically at zero temperature; and (iv) the shear viscosity takes a finite value on inclusion of dynamic xc effects and this value rapidly decreases with increasing r_s.

4.2. THE DILUTE BOSON SUPERFLUID

The visco-elastic spectra generalizing sound wave attenuation and shear viscosity to finite frequency are related by Eqs. (16) and (18) to the longitudinal and transverse current-current response functions of the homogeneous fluid. Precisely, these relations are

$$\mathrm{Re}[\zeta_2(\omega) + \frac{4}{3}\eta(\omega)] = \lim_{k\to 0} \frac{-\omega m^2}{k^2} \mathrm{Im}\,\chi_{j_L j_L}(k,\omega) \qquad (28)$$

and

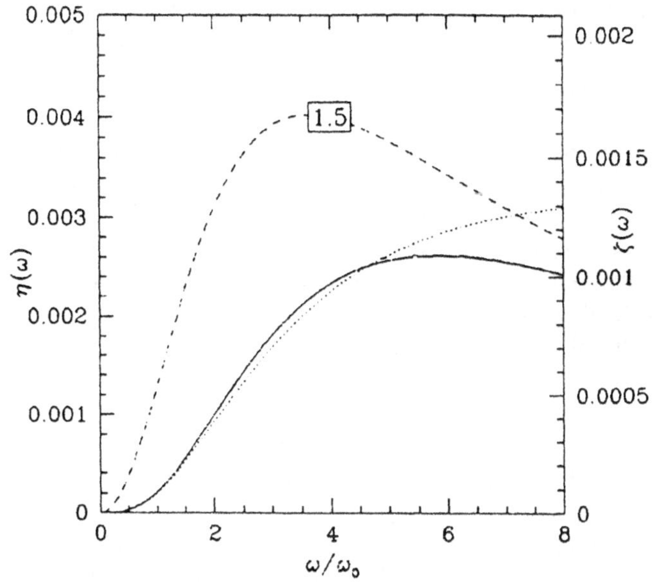

Figure 2. Spectra of shear viscosity (left scale) and bulk viscosity (right scale) as functions of ω / ω_0 at $T = 0$ (full curve) and $T = 1.5\omega_0$ (dashes). The dots show the result obtained at $T = 0$ without a cutoff in the potential.

$$\mathrm{Re}\,\eta(\omega) = \lim_{k\to 0} \frac{-\omega m^2}{k^2} \mathrm{Im}\,\chi_{j_T j_T}(k,\omega) \quad . \tag{29}$$

Of course, the imaginary parts of $\zeta_2(\omega)$ and $\eta(\omega)$, which are finite-frequency bulk and shear elastic moduli, are related to the real parts by Kramers-Kronig relations. Equations (28) and (29) are evidently analogous to Eq. (25) and can be evaluated in the same approach which led to Eq. (27) for the electron fluid. For the boson superfluid this approach may also be viewed as amounting to an expansion of the single-particle self-energy to second order in perturbation theory, followed by renormalization of the single-particle propagators into Bogolubov phonons. The extension of the theory to finite temperature is easily achieved, by introducing in the calculation of the response functions the thermal occupation factors $B(\omega) = [\exp(\beta\omega) - 1]^{-1}$ as required by the fluctuation-dissipation theorem.

The case of present interest concerns the confined condensates of Na or Rb atoms inside magnetic traps, in which the atomic interactions can be modelled by a repulsive Fermi model potential involving an s-wave scattering length a as a single parameter. The interatomic potential v_k needs to be regularized at large momenta $\mathbf{k}$, and this is most simply done by introducing a cut-off in the form

$$v_k = (4\pi a / m)\exp(-k^2 / 2\sigma^2) \quad .$$

The corresponding Bogolubov spectrum is

$$\omega_k = \left[(n v_k k^2 / m) + (k^2 / 2m)^2\right]^{1/2} \quad . \tag{30}$$

A lengthy but straightforward calculation, whose details will be reported elsewhere [25], yields for the shear and bulk viscosity spectra in the collisionless regime the result

$$\mathrm{Re}[\eta(\omega)] = \frac{12}{5}\mathrm{Re}[\zeta_2(\omega)] = \frac{v_k^2 k^6}{240\pi m^2 \omega_k^3 \omega_k'}\frac{\sinh(\beta\omega_k)}{\cosh(\beta\omega_k)-1} \quad , \tag{31}$$

in the absence of cut-off. Here, k is defined by $2\omega_k = \omega$ and $\omega_k' = d\omega_k / dk$ The high-momentum cut-off introduces a corrective factor which is somewhat different for the two spectra. At low frequency we find $\eta(\omega) \propto \omega^3$ if $\beta\omega \gg 1$ and $\eta(\omega) \propto \omega^2 T$ if $\beta\omega \ll 1$.

Figure 2 shows these visco-elastic spectra as functions of frequency at T = 0 and at a higher temperature on a scale set by the frequency $\omega_0 = 4\pi na/m$. A further result is obtained by evaluating the longitudinal spectrum at the sound-wave frequency ck, whereby one obtains the width $\Delta(k)$ of the phonon as

$$\Delta(k) = 3k^5 / (640\pi mn) \tag{32}$$

at T = 0 and

$$\Delta(k) = 3\pi^3 k (k_B T)^4 / (40 mc^4) \tag{33}$$

for $\beta\omega \ll 1$. These results for the phonon linewidth are well known from earlier work of Belyaev and Popov [26].

It can be shown that the results reported above for the visco-elastic current spectra at T = 0 correspond to the so-called one-loop approximation introduced in the context of Bose gases by Wong and Gould [27]. We have calculated the other spectra of relevance to Eqs. (16) - (19) by the same method, starting from the finite-frequency Kubo relations

$$\mathrm{Re}[\zeta_3(\omega)] = \lim_{k\to 0}\frac{-\omega}{k^2}\mathrm{Im}\,\chi_{u_s u_s}(k,\omega) \tag{34}$$

and

$$\mathrm{Re}[\zeta_1(\omega)] = \mathrm{Re}[\zeta_4(\omega)] = \lim_{k\to 0}\frac{-\omega m}{k^2}\mathrm{Im}\,\chi_{j_L u_s}(k,\omega) \quad . \tag{35}$$

It turns out that within the one-loop approximation all these dissipation spectra can be expressed in terms of four exchange-correlation building blocks: a proper condensate kernel related to the shift in local chemical potential due to interactions in the fluid away from equilibrium, two noncondensate kernels (the irreducible proper parts of the longitudinal and transverse current-current response) and a cross condensate-noncondensate vertex function. Actual calculation of these spectral functions shows that in a collisionless regime they take the same value aside from simple multiplying factors:

$$\frac{12}{5}\mathrm{Re}[\zeta_2(\omega)] = \frac{4}{5}\mathrm{Re}[mn\,\zeta_1(\omega)] = \frac{4}{15}\mathrm{Re}[m^2 n^2 \zeta_3(\omega)] = \mathrm{Re}[\eta(\omega)] \quad . \tag{36}$$

5. Summary and concluding remarks

We have reviewed the progress recently made in developing the theory of the dynamics of weakly inhomogeneous quantal fluids in the linear response regime from knowledge of the dynamics of the corresponding homogeneous fluid. The present focus in this area is on the degenerate electron gas and on the dilute Bose superfluid, for which the basic theoretical concepts and some inputs for applications have been developed. The fruitfulness of this approach remains to be tested through applications to specific physical problems. Such applications are already under way in the case of electron fluids [28].

Acknowledgments

This work is supported by the Istituto Nazionale di Fisica della Materia through the Advanced Research Project on BEC. We acknowledge very fruitful collaborations with Dr S. Conti.

References

1. Jin, D.S., Ensher, J.R., Matthews, M.R., Wieman, C.E., and Cornell, E.A. (1996) Collective excitations of a Bose-Einstein condensate in a dilute gas, Phys. Rev. Lett. 77, 420-423.
2. Mewes, M.-O., Andrews, M.R., van Druten, N.J., Kurn, D.M., Durfee, D.S., Townsend, C.G., and Ketterle, W. (1996) Collective excitations of a Bose-Einstein condensate in a magnetic trap, Phys. Rev. Lett. 77, 988-991.
3. Jin, D.S., Matthews, M.R., Ensher, J.R., Wieman, C.E., and Cornell, E.A. (1997) Temperature-dependent damping and frequency shifts in collective excitations of a dilute Bose-Einstein condensate, Phys. Rev. Lett. 78, 764-767.
4. Andrews, M. R., Kurn, D.M., Miesner, H.-J., Durfee, D.S., Townsend, C.G., Inouye, S., and Ketterle, W. (1997) Propagation of sound in a Bose condensate, Phys. Rev. Lett. 79, 553-557.
5. Stamper-Kurn, D.M., Miesner, H.-J., Inouye, S., Andrews, M. R., and Ketterle, W. (1998) Excitations of a Bose-Einstein condensate at non-zero temperature: a study of zeroth, first, and second sound, cond-mat/9801262.
6. Minguzzi, A., Chiofalo, M. L., and Tosi, M. P. (1997) Generalized quantum hydrodynamics of a trapped dilute Bose gas, Phys. Lett. A 236, 237-244.
7. Hohenberg, P.C. and Martin, P.C. (1965) Microscopic theory of superfluid Helium, Ann. Phys. (NY) 34, 291-359.
8. Griffin, A. (1993) Excitations in a Bose-condensed Liquid, University Press, Cambridge.
9. Gross, E.K.U. and Kohn, W. (1990) Time-dependent density-functional theory, Adv. Quantum Chem. 21, 255-291.
10. Gross, E.K.U., Dobson, J.F., and Petersilka, M. (1996), Density functional theory of time-dependent phenomena, in R. F. Nalewajski (ed.), Topics in Current Chemistry, Springer Verlag, Berlin.
11. Vignale, G. and Kohn, W. (1996) Current-dependent exchange-correlation potential for dynamical linear response theory, Phys. Rev. Lett. 77, 2037-2040.
12. Vignale, G. and Kohn. W. (1997) Current density functional theory of time-dependent linear response: the local density approximation, in J. Dobson, M. P. Das and G. Vignale (eds.), Electronic Density Functional Theory, Plenum Press, New York.
13. Conti, S., Nifosí, R., and Tosi, M. P. (1997) The exchange-correlation potential for current-density functional theory of frequency-dependent linear response, J. Phys.: Condens. Matter 9, L475-L482.
14. Vignale, G., Ullrich, C. A., and Conti, S. (1997) Time-dependent density functional theory beyond the adiabatic local density approximation, Phys. Rev. Lett. 79, 4878-4881.
15. Nifosì, R., Conti, S., and Tosi, M.P. (1998) Dynamic exchange-correlation potentials for the electron gas in dimensionality D = 3 and D = 2, Phys. Rev. B (in press).
16. Chiofalo, M. L., Minguzzi, A., and Tosi, M.P. (1998) Time-dependent linear response of an inhomogeneous Bose superfluid, to be published.
17. Minguzzi, A., Conti, S., and Tosi, M.P. (1997) The internal energy and condensate fraction of a trapped interacting Bose gas, J. Phys.: Condens. Matter, 9, L33-L38.
18. Huang, K. and Klein, A. (1964) Phonons in liquid Helium, Ann. Phys. (NY) 30, 203-234.
19. Forster, D. (1975) Hydrodynamic Fluctuations, Broken Symmetry, and Correlation Functions, Benjamin, Reading.

20. Hasegawa, M. and Watabe, M. (1969) Theory of plasmon damping in metals, J. Phys. Soc. Japan 27, 1393-1414.
21. Glick, A.J. and Long, W.F. (1971) High-frequency damping in a degenerate electron gas, Phys. Rev. B 4, 3455-3460.
22. Gross, E.K.U. and Kohn, W. (1985) Local density-functional theory of frequency-dependent linear response, Phys. Rev. Lett. 55, 2850-2853.
23. Böhm, H.M., Conti, S., and Tosi, M.P. (1996) Plasmon dispersion and dynamic exchange-correlation potentials from two-pair excitations in degenerate plasmas, J. Phys.: Condens. Matter 8, 781-797.
24. vom Felde, A., Sprösser-Prou, J., and Fink, J. (1989) Valence-electron excitations in the alkali metals, Phys. Rev. B 40, 10181-10193.
25. Conti, S., Minguzzi, A., and Tosi, M. P. (1998) Dissipation spectra at long wavelength in dilute Bose gases, to be published.
26. Popov, V.N. (1972) Hydrodynamic Hamiltonian for a nonideal Bose gas, Theor. Math. Phys. 11, 478-486.
27. Wong, V.K. and Gould, H. (1974) Long-wavelength excitations in a Bose gas at zero temperature, Ann. Phys. (NY) 83, 252-302.
28. Ullrich, C.A. and Vignale, G. (1998) Linewidths of collective excitations of the inhomogeneous electron gas: application to two-dimensional quantum strips, Phys. Rev. B (to appear).

SUBJECT INDEX

(each number refers to the first page of the work related to the entry)

LIST OF PARTICIPANTS

DIRECTOR

Carlo CACCAMO
Dipartimento di Fisica
Università di Messina
Contrada Papardo, CP 50, 98166 Messina (Italy)
caccamo@vulcano.unime.it

CO-DIRECTORS

Jean-Pierre HANSEN
Department of Chemistry
Lensfield Road, Cambridge CB2 1EW (UK)
jph32@cam.ac.uk

George STELL
Department of Chemistry
State University of New York
Stony Brook, New York 11794-3400 (USA)
gstell@sbchm1.chem.sunysb.edu

LECTURERS

Thierry BIBEN
Laboratoire de Physique
Ecole Normale Supérieure de Lyon
46 Allée d'Italie, 69364 Lyon 07 (France)
biben@physique.ens-lyon.fr

Siegfried DIETRICH
Fachbereich Physik
Bergische Universität Wuppertal
D-42097 Wuppertal (Germany)
Dietrich@wptsb.physik.uni-wuppertal.de

Robert EVANS
H.H. Wills Physics Laboratory
Tyndall Avenue, Bristol BS8 1TL (UK)
bob.evans@bristol.ac.uk

Michael E. FISHER
Institute for Physical Science and Technology,
University of Maryland
College Park, Maryland 20742 (USA)
Fax: +301 314 9404

Johan Skule HØYE
Institutt for Fysikk
NTNU - Norwegian University of Science and Technology
N-7034 Trondheim (Norway)
hoye@phys.ntnu.no

Michael L. KLEIN
Laboratoty for Research on the Structure of Matter
University of Pennsylvania
Philadelphia, PA 19104 - 6323 (USA)
klein@lrsm.upenn.edu

Fred LADO
Department of Physics, Box 8202
North Carolina State University
Raleigh, NC 27695-8202 (USA)
lado@gibbs.physics.ncsu.edu

Dominique LEVESQUE
Laboratoire de Physique Théorique et Hautes Energies
Bâtiment 211, Université Paris-Sud XI
91405 Orsay Cedex (France)
dl@stat.th.u-psud.fr

Giorgio PARISI
Dipartimento di Fisica and INFM
Università di Roma "La Sapienza"
Piazzale Aldo Moro 2, 00185 Roma (Italy)
giorgio.parisi@roma1.infn.it

Luciano REATTO
Dipartimento di Fisica
Università di Milano
Via Celoria 24 Milano (Italy)
luciano.reatto@mi.infn.it

Martin L. ROSINBERG
Laboratoire de Physique Théorique des Liquides
Université Pierre et Marie Curie
4 Place Jussieu, 75252 Paris Cedex 05 (France)
mlr@lptl.jussieu.fr

Pedro TARAZONA
Departamento de Física Teórica de la Materia Condensada
Universidad Autónoma de Madrid
E-28049 Madrid (Spain)
pedro@fluid5.fmc.uam.es

Arun YETHIRAJ
Department of Chemistry
University of Wisconsin
Madison, WI 53706 (USA)
yethiraj@chem.wisc.edu

SEMINAR SPEAKERS

Joseph O. INDEKEU
Laboratorium VSM - K.U. Leuven
Celestijnenlaan 200 D, B-3001 Leuven (Belgium)
joseph.Indekeu@fys.kuleuven.ac.be

Dominik MARX
Max-Planck Institut für Festkörperforschung
Heisenbergstraße 1, 70569 Stuttgart (Germany)
marx@prr.mpi-stuttgart.mpg.de

Bela M. MULDER
Theory of Complex Fluids, Condensed Matter Division
FOM Institute for Atomic and Molecular Physics
Kruislaan 407, 1098 SJ Amsterdam (The Netherlands)
mulder@amolf.nl

Lawrence R. PRATT
Los Alamos National Laboratory
Los Alamos, NM 87545 (USA)
lrp@t12.lanl.gov

Yasha ROSENFELD
Nuclear Research Center Negev
P.O. Box 9001, Beer-Sheva 84190 (Israel)
yasha@bgumail.bgu.ac.il

Mario P. TOSI
Scuola Normale Superiore
Piazza Cavalieri 7, Pisa (Italy)
tosim@bib.sns.it

PARTICIPANTS

Maria C. ABRAMO
Dipartimento di Fisica
Università di Messina
Contrada Papardo, Salita Sperone 31, 98166 Messina (Italy)
abramo@vulcano.unime.it

Zehra AKDENIZ
Physics Department, Science Faculty
University of Istanbul
Vezneciler, 34459 Istanbul (Turkey)
zakdeniz@istanbul.edu.tr

Aleksij AKSIMENTIEV
Institute of Physical Chemistry
Polish Academy of Science, III Department
Kasprzaka 44/52, 01-224 Warsaw (Poland)
alek@niut.ichf.edu.pl

Noe G. ALMARZA
Departamento de Química Física I
Universidad Complutense de Madrid
E-28040 Madrid (Spain)
noe@olivia.quim.ucm.es

Marta ALVAREZ FERNANDEZ
Instituto de Estructura de la Materia e Instituto "Rocasolano" - C.S.I.C.
Serrano 119 - 28006 Madrid (Spain)
marta@iqfr.csic.es

Juan Antonio ANTA MONTALVO
Physical and Theoretical Chemistry Laboratory (PCL)
South Parks Road, Oxford OX1 3QZ (UK)
montalvo@hoover.pcl.ox.ac.uk

Vincent BALLENEGGER
Institut de Physique Théorique
Ecole Polytechnique Fédérale de Lausanne
1015 Lausanne (Switzerland)
Vincent.Ballenegger@epfl.ch

Vytautas BALEVICIUS
Faculty of Physics
University of Vilnius
University Str. 3, 2074 Vilnius (Lithuania)
Vytautas.Balevicius@ff.vu.lt

Christoph BAUER
Department of Physics
University of Wuppertal
42097 Wuppertal (Germany)
bauer@wptsb.physik.uni-wuppertal.de

Mourad BENAKLI
University of Alabama
MINT Center, Box 870209
Tuscaloosa, Alabama 35487-0209 (USA)
mourad@co.mint.ua.edu

Mikael BJÖRLING
Department of Chemistry, Physical Chemistry
Royal Institute of Technology
SE-100 40 Stockolm (Sweden)
mikaelb@physchem.kth.se

Nikolai BRILLIANTOV
Humboldt University at Berlin
Invalidenstraße 100, D-10115 Berlin (Germany)
nbrillia@summa.physik.hu-berlin.de

Ezio BRUNO
Dipartimento di Fisica
Università di Messina
Contrada Papardo, Salita Sperone 31, 98166 Messina (Italy)
eb@ginestra.unime.it

João Manuel CORDEIRO
Departamento de Física e Química
Facultade de Engenharia de Ilha Solteira - UNESP
Av. Brasil 56, 15385-000 Ilha Solteira, SP (Brasil)
cordeiro@fqm.feis.unesp.br

Dino COSTA
Dipartimento di Fisica
Università di Messina
Contrada Papardo, Salita Sperone 31, 98166 Messina (Italy)
costa@vulcano.unime.it

Tamara COUSSAERT
Université Libre de Brussels
CP 231, Campus Plaine, Bvd du Triomphe, 1050 Brussels (Belgium)
Tcouss@ulb.ac.be

Bruno D'AGUANNO
CRS4
Via N. Sauro 10, I-09123 Cagliari (Italy)
bruno@crs4.it

Serap DALGIC
Physics Department, Science and Letters Faculty
Trakya University
22030 Edirne (Turkey)
serapd@aix.trakya.edu.tr

Seyfettin DALGIC
Physics Department, Science and Letters Faculty
Trakya University
22030 Edirne (Turkey)
seyfed@aix.trakya.edu.tr

Dung DI CAPRIO
Laboratoire SRSI
Université Paris VI
4 Place Jussieu 75252, Paris Cedex 05 (France)
Dung@sri.jussieu.fr

Marjolein DIJKSTRA
H.H. Wills Physics Laboratory
Tyndall Avenue, Bristol BS8 1TL (UK)
M.Dijkstra@bristol.ac.uk

Maria G. DONATO
Dipartimento di Fisica
Università di Messina
Contrada Papardo, Salita Sperone 31, 98166 Messina (Italy)
mdonato@vulcano.unime.it

Jean-Francois DUFRÊCHE
Laboratoire Liquides Ioniques et Interfaces Chargées
Université Pierre et Marie Curie
4 Place Jussieu, 75252 Paris Cedex 05 (France)
dufreche@ccr.jussieu.fr

Marie-Pierre GAIGEOT
CEA/Bruyeres le Chatel
DCSA/CPA BP 12
91680 Bruyeres le Chatel (France)
gaigeot@bruyeres.cea.fr

George GEORGOUSSIS
Department of Physics
National Technical University of Athens
Zografou Campus, Gr. 15780 Athens (Greece)
geogeo@central.ntua.gr

Achille GIACOMETTI
Dipartimento di Scienze Ambientali
Università di Venezia
Calle Larga Santa Marta, DD2137, I-30123 Venezia (Italy)
achille@unive.it

Pietro GIANINETTI
Dipartimento di Fisica
Università di Milano
Via Celoria 16, 20133 Milano (Italy)
gianinetti@mi.infn.it

Paolo V. GIAQUINTA
Dipartimento di Fisica
Università di Messina
Contrada Papardo, Salita Sperone 31, 98166 Messina (Italy)
giaquint@cacyparis.unime.it

Sonia JORGE-GONZÁLES
Departamento de Química Física
Universidad Complutense de Madrid, Ciudad Universitaria s/n
E-28040 Madrid (Spain)
sonia@percus0.quim.ucm.es

David GOULDING
Theoretical Chemistry Group, Chemistry Department
University of Cambridge
Lensfield Road - Cambridge CB2 1EW (UK)
dave@theor.ch.cam.ac.uk

Benito GROH
FOM Institute for Atomic and Molecular Physics
Kruislaan 407, 1098 SJ Amsterdam (The Netherlands)
groh@COMFL1.amolf.nl

Sylvain GROLLAU
Laboratoire de Physique Théorique des Liquides
CNRS - Université Pierre et Marie Curie
4 Place Jussieu, 75252 Paris Cedex 05 (France)
grollau@lptl.jussieu.fr

Noé HERRERA-PACHECO
Benemérita Universidad Autónoma de Puebla
Facultad de Ciências Físico-Matemáticas
Apartado Postal 1152, Puebla PUE 72001 (Mexico)
nherrera@fcfm.buap.mx

Pál JEDLOVSZKY
Department of Physiology and Biophysics
The Mount Sinai School of Medicine of the CUNY
New York, NY 10029 (USA)
pali@inka.mssm.edu

Gerhard KAHL
Institut für Theoretische Physik
Technische Universität Wien
Wiedner Hauptstraβe 8-10, A-1040 Wien (Austria)
gkhal@tph.tuwien.ac.at

Edouard KIERLIK
Laboratoire de Physique Théorique des Liquides
Université Pierre et Marie Curie
4 Place Jussieu, 75252 Paris Cedex 05 (France)
kierk@lptl.jussieu.fr

Sabine KLAPP
Institut für Theoretische Physik
Freie Universität Berlin
Arnimalle 14, 14195 Berlin (Germany)
klapp@archimedes.physik.fu-berlin.de

Kenichiro KOGA
Department of Chemistry
University of Nebraska-Lincoln
Lincoln, NE68588 (USA)
koga@phase2a.unl.edu

Lloyd L. LEE
School of Chemical Engineering and Material Science
100 East Boyd, Energy Center T-325, Norman, Oklahoma 73019-0628 (USA)
lle@ou.edu

Raul J. F. LEOTE DE CARVALHO
Laboratoire de Physique
Ecole Normale Supérieure de Lyon
46 Allée d'Italie, 69364 Lyon 07 Cedex (France)
rlcarval@physique.ens-lyon.fr

Yan LEVIN
Instituto de Fisica UFRGS
Caixa Postal 15051, 91501-970 Porto Alegre, RS (Brazil)
levin@if.ufrgs.br

Hartmut LÖWEN
Institut für Theoretische Physik II
Heinrich-Heine Universität Düsseldorf
Universitätsstraße 1, D-40225 Düsseldorf (Germany)
hlowen@thphy.uni-duesseldorf.de

Gianpiero MALESCIO
Dipartimento di Fisica
Università di Messina
Contrada Papardo, Salita Sperone 31, 98166 Messina (Italy)
malescio@vulcano.unime.it

Anatol MALIJEVSKY
Department of Physical Chemistry
Prague Institute of Chemical Technology
166 28 Prague 6 (Czech Republic)
anatol.malijevsky@vscht.cz

Francesco MALLAMACE
Dipartimento di Fisica
Università di Messina
Contrada Papardo, Salita Sperone 31, 98166 Messina (Italy)
mallamac@imeuniv.unime.it

Yuri MARTINEZ-RATON
Departamento de Matematicas, Escuela Politecnica Superior
Universidad Carlos III de Madrid
Butarque 15, 28911 Leganes Madrid (Spain)
yuri@elrond.alum.math.uc3m.es

Georgii A. MARTYNOV
Institute of Physical Chemistry, Math Simulation Laboratory
Russian Academy of Sciences
Leninski Prospect 31, Moscow (Russia)
Martynov@lmm.phyche.msk.su

Giorgio PASTORE
Dipartimento di Fisica
Università di Trieste
Strada Costiera 11, 34014 Trieste (Italy)
pastore@trieste.infn.it

Vesselin PAOUNOV
Surfactant Science Group
University of Hull
Hull, HU6 7RX (UK)
v.n.paounov@chem.hull.ac.uk

Giuseppe PELLICANE
Dipartimento di Fisica
Università di Messina
Contrada Papardo, Salita Sperone 31, 98166 Messina (Italy)
pellican@vulcano.unime.it

Aurelien PERERA
Physique Théorique des Liquides
Université Paris 6
4 Place Jussieu, Tour 16, 75252 Paris Cedex 05 (France)
aup@lptl.jussieu.fr

Davide PINI
Department of Chemistry
State University of New York at Stony Brook
Stony Brook, NY 11794-3400 (USA)
dpini@sbchm1.chem.sunysb.edu

Giovanni PIZZIMENTI
Dipartimento di Fisica
Università di Messina
Contrada Papardo, Salita Sperone 31, 98166 Messina (Italy)
pizzimen@vulcano.unime.it

Yuri PLEVACHUK
Institute of Applied Physics
Ivan Franko State University
49 General Chuprynka str., 290044 Lviv (Ukraine)
Plevachk@iap.franko.lviv.ua

Silvia RAMOS
H.H. Wills Physics Laboratory
Tyndall Avenue, Bristol BS8 1TL (UK)
silvia.ramos@bristol.ac.uk

Manuel REINALDO FALAGÁN
Departamento de Física Teórica de la Materia Condensada
Universidad Autonoma de Madrid
CU 512, E-28049 Madrid (Spain)
manuel@fluid3.fmc.uam.es

Albert REINER
Institut für Theoretische Physik
Technische Universität Wien
Wiedner Hauptstraße 8-10, A-1040 Wien (Austria)
areiner@tph.tuwien.ac.at

Alberto ROBLEDO
Instituto de Fisica
Universidad Nacional Autónoma de Mexico
Apartado Postal 20-364, Mexico 01000 D.F. (Mexico)
robledo@fenix.ifisicacu.unam.mx

Maria Immaculada RODRIGUEZ PONCE
Departamento de Física Atómica, Molecular y Nuclear
Universidad de Sevilla
Apartado 1065, E-41080 Sevilla (Spain)
inma@likix2.us.es

Jose Manuel ROMERO ENRIQUE
Departamento de Física Atómica, Molecular y Nuclear
Universidad de Sevilla
Apartado 1065, E-41080 Sevilla (Spain)
romero@likix2.us.es

Roland ROTH
University of Wuppertal
42097 Wuppertal (Germany)
Roth@wptsb.physik.uni-wuppertal.de

Mauro ROVERE
Dipartimento di Fisica "E. Amaldi"
Università di Roma Tre
Via della Vasca Navale 84, I-00146 Roma (Italy)
rovere@fis.uniroma3.it

Marie-Louise SABOUNGI
Argonne National Laboratory
9700 South Cass Avenue, Argonne Ill 60439 (USA)
mls@anl.gov

Franz SAIJA
Istituto Tecniche Spettroscopiche - CNR
Contrada Papardo, Salita Sperone 31, 98166 Messina (Italy)
fsaija@vulcano.unime.it

Klaus Peter SCHNEIDER
University of Leipzig
Linnestraße 5, D-04103 Leipzig (Germany)
kps@rz.uni-leipzig.de

Richard P. SEAR
Department of Chemistry and Biochemistry
University of California
Los Angeles, California 90024 (USA)
sear@chem.ucla.edu

Vladimir SOKHAN
Department of Chemistry
University of Southampton
Southampton S017, 1BJ (UK)
vps@soton.ac.uk

Mycola TASINKEVYCH
Institute of Physical Chemistry PAS
Kasprzaka 44/52, 01-224 Warsaw (Poland)
Miko@fuks.ichf.edu.pl

Margarida TELO DA GAMA
Departamento de Física, Faculdade de Ciências
Universidade de Lisboa
Av. Prof. Gama Pinto 2, P-1699 Lisboa Codex (Portugal)
margarid@alf1.cii.fc.ul.pt

John VALLEAU
Chemical Physics Theory Group, Lash Miller Chemical Laboratories
University of Toronto
St. George Street, Toronto, Ontario M5S 1A1 (Canada)
Jvalleau@taurasi.chem.utoronto.ca

Martin VAN DER HOEF
Department of Chemical Engineering
University of Twente
P.O. Box 217, 7500 AE Enschede (The Netherlands)
M.A.vanderhoef@ct.utwente.nl

René VAN ROIJ
H.H. Wills Physics Laboratory
Tyndall Avenue, Bristol BS8 1TL (UK)
R.Vanroij@bristol.ac.uk

Martin WATZLAWEK
Institut für Theoretische Physik II
Heinrich-Heine Universität Düsseldorf
Universitätsstraße 1, D-40225 Düsseldorf (Germany)
martin@thphy.uni-duesseldorf.de

Volker WEISS
Institut für Anorganische und Physikalische Chemie (FB 2)
Universität Bremen
NW 2 Leobenertraße, 28359 Bremen (Germany)
volker@calvados.chemie.uni-bremen.de

Nigel B. WILDING
The University of Edinburgh
James Clerk Maxwell Building, Mayfield Road, Edinburgh EH9 3JZ (UK)
N.B.Wilding@ed.ac.uk

Hong XU
Département de Physique des Matériaux
Université Claude Bernard, Lyon1
43 Boulevard du 11 nov. 1918, 69622 Villeurbanne Cedex (France)
hxu@dpm.univ-lyon1.fr

Xiao C. ZENG
Department of Chemistry
University of Nebraska-Lincoln
Lincoln, NE 68588 (USA)
xzeng@unlinfo.unl.edu